Proceedings of the II International Conference on

APPLICATIONS OF PHYSICS TO MEDICINE AND BIOLOGY

Proceedings of the
II International Conference on

Applications of Physics to Medicine and Biology

Giorgio Alberi Memorial

Organized by
International Centre for Theoretical Physics and
Associazione Italiana di Fisica Biomedica

Trieste, Italy
7 - 11 November 1983

Editors: Ž Bajzer
P Baxa
C Franconi

World Scientific

Published by

World Scientific Publishing Co Pte Ltd.
P O Box 128
Farrer Road
Singapore 9128

Proceedings of the II International Conference on Applications of Physics to Medicine and Biology

ISBN 9971-966-81-6

Printed in Singapore by Continental Press

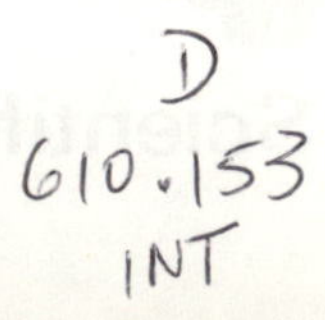

INTERNATIONAL ADVISORY COMMITTEE

C. Franconi (Conference Coordinator) - Italy

J.R. Cameron - U.S.A.

J. Clifton - U.K.

A. Kaul - Fed. Rep. Germany

S. Mascarenhas - Brazil

LOCAL ORGANIZING COMMITTEE

Ž. Bajzer - Yugoslavia

P. Baxa - Italy

L. Dalla Palma - Italy

S. Lin - Italy

P. Schiavon - Italy

UNDER THE AUSPICES OF

International Union for Pure and Applied Physics

International Organization for Medical Physics

European Federation of Organizations for Medical Physics

Società Italiana di Fisica

SPONSORS

Commissariato del Governo nella Regione Friuli-Venezia Giulia

Regione Friuli-Venezia Giulia

The City of Trieste

ENEA Comitato Nazionale per la Ricerca e per lo Sviluppo dell'Energia Nucleare e dell'Energie Alternative

AMPLISILENCE

ELSCINT

SORIN BIOMEDICA

EDITORIAL PREFACE

The Second International Conference on Applications of Physics to Medicine and Biology was organized by the International Centre for Theoretical Physics and Associazione Italiana di Fisica Biomedica (A.I.F.B). It was dedicated to the memory of Professor Giorgio Alberi, who devoted the last years of his life promoting the idea of applying the knowledge accumulated in physics to medicine and biology. For this purpose, he conceived such conferences of a very general format and was the organizer of the first one.

The main scope of this second conference remained the same as that of the preceding one, i.e. to encourage the application of contemporary physical methods to medicine and biology throughout the world and especially in developing countries. It followed the line of keeping physicists, engineers, biologists and medical doctors abreast with the main developments in the various applications of physics to biomedical sciences, and also providing basic information on new methods and techniques in biomedicine which are under assessment or being put into clinical practice.

The core of the conference were invited talks, organized in four symposia covering four main topics: technologies for the cardiovascular system, NMR in biomedicine, nuclear methods and techniques and hyperthermia and cancer. Other topics were included through invited talks devoted to biomagnetism, biological effects of electromagnetic fields, computerized radiometry, telemedicine, and to human speech and communication aids, which pointed out interesting possibilities and developments in the wide range of potential applications of physics and engineering to medicine and biology.

In addition to the invited papers, more than one hundred contributed papers were presented in three poster sessions. These covered a broad field of applications of physics to biomedicine giving the latest results in related research activities throughout the world. The same poster sessions included also "Work in progress in medical physics in Italy 1983", this being the scientific part of the Second Annual A.I.F.B. Meeting.

This book of proceedings contains all invited papers being prepared as manuscripts in camera-ready form and received up to 28 February 1984. The contributed papers included in this volume were reviewed and accepted for publication by the Scientific Advisory Board. The author index and programme of the conference are included at the end.

We should like to acknowledge the continuous support and encouragement of Professor Abdus Salam, Director of the International Centre for Theoretical Physics, his staff and collaborators. In particular we appreciated very much the effort of Mrs. Janet Varnier who patiently helped us in the preparation of the manuscripts and eased our editorial work.

We also wish to acknowledge the invaluable scientific advice of the International Advisory Committee. Finally, thanks are due to World Scientific Publishing Co. Pte. Ltd., Singapore, for the rapid publication of this volume.

The Editors

CONTENTS

INVITED PAPERS

CONTRIBUTED PAPERS

Physiological Models and Measurements
Radioisotopes

Biological Materials
Technology, Instrumentation and Prosthetic Devices
Non-ionizing Radiation

Ionizing Radiation
Risk Assessment and Control
Hyperthermia
Education and Training

INVITED PAPERS

PROCEEDINGS OF THE II INTERNATIONAL CONFERENCE ON APPLICATIONS OF PHYSICS TO MEDICINE AND BIOLOGY
edited by Ž. Bajzer, P. Baxa & C. Franconi

HUMAN SPEECH AND COMMUNICATION AIDS

Gunnar Fant
Dept. of Speech Communication & Music Acoustics
Royal Institute of Technology (KTH)
S-100 44 Stockholm
SWEDEN

ABSTRACT

This is an overview of speech analysis and synthesis work with applications in medical diagnosis and handicap aids developed at the Royal Institute of Technology (KTH) in Stockholm. An introduction to acoustic theory of speech production is followed by brief comments on phonatory types and analogies to cardiovascular pulse shaping. An improved electrolarynx with intonation control is described. Speech synthesis techniques have been developed for translating unlimited vocabulary texts in seven languages, specified in normal orthography to intelligible speech. Personal speaking aids for people with complete loss of speech motor function are now coming into use. The user can input his text through a computer keyboard or through a special Bliss symbol communication board which suits those who do not have command of the normal written language. A brief summary is made of problems associated with hearing loss and the present development of cochlear implants and tactile aids.

1. Introduction

Research into human speech covers a large area of activites in communiation engineering, linguistic phonetics, psychology, physiology, medical sciences, and special education. In recent years we have experienced a steady growth of activities in this interdisciplinary area.[1] Many new specialities have developed prompted by the needs of the society and of rapidly expanding technological potentials.

There is a growth of both applied work and basic research, of system developments as well as of speech communication theory. The trend of specialization on the applied end combines with an opposite trend, that of integration of concepts, models, methodology, instrumentation, and even of research objectives within the major disciplines.

Thus, we may find developments of methods and instrumentation for acoustic diagnosis of voice disorders in both medical, technical, and phonetic-linguistics departments. The main body of basic research into speech production, speech analysis, and speech perception is found within linguistics, communication engineering and psychology, rather than in medical science.

An apparent growth problem in the applied field is the technological overoptimism, a mismatch between high expectations and our present incomplete insight in human functions.[2)] The advance of microelectronics and computer technology has made possible the implementation of extremely complex systems into compact chip designs. At the same time, there has developed very powerful signal processing algorithms. Speech synthesis as well as automatic recognition of speech are already being exploited in talking chips and simple voice-controlled devices. Speech technology works fine for economical storage and replay of human utterances but encounters limitations when it comes to more advanced objectives such as translation of any text from usual orthographic representation to a readily acceptable synthetic speech output. Such text-to-speech conversion systems for unlimited vocabulary retain obvious synthetic qualities but have already achieved a level of intelligibility where they can be of considerable use in handicap applications and special information retrieval systems. Improvements are steadily under way. However, a perfect automatic reading of a text implies that the computer should know what it is reading, i.e., semantic and situational information has to be included.

The bottle neck is not technology but basic knowledge. We need a fifth generation of speech scientists rather than a fifth generation of computers. With an improved insight in human language, speech, and perception we will eventually pave the way for a society where humans and computers interface through the most basic and direct mode of communication - that of spoken language. You speak to the computer instead of manipulating it from a keyboard and the computers may deliver its response in spoken form as an alternative to the graphic terminal. This principle and associated research is illustrated by Fig. 1.

2. The Speech Code

Our major source of insight in the speech process is through analysis of the speech wave as recorded from a microphone. At this stage we have, at least theoretically, a complete insight in the spoken utterance whereas an articulatory description in terms of the dynamics of the speech articulators is less precise. High-speed moving pictures of the vocal folds and x-ray cineradiography cannot capture all the elements involved in the production. E.m.g. analysis provides but a supplementary source of insight. Our insight in the peripheral auditory system is also incomplete and we have but vague qualitative models of the complexity of the brain functions associated with the formulation, the motor control, and the perception of an utterance.

At this stage there is an obvious divergence of research interests. The medical specialist is primarily engaged in pathology and possibly also in overall functional mechanisms, whereas we speech scientists engaged in studies of language and speech universals have a far-reaching ambition to create abstract models of the encoding of a message in all successive stages within the speaker and the listener. Eventhough this is an impossible task in details, we need an insight in basic principles to guide our understanding of the nature of speech and the constraints imposed by language and by human production and perception.[1)2)]

The situation is similar to that of breaking the code of an unknown signaling system. If we want to teach computers to talk and to understand speech, we must have access to this code in all its variations depending on speaker type, dialect, stylistic variations in addition to the more apparent rules of how the position of a sound within an utterance and overlayed stress and intonation pattern influence its acoustic pattern. This is a theme of variability and invariance of searching for information-bearing elements related

MAN-MACHINE COMMUNICATION SYSTEMS

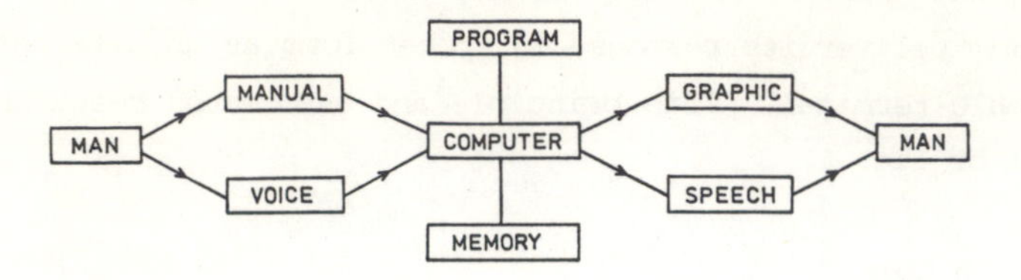

RESEARCH		APPLICATIONS
SPEECH PRODUCTION	SYNTHESIS	INFORMATION SYSTEMS
SPEECH ANALYSIS	RECOGNITION, CODING	DIGITAL TRANSMISSION
SPEECH PERCEPTION	EVALUATION	HANDICAP AIDS
MUSICAL ACOUSTICS	SINGING	DIAGNOSIS
THEORY OF MUSIC	MUSICAL INSTRUMENTS	SPEECH AND LANGUAGE TRAINING

Fig. 1. Man-computer interface with speech in and out. Speech research.

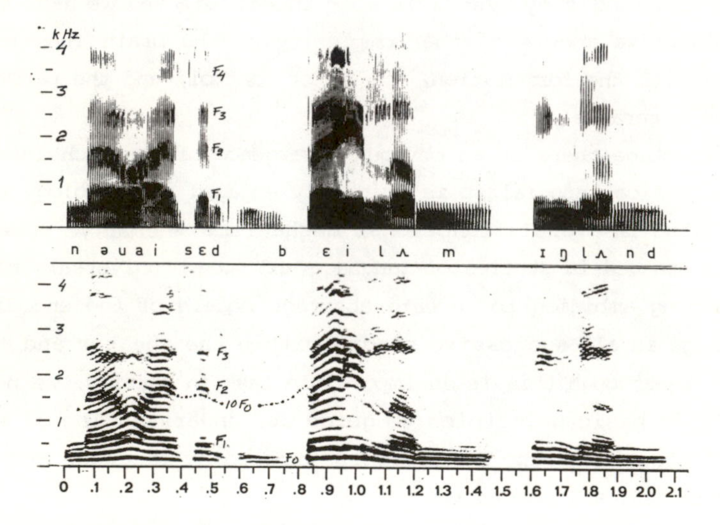

Fig. 2. Broadband and narrowband spectrograms of a sentence: "No, I said Baleham England". Intonation may be quantified by the 10th harmonic as indicated in the narrowband (lower) version.

not only to a formal message content but also to speaker characteristics and to specific denotations and emphasis.

3. Speech Analysis. Spectrographic Representation

The most direct access we have to the speech code is, accordingly, through studies of the speech wave.[3] We must learn the "language of visible speech", or more concrete, we must learn how to read speech spectrograms. Fig. 2 is an example of conventional spectrograms. The text is: "No, I said Baleham England". The upper spectrogram was produced with so-called broad-band analysis which brings out the "formant" structure, i.e., spectral peaks, characterizing successive sounds. The bottom graph is a narrow-band analysis of the same utterance which shows the frequency variation of the voice fundamental and its harmonics within the sentence as a guide to intonation studies.

The spectrogram shows a continuity of formant movements which mirror the continuous movements of the articulators. Adjacent sounds blend into each other and there are not always clear boundaries between adjacent speech sounds.

As an aid to establish systematic relations between articulations and sound, there has developed an acoustic theory of speech production.[4] As shown in Fig. 3, the articulation of a vowel may be specified by an area function describing the variation of the cross-sectional area of the aircolumn within the vocal tract from the vocal folds to the lips. With the tongue in a fronted position appropriate for a vowel [i], the first formant occupies a low position close to the bottom of the frequency scale, whereas the second formant is found at a high position closer to the third formant.

Vocal tract area functions and corresponding spectra of a few vowels are shown in Fig. 4. These are amplitude versus frequency sections at a given time, whilst those of Fig. 2 are frequency versus time spectra with intensity represented by the gray scale.

The identity of speech sounds is related to their formant patterns. There is no absolute invariance. Because of smaller vocal tracts and laryngeal dimensions, females and children have proportionally higher scaled formant frequencies than men. When a vowel is sung on different notes or when in speech an intonation contour

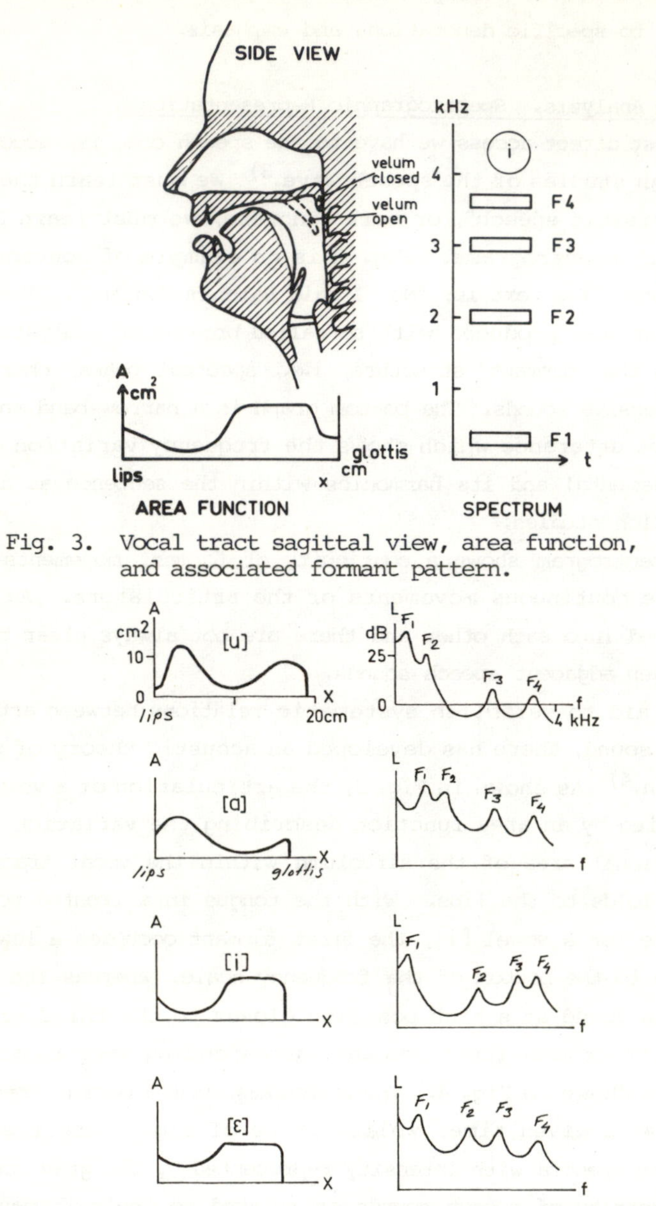

Fig. 3. Vocal tract sagittal view, area function, and associated formant pattern.

Fig. 4. Vocal tract area functions and corresponding spectral sections with formant peaks indicated.

is superimposed on a fixed vowel, the formant pattern stays invariant whilst the fundamental frequency varies.

These relations are best described by a source filter analysis of the production process, as in Fig. 5. The understanding is enhanced by applying both time-domain and frequency-domain concepts. During phonation, at appropriate subglottal air pressure and adjustment of laryngeal musculature, the vocal cords execute aerodynamically induced closing and opening movements which produce a periodic sequence of emitted air pulses through the glottis. The inverse of the distance between two successive air pulses is the voice fundamental frequency.

The time-domain correspondence of a formant is a damped oscillation evoked in the supraglottal cavity system at the instant of glottal closure. One important technical method of studying the voice source characteristics is by means of inverse filtering. The process is conceptually simple. Since the speech wave is a product of a source and a filter, an additional filter inverse to that of the vocal tract will recover the underlying glottal flow pattern.[5)6)7)]

4. Clinical Applications of Speech Analysis

Inverse filtering provides a non-invasive technique for studying the function of the vocal cords.[8)] Typical pulse shapes are illustrated in Fig. 6. A breathy voice is characterized by a smooth return of the pulse to a minimum value which is superimposed on a steady DC air component representing "leakage". The spectrum shows a low-frequency dominance.[9)] In the normal voice, on the other hand, the abrupt decay of the pulse causes a significant excitation of formants. This excitation is proportional to the derivative of the pulse at closure or to the "declination" time of the closing branch which is the ratio of the pulse peak amplitude to the derivative at closure.

The analogy of the human voice to the cardiovascular system is apparent. F. Lund in 1949 introduced the parameter "inclinication time" for analysis of pulse curves from fingers and toes.[10)] It would be interesting to see to what extent models from speech analysis could promote developments within cardiovascular system analysis

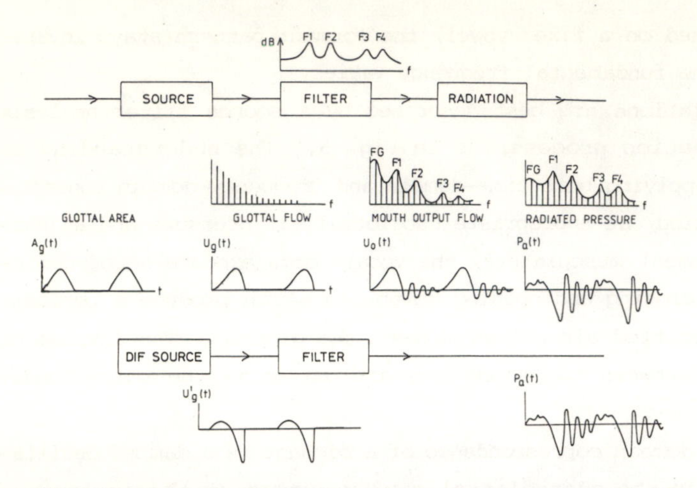

Fig. 5. Frequency and time domain representation of the production of voiced sounds.

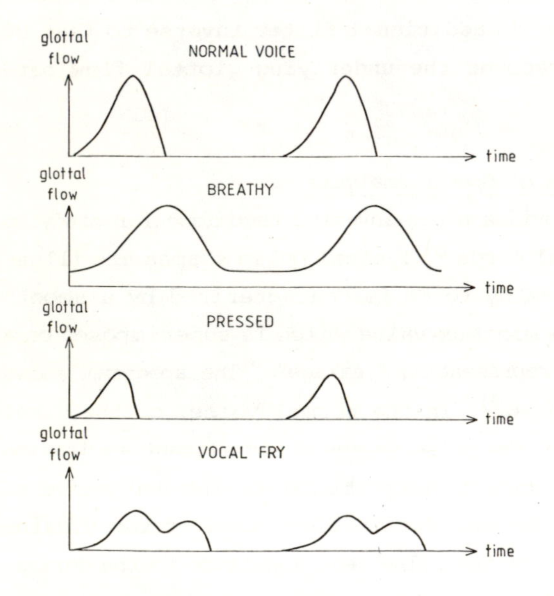

Fig. 6. Typical glottal flow pulse shapes for different phonatory types.

and vice versa. In both systems the pulse curve conveys important diagnostic information concerning a valve functioning as a source and the properties of a transmission system. In a vocal fry register at low voice fundamental frequencies, the flow often attains a "dicrotic" pattern with one large peak followed by a smaller peak within the same fundamental period, see Fig. 6. In peripheral pulse curves, on the other hand, a secondary shallow hump is a normal attribute associated with the elasticity of blood vessels. However, these analogies can be brought to a more stringent base by the process of separating source and transmission properties in the analysis.

I shall not engage in the details of instrumental methods for diagnosis of voice disorders but merely mention some possibilities. Various measures of aperiodicity and noise components have been suggested, e.g., for the early diagnosis of laryngeal cancer. Such methods are still under development and have not yet penetrated clinical practice. It is hard to instrumentally compete with a trained ear. On the other hand, we could gain some additional evidence.

The intonation contour in connected speech is highly correlated to states of depression. Åsa Sjölin-Nilsonne, a Swedish psychiatrist, has demonstrated a leveling of the intonation contour associated with a narrowing of the span of the voice fundamental frequency histogram and a decrease of the rate of change in intervals of inflection.[11] These voice changes vary with the patients' mental states.

5. Electrolarynx

In our department at KTH in Stockholm we have developed an artificial larynx for those who have had their larynges removed after throat cancer. Conventional electrolarynxes produce a monotonous sound. The novelty with our device[12] is that it allows its user to produce a fairly natural intonation. This is accomplished by means of a manual control of the fundamental frequency of the vibrator which is attached to the neck. The frequency control is contained in a small hand-held unit manipulated by opening and closing movements of one or two fingers, see Fig. 7. I shall play you a demonstration tape in which the device first will be operated in a monotone mode and then with the intonation control. I may add that the demonstration pertains to a phonetically trained normal subject. A conclusion is that the fine adjustments of intonation pattern, normally involving the cricothyroid and cofunctioning laryngeal muscles, may be transferred by learning to the fingers.

The electrolarynx is a substitute for the laryngeal voice source in speech whilst the subject executes normal dynamic control of his articulation. People with complete loss of speech motor functions can now use a complete synthesizer. This is a line of development in which we are heavily engaged.

6. Speech Synthesis Aids

Let me first point out that our approach is to produce synthetic speech from a computerized analog of the human vocal tract, represented by a voice source and a noise source and appropriate filter circuits to continuously shape speech sounds according to a set of control functions which, in essence, are the analog of human neuromuscular patterns. The input to the synthesizer is ordinary spelled text. The translation from common orthograph to phonetic symbols and to phonetic parameters is performed within the system, see Fig. 8.[13] The user can, thus, type his text on an ordinary keyboard or the synthesizer may be connected to the text read-out of a computer or to a tape record containing the printing instructions for any text, e.g., a book or newspaper article.

The systems make use of all available knowledge of spelling rules in a language and it can store phonetic representations and control parameters of the most common words to be used or some special vocabulary frequently employed by the user. The core of the control system is a large set of rules for translating the spelled forms to synthesis parameters. There is also special routines employed for reading sequences of numbers and special symbols.

This device, which I now will demonstrate to you, was developed in our institution and is now sold by a Swedish firm. It is provisionally programmed to speak in seven different languages: Swedish, English, French, German, Italian, Spanish, and Chinese. It has already found a use as a speech prothesis for nonverbal handicapped

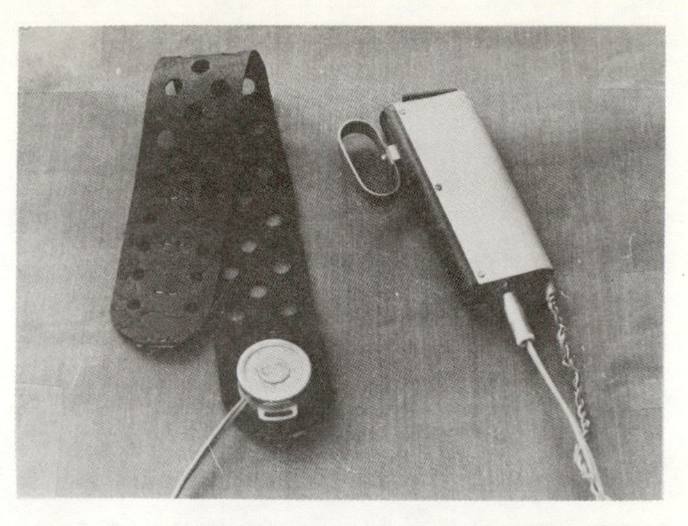

Fig. 7. Electrolarynx with separate intonation control.

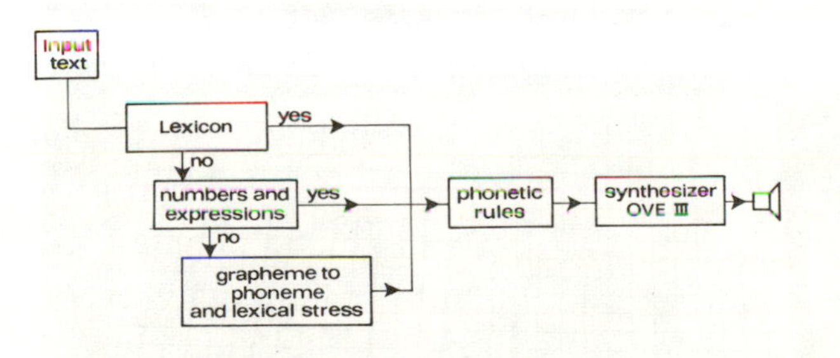

Fig. 8. Text-to-speech conversion.

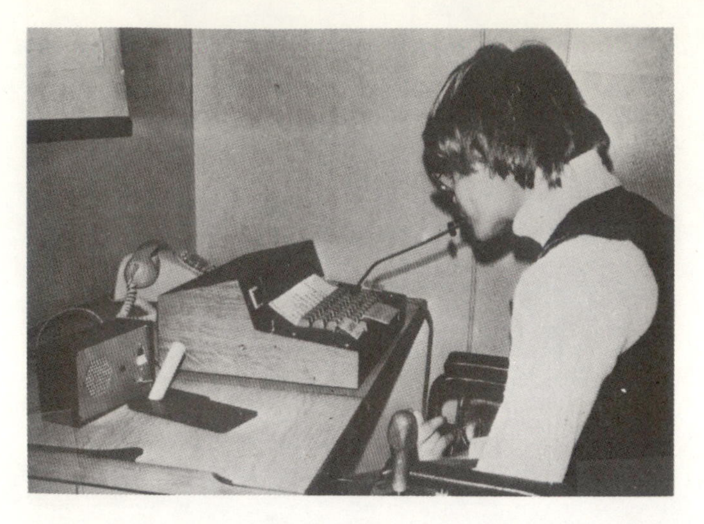

Fig. 9. CP-handicapped boy using mouthstick for keyboard control of personal synthesizer.

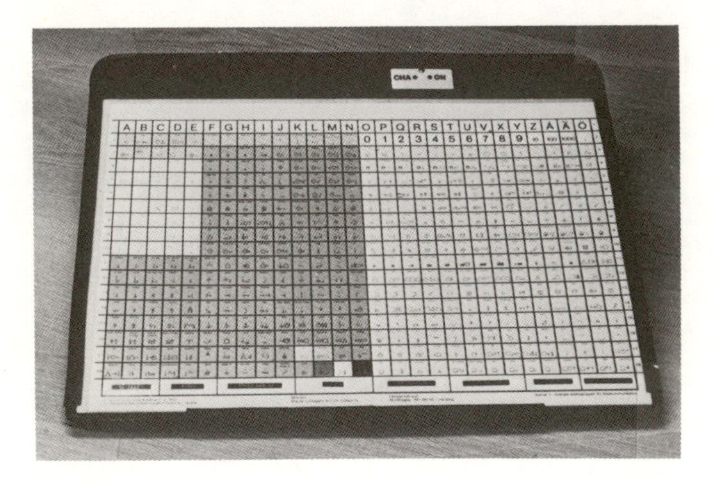

Fig. 10. Bliss-board communication table.

and as a reading device for the blind. A larger market is expected in information retrieval systems for public use and in office automation.

Fig. 9 illustrates how the synthesizer is used as a communication aid by a CP-damaged boy who manipulates the keyboard with a mouthstick.[14] He has a good general knowledge of normal language. The access to a speaking output provides him with a more direct mode of communication than by writing or typing out a text. The synthesizer provides unlimited access to any text and can be preprogrammed to store commonly used words and parts of sentences.

For those who lack both speech and command of written language, we have developed a speaking attachment to a Bliss board, Fig. 10.[14] Instead of or as a supplement to normal orthography, a Bliss board contains a matrix of symbols that stand for objects, actions, grammatical modifications, and emotions. This symbol system is by now well established in many countries. The originator, Charles Bliss, was inspired by the Chinese sign system. An example of a sentence in Bliss language is shown in Fig. 11. With our speaking Bliss board, the synthesis of words may be executed one at a time as they are selected. The main feature, however, is that as the user activites the symbol "period" [.] for sentence ending, the internal program rearranges and transforms the sentence into a grammatically correct form. I shall illustrate this with a tape-recording.

Our first years of experience with speaking aids indicated that they promote language development and command of normal written language and spelling. There is a need to make the synthesizers adaptable to a large range of possible voice qualities but this is under way.

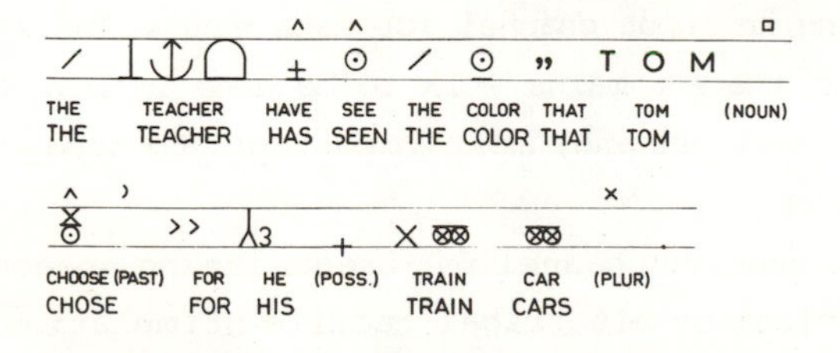

Fig. 11. Example of a sentence in Bliss symbols.

7. Aids for Profound Hearing Loss

It is a well-known fact that hearing aids do not have the same sensory restoring potential as eyeglasses. Loss of hearing is in limited cases only a matter of attenuation that can be compensated by amplification. The usual hearing loss is sensorineural. We are all more or less confronted with this at old age. It involves reduced frequency discriminability, reduced intensity handling range, loss of coordination between left and right side auditory processing, impaired short-time memory, and difficulties in concentrating on a single speaker in a group. For some people with a substantial hearing loss, a hearing aid is of little or occasional help only.

It would bring me too far in this article to review major problems of deafness and selection of hearing aids. I have a few comments to make about recent developments for profoundly hard of hearings.[15] There are great hopes attached to the so-called cochlear implant techniques for electrical stimulation of the inner ear through a single or a multichannel electrode inserted into the cochlea or at the round window. Patients hope to get their hearing back and many otology specialists feel the challenge of taking up a new fashinable operational technique. However, implant aids provide a very crude auditory sensation that is a useful supplement to lipreading but can generally not convey sufficient speech information when used alone. However, exceptions have been reported with aids employing a complex recoding of the speech signal to provide an optimum interface by spatially distributed formant pattern information through a multielectrode system. Implant aids are generally not used for deaf-born patients. A critical analogy to implant techniques would be that of driving a long nail into a computer with the hope of creating an input channel for data entry. The situation is not that bad but there remains work to be done in investigating various implant techniques and, most important, to study various speech coding schemes.

An alternative sensory channel for transmitting speech to the deaf is via mechanical or electrical tactile stimulation on the skin, e.g., by means of a single vibrator held in the hand which is the simplest technique. One of my colleagues, Karl-Erik Spens, has evaluated various techniques including the use of a multi-element

Optacon vibrator of the type used by the blind in connection with a photocell for optical scanning of a text.[16)] The tactile transducer is applied to a finger. By a proper selection of spectrum representation in the rows of the vibrator matrix and a few time slots for successive columns of the pattern, he learned to recognize the ten digits with a fair degree of accuracy. However, the training time needed was substantial. Other people have designed laborious systems with spatially distributed vibrators carried in a belt around the waist or on the legs or arms, see Fig. 12. We may expect to improve tactual aids by more complex coding strategies and sensory capabilities and by radical miniaturization. Much of present development work is directed towards simple aids. One such aid, designed by Spens, has the vibrator built into a ring or a bracelet or an earmould. It has been very well received by patients.

A review of these techniques is given by Spens.[16] An extensive survey of speech-coding techniques including cochlear implants is given by Risberg.[15)]

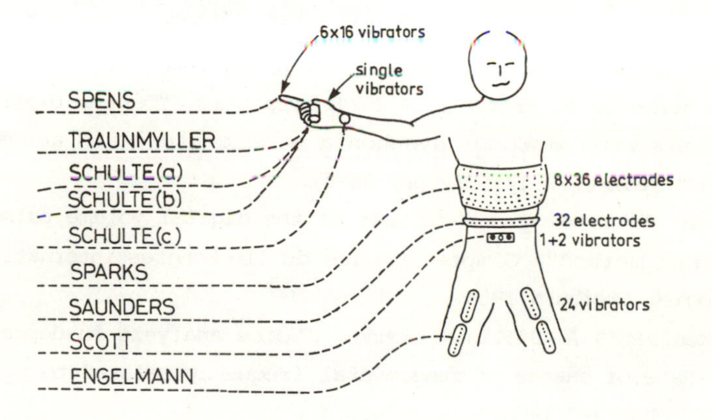

Fig. 12. Various tactile aids developed for communication with deaf subjects, Spens [15)].

References

1) G. Fant: "Perspectives in speech research", STL-QPSR 2-3/1980, pp. 1-16.

2) G. Fant: "Phonetics and speech technology - a revisit", STL-QPSR 2-3/1983, pp. 20-35.

3) G. Fant: "Analysis and synthesis of speech processes", pp. 173-277 in Manual of Phonetics (B. Malmberg, ed.), North-Holland Publ. Co., Amsterdam 1968.

4) G. Fant: Acoustic Theory of Speech Production, Mouton, Haag, 1960 (2nd edition 1970).

5) J. Sundberg & J. Gauffin: "Waveform and spectrum of the glottal voice source", pp. 301-320 in Frontiers of Speech Communication Research (B. Lindblom & S. Öhman, eds.), Academic Press, London 1979.

6) G. Fant: "The voice source - acoustic modeling", STL-QPSR 4/1982, pp. 28-48.

7) T.V. Ananthapadmanabha & G. Fant: "Calculation of true glottal flow and its components", Speech Communication 1 (1982), pp. 167-184.

8) J. Gauffin & J. Sundberg: "Data on the glottal voice source behavior in vowel production", STL-QPSR 2-3/1980, pp. 61-70.

9) B. Hammarberg, B. Fritzell, & H. Schiratzki: "Teflon injection in 16 patients with paralytic dysphonia - perceptual and acoustic evaluations", STL-QPSR 1/1981, pp. 38-57.

10) F. Lund: "Morphological analysis of the digital volume pulse as a diagnostic method", Comptes rendus du IIe Congres international d'Angeiologie, Freiburg 1955.

11) A. Askenfelt & Å. Sjölin-Nilsonne: "Voice analysis in depressed patients: Rate of change of fundamental frequency related to mental state", STL-QPSR 2-3/1980.

12) G. Fant, K. Galyas, P. Branderud, S-G. Svensson, & R. McAllister: "Aids for speech-handicapped", Scand.J.Rehab.Med. 8 (1976), pp. 65-66.

13) R. Carlson, K. Galyas, B. Granström, S. Hunnicutt, B. Larsson, & L. Neovius: "A multi-language, portable text-to-speech system for the disabled", STL-QPSR 2-3/1981, pp. 8-16.

14) R. Carlson, B. Granström, & S. Hunnicutt: "A multi-language

text-to-speech module", STL-QPSR 4/1981, pp. 18-28.
15) A. Risberg: "Speech coding in aids for the deaf: An overview of research from 1924 to 1982", STL-QPSR 4/1982, pp. 65-98.
16) K-E. Spens: "Tactile speech communication aids for the deaf: A comparison", STL-QPSR 4/1980, pp. 23-39.

text to speech modular." BIT, 1986, 4(34), pp. 16-28.

(b) R. Schafer. "Speech coding in the Festtext text." An overview of research from 1974 to 1987", ICASSP, 47303, p. Knox.

(c) L.R. Rabin. "Real-time speech communication aids for the deaf: a communicate." ICASSP, I.360, pp. 33-39.

PROCEEDINGS OF THE II INTERNATIONAL CONFERENCE ON APPLICATIONS OF PHYSICS TO MEDICINE AND BIOLOGY
edited by Ž. Bajzer, P. Baxa & C. Franconi

TECHNOLOGIES OF THE TOTAL ARTIFICIAL HEART

E. Hennig
Chirurgische Klinik, Klinikum Charlottenburg, Freie Universität Berlin, Spandauer Damm 13o, 1ooo Berlin 19

Introduction

Comparing the natural heart with the total artificial heart system, an engineer has to confess that the biological design is a marvellous piece of workmanship that neven can be fully replaced by a technical system.

To replace the pumping function of the heart, we do need two pumps replacing the action of both cardiac chambers, two actuators to replace the muscle of the heart, a sophisticated control system to replace the physiological control of the left and right cardiac output with respect to the pre- and afterload as well as the metabolic demands, and we do need an energy source replacing the metabolism of the myocardium.

A schematic of a pneumatic artificial heart system where the relatively bulky energy converter, the automatic control system for the driving parameters and the energy supply have to be placed outside of the body, is shown in Fig. 1. Only the two blood pumps are implanted into the thoracic cavity. With a system like this we do need percutaneous devices allowing the infection-free entrance of the driving lines and measurement leads into the body.

The ultimate goal is the development of an artificial heart system that is fully implantable (Fig. 2). All components are kept inside the body, the necessary energy will be provided by an internal energy source. As long as no long lasting internal energy source will be available, a system consisting out of an external energy supply and electrical transmitters, e.g. by inductively coupling superficial and implanted coils of wire, recharging an implanted battery, would be acceptable. This fully implantable system would allow free mobility for the patient, but from the technical point of view it creates a number of problems that have not been solved as yet.

We have to keep in mind that the requirements for implantable components of the system are much higher than for those kept outside the body. External components can be easily supervised, periodical maintenance is possible or, if necessary, the planned exchange of parts subjected to high load. With additional back-up systems, this will lead to long-term high reliability. It is not necessary to absolutely minimize weight and volume of these components, thus increasing the long-term durability of the mechanical parts.

Implanted components have to be extremely small and light weight to fit in the anatomical space and to be physiologically compatible. They have to be hermetically sealed, a safe maintenance-free operation for at least five years must be assured. These systems are inherently complex and, as we all know, with an increasing number of moving parts the reliability of the system decreases. Current foci of research activities are the problems of displacement volume compensation, fluid penetration across the polymer blood pump diaphragm and the demonstration of long-term durability and reliable function in vitro and in vivo. The engineer faces different technical problems, depending on the kind of system components and their design, especially whether they are meant to be implantable or kept extracorporeal (Tab. 1).

COMPONENTS OF THE TAH-SYSTEM

- Energy converters
- Blood pumps
- Percutaneous devices
- (Transcutaneous devices)
- Measurement, surveillance
- Control systems
- Energy source and storage

ENERGY CONVERTERS

Many different approaches for the development of energy converters for circulatory assist devices or total artificial hearts are currently under investigation, in the stage of in vitro or in vivo testing.

Extracorporeal driving systems:

Most experience has been gathered with extracorporeal driving systems. In all of these systems gases are used for the transmission of a pressure and flow pulse to avoid inertial forces. In the "open" pneumatic system, the energy is stored in pressure- and vacuum tanks. The pulse wave is generated by valves,activated by an electronic pulse generator.Low and high pressure systems have been developed,the latter using a conversion piston for the necessary low pressure activating diaphragm blood pumps.

To provide controlled systolic pressure and diastolic vacuum as well as a preset systolic and diastolic timing is much easier with an electropneumatic driving system utilizing a linear piston driven by a reversing electromotor[1)] (Fig. 3). The rotation of the motor is converted into the translatory motion of the pneumatic piston, using a rack and pinion. Driving frequency, systolic pressure, diastolic pressure and length of systole can be controlled electronically by modulating the electromotor's voltage.

Instead of the linear moving piston, a rotary piston can be used, the pulsatile air flow is generated by driving a Wankel-compressor in a forward/reverse mode by a DC electric motor (Fig. 4). This concept permits a reduction of the energy converter's size and weight, resulting in a portable driver[2)]. The family of electropneumatic extracorporeal driving systems used in Berlin for animal experiments with the total artificial heart (TAH) and ventricular assist devices (VAD) is shown in Fig. 5. The stationary units have been developed by AEG-Telefunken. The small portable driver consisting out of a Wankel-compressor, a battery pack and the control electronics, allowing an independent operation for two hours, is an inhouse design.

Implantable driving systems:

Implantable energy converters for the TAH are still in the beginning stage of their development. At present, the research is concentrating on the development of implantable driving systems for left ventricular assist devices. The National Heart, Lung and Blood Institute of the United States (NHLBI) is sponsoring a program in which two general categories of implantable energy systems - electrical energy systems and thermal energy systems - are under development. Main goals are selection of a source of energy to provide mobility, selection of the appropriate energy conversion technique to translate the energy from the source to the form required to actuate the blood pump and the definition of methods to control the blood pump to meet physiological needs. In an additional program, the NHLBI[3)] supports several research groups who develop the necessary percutaneous leads or transcutaneous transformers to transfer electrical energy and signals into the body from the external battery pack or out of the body for control and supervision of the implanted system.

The thermal engines under development have the potential of operating with a thermal storage capsula heated by an electric heater or by a nuclear heat source like a PU 238 capsula. The funding on the nuclear fuel powered artificial heart has discontinued in the United States as well as in Italy and Germany, where similar programs where running[4,5,6)].

Some of the current concepts for implantable energy converters for VAD are listed in Tab. 2, explaining the program sponsored by the NHLBI 1982[7)].

Implantable energy converter (NIH Program 1982)

Developer (system)	Description	Weight(g)	Volume(cc)
Novacor Medical Corp. (electromagnetic)	short stroke solenoid energizes spring (pusher plate pump)	675	25o (without pump)
ABIOMED (electrohydraulic)	unidirectional axial flow pump driving hydraulic fluid, four way sliding sleeve valve (double chambered free diaphragm blood pump)	6oo	6oo
NIMBUS INC. Cleveland Clinic Foundation (electrohydraulic)	high speed brushless motor, gear pump, double acting piston (pusher plate pump)	1.ooo	65o
THERMO ELECTRON Corp. Children's Hospital Medical Center (electromechanical)	low speed torque motor, pair of cam followers (pusher plate pump)	41o	87o
GOULD INC. THI (electromechanical)	high speed brushless motor, gear reduction, face cam actuates levers (pusher plate pump)	49o	147 (without pump)
NIMBUS INC. University of California (thermocompressor)	Stirling cycle thermocompressor with heat pipe	1.22o	62o
University of Washington Cleveland Clinic Foundation (thermal converter)	Stirling/hydraulic converter	93o	41o

The electromagnetic system consists practically out of only two moving parts, directly coupled with a dual pusher plate blood pump. It has demonstrated impressive reliability.

The electrohydraulic converter with a dual chambered pump and integrated tri-leaflet valves solves the displacement problems existing in all fully implantable systems. The stroke volume of the pump will be compensated during ejection and filling even with the hermetical sealing of energy converter and pump.

The electrohydraulic energy converter with a high speed continuous running motor driving a hydraulic gear pump supplies the hydraulic fluid to a double acting piston which drives the pusher plate pump. For the volume compensation, an implanted flexing bag is used. The electromechanical energy converter with a low speed torque motor activates a pusher plate pump at one eject cycle per evolution, using a pair of cam followers.

The electromechanical energy converter with a high speed brushless DC motor uses a gear reduction mechanism to drive a face cam that actuates levers for the movement of the pusher plate.

The two different approaches for thermal converters use electrical heat sources and thermal salt energy storage, allowing up to ten hours of tetherfree operation. They are also directly coupled with a pusher plate pump, both engines are tested with the same parathoracic pump (12A and 12C, developed by Thermo Electron Corp.).

Some of these concepts have the potential to be used after modification as implantable energy converters for the TAH system, but today their clinical readiness is not foreseeable.

Other concepts for electromechanically driven total artificial hearts, such as the early design of M. Arabia[8)] and the current design under development by G. Rosenberg and W. Pierce[9)] are also not ready for the clinical use. The implantable electrohydraulic energy converter developed by R. Jarvik[10)] is in the animal experiment stage, its clinical application cannot be expected in the near future.

Taking today's technical status of implantable energy converters into consideration, the practical application of the total artificial heart in humans will be based on percutaneous systems with implanted blood pumps and external electropneumatic energy converters[11]. With such systems in more than 12oo animal experiments worldwide with maximal survival times of up to 3oo days, sufficient experience has been achieved to start the first clinical experiments. To achieve comparable experience with implantable energy converters will take, for my opinion, at least three to five years of extensive animal experimentation. But the implantable systems will evolve if the necessary financial support for the research will be provided[12]. First steps in this field of research are animal experiments with modified implantable ventricular assist devices that already have proven their applicability[13].

BLOOD PUMPS

Different designs of blood pumps have evolved within the past two decades of development. Their special features are influenced by the kind of the respective driving system utilized, the blood/material contact hypothesis of the designer and fabrication methods and materials available. An incomplete list of implantable blood pumps in experimental animals or clinical use today is summarized in Tab. 3 [14,15,16,17]. Almost all of the designs are using separate blood pumps for the lung and systemic circulation. The blood driving chambers are separated by a flexible elastomer diaphragm allowing in connection with a relatively rigid housing the conversion of the activating air pulse into a corresponding blood pulse.Today,two main concepts are favored: the free diaphragm pumps where the housing and the diaphragm forms a continuous blood contacting surface, and the tethered-sac design where a separate blood sac is enclosed within a rigid housing. Most pumps have tilting disc valves to achieve unidirectional flow.

Problems of sufficient anatomical fit within the thoracic cavity of the respective recipients and the internal flow pattern of the blood allowing the necessary wash-out with each pumping cycle to avoid regions of stagnating flow and such preventing thrombus formation, have been adequately solved. Today's most severe problem is the fatigue behaviour of the moving parts of the pumps, especially of the diaphragm in regions of high flexion in connection with mineralization of the polymers.

Implantable Blood pumps - TAH (1983)

Developer	Description	Blood Contact
Pennsylvania State University (Thoratec)	sac-type, pneumatic-ally activated, tilting disc valve	smooth continuous Biomer^R
Pennsylvania State University	sac-type, pusher-plate electromechanical tilting disc-valves	smooth continuous Biomer^R
University of Utah Salt Lake City	free multilaminar diaphragm, pneumatic-ally activated, tilting disc valves	smooth continuous Biomer^R
University of Utah Salt Lake City	multilaminar diaphragm electrohydraulical tilting disc valves	smooth continuous Biomer^R
Cleveland Clinic Foundation	polyolefin rubber diaphragm, pusher-plate, electrohydrau-lically activated tri-leaflet valves (Duramater)	textured surface, coated with gelatine, cross-linked with glutaral-dehyde (biolized)
Klinikum Charlot-tenburg, FU Berlin	free diaphragm, pneum-atically activated, tilting disc valves	smooth continuous Pellethane^R
Klinikum Charlot-tenburg, FU Berlin	multilaminar dia-phragm, pneumatically activated, tilting disc-valves	smooth continuous Pellethane^R
University of J.E. Purkinje KUNZ Brno	free diaphragm pneumatically acti-vated, flap-valves	smooth Polyurethane
University of Tokyo	sac-type, pneumatical-ly activated, tilting disc-valves	smooth continuous Polyurethane
University of Hiroshima	free diaphragm pneumatically activ-ated, tilting disc-valves	smooth continuous Polyurethane
National Cardio-Vascular Research Center Osaka	free diaphragm pneumatically activ-ated, tilting disc-valves	smooth continuous Avcothane 51^R
Chirurgische Klinik Innsbruck	free diaphragm, pneum-atically activated, disc-valves	smooth continuous Avcomate 61o

Design:

I would like to discuss these fatigue problems on the basis of our experience with free diaphragm polyurethane blood pumps, driven by the AEG-Telefunken electropneumatic energy converter[18)]. In all of our long surviving animals (up to 213 days), blood pumps made out of polyurethane, fabricated by a dip-coating process[19,2o)], have been implanted (Fig. 6). The most common reason for the termination of the experiments was fatigue failure of these pumps. Even after a relatively short functional time of less than 3o days, micro cracks formed around the diaphragm housing junction (DHJ), leading to thrombus formation, followed later by calcification (Fig. 7). The rigid calcium crystals attached at the blood contacting surface in regions of maximum flex and high buckling stress resulted in secondary wear, puncturing the diaphragm, separating the blood- and air chamber[21)]. Apart from the durability, the reproducibility has been an unsolved problem because the fabrication process was very difficult to control.

The production was extremely time consuming, the many necessary steps led to a high number of rejects.

To solve these problems, a completely new pump has been developed, basing on the Jarvik design of pneumatically driven blood pumps with multilaminar diaphragms (Fig. 8). The new pump consists out of a housing with an integrated diaphragm, forming a continuous blood cantacting inner surface. Two or more driving diaphragms are connected with a circular base-plate, forming the driving chamber[22].

Because of the multilaminar construction of the diaphragm, the single membranes can be very thin and high bending stresses as well as high flex in the DHJ can be avoided. A proper combination in size of the diaphragms leads to a nearly unloaded blood contacting membrane, the stresses are taken over by the air pressure loaded driving membranes. The angle of bending at the DHJ can be reduced by separating the circular flex lines for positive (systolic) and negative (diastolic) deflection by a supporting ring.

Fabrication:

The design allows a simple fabrication, using a combination of vacuum molding of parts out of prefabricated polyurethane sheetings and dip-molding of parts out of polyurethane solution (Pellethane^R). The rigid upper part of the housing with in- and outlet ports is vacuum molded out of Plathurane sheetings with a thickness of 4 mm.

The base-plate is vacuum molded out of the same material[23]. The second driving diaphragm is also vacuum molded, using Plathurane sheetings with a thickness of o.25 mm. For this technique, molds made out of "plaster of Paris" formed in basic Silicone molds have to be used. The first driving as well as the blood contacting diaphragm are fabricated out of Pellethane^R solution, using stainless steel molds with extremely high polished surface. All dip-coating is done in a class loo laminar flow bench to get particle-free air-dried blood contacting surfaces. The thickness of the membranes depends on the viscosity and the solid content of the PE-solution used. The different molds necessary for the fabrication of a blood pump for the replacement of the left ventricle are shown in Fig. 9. The parts of the pumps are fabricated separately. They can easily be inspected. Assembling the parts (Fig. 1o), is simplified by special constructional features of the pump, allowing form and force locking. The first and second driving diaphragm are glued together around the circumference, using the stainless steel mold as a tool. A lubricating medium as graphite, silicone or teflon oil prevents the membranes from sticking together. The housing of the blood pump with the integrated blood diaphragm is also glued together with the driving membrane aroung the circumference with the help of a stabilizing aluminum or titanium ring. The vacuum molded base-plate with a connector for the pneumatic driving line is attached to the housing by glueing, closing the driving chamber of the pump. The Björk-Shiley-tilting-disc-valves in inflow and outflow position are mounted with the help of a specially developed quick connect system made out of silicone coated stainless steel or titanium. The already assembled pumps with the vascular adapters are shown in Fig. 11.

This fabrication process is not only less time consuming than the dip-coating method, it also allows a simple change of the shape of the blood pumps according to the anatomical needs. Pumps with different shapes, especially with different positions of the in- and outflow ports for the implantation in calves, sheep, goats or humans, can be easily fabricated, using the same stainless steel molds. Only the simple vacuum forming molds for the upper part of the housing have to be modified.

With the same basic molds, the driving and blood diaphragms for a family of pumps can be built, usable as extracorporeal left or biventricular assist device or left respective right ventricular replacement.

For the clinical use of our TAH percutaneous system, a modified upper part of the housings of the left and right pump had been evaluated, that fits well within the average dimensions of the human thoracic cavity[24] (Fig. 12).

Performance:

The performance of the blood pumps in terms of cardiac output versus pulse rate is mainly effected by the dynamic stroke volume (DSV). The DSV is limited by the overall size of the blood pump. Its maximum is achieved in the optimal working mode of complete filling and emptying of the blood chamber. Only if the pump is operating with incomplete filling, variations of the filling pressure, the sum of atrial pressure and diastolic vacuum will cause variations of the DSV (Fig. 13). The sensitivity of the artificial heart for changes in cardiac output caused only by changes in atrial pressure is very low compared with those of the natural heart. Most of the energy filling the blood chamber of the pump is consumed by friction losses in the pneumatic drive lines where an equivalent amount of air has to be forced out of the system. Even with high flexible diaphragms and pneumatic lines with larger diameters, for higher frequencies an unphysiologically high atrial pressure is necessary for a complete filling during diastole[25]. To get the maximal possible cardiac output for a given total size of blood pump, we are operating our system always in the complete full/empty mode (maximum DSV), so the pumps are always working with the lowest possible pulse rate for a given cardiac output.

The maximal DSV is affected by the afterload and an increase in arterial and therefore systolic driving pressure results in a decrease of the actual stroke volume. This is partially due to the compliance of the pump's housing and also influenced by changes of the regurgitation of the valves. If both pumps for the replacement of the left and right cardiac chamber have the same static stroke volume (SSV), there will always be an imbalance between the left and right dynamic cardiac output because of the remarkably different pressures in the aorta and pulmonary artery and thus DSV. If the pumps are operating with the same frequency and full/empty mode, the output of the right pump will exceed this of the left for up to 15 %. The bronchial flow will add to this imbalance with 1.5 to 3 %. To compensate for this phenomena, common in all blood pumps with not completely rigid housings and compressible driving media, we use a combination of blood pumps with different static stroke volumes. The pump for the replacement of the left cardiac chamber has a 15 to 18 % larger SSV, such creating a "cardiac reserve" of the left pump to avoid a possible lung edema[26].

The complete performance diagram (Fig. 14) for a combination of a right pump with 1oo ml SSV and a left pump with 12o ml SSV shows that within the possible limits of the afterload of the pulmonary circulation (right pump) and of the body circulation (left pump), this coupling works in a good balance over a wide variety of pulse rates. This combination, fitting well within the thoracic cavity of a 45 kg Jersey calf, will over the possible pulse rate range always assure a certain cardiac reserve of the left pump (Fig. 15).

Fatigue Testing:

For the empirical optimization of the thickness of the diaphragms in the new pump design, a series of pumps with a variation in the thickness of a single membrane (5o µm to 25o µm) has been built and investigated in a fatigue test station with 24 positions. The test station (Fig. 16) runs with a pulse rate of 1oo beats/minute, the driving pressures are 1.5 time equivalent to those of the left pump under physiological conditions[27)].

Apart form some early failures because of insufficient lubrication due to impurity of the used graphite, the pumps are running for more than one year now in ongoing tests (Fig. 17). An influence of the membrane's thickness within the given limits on the durability could up to now not be observed. Until we will have the fatigue results showing a good correlation between structural dimensions of the membranes and durability, the fabrication parameters for the pump parts made out of polyurethane solutions (percent solid contents, viscosity, temperature, amount of material) are chosen such as to assure a thickness of 11o to 15o µm around the DHJ and the circumference of the driving diaphragm. Samples out of the high flexed regions of the diaphragm were taken for microscopic investigations, no structural damage at the blood-contacting surface could be detected even after more than 1oo x 10^6 load cycles.

CONNECTORS

The rigid silicone coated stainless steel parts mounted at the in- and outflow ports have to be connected with the atrial and arterial vessels. A quick connect system has been developed, using adapters made out of Dacron reinforced silicone rubber prepared at one end for the end to end anastomosis with the natural tissue and the other end with a stainless steel spring for a safe and tight fixation at the pump. With the help of a modified clamp, the spring can be spread to facilitate an easy insertion of the pump.

Integrated in these four adapters, shown in Fig. 18, are the air capsulae for the noninvasive continuous measurement of the blood pressures. The blood contacting surface of the adapters is extremely smooth. Steps at the connection site are avoided as far as possible because they are known as starting points for thrombus formation and could induce the growth of the so called pannus. Pannus formation occured especially in the left or right atrium, but only if an infection around the suture lines was present. In these cases, an inflow obstruction developed very rapidly, leading to the termination of the experiment within some days[28)].

PERCUTANEOUS DEVICES

As long as fully implantable artificial heart prostheses are not available, percutaneous devices are essential parts of the total artificial heart system. They serve as percutaneous conduits for the transmission of energy and signals. With our pneumatically driven system, we do need the percutaneous device for the entrance of the two pneumatic driving lines into the thoracic cavity to be connected with the implanted blood pumps and for the leads transferring pneumatical or electrical signals. To achieve this entrance, an appropriately sized defect has to be created in the skin, kept open by the presence of a percutaneous device, preventing the reforming of the integumental continuity[29]. For the individual design of the PD, many factors have to be taken into consideration, such as the size of the implant, selected material, mechanical stresses and implant site.

In our lab, a skin button has been developed, allowing the minimization of mechanical stresses[3o)](Fig. 19). A soft air chamber (a) interposed between skin and implant absorbes the surface near stresses. Its Dacron velour cover, acting as micro anchor, allows epithelial tissue ingrowth. A Dacron velour flange (b), positioned just below the dermis, allows fibrous tissue ingrowth, a fixation plate (c) absorbes the stronger pulling forces on the driving lines. This device for higher acting forces could be maintained for up to seven months in a calf under total artificial heart conditions. The percutaneous device for the measurement lines is not subjected to high mechanical load, a simple dacron velour flange with a soft air chamber placed between the PD and the epidermal attachment surface has proven to be functional for more than six months. The skin buttons used in the Berlin artificial heart project are shown in Fig. 2o.

MEASUREMENTS

Driving pressures:

Since the blood pumps are activated by a pressure controlled pneumatical driving system, the knowledge of the pressure pulse curve within the pump is essential for manual or automatical setting of the driving parameters. The pressure pulse curve gives the necessary informations for an optimal endsystolic and enddiastolic timing when the blood pump is completely emptied or filled. Because of the known difficulties in measuring the blood pressure invasively, we monitor the noninvasively measured air pressure in the driving chamber of the pump.

To measure the pressure within the blood pumps, it is not necessary to have a transducer there. An additional catheter fed through the driving tube close to the air chamber can measure this pressure very easily, the transducer is positioned outside of the animal. The curvature of the air pressure pulse in the pump is very similar to that of the blood pressure, it allows a very delicate analysis of the filling and ejecting cycles[31]. While the diaphragm is at its endsystolic fully expanded position, the shape of the pneumatic pressure pulse shows an additional peak. A suction spike is detectable when the diaphram is in the fully expanded enddiastolic position. The detection of these peaks within the air chamber is very sensitive and permits an optimal setting of the driving parameters, so that the time of diaphragm rest in either endposition can be greatly reduced.

Cardiac output:

The cardiac output is continuously calculated out of the air pressure and the air flow within the driving tube[32]. The air flow is indirectly derived from the velocity of the piston drive, given by the tacho signal. This computation of cardiac output allows the estimation of the beat-by-beat values[33]. If only the mean cardiac output is required, the calculation can be done by a simple analog computer working with a satisfying acuracy[34].

Blood Pressures:

For automatic control used for physiological studies and for the surveillance of the complete total heart system, the long term measurement of the circulatory parameters such as atrial and arterial pressures, is desirable. A semi-noninvasive measurement system using small air capsulae has been developed, allowing the continuous long-term reliable measurement with extracorporeal pressure transducers [35,36]. The air capsulae are integrated in the four adapters already shown in Fig. 18. The necessary hardware for this pressure measurement system, a small pneumatic rollerpump, an automatic timer for its activation, a pressure transducer and amplifier form a compact miniaturized self-containing unit. In addition to the driving and circulatory pressures, the p-wave of the remaining right atrium is documented continuously by implanted electrodes.

A complete list of the signals for control and monitoring is given in Fig. 21. Fig. 22 shows an example of a documentation of an episode of an artificial heart experiment. All data derived in the animal experiments are online stored and processed in a computer (PDP 11/4o) with the help of appropriate software[37].

Control of the Pneumatic Total Artificial Heart System

This problem can be subdivided into three intermashing problems:

- the adaptation of the driving unit to the blood pump
- balancing the cardiac output of the right pump with that of the left
- adapting the total cardiac output according to the physiological needs.

Adaptation of the driving unit to the blood pump:

The blood pumps have to be always driven in the optimal working mode, in which the blood chamber is completely filled and emptied within each cycle. The filling of the pump is effected by the pressure in the atrium (preload) and the diastolic vacuum produced by the unit. The emptying of the pump depends on the arterial pressure (afterload) and the acting systolic air pressure of the driving unit.

Pressure rise and fall (dP/dT) are assumed to be constant. Minor influences such as pressure losses because of the artificial heart valves can be neglected, they will be compensated by the respective driving pressures. Four different possibilities to achieve the optimal working mode have been investigated, resulting in different system behaviour.

a) Systolic and diastolic driving pressures are kept constant: The optimal systolic and diastolic duration is detected by the filling spikes in the pressure pulse curve and the reversing points of the air flow. This method results in a change of frequency and cardiac output, depending on pre- and afterload (Fig. 23).

Increasing afterload will cause a decreasing frequency and cardiac output because the systolic duration will be longer. By an increase in preload with constant afterload, the diastolic duration will be shortened, the frequency and thus cardiac output will increase. This method, defined as full stroke triggering, assures always the optimal working mode, but the sensitivity of the blood pump for changes in pre- and afterload will be very poor.

b) The frequency is kept constant and systolic and diastolic driving pressures are controlled according to the resptective preset duration. In this case, the cardiac output will not be influenced by changes of pre- and afterload and will be constant as long as the venous return allows a complete filling of the pump.

c) To achieve a higher sensitivity of the system, we combined a full stroke trigger with a second controller that keeps the relative systolic/diastolic duration constant by changes of the systolic driving pressure. The system acts similar as the full stroke trigger, but in cases of an increase in afterload (increase of peripheral resistance with constant flow), the cardiac output is kept constant instead of decreasing. Changes in preload result in a shorter diastolic duration, to keep the relation systole/diastole constant, the controller will increase the systolic driving pressure, thus the increase in frequency will be higher than with only the full stroke trigger (Fig. 24).

d) The combination of full stroke trigger, positive driving pressure control and an additional vacuum controller will give the system a very high sensitivity for changes in pre- and afterload. The "Starling curve of the artificial heart" can be changed near to that of the natural heart. The working principle of this vacuum controller depends on the influence of changes in atrial pressure on the air pressure pulse curve. Increasing atrial pressure can be detected by the increase of the stable value of the diastolic plateau. The vacuum controller is designed to keep a preset value in a certain diastolic interval constant. Increasing preload results in additional increasing vacuum and thus in higher cardiac output. Fig. 25 shows an example of a combined control function necessary to produce the desired Starling curve.

As controller input for all of these different modes, only noninvasively derived values measured in the air chamber of the pump or the driving line have to be used. Changes in the venous or arterial hemodynamics are estimated out of the curvature of the air pulse.

The combined system (full stroke trigger, driving pressure control, vacuum control) resembles the inotropic effect of afterload according to Starling's law and the chrono- and inotropic effect of preload similar to neurosympathetic stimulation, thus decorating the artificial heart with a "physiological behaviour". It can be used for left heart assist pumps and for either the right or left pump of the artificial heart. It adapts driving system and pump, keeping the optimal working mode even under extreme conditions and changes pump frequencies and CO according to the venous return.

Balance of the cardiac output of the left and right pump:

In the circulatory system, both pumps are working in series. If one of the pumps has a slightly higher cardiac output than the other, the vascular bed in the following part of the circulation will get overfilled. Especially dangerous is this if the cardiac output of the right pump exceeds this of the left. The pressure in the lung circulation will increase beyond its critical value, a hemorrhagic lung edema will occur. We are trying to take care of this problem by the design of the blood pump itself, by the control mode and an additional emergency control for excessive left atrial pressure.

As discussed before, the performance of the pneumatic blood pumps with a certain elastic compliance is greatly influenced by the afterload. As shown in Fig. 14, with constant pulse rate and optimal working mode, the CO can decrease up to 15 % within the physiologically possible limits of an aortic pressure increase. Therefore, we designed the pump for the replacement of the left cardiac chamber with an adequate higher static stroke volume than that of the right pump to achieve - as far as possible - an equal effective stroke volume (ESV) of both pumps under the given circulatory conditions. If we have to use pumps with equal SSV, we can only keep the balance between left and right by different pulse rates for both pumps or by incomplete filling or emptying of the right pump.

To avoide two independent controllers for both pumps, we chose the design consideration with different static stroke volumes and an additional emergency control. As emergency controller input we can use either the diastolic air pulse plateau of the left pump or the left atrial pressure measured continuously with the air capsula method. If a certain preset level of the left atrial pressure has been reached, the controller decreases the driving pressure of the right pump and thus its cardiac output. An example of the function is shown in Fig. 26. If this controller is activated, an acoustic alarm indicates a possible malefunction of the total system.

Controlling the cardiac output according to the needs of the organism:

Which physiological parameters reflect the momentary demand for cardiac output best and are at the same time reliable and long term stable measurable? Up to now, this question could not be answered unanimously. Metabolic parameters (e.g. mixed venoarterial oxygen tension) would be best suited because their information is combined out of the gas exchange in the lungs and the cardiac output, but sensors producing reliable signals usable as controller input are not yet developed. Control systems basing on different hemodynamic pressures such as aortic pressure or the peripheral resistance of the body circulation, have been discussed as well as the sinus node frequency of the remaining right atrium.

Within this paper, it is impossible to present and discuss our results. The reader may be referred on the literature[38,39,4o,41].

Now, we use changes in right atrial pressure or its noninvasively detected derivatives as leading value for the cardiac output controller. The above described combined controller has been tested in animal experiments. One example of the results is shown in Fig. 27. With constant cardiac output over the first day, the typical daily profile of the right atrial pressure developed. With onset of cardiac output control according to the venous return, one day of adaptation was necessary to get the complete inverse relation, shown at the 3rd and 4th day. The right atrial pressure decreases remarkably to normal values, a daily profile of cardiac output very similar to this of the animal with natural heart developed. The mean cardiac output necessary to keep this normal level of atrial pressure, was much lower than before with constant cardiac output. After the controller had been turned off, the former unphysiological relation developed again.

It is obvious that the artificial heart control has advantages for the technical as well as the biological system. Technical aspects are:

- reduced mechanical load (driving pressures, frequency)
- reduced energy consumption
- efficient use of the available space (optimal full/empty mode)
- safety in case of system failure, increased reliability.

Important for the biological system seems to be:

- reduced mechanical load on the blood
- balance of lung and systemic circulation (left atrial pressure)
- adaptation to changes in afterload
- CO according to the metabolic demands (venous return, right atrial pressure)

Additional positive results of the perfusion with demand controlled CO are still in discussion (reduced mean cardiac output, reduced mean systemic pressure, normal amount of circulating blood volume).

The new compact control system using seven micro processor boards for the automatic control of cardiac output with different biological parameters as input values and the functional surveillance of the complete total artificial heart system is now in its final preparation. Aside of the automatic control, it allows the predictive failure analysis; out of the comparison of technical parameters such as temperature, current or voltage of the motors or others it will give a prognosis for the further function of the system. A predictive failure will be detected and signalized by the computer, additionally the computer acts with logical decisions, e.g. activating the failsafe system.

Summary

With the pneumatic artificial heart system where only the blood pumps are implanted and all other components are external, most experience in animal experiments has been gained over the last 2o years. The results have shown that this system has a good potential for the clinic al application, at least for a temporary use.

Fully implantable total artificial heart systems will not be available within the foreseeable future. The technical problems are numerous, especially with respect to the durability and reliability because of the system's complexity. Maintenance or replacement of a failing component or a periodic exchange of parts requires surgery.

Non of the electrically driven artificial hearts have up to now achieved an acceptable functional time in animal experiments or in bench testing. A similar if not longer time for the development of a fully implantable TAH system can be expected than that already spent for the pneumatic artificial heart.

Therefore, we are concentrating on further improvement of pneumatic systems during long-term animal experiments (Fig. 28) to prove sufficient in vivo reliability, giving confidence for the clinical use.

These long-term experiments are only possible if we change over to another animal species because the growing calf limits the duration of the experiment by its weight gain. We are hopeful that with the improved total artificial heart system survival times of more than one year will be possible within the next future.

In preparation for the clinical use, the development and improvement of portable electropneumatic driving units is of high importance (Fig. 29). They will allow the patient a tetherfree life at least for some hours during the day. In animal experiments their reliable performance could be proven (Fig. 35).

Only if the durability of the components of the system and the functional reliability of the total system have been proven in extensive in vitro and in vivo tests, the clinical application will be justified with regard to the technical aspects.

But taking the implantation in patients into consideration, not only the engineering aspects for a safe and functional system have to be fulfilled, several human factors such as patient confidence, quality of life and the risk-benefit relation, will play an important role in qualifying the TAH as ready for the clinical use.

Acknowledgement

The research work in Berlin, presented only in parts in this paper, was conducted in the Department of Surgery under the supervision of Prof. Dr. med. E.S. Bücherl in the Klinikum Charlottenburg of the Free University Berlin. It is a result of the joint effort of many technical and medical coworkers.

Our research was supported first by the Bundesministerium für Forschung und Technologie (BMFT) and later by the Deutsche Forschungsgemeinschaft (DFG).

References

1. U. Nemsmann, R. Neidhold, J. Wagner:
   Erzeugung von variablen Solldruckkurven für implantierbare Blutpumpen.
   Biomed. Tech. 22, 327, 1977b

2. J. Frank, K. Affeld, A. Mohnhaupt, P. Baer, F. Zartnack, E.S. Bücherl:
   First experiences with a totally mobile artificial heart system.
   Trans. Am. Soc. Artif. Intern. Organs (ASAIO), 26, 72, 198o.

3. F.D. Altieri:
   Status of implantable energy systems to actuate and control ventricular assist devices.
   Artif. Organs, 7(1), 5, Raven Press, New York, 1983.

4. D.A. Hughes, R.J. Faeser, B.D.T. Daly, C.H. Edmonds, L.V. Feigenbutz, S.R. Igo, K. Hagen, J.J. Migliore, A.E. Ruggles, W.J. Robinson, F.N. Huffman, and J.C. Norman:
   Nuclear-fueled circulatory support systems XII: Current status.
   Trans. Am. Soc. Artif. Intern. Organs (ASAIO), XX, 737, 1974.

5. R.D. v.Reth:
   Development of an implantable thermal engine as power source for artificial blood pumps.
   Proc. Europ. Soc. Artif. Organs (ESAO), I, 151, 1974.

6. S. Bevilacqua, A. Famulari, G. Cucciara, and R. Cortesini:
   Electrically driven nuclear artificial heart.
   Proc. Europ. Soc. Artif. Organs (ESAO), II, 133, 1975.

7. Abstract Contractors Meeting, Division of Heart and Vascular Deseases, National Heart, Lung, and Blood Institute, December 6, 7, and 9, 1982.

8. M. Arabia, C. Franconi, M. Guerrisi, A. Magrini, N. Yamamoto, A. Vakamudi, M. Drummond, and T. Akutsu:
   A new automatically controlled electric TAH.
   Trans. Am. Soc. Artif. Intern. Organs, XXVI, 6o, 198o.

9. G. Rosenberg, A. Snyder, W. Weiss, D. Landis, D. Geselowitz, and W. Pierce:
   A roller screw drive for implantable blood pumps.
   Trans. Am. Soc. Artif. Intern. Organs, XXVIII, 123, 1982.

1o. R.K. Jarvik:
   The total artificial heart.
   Sci. Am. 244, 74, 1981.

11. R.K. Jarvik:
   Electrical energy converters for practical human total artificial hearts - An opinion in support of electropneumatic systems.
   Artif. Organs, 7(1), 21, Raven Press, New York, 1983

12. W.S. Pierce, J.L. Myers, J.H. Donachy, G. Rosenberg, D.L. Landis, G.A. Prophet, and A. Snyder:
Approaches to the artificial heart.
Surgery, 9o(2), 137, 1981.

13. Y. Nosé, G. Jacobs, R.J. Kiraly, L. Golding, H. Harasaki, S. Takatani, S. Murabayashi, R.W. Sukalac, H. Kambic, and J. Snow:
Experimental results for chronic left ventricular assist and total artificial heart development.
Artificial Organs 7(1), 55, Raven Press, New York, 1983.

14. J. Vasku:
Artificial heart-pathophysiology of the total artificial heart and of cardiac assist devices.
J.E. Purkyne University Brno, Medical Faculty, 1982.

15. O.H. Frazier, T. Akutsu, and D.A. Cooley:
Total artificial heart (TAH) utilization in man.
Trans. Am. Soc. Artif. Intern. Organs, XXVIII, 534, 1982.

16. K. Atsumi, I. Fujimasa, K. Imachi, H. Miyake, N. Takido, M. Nakajima, A. Kouno, T. Ono, S. Yuasa, Y. Mori, S. Nagaoka, S. Kawase, and T. Kikuchi:
Three goats survived for 288 days, 243 days and 232 days with hybrid total artificial heart (HTAH).
Trans. Am Soc. Artif. Intern. Organs, XXVII, 77, 1981.

17. F. Unger:
The ellipsoidheart in total artificial heart replacement.
Assisted Circulation, Ed. F. Unger, Springer-Verlag Berlin-Heidelberg-New York, 353, 1979.

18. E.S. Bücherl, E. Hennig, P. Baer, A. Bücherl, J. Frank, G. Grötzbach, I. Jannek, H. Keilbach, R. Langer, W. Lemm, H. Pannek, H. Weidemann, F. Zartnack:
The artificial heart program in Berlin - past, present, future.
Heart Transplantation 1, 4, 1982.

19. F. Rennekamp, K. Affeld, W. Lemm, E.S. Bücherl:
Neue Technik zur Herstellung von Blutpumpen für den Totalherzersatz aus Polyurethan mit nahtloser Innenoberfläche.
Biomed. Technik 23 (Erg.Bd.), 198, 1978.

2o. F. Rennekamp, W. Lemm, F. Zartnack, E. Hennig, H. Keilbach, H.D. Clevert, W. Krautzberger, Ch. Große-Siestrup, K. Gerlach, K. Affeld, V. Unger, J. Frank, F. Kuhlmann, A. Mohnhaupt, E.S. Bücherl:
Long term results with seamless blood pumps out of polyurethanes for the replacement of the heart.
Proc. Europ. Soc. Artif. Organs (ESAO) VI, 94, 1979.

21. E. Hennig, H. Keilbach, D. Hoder, E.S. Bücherl:
Calcification of artificial heart valves and artificial hearts.
Proc. Europ. Soc. Artif. Organs (ESAO) VIII, 76, 1981.

22. F. Zartnack, E. Hennig, F. Ott, A. Schiessler, E.S. Bücherl:
Development and fatigue testing of a new blood pump.
Proc. Europ. Soc. Artif. Organs (ESAO), Life Support Systems, Saunders Comp. Ltd., Longon, 1983.

23. F. Zartnack, K. Affeld, E.S. Bücherl:
The vacuum molding technique, a new method for fabricating polyurethane blood pumps.
Proc. Europ. Soc. Artif. Organs (ESAO), VI, 99, 1979.

24. H.W. Pannek, E.S. Bücherl:
Design and realization of TAH-pumps for human application.
Proc. Europ. Soc. Artif. Organs (ESAO), VIII, 33, 1981.

25. E. Hennig, A. Mohnhaupt, E.S. Bücherl:
The influence of the filling pressure on the output of pneumatically driven blood pumps.
Proc. Europ. Soc. Artif. Organs (ESAO), III/1976, 19, 1978.

26. E. Hennig, K. Affeld, H. Kleß, A. Mohnhaupt, R. Mohnhaupt, H.D. Clevert, H. Keilbach, W. Krautzberger, H.O. Kleine, H. Weidemann, E.S. Bücherl:
Controlling of artificial blood pumps after total heart replacement - an example of disregulation.
Proc. Europ. Soc. Artif. Organs (ESAO), II/1975, 138, 1976.

27. F. Zartnack, W. Dunkel, K. Affeld, E.S. Bücherl:
Fatigue problems in artificial blood pumps.
Trans. Am. Soc. Artif. Intern. Organs (ASAIO), 24, 6oo, 1978.

28. H. Weidemann, Ch. Große-Siestrup, K. Gerlach, A. Kaufmann, E.S. Bücherl:
Pathological-anatomical findings in calves after total artificial heart replacement.
Proc. Europ. Soc. Artif. Organs (ESAO), VIII, 12, 1981.

29. A.F. v.Recum, J.B. Park:
Permanent percutaneous devices.
CRC Critical Review on Bioengineering, 5(1), 37, 1981.

3o. Ch. Große-Siestrup, K. Affeld:
Design criteria for percutaneous devices.
Journal Biomed. Mat. Res., 1983 (in print).

31. E. Hennig, A. Mohnhaupt, W. Krautzberger, E.S. Bücherl:
Die Druckmessung in der Luftkammer pneumatisch angetriebener künstlicher Blutpumpen, ein Mittel zur Abstimmung der Parameter des extrakorporalen Antriebs.
Ergebnisbericht zum Vorhaben BMFT/KUE 4o1, Berlin, 1976.

32. K. Affeld, R. Mohnhaupt, A. Mohnhaupt, E.S. Bücherl:
Antriebsseitige Messungen von Kreislaufgrößen.
Langenbecks Arch. Chir. 335, 47, 1974.

33. H. Kleß, E.S. Bücherl:
Extracorporeal beat-by-beat evaluation of right ventricular stroke volume for the artificial heart.
Proc. Europ. Soc. Artif. Organs (ESAO), 1975(2), 136, 1976.

34. E. Hennig, H.D. Clevert, E.S. Bücherl, F. Wallner:
A novel control system for pneumatic left-heart assist devices (LHAD).
Proc. Europ. Soc. Artif. Organs (ESAO), IV/1977, 342, 1978.

35. G. Rosenberg, W.S. Pierce, J.A. Birghton, W.M. Philips, D.L. Landis, J. Lenker:
A pressure-measuring system for long-term in vivo pressure measurement.
Surgery 79, 1976.

36. K. Affeld, M. Echt, A. Mohnhaupt, P. Baer, E. Hennig, E.S.Bücherl
A new method for drift-free measurement of venous pressures.
Proc. Europ. Soc. Artif. Organs (ESAO), 1975, 2, 142, 1976.

37. H. Kleß:
Zur Strukturierung der Datenerfassung und -verarbeitung bei datenintensiven Langzeitexperimenten.
EDV in Medizin und Biologie 2, 1978.

38. E. Hennig, Ch. Große-Siestrup, W. Krautzberger, H. Kleß, E.S. Bücherl:
The relationship of cardiac output and venous pressure in long surviving calves with total artificial heart.
Trans. Am. Soc. Artif. Intern. Organs (ASAIO), 24, 616, 1978.

39. K.J.G. Schmailzl, E. Hennig, E.S. Bücherl:
Comparison of biological signals for the automatic control of the total artificial heart.
Europ. Soc. Artif. Organs (ESAO), IX Annual Meeting, Brussels, 1982.

4o. E. Hennig, A. Mohnhaupt, E.S. Bücherl:
Flow regulation by venous return in TAH experiments.
ASAIO 198o (Abstract).

41. A. Bücherl, E. Hennig, E.S. Bücherl:
The p-wave of the atrium after total heart replacement.
2nd International Workshop, ISAO, Innsbruck, 1981 (in print).

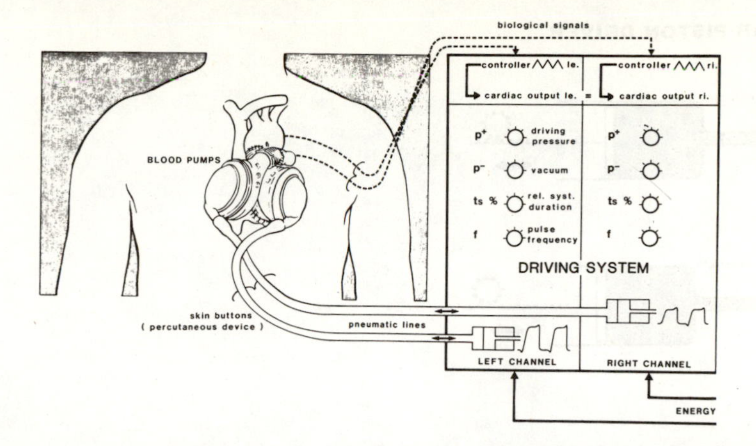

Fig. 1: Schematic of a percutaneous artificial heart system. Two implanted blood pumps are pneumatically activated by an external driving system via two pneumatic lines, transmitted into the thoracic cavity by percutaneous devices.

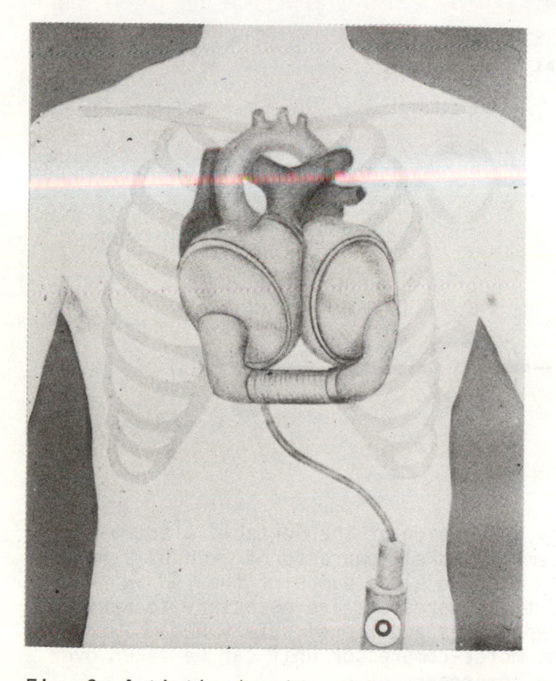

Fig. 2: Artist's drawing of the ultimate goal of the artificial heart research: Blood pumps and all other system components, including energy source, are implanted.

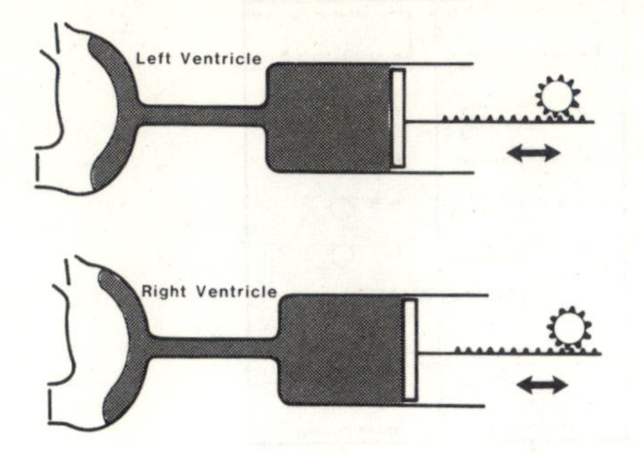

Fig. 3: Schematic of the linear piston electropneumatic energy converter, developed by AEG-Telefunken. The reversing rotational movement of the electromotor is conversed in the back- and forward movement of the pneumatic piston by a rack and pinion.

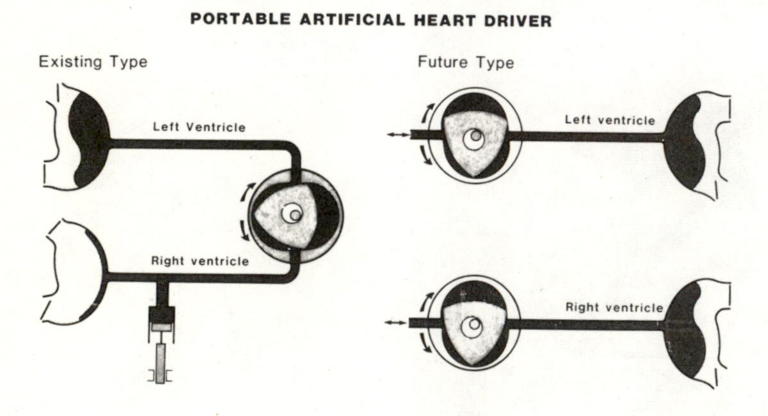

Fig. 4: Schematic of the energy conversion of the portable electropneumatic driving system. A Wankel-compressor driven by a reversing DC motor can supply the two blood pumps of the TAH in counterpulsation mode, two compressors are necessary to achieve copulsation. This concept gives additional safety because the function of one failing motor-compressor unit can be taken over by the second in case of emergency.

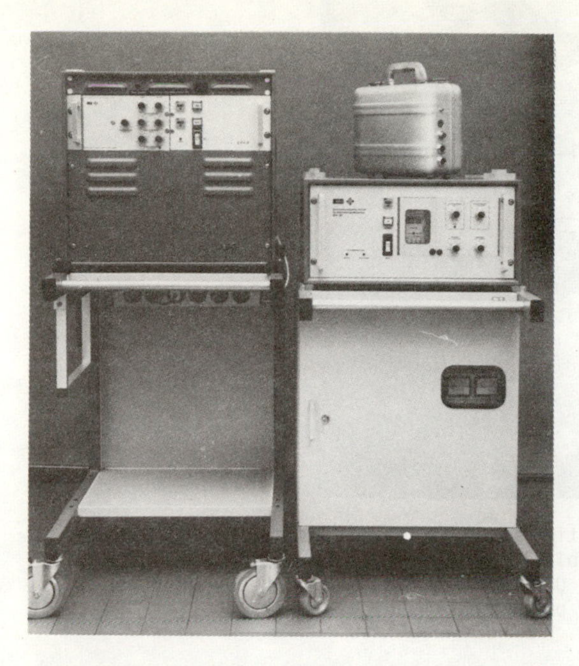

Fig. 5:
Electropneumatic energy converters used by the Berlin artificial heart research group. Left: Driving system for the total artificial heart, consisting out of two separate motor-piston units, electronic control and a battery pack for two hours performance without external energy. Right: Driving unit for ventricular assist devices, one motor-piston combination also with control and battery pack. This unit serves as back-up system for a failing channel in TAH experiments. Left on top: Electropneumatic portable driving unit with Wankel compressor, control and battery pack.

Fig. 6: Implantable blood pumps for the replacement of the left and right ventricle made out of Pellethane^{R} fabricated with the dip-coating technique. The inner surface is smooth and continuous with one seamless diaphragm. Björk-Shiley-tilting-disc-valves are used in inlet- and outlet positions.

Fig. 7: Calcified blood contacting side of the diaphragm of a PE blood pump 154 days after implantation in a calf. Mineralization is concentrated along the DH junction, because of secondary wear the diaphragm has been punctured, leading to air embolism.

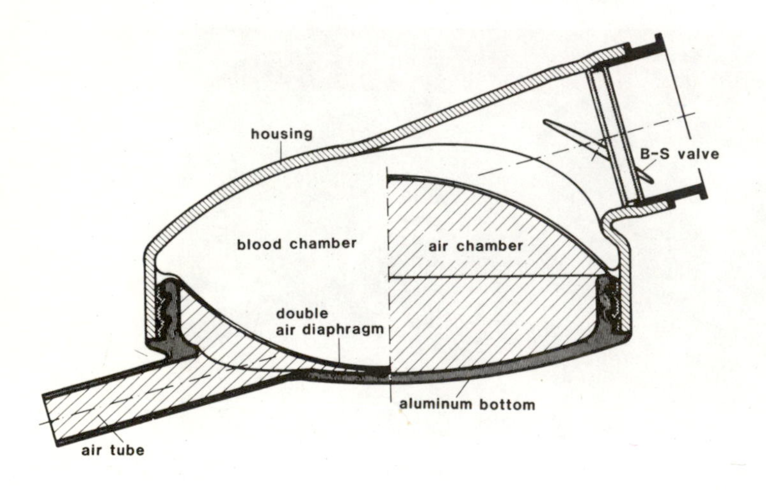

Fig. 8: Cross-section of the blood pump with tri-laminar diaphragm. Blood-and air chamber are separated by a combination of three diaphragms, the blood contacting diaphragm forms a continuous surface with the housing, vacuum molded out of polyurethane. The bottom with supporting ring allows two separated areas of bending flex for systole and diastole. The pneumatic load is taken over by the double-air-diaphragm (left: diastolic position, right: systolic position).

Fig. 9: Set of molds necessary for the fabrication of a left pump. Stainless steel molds (top) for the dip-coating technique (blood diaphragm, first air diaphragm) and molds out of "plaster of Paris" for the vacuum forming (housing, air diaphragm, bottom).

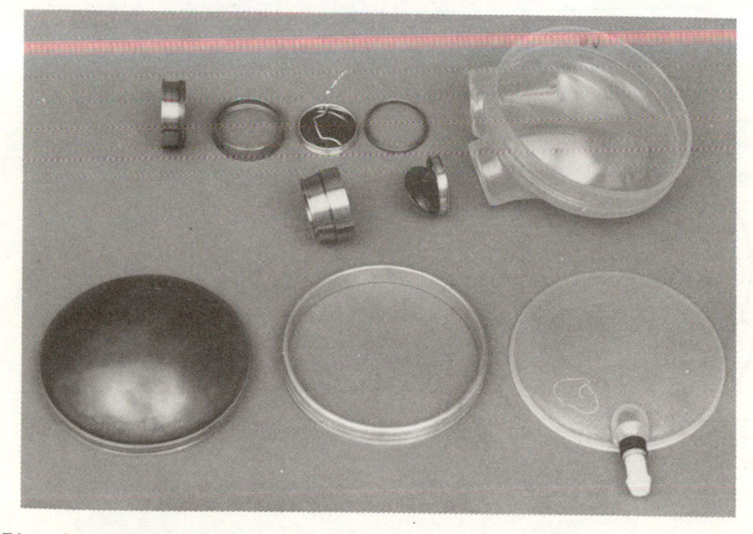

Fig. 1o: Parts of the left pump before assembling: In- and outlet connectors with Björk-Shiley valves, housing with integrated blood contacting diaphragm, two-laminar driving diaphragm, supporting ring, bottom.

Fig. 11: Multilaminar diaphragm blood pumps for the replacement of the left and right ventricle in calves. The quick-connect system for the anastomosis to the vascular system has been already mounted for demonstration.

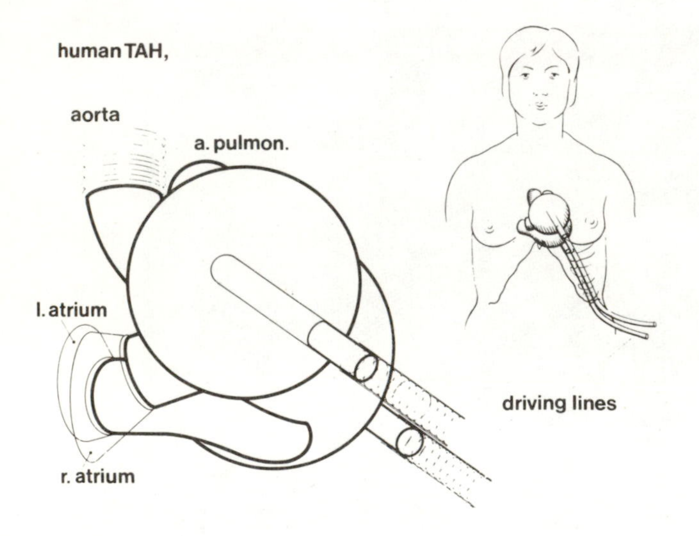

Fig. 12: Shape and position of the in- and outflow ports of blood pumps for the total artificial heart replacement in humans.

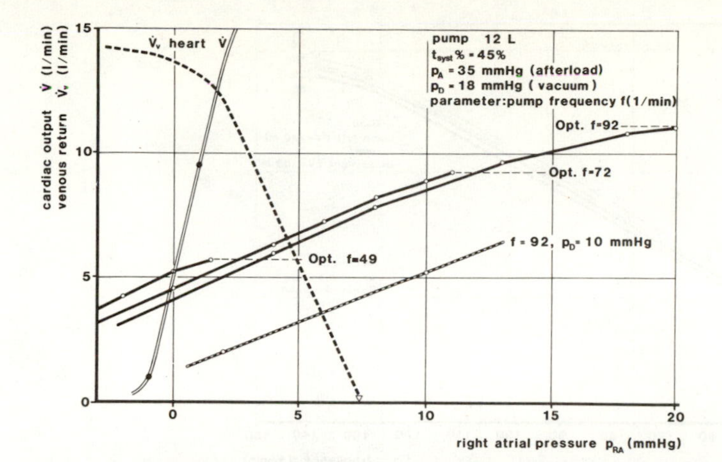

Fig. 13: Family of Starling curves of a MLD-pump with different preset constant frequencies. Afterload and vacuum are kept constant. The flat inclination demonstrates the insensitivity for changes in right atrial pressure due to the high air flow losses in the driving lines. An optimal cardiac output exists for each frequency when the full/empty working mode has been reached.

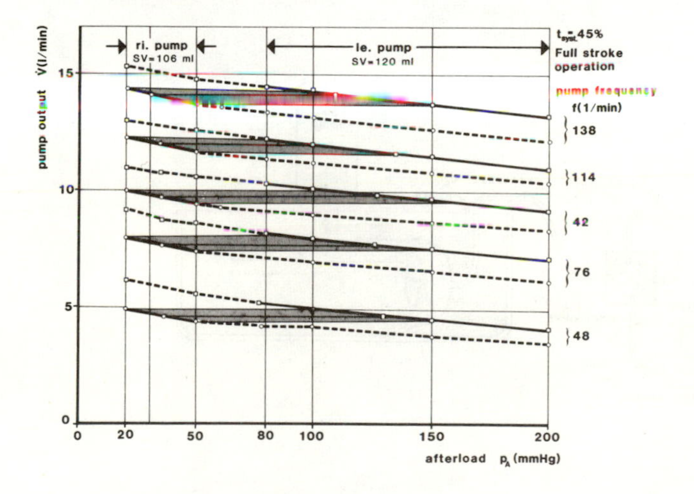

Fig. 14: Performance diagram of two blood pumps with different static stroke volume (SSV). The pumps are driven in the full/empty mode, parameter for each curve is a certain constant frequency. With increasing afterload and unchanged frequency, the CO decreases, mostly due to the elastic compliance of the housing of the pump. Different SSV (right pump: 1o6 ml, left pump: 12o ml) allows a balance of the output of the right and the left pump within the limits of the respective biological afterload, keeping the same frequency.

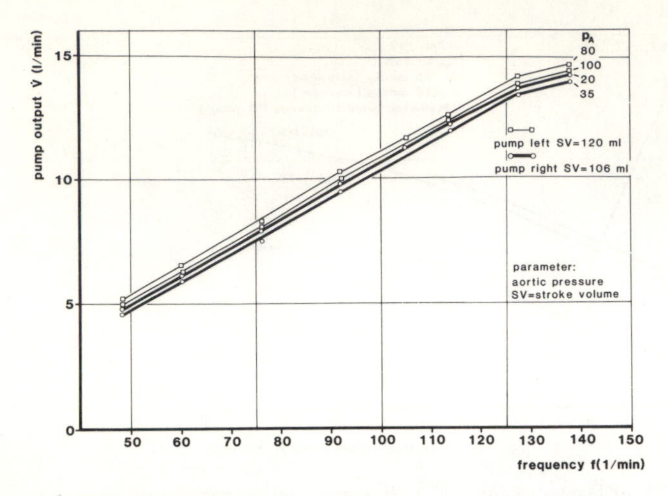

Fig. 15: Pump output versus frequency, demonstrating the balance between the left and right pump's ejection with different afterloads; a small "cardiac reserve" of the left pump is always kept to compensate for example for the bronchial shunt.

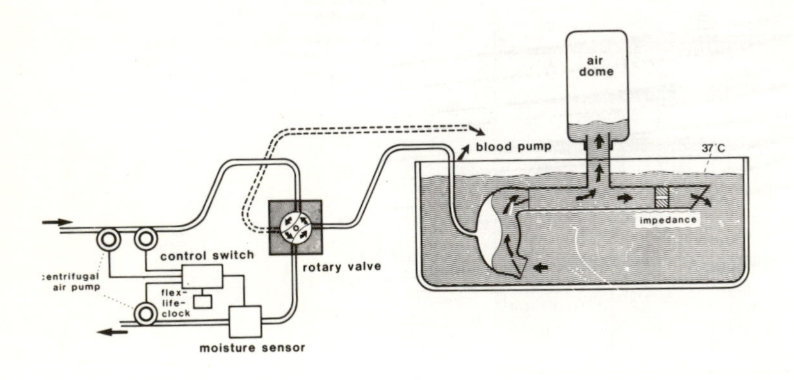

Fig. 16: Schematic of the in vitro fatigue test station, 24 places are activated by a central air pressure and vacuum supply controlled by a rotary valve.

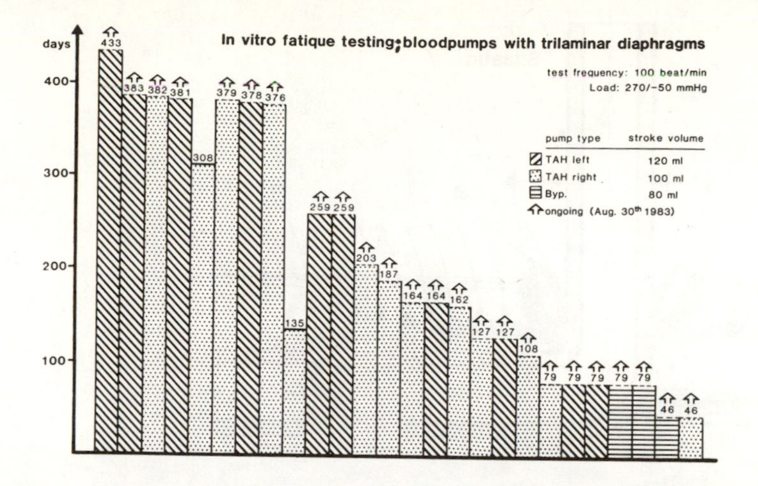

Fig. 17: Recent in vitro fatigue test results (August, 1983) for three different types of MLD-pumps. The stepwise increase in days performed are a result of the increase of the number of test stations available within the last year. Two pump failures occured: Rupture of the blood diaphragm after 3o8 days - diaphragm thickness: 4o µ , leakage between bottom and housing after 135 days - improper mounting.

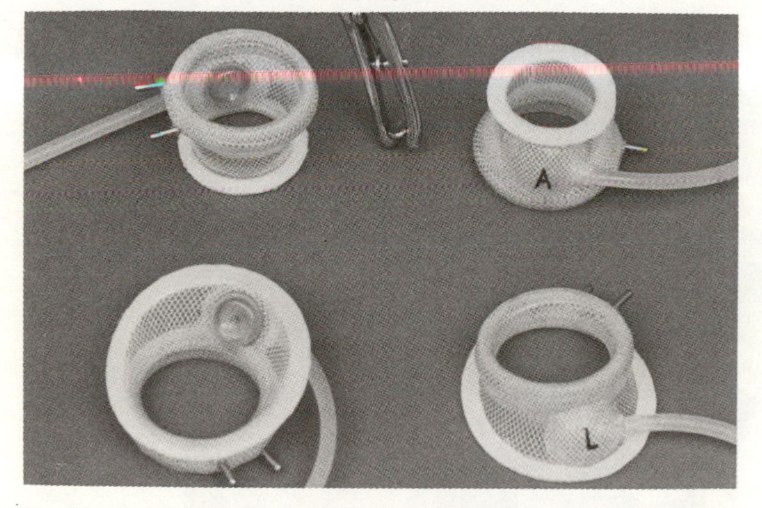

Fig. 18: Quick- connect system for the left and right pump (top: pulmonary artery, aorta; bottom: right and left atrium). The adapters are made out of reinforced silicone rubber, air capsulae for the noninvasive measurement of the respective blood pressures are integrated. A modified clamp has to be used for spreading up the stainless steel spring, giving a tight and firm fixation with the in- and outflow ports of the pump.

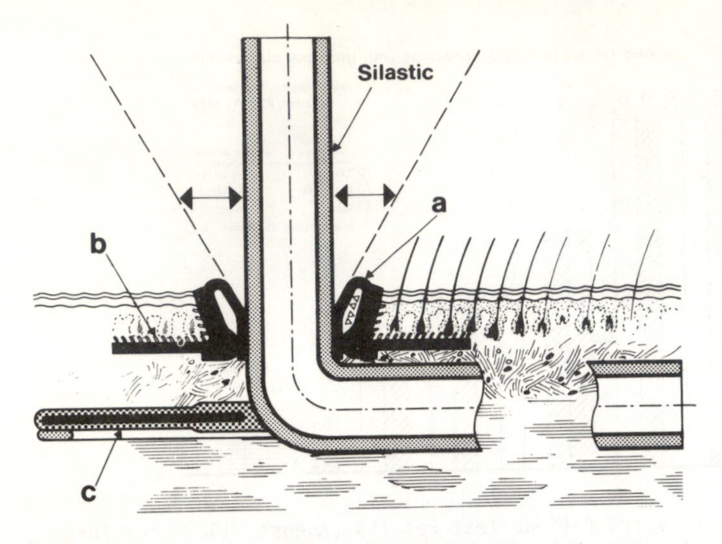

Fig. 19: Schematic of the percutaneous device, necessary for the permanent infection-free entrance of the pneumatic driving lines into the thoracic cavity (explanation of the special sections a, b, c - see text).

Fig. 2o: Set of PDs (skin buttons) necessary for the total artificial heart system. Two PDs, capable of taking higher mechanical load, are used for the transmission of pneumatic energy to the implanted pumps, the small PD is used for the transmission of small air tubes (blood pressure) and wires (electrical/biological signals).

Signals for control and Monitoring

Presently	VENTRICULAR PRESSURE
	AIR FLOW
	ATRIAL PRESSURE
	AORTIC PRESSURE
	CARDIAC OUTPUT
	P-WAVE
In Future	POWERSTAGE TEMPERATURE
	LEAKAGE RATE
	TRANSDUCER STABILITY

Fig. 21:
Measurement of biological an technical parameters, continuously monitored for the surveillance of the TAH system and the technical and physiological control. Some of these parameters serve as input for a micro processor, predictive failures will be detected and signalized, in case of emergency the back-up system will be activated automatically.

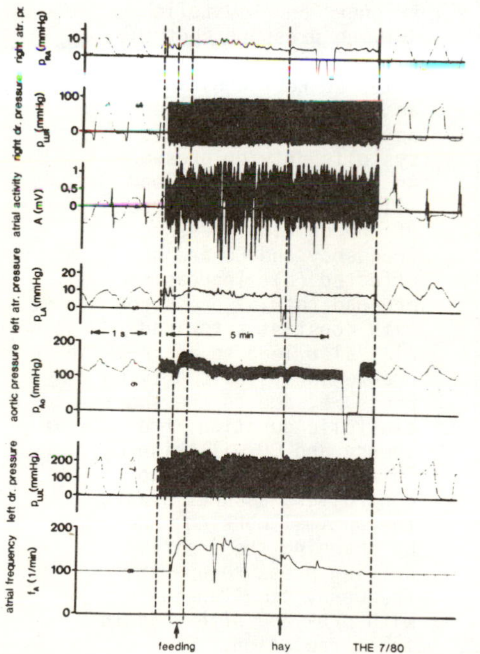

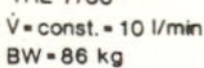

Fig. 22:
Feeding protocol of a TAH experimental animal, showing the parameters that are monitored continuously in our research laboratory.

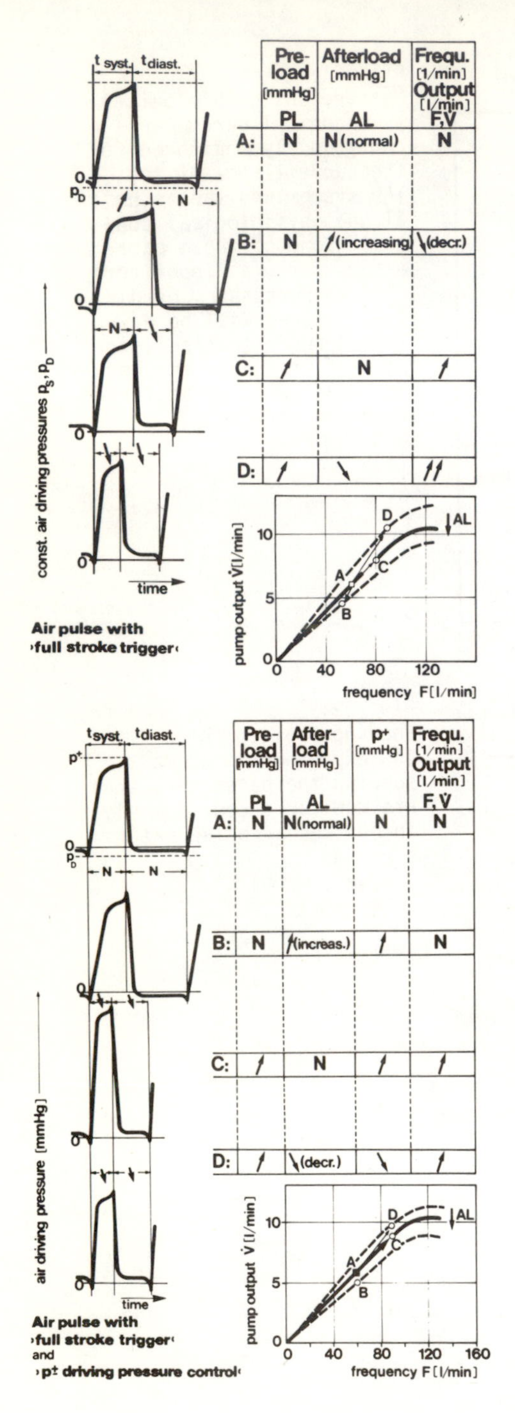

Fig. 23:
Curvature of the air-pressure-pulse in the blood pump, influenced by changes of the pre-and afterload. The endsystolic and diastolic spikes are indicating the complete emptying and complete filling of the blood chamber, the diaphragm is resting. These points have to be detected and used for the initiation of the next pumping cycle to keep the optimal full/empty working mode.

Fig. 24:
Combination of "full stroke trigger" and "systolic driving pressure control". The relative systolic duration is kept constant by an automatic controller acting on the pneumatic systolic driving pressure. An increase in afterload will result in an increase of the driving pressure, frequency and CO are not affected (b). Increasing preload (atrial pressure) with constant afterload will also lead to an increase in systolic driving pressure, due to the short diastolic duration, frequency and CO will rise(C). This combination of two controllers assures always the optimal working mode by changing the positive driving pressure and the frequency simultaneously with pre- and afterload in the circulation.

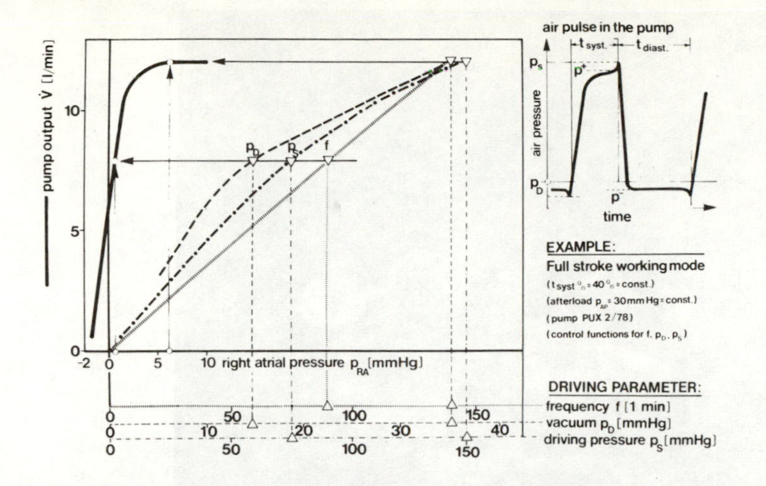

Fig. 25: Example of automatic control, showing the necessary function curves for systolic and diastolic driving pressure and pump frequency to keep the preset Frank Starling curve. These curves are achieved by using a combination of three controllers (full stroke trigger, positive driving pressure control, vacuum control).

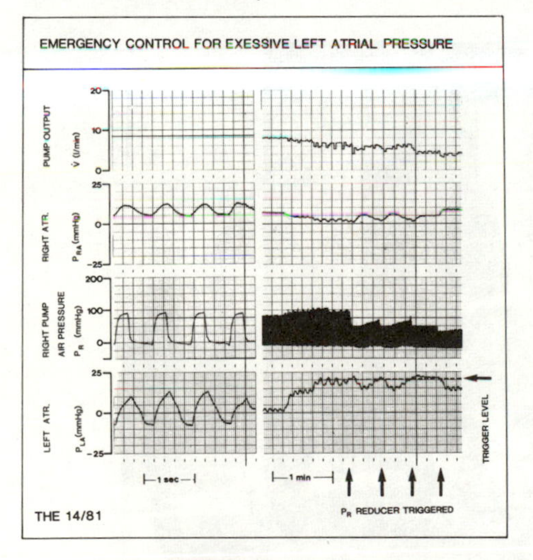

Fig. 26: In vivo example of the performance of the emergency control that is triggered by excessive left atrial pressure.

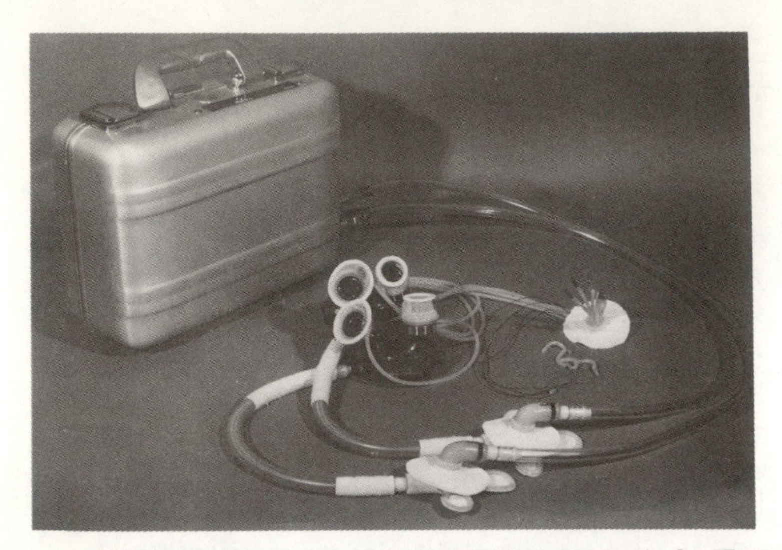

Fig. 29: Components of the wearable, battery powered electropneumatic total heart system, developed in Berlin. Only the blood pumps of this percutaneous system have to be implanted, the other technical components can be kept outside of the body, allowing easy maintenance and, if necessary, the exchange of parts without surgery.

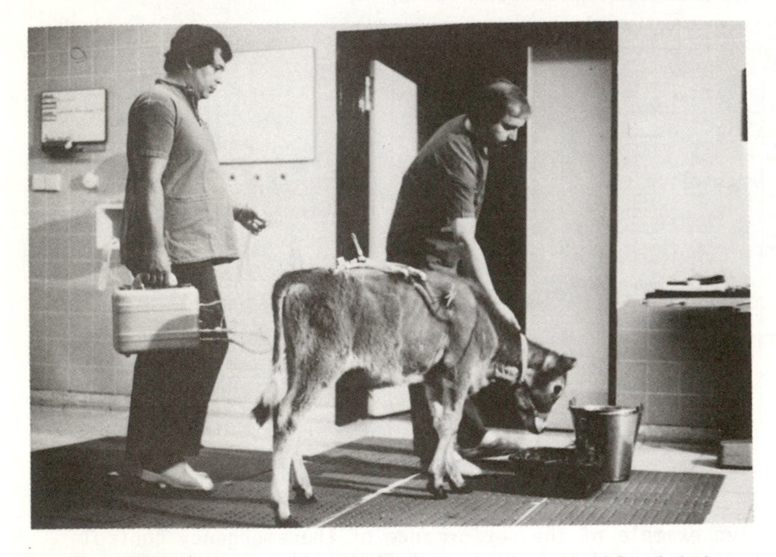

Fig. 3o: TAH-calf with an electropneumatic portable driving system for tetherfree mobility.

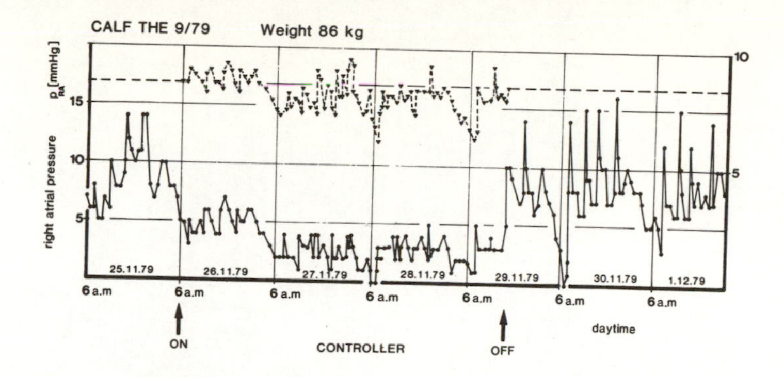

Fig. 27: Comparison of the daily profile of right atrial pressure and pump output with and without automatic control. The mean daily value of CO and the right atrial pressure will be lower if the controller is adapting the perfusion according to the venous return. The typical daily profile of the right atrial pressure with constant cardiac output (left) changes to the typical natural daily profile of CO with near to constant right atrial pressure with automatic control.

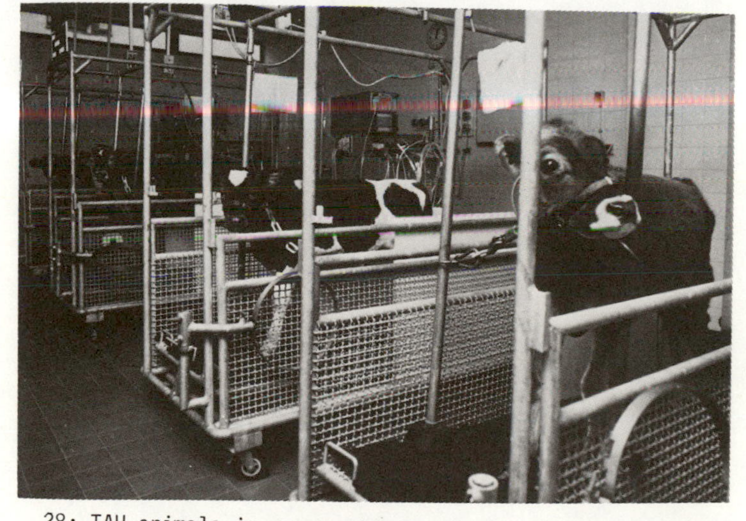

28: TAH animals in our postoperative-care unit. Additionally to the in vitro testing, in vivo reliability and fatigue testing of all components of the system has to be performed in order to achieve the clinical readiness.

PROCEEDINGS OF THE II INTERNATIONAL CONFERENCE ON APPLICATIONS OF PHYSICS TO MEDICINE AND BIOLOGY
edited by Ž. Bajzer, P. Baxa & C. Franconi

CARDIOVASCULAR SYSTEM MODELLING FUNDAMENTALS

Enzo Belardinelli

Department of Elettronics, Informatics
and System Engineering

Viale Risorgimento n.2, 40136 Bologna

Italy

ABSTRACT

Descriptive and interpretative models are defined; some examples of both models are given with reference to cardiovascular system. Since simplification procedures are in the interpretative modelling often required, some reduction algorithms are produced and examples of their use in cardiovascular modelling are given.

1. Introduction

"The status of a science is commonly measured by the degree to which it makes use of mathematics". Not everyone shares this opinion of the psychologist S.S. Stevens. Physicists and engineers have always used the mathematical tool to represent the real world; on the contrary, strong resistance to its use has been offered by some biologists. This opposition is generally justified by extreme complexity of the biologic world, which does not permit schematisms or simplifications. It is true that a biologic system has a high degree of complexity; the inference that physical and mathematical analysis brings to an unavoidable distortion of the biologic world is false.

Certainly every physical and mathematical search implies a suitable schematism, but in order to conclude that such schematisms are unacceptable it is necessary to prove the existence of methods which do not employ schematisms or utilize less deforming models. Since such methods do not exist. rejection of the mathematical tool should result in a consistent refusal of any other method.

The aim of this lecture is to show that mathematical models are very useful to study the cardiovascular system and provoke some methodological considerations which can help to understand the essential characteristics of models increasingly used in cardiovascular research. Some fundamental problems arising in the course of modelling are presented and some possible solutions proposed.

2. Descriptive (or representative) and interpretative models

The difference between the two classes of models comes out clearly from the following example. The behavior of a cardiac ventricle is represented in different ways; one of these [1] [2] is based on the linear relationship between ventricular pressure, $P_r(t)$, and ventricular volume, $V_v(t)$:

$$V_v(t) = C_v(t)\ P_v(t) + V_c \qquad (1)$$

where V_c is a constant and $C_v(t)$, ventricular compliance, is a time function which, on a first, rough approximation [3], does not depend on pressure and volume.

Another model is presented schematically in fig. 1, which is self-explanatory [4].

As appears clearly, the second model goes into the details of ventricle anatomy and muscle dynamics. The dynamics of the muscle, in turn, may be interpreted by a model of the contractile process, which is based on the interaction of the sarcomere filaments through a complex mechanism of forming and destroying elastic links [5].

The first model has the aim of describing how the ventricle acts as a pump which supplies the vascular circulation; the purpose of the second model is to interpret the ventricle inside in order to understand how pump movements are generated.

It is easy to realize that the second model is much more complex than the first.

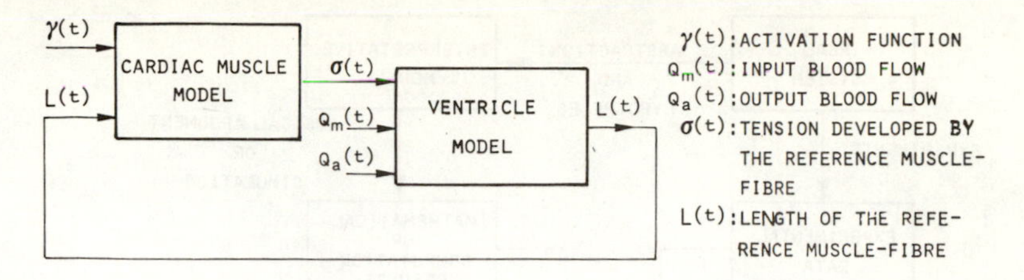

Fig.1 - Schematic diagram of the left ventricle interpretative model.Reference muscle-fibre represents the average behavior of all ventricular wall fibres.Input and output flows are the flows through mitral valve and aortic valve, respectively.

The above-mentioned models can be interpreted by the general schematic diagrams of figs. 2,3.

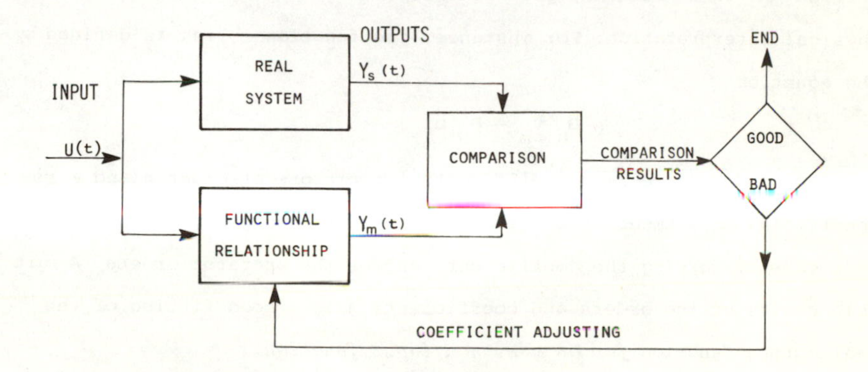

Fig.2 - Schematic diagram of representative model.

With reference to fig.2, the functional relationship between input u and output y_m (*) is determined by empirical criteria resorting to in-

(*) Input and output functions means the independent and dependent variables, respectively.

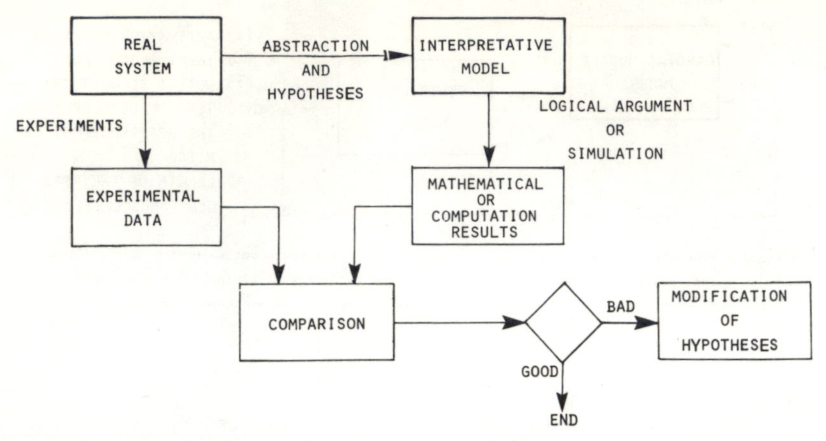

Fig.3 - Schematic diagram of interpretative models.

tuition or to a mathematical formalism, which has nothing to do with physical interpretation. For instance, the functional link is defined by the equation

$$D_n \, y_m = D_\nu \, u$$

where D_n and D_ν are linear differential operators of order n and ν respectively. Adjustment of the functional relationship may be made in this case by varying the coefficients and/or the operator orders. A suitable choice of the orders and coefficients allows good fitting of the real output function *for an assigned input function*.

Quite different is the procedure of the interpretative model (fig. 3). The abstraction process is the most important and the most difficult. Some authors think it impossible to write down precise rules for it and assert that it is pure art. On the contrary several modellers believe that well-defined criteria can and must be used. They are: a) exact formulation of the problem, which clears up the purpose of the model; b) previous analysis of system behavior on the whole and its decomposition, when possible, on subsystems; c) writing relations between subsystems; d) identification of independent and dependent variables on the subsystems; e) writing the functional equations on the variables.

In step e) physical knowledge is used in *all* cases, *except* when the actual state of knowledge does not allow "a priori" determination of the functional relationships between variables. In these cases it is necessary to use empirical equations.

Whenever comparison with the experiment gives no adequate results, the hypotheses in steps b) and c) must be reviewed.

It is clear that, whenever the comparison is satisfactory, the model is not only an abstract system whose response is like that of the real system, but it is a tool for interpreting the physical nature of the system.

The use of physical laws to write down the equations gives a solid foundation to the search and strengthens the hope that the phenomenon has been explained.

The interpretation of some fundamental hemodynamical laws attests the above-mentioned assertions.

In the following the concept of input impedance is often used. Since such a concept, widely used in electrical circuit theory, is not commonly employed outside this technical field, it is useful to give its definition. Given a vascular compartment without internal sources of energy and supplied by single vessel, the input impedance in a section of the vessel is the ratio of the L.T. of the pressure and the flow in the considered section (fig. 4).

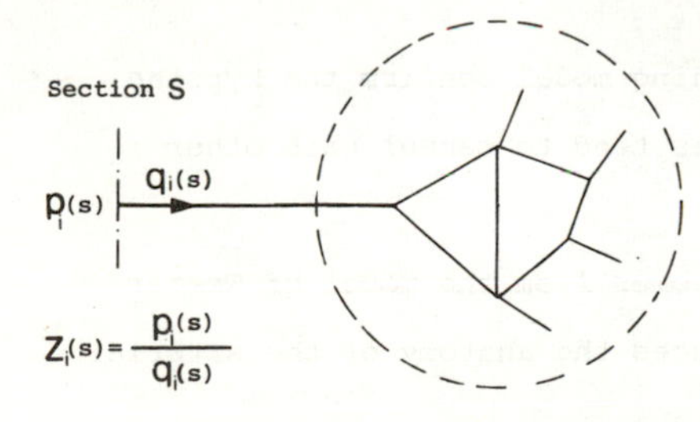

Fig.4 - Definition of input impedance.

It is widely confirmed that the input impedance of the arterial systemic tree is very well represented by the so-called three element windkessel model [6] (fig. 5). Analysis of such a simple model brings up a question. Why can so simple a model represent such a complex systemic vascular circulation? In other words why does a hemodynamical sys-

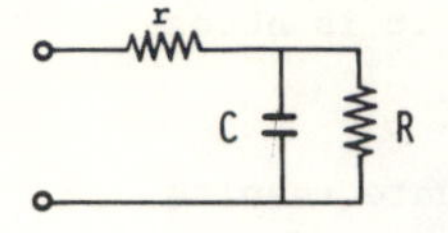

C :COMPLIANCE

R :PERIPHERAL RESIST.

r :CHARACTERISTIC IMPEDANCE AT HIGH FREQ.

Fig.5 - Three-element windkessel model.

tem with hundreds of vessels, each consisting of a tube with pulsatile walls, lead to so a simple relation between input pressure and flow? It is well known that an artery may be regarded as a transmission line along which the sinusoidal waves travel with a frequency-dependent velocity [7]. When the waves reach a mismatched load they are reflected back to the source. Wave attenuation is very small [7]; consequently, a wave which leaves the heart must come back to the heart with a non-negligible amplitude. Reflections at the distal extremities of the arteries and the weak attenuation seem to contrast with the experimental evidence that the effect of the reflected waves on the heart is negligible.

In 1964 Taylor [8] proposed a model of the whole arterial system consisting of many segments, which are arranged in a tree similar to the one formed by the systemic arteries. In this model every segment divides into two branches and at each level of dichotomization the lengths of the segments differ from one another. Taylor assumed that the mean length of all vessels at the nth level is the length of the first segment divided by n+1, while individual length at any level are chosen to be distributed around their mean according to a second-order gamma probability distribution.

The results of this randomly branching model confirm the hypothesis that reflections from different sites tend to cancel each other because of their random separation.

A further test of this hypothesis comes from the model of Westerhof, Noordergraf et al.[9], which reproduces the anatomy of the arterial systemic circulation.

This model is meaningful since it clarifies the essential characteristics and the limits of an interpretative model. The anatomy of the system is assumed as in fig.6, where each block represents an artery.

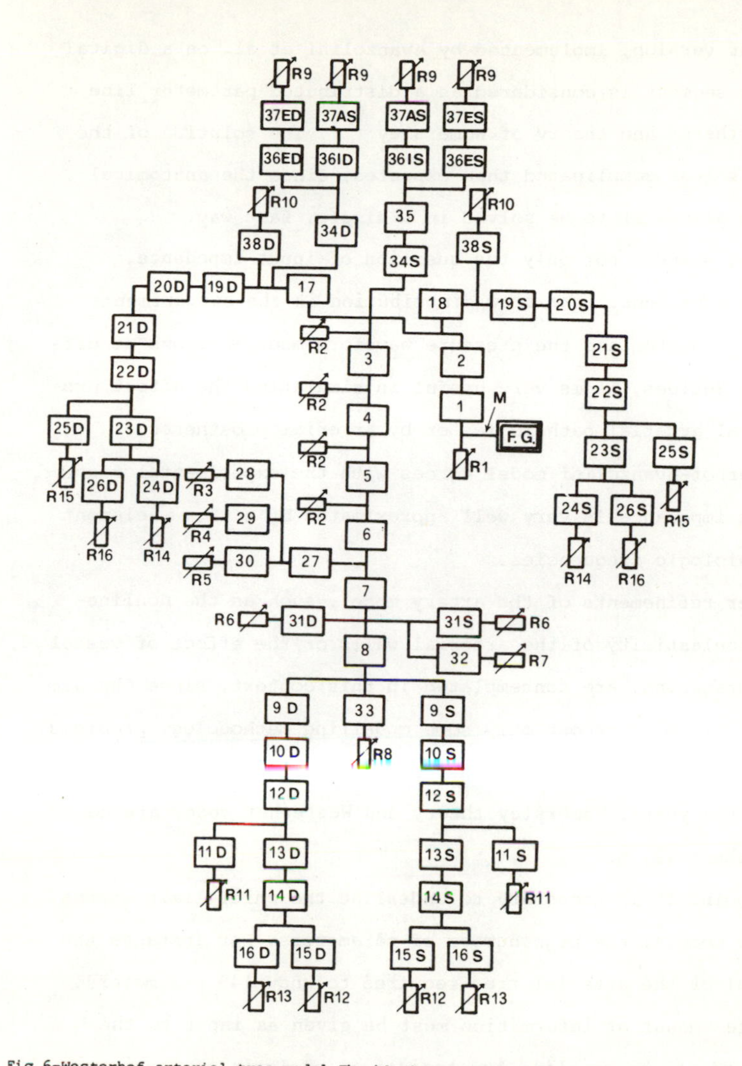

Fig.6-Westerhof arterial tree model.The blocks represent an artery or an arterial tree segment and are arranged according to anatomic configuration of the systemic arterial tree.F.G.:function generator which simulates the cardiac output.

odel includes all the chief arteries of radius greater than 1 mm. th,radius,wall thickness and Young modulus are given in a table not reproduced in this text.In Westerhof's original work each ar ıent is represented by lumped-parameter model and realized by a- ctrical circuits.

In a later version, implemented by Avanzolini et al. on a digital computer, each segment is considered as a distributed parameter line using the hypotheses and theory of Womersley [10]. The solution of the equation set is less complicated than expected, since the anatomical tree structure allows it to be solved in a simple, fast way.

This model answers not only the question of input impedance, but also other questions, like flow distribution in the compartments of the vascular system and the pressure waveform modifications at different points. Besides, it is very useful in simulating the effect produced by several arterial pathologies or by arterial prostheses.

The Westerhof-Avanzolini model agrees with the experimental result that the input impedance is very well approximated by a three-element model for physiologic frequencies.

No further refinements of the artery model, such as the nonlinearity and viscoelasticity of the arterial walls or the effect of vessel tapering and branching, are contemplated in this context, since the aim of the lecture is to point out only some modelling methodology problems.

It is obvious that each theory and model undergoes refinements in the course of the years; Womersley theory and Westerhof model are no exception to the rule. e.

At this point it is necessary to underline the chief limit of the interpretative models: the high number of parameters. For instance the Westerhof model of the arterial tree requires to know 145 parameters.

This large amount of information must be given as input to the model system and can be supplied by statistical analysis of large experimental data sets.

An interpretative model is therefore much more closely connected with experimental work than a descriptive one; it needs experimental data not only in the final validation step but also in the computer simulation phase. Owing to the large number of parameters very few of them can be adjusted by a best fitting procedure; most of them must be previously known.

The experimental data necessary to specify the parameters of an interpretative model are different from those required for model validation. The first concern the elementary properties of the biologic structures, the second the behavior of the system as a whole.

The first kind of experimental work should be further increased; for instance, the viscoelastic behavior of the arterial walls requires new experimental investigations, since the frequency dependence of sinusoidal wave attenuation, in the physiologic range of frequencies, is not yet clear.

The interpretative models are in general quite involved and are, therefore, not suitable for an intuitive and simple insight of the phenomenon.

Model complexity is an unavoidable consequence of that of the biologic system. A rough simplification of the model produces a distortion of reality whereas a careful and limited reduction process could lead to a more intelligible model.

This subject is discussed in the next section.

3. Some model reduction methods

First of all it is necessary to explain the meaning of a reduced or simplified model. An example should clear up the concept.

Very often in problems of vascular dynamics transcendental functions of a complex variable appear, as in the formula of input impedance Z_i of an artery loaded at the distal extremity with an impedance Z_L:

$$Z_i = \frac{Z_L + z\,\ell \tanh(\Gamma\ell)/(\Gamma\ell)}{1 + Z_L\, y_t\, \ell \tanh(\Gamma\ell)/(\Gamma\ell)} \tag{2}$$

$$\Gamma := \sqrt{z\, y_t}$$

$$z := r_p\, x^2 I_o(x)/(8 I_2(x))$$

$$x := R\sqrt{\rho s/\mu} \qquad\qquad y_t := sc;$$

ℓ : artery length; ρ : blood density; μ : blood viscosity; R : inner artery radius; r_P: Poiseuille resistance (for unit artery length); c : artery compliance (for unit length); s : complex variable of L.T.; I_ν : modified Bessel functions.

Formula (2) derives from the Womersley theory which supposes the artery to be cylindrical; for this reason the tapering parameters do not appear.

An essential simplification of function (2) is produced by replacing in it the meromorphic functions z and $\tanh(\Gamma\ell)/(\Gamma\ell)$ by rational approximants. In this way the function Z_i, which has an infinite number of zeroes and poles, is replaced by a function with a finite number of zeroes and poles.

Moreover, a rational function, expressed as ratio of two polynomials of m and n degrees, is simplified if it is approximated with a quotient of polynomials of degrees $\mu<m$ and $\nu>n$.

Obviously the approximation is valid only in a limited domain of the complex plane.

Now we see some methods which can be used to approximate the meromorphic functions. An example of this procedure is due to Jager et al.[11]; it concerns simplification of the longitudinal impedance z. This function may be expanded in an infinite continued fraction:

$$z(x) = r_P + x_I + \frac{1}{3/x_I} + \frac{1}{2r_P} + \frac{1}{5/x_I} + \frac{1}{3r_P} + \cdots \qquad (3)$$

$$x_I := r_P x^2/8 = s\frac{\rho}{R^2} =: s\lambda; \qquad \lambda : \text{inertance for unit length.}$$

The other symbols have the above-specified meaning.

Expansion (3) may be interpreted by means of a nice staircase equivalent circuit. Cutting off the expansion at the Nth term gives the Nth function's approximant, which is rational.

In a similar way an approximant of the function $\tanh(y)/(y)$ may be found, which has the following continued fraction expansion:

$$\tanh(y)/(y) = \frac{1}{1} + \frac{y^2}{3} + \frac{y^2}{5} + \frac{y^2}{7} + \cdots \qquad (4)$$

The use of formula (4) allows interpretation of the input impedance of an artery (see (2)) with the equivalent circuit of fig. 7 [12].

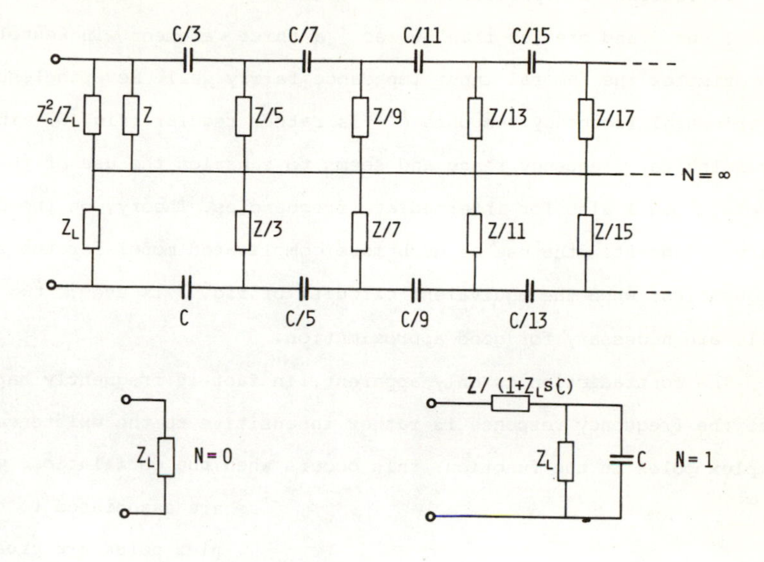

Fig.7 - Equivalent circuit for the input impedance of an artery loaded with Z_L. C : compliance of the artery ; Z : longitudinal impedance ($z\ell$).

For N = 0 the approximant is simply Z_L; for N = 1 the approximant is:

$$Z_{i,1} = \frac{Z_L}{1+Z_L y_t \ell} + \frac{z\ell}{1+Z_L y_t \ell} . \qquad (5)$$

When Z_L is a resistance or a windkessel impedance, i.e. a resistance in parallel with a compliance, the first term at the right hand side of (5) is also a windkessel impedance.

With reference, for instance, to the superficial femoral artery, the 1st approximant of the input impedance approximates the function with an error less than 10% for frequencies lower than 1 sec^{-1}; moreover, for these frequencies the second term on the right of (5) is negligible.

We may conclude that in this case a windkessel model approximates the input impedance of the femoral artery fairly well.

Only for frequencies greater than 20 sec^{-1} can the input impedance of the femoral artery be approximated with the characteristic impedance, giving an error of less than 10%.

It follows from previous considerations that for frequencies less than 1 sec^{-1} and greater than 30 sec^{-1} a three - element windkessel model approximates the femoral input impedance fairly well. Nevertheless the experimental frequency response [13)] is rather regular (fig. 8) in the physiological frequency range and seems to sanction the use of the a-bovesaid model also for intermediate frequencies. Theory, on the contrary, suggests the use of much more complicated model for the same frequencies; when the equivalent circuits of fig. 7 is used, two cells are necessary for good approximation.

The contradiction is only apparent. In fact it frequently happens that the frequency response is rather insensitive to the existence of complex poles in the function; this occurs when the oscillations which are associated to complex poles are greatly damped. Therefore, if the problem is only to represent the frequency response in a limited frequency range, a very simple model may be used, for example the three-element windkessel model.

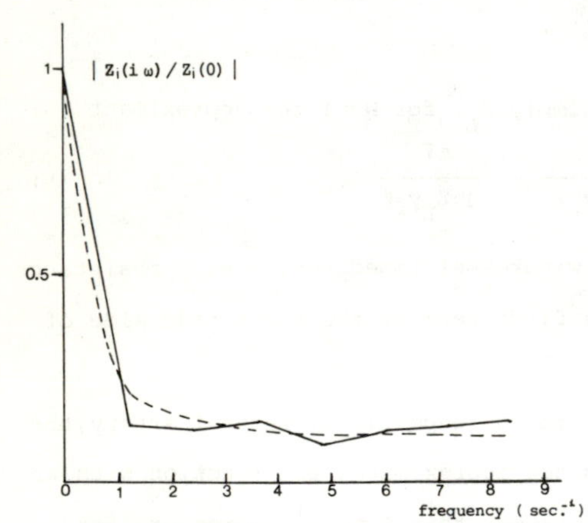

Fig.8 - Frequency response of the sup. femoral artery (solid line). Frequency response obtained by an optimal three-element model (dashed curve)

But in all problems where the knowledge of all dominant poles is necessary, the use of a more complete model becomes unavoidable. This is the case of frequency input impedance responses with conspicuous maxima, as in the dog abdominal aorta and brachiocephalic artery [14]. Besides, also for arteries with roughly monotone decreasing response it may be convenient to point out complex poles, since they are highly sensitive to some pathologies as we shall see later.Only dominant poles have to be usually determined.

For the sake of clarity, it should be stated that the dominant poles are the singularities which mainly influence the frequency or impulse responses. Other possible poles can be neglected if that implies an error on the responses less than an assigned value. Therefore the concept of dominant poles is strictly connected with that of acceptable error in a specific problem. A more exhaustive definition can be given with reference tc Mittag-Lefler partial-fraction expansion of meromorphic functions.

At this point it must be clarified whether the use of a simplified model like those obtained by the theory of approximation of function, is convenient or not. For example, one might think of computing the poles of the input impedance (2) of an artery directly by computing the zeroes of the denominator, i.e. by solving the equation:

$$1 + Z_L\, y_t \ell \tanh(\Gamma\ell)/(\Gamma\ell) = 0 \; . \qquad (6)$$

Unfortunately the numerical methods for solving equations like (5) are terribly involved and computer time consuming. When the function on the left hand side of (6) is approximated by a rational function, the solution of the equation becomes much simpler. Under suitable but quite broad conditions the solution of the approximated equation gives an approximation of the zeroes as follows from a Tricomi's theorem.[15]

When the function admits a continued fraction expansion, cutting off the expansion is a method to obtain a rational approximant.

Closely connected with the continued fractions are the Padè approximants when the functions are holomorphic in the origin of the complex plane [16]. Given such a function G(s), the Padè method approximates it with a rational function

$$G_a(s) = \frac{\sum_{o}^{L} {}_i\ \beta_i s^i}{\sum_{o}^{M} {}_i\ \alpha_i s^i} \qquad (7)$$

in such a way that an equal number of Mac Laurin series expansion coef ficients agree with the corresponding ones of G(s). If the number of terms considered in the Mac Laurin series expansion is N+1, then it fol lows that

$$N = L + M . \qquad (8)$$

For every choice of the integers L,M satisfying (8), one has an approximant generally indicated with the symbol L/M .

The Padè approximants are a generalization of the continued fraction approximants, in fact it is proved that these are particular sequences of the Padè approximants.

A common view is that the Padè method is simple computationally. Indeed simple algorithms exist for computing the approximant coefficients *starting from MacLaurin expansion coefficients*, but it is not always easy to compute these for an involved function like artery input impedance. When the approximant requires a large number of MacLaurin coefficients the procedure for their computation is quite cumbersome. For this reason it is better, in many cases, to approximate a compound function through approximation of its elementary functions.

The following example illustrates a problem which requires function approximation. The problem arises from the need to evaluate the severity of an arterial stemosis by measuring the blood flow.

An arterial stenosis, whenever it is, bears on the whole cardiovascular system; for example, a bilateral occlusion of the femoral arteries [17)] produces an increase of 10 mmHg of the systolic pressure in the ascendens aorta.

Obviously the more marked effects are produced close to the steno sis, especially on the circulation compartment downstream of the stenosis.

A way to evaluate the effect of the stenosis is to use the Westerhof arterial tree simulator, previously mentioned, provided the stenotic segment is represented with a modified model to account for section reduction. But this approach is not suitable for diagnostic use.

A more effective method must be founded on measurements made on the patient. In the following it is supposed that invasive techniques such as angiography are excluded.

The most direct method is to measure the mean velocity(*) in a section upstream of the stenosis and in a section of the stenotic segment (fig. 9a).

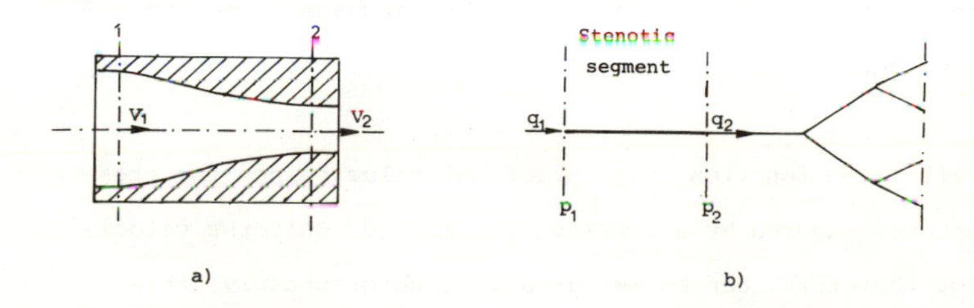

Fig.9 - Different methods for evaluation of arterial stenosis (see text).

If A_1 e A_2 are the respective areas, then

$$\frac{V_1}{V_2} = \frac{A_2}{A_1} := k \quad ,$$

on the hypothesis that there are no branches between the sections. The use of ultrasound Doppler velocimeters allows approximate evaluation

(*) Mean velocity is the average of the fluid velocities across the section

of the mean velocity.

A critical analysis of this method is not made here; suffice to say that the results are affected by not easily predictable errors.

Another method is based on analysis of the velocity/time waveforms.

Measurements on patients [18] show that the velocity/time waveform at the output of the stenotic arterial segment is modified with respect to the normal situation. To correlate this alteration with the area ratio k, is a good idea.

The problem is to quantify by means of suitable parameters the waveform distortion produced by the stenosis and to evaluate the sensitivity of the parameters with respect to the reduction of k. We can schematize the problem in this way. First of all we need to define some functions which are dependent *only* on the parameters of the arteries. With reference to fig. 9b we can consider three ratios which have this property: $p_2(s)/p_1(s)$, $p_1(s)/v_1(s)$, $v_2(s)/v_1(s)$.

Of these transfer functions, using the language common to system theory, the second is the input impedance of the stenotic segment.

Not all these functions have practical value, since the pressure cannot be measured by a non-invasive method. Only the velocities ratio, therefore, can be measured but, unfortunately, this function is less sensitive to the variation of k than the other, as we shall see later.

The above-defined transfer functions have the following expression:

$$G_1(s) := \frac{p_2(s)}{p_1(s)} = \frac{Z_L/\cosh(\Gamma\ell)}{Z_L+z\ell\,\tanh(\Gamma\ell)/(\Gamma\ell)}$$

$$Z_i(s) = \frac{Z_L+z\ell\,\tanh(\Gamma\ell)/(\Gamma\ell)}{1+Z_L\,y_t\,\ell\,\tanh(\Gamma\ell)/(\Gamma\ell)}$$

$$G_2(s) := \frac{1}{k} \frac{1/\cosh(\Gamma\ell)}{1+Z_L\, y_t\, \ell \tanh(\Gamma\ell)/(\Gamma\ell)}$$

$$k := \frac{A_2}{A_1} = \frac{\text{reduced area}}{\text{normal area}}$$; Z_L : load impedance of the stenotic segment;

other symbols have been previously defined.

$Z_i(s)$ and $G_2(s)$ have the same denominator, which is different from that of $G_1(s)$. The qualitative interpretation of the formulas allows interpretation of the poles displacement as being due mainly to increasing of (the modulus of) the longitudinal impedance z for G_1 and to reduction of (the modulus of) the transversal admittance for Z_i and G_2, when the parameter k becomes smaller. In the simplest form longitudinal impedance is the series of the Poiseuille resistance and of the reactance due to inertance; transversal admittance is that due to compliance. Moreover some zeroes of $Z_i(s)$ agree with some poles of $G_1(s)$, because the numerator of Z_i is equal to the denominator of G_1.

We consider now the particular case of a stenosis in the proximal portion of the superficial femoral artery (fig. 10). The functions G_1, Z_i, G_2 have in this case a form different from those above-mentioned since there is a branch between input and output. For the sake of simplicity, these formulas are not reported. The assumptions are: a) uniform reduction of the inner radius of the stenotic segment which produces an increase in thickness, b) the Womersley hypotheses [19] for fluid motion and wall displacement, c) invariance during the stenotic process of the peripheral resistances, d) lack of collateral circulation.

Obviously all the hypotheses can be discussed and criticized, since all are schematizations of the real processes. Also for this model the validation of the hypotheses is made in the last step (see schematic diagram of fig. 3) by comparison with experimental data.

The problem is: to evaluate the zeroes and poles displacement of the functions G_1, Z_i, G_2 when the value of K goes from 1 (normal situation) to 0 (complete occlusion).

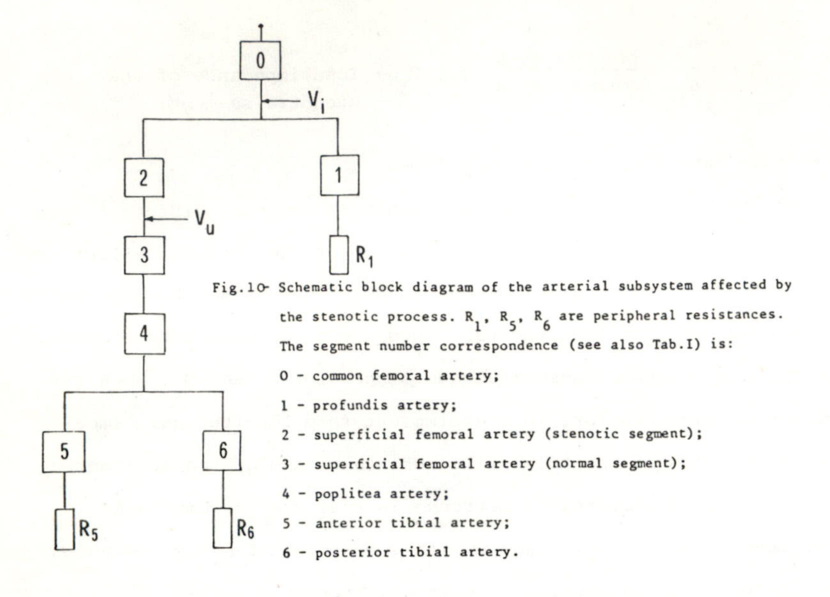

Fig.10- Schematic block diagram of the arterial subsystem affected by the stenotic process. R_1, R_5, R_6 are peripheral resistances. The segment number correspondence (see also Tab.I) is:

0 - common femoral artery;
1 - profundis artery;
2 - superficial femoral artery (stenotic segment);
3 - superficial femoral artery (normal segment);
4 - poplitea artery;
5 - anterior tibial artery;
6 - posterior tibial artery.

Numerical methods have been used in order to compute directly the zeroes and poles of the function in the original form for each value of k, but they are too involved and not reliable owing to the particular nature of the function. Such methods use standard subroutines to generate the transcendental elementary functions which are not suitable for this kind of problem.

In order to evaluate the zeroes and poles quickly and surely, the only procedure is to approximate the elementary functions with rational approximants. The approximation must be made on a circular domain around the origin of the complex plane and not only on an interval of the imaginary axis. The elementary functions to be approximated are $\tanh(y)/y$ and z; for the first a continued fraction approximant has been used and for the second a rational approximant obtained by cutting off an infinite product [20]. In this way the original functions of the model, impedances and transfer functions, become ratios of polynomials on s, the independent variable. These polynomials are of high degree and therefore must be approximated with polynomials with lower degree. This problem can be solved again by Padè approximants or other techniques like the method of dominant zeroes retention [21)22)].

The used valuesof the parameters are reported in Table I. Referring to function $G_7(s)$, the results are shown in fig. 12 where the dominant pole locus is reported[23)]. The length of the stenotic segment is in this case equal to the length of the superficial artery (distributed stenosis).

Table I. Anatomical data for the arterial subsystem of Fig.10.

Segment number	Artery	Radius	Length	Wall thickness	Young's modulus
n		r_n (cm)	ℓ_n (cm)	h_n (cm)	E_n ($\times 10^5$ Kg m^{-1} s^{-2})
0	Femoralis communis		unused		
1	Profundis	.225	12.6	.049	16
2	Femoralis superficialis	.242	5.0 (°) 25.4 (^)	.05	4
3	Femoralis superficialis	.242	20.4 (°) unused (^)	.05	4
4	Poplitea	.2	18.9	.047	8
5	Tibialis anterior	.13	34.3	.039	16
6	Tibialis posterior	.18	32.1	.045	16

Peripheral resistances ($\times 10^5$ m^{-4} Kg s^{-1})

R_1 = 55.000 $\quad$ R_5 = 94.000 $\quad$ R_6 = 46.000

Blood kinematic viscosity $\nu = .038 \times 10^{-4}\ m^2\ s^{-1}$

(°) Case a) of section 1 (short stenosis).

(^) Case b) of section 1 (long stenosis).

We see that a pair of initially complex dominant poles become real for a very small value of K = 0.07. Substantial displacement of the same poles occurs for a value of K = 0.2.

Therefore, the method is *not* sensitive enough to the reduction of k. Better results are obtained from the transfer function $G_1(s)$;

in fact, the same effects are in this case produced by values of k about twice as those mentioned above. The same result has been obtained for dominant zeroes displacement of Z_i. The shape of the dominant poles locus is the same as that obtained by measurements made on the patient [18]. Experimental work is in progress in the Italian Cardiovascular Project, in order to evaluate the sensitivities of the singularities of the functions by measurements on the dog.

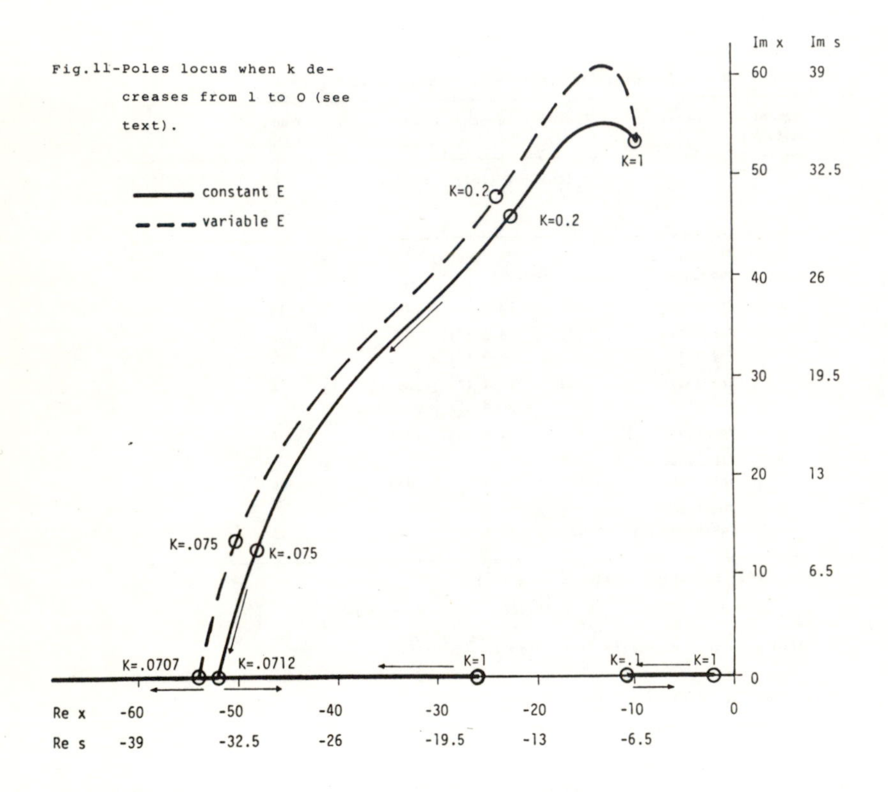

Fig.11-Poles locus when k decreases from 1 to 0 (see text).

The foregoing examples allow us to conclude that the approximation theory is an effective tool for interpreting the behavior of vascular subsystems and for solving the problem of approximating zeroes and poles of haemodynamical functions.

Conclusion

Representative and interpretative models are both useful tools for cardiovascular research. They have different characteristics and aims; it is important that difference be well understood.

Representative models synthetize the results of the experiments; interpretative models attempt to explain them by means of physical theories.

To ascribe an interpretative function to a descriptive model can produce a distortion of reality; to claim divination faculties by using interpretative models is an illusion.

The progress of the knowledge requires, in fact, a harmonic and interacting development of experimental and theoretical research, as the history of all sciences teaches.

References

1. Warner H.R.: The use of an analog computer for analysis of control mechanisms in the circulation. Proc. IRE, 47, 1959
2. Suga H.: Time course of left ventricular pressure-volume relation ship under various extents of aortic occlusion. Jap. Heart Journ. 11, 1970
3. Westerhof N., Elzinga G.: The apparent source impedance of heart. Ann. Biomed. Engng. 6, 1978
4. Beneken, J.E.W.: A mathematical approach to cardiovascular function. Report n. 2-4-5/6. Institute of medical physics. Utrecht, May 1965
5. Huxley A.F.: Muscle structure and theories of contraction. Prog. Biophys. and Biophys. Chem. 7, 257-318, 1957
6. Westerhof N., Elzinga G., Sipkema P.: An artificial arterial system for pumping hearts. J. Appl. Physiol. 31, 1971
7. Pedley T.J.: The fluid mechanics of large blood vessels. Cambridge University Press. Cambridge 1980
8. Taylor M.G.:Wave travel in arteries and the design of the cardio-

vascular system. In "Pulsatile blood flow", E.D. Attinger (Ed.). pp. 343-372, Mc Graw Hill, New York, 1964

9. Westerhof N., Bosman F., De Vries C.J., Noordergraaf A.: Analog studies of the human systemic arterial tree. J. Biomechanics, 2, 1969

10. Avanzolini G., Belardinelli E., Capitani G., Passigato R.: Steady-state numerical model of the human systemic arterial tree. IFAC Symposium on Automatic Control and Computers in the Medical Field. Bruxelles, 1971

11. Jager G.N., Westerhof N., Noordergraaf A.: Oscillatory flow impedance in electrical analog of arterial system. Circulation Res., 16, 1965

12. Belardinelli E.: Expansion of the arterial input impedance. 3rd International Conference on mechanics in medicine and biology. Compiegne, 1982

13. Farrar D.J., Malindzak G.S., Johnson G.: Large vessel impedance in peripheral atherosclerosis. Cardiovascular surgery, Supp. 2, Circulation, 56-3, 1977

14. O'Rourke M.F.: Pressure and flow waves in systemic arteries and the anatomical design of the arterial system. J. Appl. Physiol. 23, 1967

15. Tricomi F.: Sugli zeri delle funzioni di cui si conosce una rappresentazione asintotica. Annali di Matematica, Serie IV, Tomo 26, 1947

16. Baker G.A.: Essentials of Padè approximants. Academic Press, New York, 1975

17. Murgo J.P., Westerhof N., Jolma J.P., Altobelli S.A.: Aortic input impedance in normal man: relationship to pressure wave forms. Circulation, 62, 1, 1980

18. Brown J.M., Nahorski Z.T., Woodcock J.P., Morris S.J.: Transfer-function modelling of arteries. Med. and Biol. Eng. and Comput., 16, 1978

19. Womersley J.R.: An elastic tube theory of pulse transmission and oscillatory flow in mammalian arteries. WADC Tech. Rep. TR 56-614, 1957

20. Belardinelli E.: Sull'impedenza longitudinale del modello lineare di arteria. Parte I, II. Atti Accademia delle Scienze, Serie XIII, Tomo 8, Bologna 1981

21. Gutman P.O., Mannerfelt C.F., Molander P.: Contributions to the model reduction problem. IEEE Trans. Aut. Contr., 27, 1982

22. Shamash Y.: Linear system reduction using Padè approximation to allow retention of dominant modes. Int. J. Control, 21, 1975

23. Belardinelli E.,Gnudi G.,Evangelisti A.: Modello matematico delle stenosi arteriose periferiche.Atti Accademia delle Scienze,Serie XV,Bologna 1983.

PROCEEDINGS OF THE II INTERNATIONAL CONFERENCE ON APPLICATIONS OF PHYSICS TO MEDICINE AND BIOLOGY
edited by Ž. Bajzer, P. Baxa & C. Franconi

THE DESIGN, DEVELOPMENT AND ASSESSMENT OF HEART VALVE SUBSTITUTES

H. Reul* and M. M. Black**

* Helmholtz-Institute for Biomedical Engineering at the RWTH Aachen, D-5100 Aachen, West-Germany

** Department of Medical Physics and Clinical Engineering, Royal Hallamshire Hospital
Sheffield S10 2JF, England

ABSTRACT

The different designs as well as the advantages and disadvantages of mechanical and tissue valves are discussed in the first section of the paper. A brief historical review on the major trends in heart valve developments is included.
The second part concentrates on the various methods of in-vitro testing. Starting with steady flow studies, this section includes also a variety of pulsatile flow testing methods such as hot-film-anemometry for wall-shear-stress measurements, Laser-Doppler anemometry for the assessment of velocity fields and heart valve fatigue testing.
Future developments in heart valve design and the need for standardized testing methods are discussed in the final section.

1. INTRODUCTION

The replacement of malfunctioning heart valves with prosthetic counterparts has been an accepted routine clinical precedure for over tewnty years. Hufnagel[21] showed that manmade materials could be tolerated in the blood stream and his caged-ball prosthesis placed in the descending aorta was able to funtion in many patients for reasonably long periods. Total or partial valve replacement became a viable therapeutic option subsequent to the development of extracorporeal circulation, as described by Gibbon[19].

Since these early days there have been many developments in the design and manufacture of artificial valves, and it is no exaggeration to say that several hundred different configurations have been considered. The vast majority of these designs have not gone beyond initial laboratory evaluation and, indeed, some have not even reached that stage. This may appear to suggest that much time and effort has been expended on valve development which has yielded a disproportionately small number of viable prostheses. Such an assessment is, however, quite unjustified since it neglects the immense wealth of knowledge that this research has created on a wide range of topics which are complementary to the ultimate development of an optimum valve design. For example, the need to test valves in the laboratory necessitated the development of sophisticated pulse duplicators which could reproduce the physiological and haemodynamic characteristics of the cardiac cycle. The design of such in vitro test systems has led to a greater understanding of normal valve performance in vivo. Furthermore, studies of the problems associated with the interaction of blood and foreign materials have yielded important information on anticoagulation techniques as well as encouraging the search for materials with improved properties of biocompatibility. This latter topic is of relevance to the whole field of prosthetic implants.

The range of research projects which has been created as a result of prosthetic heart valve development is very extensive and, more importantly, has led to the involvement of non-medical researchers from many different scientific backgrounds. This intimate collaboration of clinicians and scientists is an essential ingredient if optimum valve designs are to be obtained. The first valves to be used were almost entirely the result of surgical innovation. Whilst not wishing to detract from their initial success, it is clear that even today there is no ideal valvular replacement. To achieve the optimum design will require the continued collaboration of medical and non-medical scientists.

In the sections which follow an account will be given of the development of various valve forms and the techniques used to test them in the laboratory. Since the autors are non-medical scientists, there will be a tendency to place more emphasis on the physical aspects of valve design. However, in the final analysis, what matters are the long-term follow-up results of clinical implantation. Such data, based on over 3000 cases, are available from the Multicentre Valve Trial set up by the Cardiac Surgical Research Club in Britain.

2. MECHANICAL VALVES

The use of a caged-ball valve in the descending aorta became obsolete with the development of what today is still referred to as the Starr-Edwards ball and cage valve, as illustrated in Figure 1. In concept it was similar to the original Hufnagel valve but was designed to be inserted in place of the excised diseased natural valve. This form of intracardiac valve replacement in the mitral position was reported by Starr[43] and for aortic and multiple valve replacements by Cartwright et al.[13]. Since 1962, the Starr-Edwards valve has undergone many modifications in order to improve its perfomance in terms of reduced haemolysis and thromboembolic complications. However, the changes have involved materials and techniques of construction and have not altered the overall concept of the valve design in any way. Attempts to reduce the oberall height of the valve by using a disc occluder in a flattened cage were not successful and the still popular current models continue to use a spherical occluder.

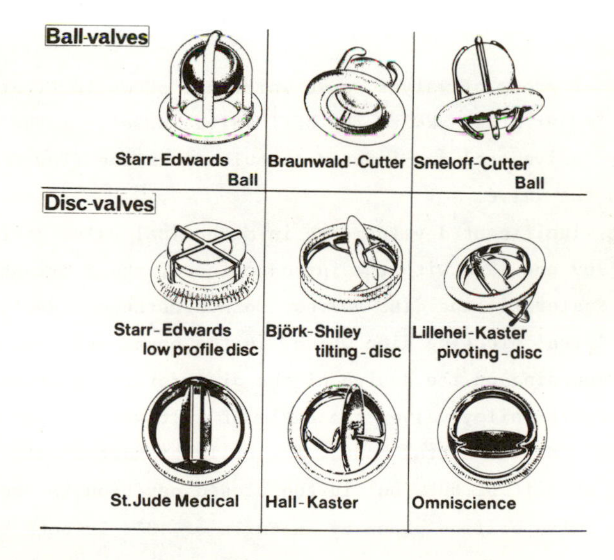

Figure 1: Some types of currently used mechanical heart valve prostheses

Other manufacturers have produced variations of the ball and cage valve, notably the Smeloff-Cutter valve and the Magovern prosthesis. In the case of the former, the ball is slightly smaller than the orifice. A subcage on the proximal side of the valve retains the ball in the closed position with its equator in the plane of the sewing ring. A small clearance around the ball ensures easy passage of the ball into the orifice. This clearance also gives rise to a mild regurgitation which was felt, though not proven, to be beneficial in preventing thrombus formation. The Magovern valve is a standard ball and cage format which incorporates two rows of interlocking mechanical teeth around the orifice ring. These teeth are used for inserting the valve and are activated by removing a special valve holder once the valve has been correctly located in the prepared tissue annulus. The potential hazard from dislocation from a calcific annulus due to imperfect placement was soon observed. This valve is no longer in use.

Many other variations of caged occluder valves could be listed but none have been able to displace the Starr-Edwards as the preferred valve in this category. Even after twenty years of valve development, the ball and cage format remains a valve of choice amongst some surgeons. However, it is no longer the most popular mechanical valve, having been superseded, to a large extent, by tilting disc or hinged leaflet valves.

This latter group of valves, some which are shown in Figure 1, overcomes two major drawbacks to the ball valve, namely, large profile height and excessive occluder-induced turbulence in the flow through and distal to the valve.

The most significant developments in mechanical valve design occurred in 1969 and 1970 with the introduction of the Björk-Shiley and Lillehei-Kaster tilting disc valves. Both prostheses involve the concept of a 'free' floating disc which, in the open position, tilts to an angle depending on the design of the disc retaining struts. In the original Björk-Shiley valve, the angle of tilt was 60° for the aortic, and 50° for the mitral model. The Lillehei-Kaster valve has a greater angle of tilt of 80°, but in the closed position is pre-inclined to the valve orifice plane by an angle of 18°. In both cases the closed valve configuration permits the occluder to fit into the circumference of the inflow ring with virtually no overlap, thus reducing mechanical damage to erythrocytes. A small amount of regurgitation backflow induces a 'washing out' effect of 'debris' and platelets and theoretically reduces the incidence of thromboemboli (cf. Smeloff-Cutter valve description).

The obvious advantage of the tilting disc valve is that in the open position it acts like an aerofoil in the blood flowing through the valve, and induced flow disturbance is substantially less than that obtained with a ball occluder. Although the original Björk-Shiley valve employed a 'Delrin' occluder, all present-day tilting disc valves use pyrolitic carbon for this component. It should also be noted that the 'free' floating disc can rotate during normal function, thus preventing excessive contact wear from the retaining components on one particular part of the disc.

Tilting disc valves remain popular (cf. Figure 3) although central orifice tissue valves, as discussed in the section which follows, are gaining in usage. Various improvements to this form of mechanical valve have been developed but have tended to concentrate on alterations either to the disc geometry as in the Björk-Shiley Concavo-Convex design, or to the disc retaining system as with the Omniscience valve.

Perhaps the most interesting development has been that of the bileaflet pyrolitic carbon valve produced by St. Jude Medical Inc. This design incorporates two semicircular hinged pyrolitic carbon occluders which in the open position provide minimal disturbance to flow. Details of in vitro studies on this valve were published by Emery & Nicoloff[14)] and early clinical evaluation of its perfomance by Emery et. al.[15)]. The evidence provided in these papers indicates some promise for this design of mechanical valve.

The informed reader will realize that this section on mechanical valves has highlighted a relatively small number of the many various forms which have been made. However, those that have been included are either the most commonly used, or are those which have made notable contributions to mechanical valve design. In a later section the methods used to test valves in the laboratory are discussed. The lack of standardization of such test procedures inevitably makes it difficult to obtain genuinely comparative data on the various different valves. Sadly, in vitro performance differences amongst the valves are often relatively small, and claims made for new designs may lack scientific integrity in the face of commercial necessity. However, the majority of mechanical valves do offer patients reasonable hope for several years of event-free survival even though they bear no morphological comparison to natural valves. In the following section the development of 'tissue' valves which are simular or, indeed, identical to natural valves, is discussed.

3. TISSUE VALVES

One major disadvantage with the use of mechanical valves is the need for continuous anticoagulation therapy to minimize the risk of thrombosis and thromboembolic complications. Furthermore, the haemodynamic function of even the best types of mechanical valves differs significantly from that of normal heart valves. An obvious step, therefore, in the development of heart valve substitutes was the application of naturally-occurring heart valves. This was the basis of the approach to the use of treated human aortic valves removed from cadavers for implantation in place of their diseased counterparts.

The first of these procedures was carried out by Ross[38)] in 1962, and the overall results so far have been satisfactory (Wain et al.[46)]). This is, perhaps, not surprising since the replacement valve is optimum both from the point of view of structure and function. In the open position these valves provide unobstructed central orifice flow and also have the ability to respond to deformations induced by the surrounding anatomical structure. As a result, such substitutes are less damaging to the blood when compared with the rigid mechanical valve. The main problem with these cadaveric allografts, as far as may be ascertained, is that they are no longer living tissue and therefore lack that unique quality of cellular regeneration typical of normal living systems. This makes them more vulnerable to longterm damage.

An alternative approach is to transplant the patient's own pulmonary valve into the aortic position. This operation was also first carried out by Ross[39)] in 1967, and his subsequent study of 176 patients followed up over 13 years showed that such transplants continued to be viable in their new position with no apparent degeneration (see Wain et al[46]). This transplantation technique is, however, limited in that it can only be applied to one site.

The next stage in the development of tissue valve substitutes was the use of autologous fascia lata either as free or frame-mounted leaflets. The former approach for aortic valve replacement was reported by Senning[41] in 1966, and details of a frame-mounted technique were published by Ionescu & Ross[23] in 1966. The approach combined the more natural leaflet format with a readily available living autologous tissue. Although early results seemed encouraging, Senning[42] expressed his own doubt on the value of this approach in 1971, and by 1978 fascia lata was no longer used in either of the above, or any other, form of valve replacement. The failure of this technique was due to the inadequate strength of this tissue when subjected to long-term cylic stressing in the heart. This situation might have been avoided if adequate analysis of the mechanical behaviour of fascia lata had been undertaken.

In parallel with the work on fascia lata valves, alternative forms of tissue leaflet valves were being developed. In these designs, however, more emphasis was placed on optimum performance characteristics than on the use of living tissue. In all cases the configuration involved a three-leaflet format which was maintained by the use of a suitably designed mounting frame. It was realized that naturally-occurring animal tissues, if used in an untreated form, would be rejected by the host. Consequently, a method of chemical treatment had to be found which obviated this antigenic response but did not degrade the mechanical strength of the tissue.

Formaldehyde has been used by histologists for many years to arrest autolysis and 'fix' tissue in the state in which it is removed. It had been used to preserve biological tissues in cardiac surgery but, unfortunately, was found to produce shrinkage and also increase the 'stiffness' of the resulting material. For these reasons, formaldehyde was not considered ideal as a method of tissue treatment.

Glutaraldehyde is another histological fixative which has been used especially for preserving fine detail for electron microscopy. It is also used as a tanning agent by the leather industry. In addition to arresting autolysis, glutaraldehyde produces a more flexible material than does formaldehyde, with improved strength characteristics due to increased collagen cross-linking. Glutaraldehyde has the additional ability of reducing the antigenicity of xenograft tissue to a level at which it can be implanted into the heart without significant immunological reaction.

In 1969, Kaiser et al.[25)] described a valve substitute using an explanted glutaraldehyde-treated porcine aortic valve which was mounted on to a rigid support frame. Following a modification described by Reis et al.[36)], in which the rigid frame was replaced by a frame having a rigid base ring with flexible posts, this valve became commercially available as the Hancock Porcine Xenograft. It remains one of the two most popular valve substitutes of this type, the other being the Carpentier-Edwards Bioprosthesis introduced commercially by Edwards Laboratories in 1976. This latter valve uses a totally flexible support frame.

In this type of prosthesis, the use of the intact biologically-formed valve obviates the need to manufacture individual valve cusps. Whilst this has the obvious advantage of reduced complexity of construction, it does require a facility for harvesting an adequate quantity of valves so that an appropriate range of valve sizes of suitable quality can be made available. This latter problem did not occur in the production of the three-leaflet calf pericardium valve developed by Ionescu et al.[24)]. The construction of this valve involves the moulding of fresh tissue to a tricuspid configuration around a support frame and, whilst held in this position, the tissue is treated with glutaraldehyde solution. The valve, marketed in 1976 as the Ionescu-Shiley Pericardial Xenograft, is currently the most popular valve of its type. Examples of mounted porcine valves and bovine pericardium valves are shown in Figure 2. Clinical results so far obtained with tissue valves indicate their superiority to mechanical valves with respect to a lower incidence of thromboembolic complications and more natural haemodynamic characteristics. For these reasons the use of tissue valves has increased significantly, as reflected in the data submitted to the Multicentre Valve Trial. This latter information is shown in Figure 3 for the period 1974 - 1982.

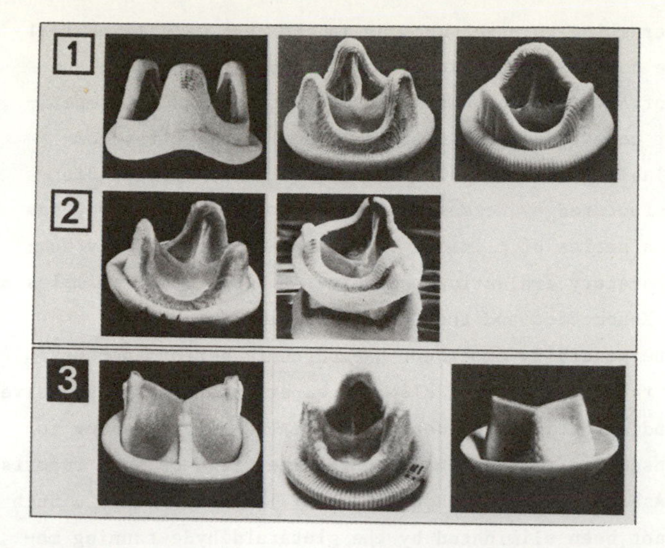

Figure 2: Tissue valves
1st row: Porcine valves, from left to right:
HANCOCK, CARPENTIER-EDWARDS, XENOMEDICA
2nd row: Porcine valves continued, from left to right:
WESSEX-MEDICAL, BIOIMPLANTAT CANADA
3rd row: Pericardial valves, from left to right:
IONESCU-SHILEY, HANCOCK, MITROFLOW

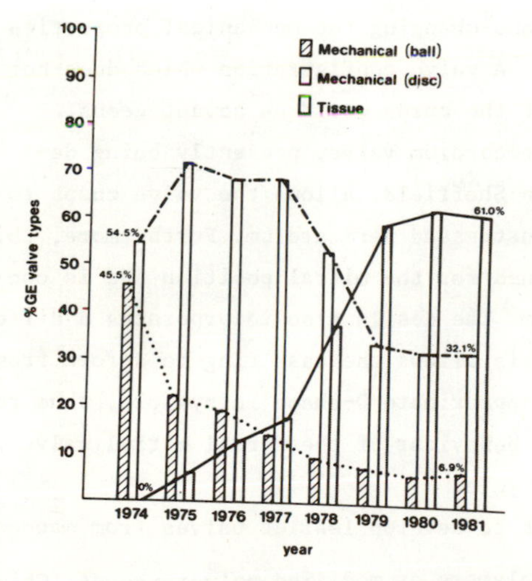

Figure 3: Percentage distribution of major valve types (values obtained from data recorded by the Multicentre Valve Trial (1974-1982) comprising a total of 3373 valve implants)

It is interesting to note that, as is the case with mechanical valves, all the popular tissue prostheses are manufactured in the United States of America and, until recently, there was no competition from other countries in this field. However, the first commercially-available British tissue valve, the Wessex Porcine Bioprosthesis manufactured by Wessex Medical Laboratories Limited, has recently begun a series of clinical trials having successfully completed both laboratory evaluation and animal trials. Other examples are the Swiss-made Xenomedica and the Canadian Bioimplant valves.

Despite the generally excellent haemodynamic performance of tissue valves, recent reports on clinical experience with these valves increasingly indicate time-dependent structural changes leading to failure and subsequent replacement. Amongst the first of such reports were those by Ashraf & Bloor[2)], Housman et al.[20)] and Broom[10)]. Such problems have not been eliminated by the glutaraldehyde tanning methods so far employed, and it is not easy to see how these drawbacks are to be overvome unless either living autologous tissue is used or the original structure of the collagen and elastin are chemically enhanced. On the latter point there is, as yet, much room for further work. For instance, the fixing of calf pericardium under tension during the moulding of the valve cusps will inevitably produce 'locked-in' stresses during fixation, thus changing the mechanical properties of the tissue (Reece et al.[35)]. A valve configuration which does not involve fixation moulding of the cusps would be advantageous.

The design of a new pericardium valve, presently being developed by Black et al.[8)] in Sheffield, allows the valve cusps to be formed using fully fixed, unstressed pericardium. Furthermore, this valve is specifically designed for the mitral position and is consequently of a bicuspid form. The design also incorporates a differentially flexible frame. This allows the base ring to deform from circular in diastole to an approximate D-shape in systole, thus reproducing the physiological behaviour of the normal mitral valve annulus during the cardiac cycle.

There are also attempts to develop leaflet valves from man-made materials such as block-copolymers or modified polyurethanes (Ghista & Reul[18)]). The major advantages of such a valve type would be better reproducibility and cost. The currently tested valve exhibits sufficient fatigue life of more than 400 Mill. cycles, excellent hemodynamics but, on the other hand, also shows severe leaflet calcification after implantation times of more than 100 days in calves. It is expected to reduce these problems by further modifications of the material and/or changing the test animal species.

4. MECHANICAL VERSUS TISSUE VALVES

It will be obvious from the foregoing sections that the different types of heart valve substitutes currently available have individual advantages and disadvantages. Certainly, none of the present prostheses could be regarded as ideal, or even optimum, replacements for natural valves. To identify in detail the good and bad points of each valve would require a publication in its own right and would not add significantly to the value of this present paper. However, in general terms, valve type advantages and disadvantages can be classified as shown in Table 1: From this table it is clear that surgeons are faced with a difficult situation when deciding on the most appropriate valve type for any particular patient. In many cases a surgeon will tend to use one particular valve type almost to the total exclusion of any other. This does not imply a narrow outlook on the part of the surgeon, but rather a natural scepticism in relation to what is often clinically unsubstantiated information provided by various manufacturers regarding new valve products.

Table 1. Advantages and disadvantages of mechanical and tissue valves

Valve type	Advantages	Disadvantages
Mechanical	Long-term durability Consistency of manufacture	Unnatural form Patient usually requires long-term anticoagulant therapy
Tissue	More natural form and function Less need for long-term anticoagulant therapy	Unproven long-term durability Consistency of manufacture is more difficult In vivo calcification

For the present, the surgeon has to make his own assessment of the potential advantages and disadvantages of any particular design. Whilst the apparent long-term durability of mechanical valves is immediately attractive, problems of maintaining large and increasing numbers of patients on optimum anticoagulation therapy can lead to logistic difficulties. Incidence of significant levels of haemolysis due to mechanical damage can also militate against their extensive use by many surgeons.

As already noted, the ultimate assessment of any valve can only be achieved through long-term clinical implantation and follow up. Individual surgeons maintain such records on their own cases. However, it is obvious that a survey involving many units offers a greater opportunity for collecting, not only much larger numbers, bur also for obtaining data on a wider range of valve types. Such a trial was initiated in 1974 under the auspices of the Cardiac Surgical Research Club. At present there are over 3000 individual valve implants recorded on computer, covering 37 valve models. Analysis of this Multicentre Valve Trial information was used to provide the data given in Figure 3 and Tables 2 and 3. It can be seen from Figure 3 that approximately 60 % of the valve implants recorded by the Trial in each of the last three years involved tissue valves. This result is, of course, dependent on the preferences of the 29 surgeons participating in the Trial, but is a reasonable guide to the overall picture in the United Kingdom at the present time.

The data given in Tables 2 and 3 represent the actuarial probabilities of 'patient survival' and 'freedom from valve complications' respectively. In the case of patient survival, the most obvious feature is the sudden drop in survival to 61.5 % after 7 years for those patients with a Hancock porcine mitral prosthesis. Although the results for 6 to 7 years follow up are, as yet, incomplete for some of the other valves listed in Table 2, those which have gone beyond 5 years do not indicate such a dramatic fall.

A simular phenomenon is observed from the figures defining freedom from valve complications. In this case the valve showing the most severe drop is the Starr-Edwards 1260 aortic model at a value of 42.5 % after 6 years, with its mitral counterpart, the 6120 model, at 52.9 % after 8 years. None of the other valves listed in Table 3

Table 2: Actuarial values of probabilities of patient survival after valve replacement: Multicentre Valve Trial 1974 - 1982 (follow-up data for years 6 to 8 not yet complete)

Valve model	Position	Years of follow up 1	2	3	4	5	6	7	8
Mechanical valves									
Björk-Shiley	Aortic	95.3	93.4	92.1	90.5	89.3	82.9	82.9	-
	Mitral	93.4	91.8	89.9	88.1	83.3	83.3	83.3	-
Lillehei-Kaster	Aortic	95.3	92.7	88.7	88.7	88.7	-	-	-
	Mitral	95.9	92.9	91.1	91.1	91.1	-	-	-
Starr-Edwards 1260	Aortic	98.2	95.0	91.2	91.2	88.9	82.1	-	-
Starr-Edwards 6120	Mitral	96.1	92.7	90.0	88.8	84.0	78.0	78.0	78.0
Tissue valves									
Carpentier-Edwards	Aortic	97.0	94.7	91.2	91.2	91.2	-	-	-
	Mitral	94.8	92.5	84.8	79.5	79.5	-	-	-
Free homograft	Aortic	100.0	97.0	97.0	97.0	97.0	97.0	-	-
Hancock	Aortic	95.1	95.1	95.1	88.8	88.8	88.8	-	-
	Mitral	94.0	92.3	92.3	92.3	92.3	61.5	61.5	-

Table 3: Acturarial values of probabilities of freedom from valve complications after valve replacement: Multicentre Valve Trial 1974-82 (follow-up data for years 6 to 8 not yet complete)

Valve model	Position	Years of follow up 1	2	3	4	5	6	7	8
Mechanical valves									
Björk-Shiley	Aortic	90.8	86.8	84.9	82.2	78.2	72.2	72.2	-
	Mitral	89.0	83.8	81.1	77.5	71.6	71.6	71.6	-
Lillehei-Kaster	Aortic	94.0	91.3	85.5	85.5	85.5	-	-	-
	Mitral	92.4	86.6	83.3	74.9	74.9	-	-	-
Starr-Edwards 1260	Aortic	85.2	74.4	69.2	66.9	60.7	42.5	-	-
Starr-Edwards 6120	Mitral	88.2	79.9	74.4	70.9	61.3	52.9	52.9	52.9
Tissue valves									
Carpentier-Edwards	Aortic	92.9	88.4	84.9	84.9	84.9	-	-	-
	Mitral	88.5	84.7	71.9	66.4	66.4	-	-	-
Free homograft	Aortic	91.2	82.1	82.1	76.2	76.2	76.2	-	-
Hancock	Aortic	92.7	90.2	90.2	83.2	83.2	83.2	-	-
	Mitral	86.4	79.3	70.3	70.3	70.3	70.3	70.3	-

5. IN VITRO TESTING OF HEART VALVE SUBSTITUTES

Although the ultimate assessment of prosthetic valve performance must come from animal and, more importantly, clinical implantation, the initial evaluation of any new design can be obtained form appropiate laboratory tests. There are several areas where in vitro testing can yield valuable information for both the cardiac surgeon and the valve designer. However, if such studies are to be useful form a 'clinical' point of view, the experiments must be performed under carefully controlled conditions and their relevance to the clinical situation must be clear. Many sophisticated laboratory studies have been undertaken which provided comparative evaluations of valve substitutes but did not give any insight into the valves' in vivo performances. The interpretation of such in vitro test results must be approached with caution by recognizing their limitations when extrapolated to the in vivo situation (Tindale et al.[45)]).

6. STEADY FLOW STUDIES

Haemodynamic testing or, more correctly, hydrodynamic testing is a useful indicator of the fluid mechanical perfomance of a valve. Steady flow studies are by far the simplest to perform since they do not involve the need to simulate the pulsatile nature of the cardiac cycle. However, it is important that the valve test section of the flow circuit be designed so as to reproduce the anatomy of the biological flow passages. Several studies, including those by Kramer et al.[26)], Naumann & Kramer[32)] and Bellhouse & Bellhouse[3)] have shown that flow passage geometry has a profound effect on the valve's performance.

The pressure drop across a valve can be measured over a range of flow rates, from which various parameters can be derived such as 'effective orifice area' and 'discharge coefficient'. This simple test procedure can also highlight defects in the opening characteristics of a tissue valve, for example, excessive leaflet stiffness or leaflet flutter, either of which may be detrimental to its subsequent in vivo performance. The sites at which pressure measurements are made both upstream and downstream from the valve must be standardized. In the case of aortic valve substitutes, several downstream pressure measurements must be made at different axial locations, since, due to the phenomenon of 'pressure recovery', an axial variation in the measured pressure drop will obtain. This phenomenon is well known to fluid mechanicists from the study of flow through constricting orifices. It can be explained by the interchanges between kinetic and potential energy of the fluid which occur due to the restrictive nature of the orifice.

An according set-up, such as used in Aachen for steady flow studies is shown in Figure 4.

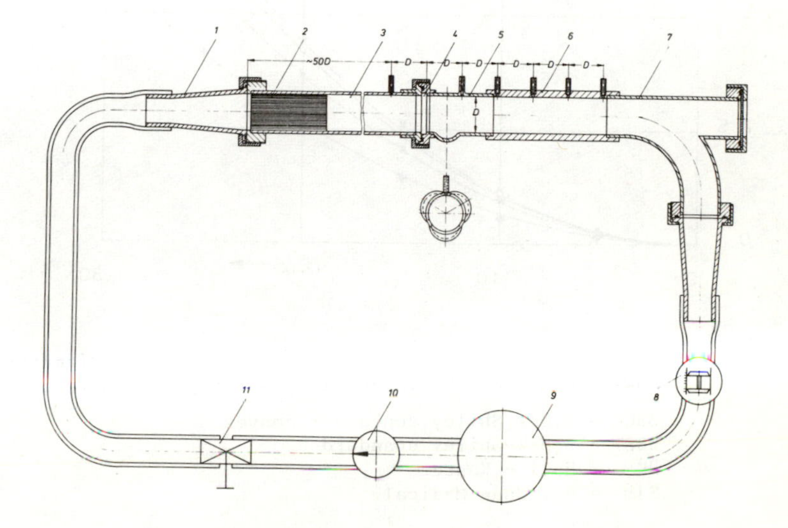

Figure 4: Set-up for the steady flow experiments: (1) flow inlet diffuser; (2) honeycomb; (3) inlet tube; (4) heart valve mounting ring; (5) model aortic root; (6) downstream measuring section; (7) bifurcation with optical observation window; (8) rotameter; (9) fluid reservoir; (10) centrifugal pump; (11) throttle valve.

The model aortic root has three symmetrically spaced (120°) sinuses. The pressure measuring sites are located one diameter upstream of the valve mounting ring and one to five diameters (in steps of one) downstream. This arrangement facilitates the measurement of pressure recovery downstream of the valve.

Figure 5 presents the measured pressure loss for four different mechanical heart valves as obtained in the above set-up.

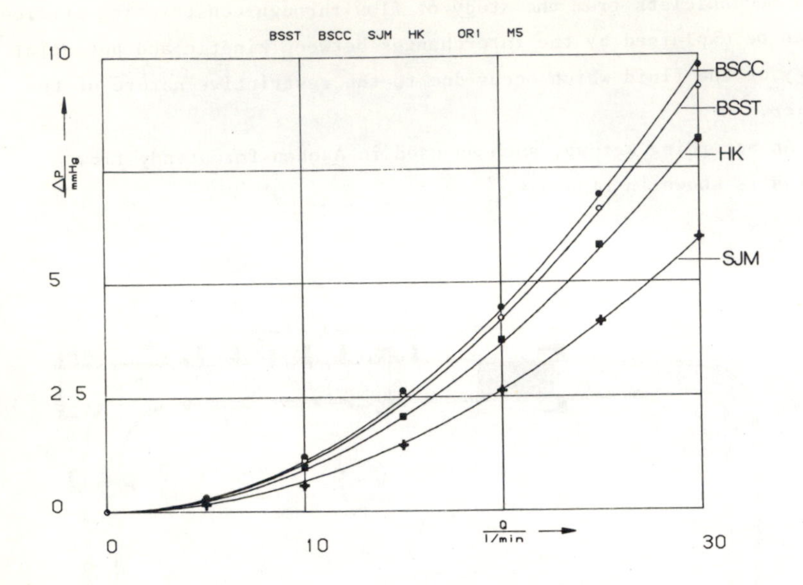

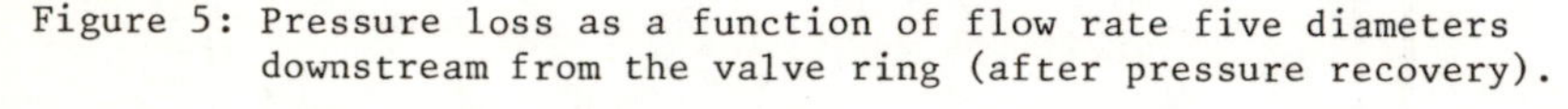

Figure 5: Pressure loss as a function of flow rate five diameters downstream from the valve ring (after pressure recovery).

BSCC = Björk-Shiley concave - convex
BSST = Björk-Shiley standard
HK = Hall - Kaster
SJM = St. Jude Medical

Figure 6 shows plots of the axial variation of the measured pressure drop for a series of 23 mm tissue annulus diameter aortic valve substitutes. The results have been grouped according to the type of valve, that is, mechanical or biological. It is of interest to note that the valves tested form two quite distinct categories, with the pressure drop at any axial location being significantly greater overall for the tissue valves. However, ist must be stressed that these data cannot be used to extrapolate to the behaviour of other valve sizes.

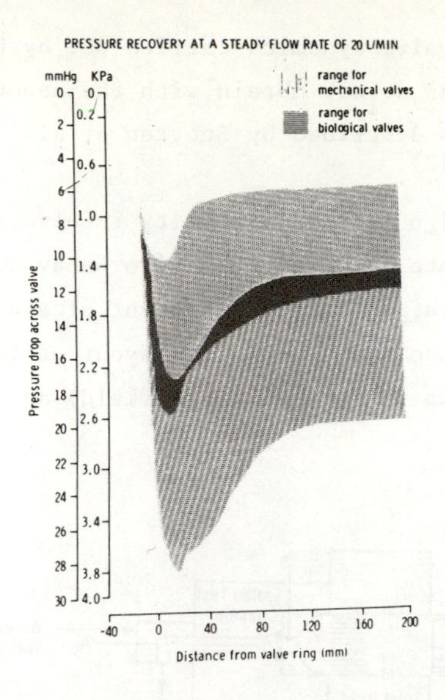

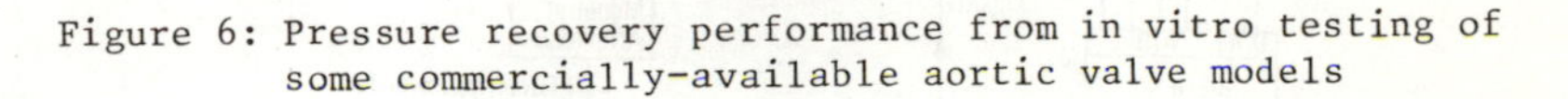

Figure 6: Pressure recovery performance from in vitro testing of some commercially-available aortic valve models

7. PULSATILE FLOW STUDIES

Whilst steady flow testing provides a useful 'first line' approach to the in vitro assessment of valves, its shortcomings are obvious. Since the normal pulsatile flow of blood is not simulated, valve dynamics cannot be studied, nor can the degree of valvular regurgitation be accurately assessed. In order to investigate valve performance in greater depth a 'pulse duplicator', which accurately simulates the pumping action of the heart, is required. Over the last thirty years many different designs of pulse duplicators have evolved. Although they all have one feature in common, namely a device for producing pulsatile flow through the test valve, the majority of the earlier designs fell short of their goal to simulate the action of the human heart. This, in turn, led to inaccuracies when results from such test systems were unwisely extrapolated to the in vivo situation. For example, the in vitro results of Mc Millan et al.[30)] suggested that the human aortic valve did not open fully at any time during systole. This was later completely contradicted by studies performed by Padula et al. (1968) on the aortic valve in an actively beating heart, and also by more refined model studies of aortic valve dynamics carried out by Bellhouse & Bell house [3]. Such inaccuracies do not obtain with the newer, more sophisticated pulse duplicators described by Scotten et al. [40], Martin et al. [28] and Bruss et al. [12].

Pulse duplicator design must of necessity involve a compromise between the need to simulate accurately in vivo behaviour so as to obtain clinically useful data, and the requirement of a system which is practicable for routine laboratory use. The layout of two such systems, currently in use, are shown in Figure 7 (Sheffield) and Figure 8 (Aachen).

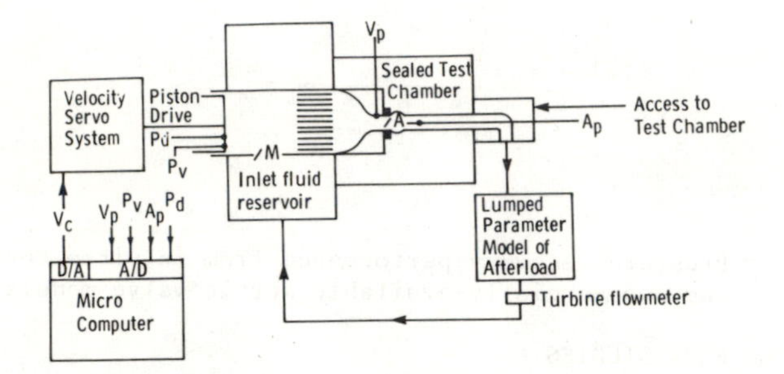

Figure 7: Schematic of Sheffield pulsatile flow test rig where M and A represent test valves in the mitral and aortic positions respectively; Vp and Ap are model ventricular and aortic pressures monitored with purpose built catheter tipped devices; Pd and Pv are the displacement and velocity of the piston at any instant and Vc it the piston control signal derived from the microcomputer

The parameters which have been used to define valve performance during pulsatile flow are many and varied. The most commonly used are pressure drop and regurgitation and, to a lesser extent, power and energy losses through the valve, as well as turbulence measurements.

Many problems and complications associated with valve prostheses are related to non-stationary flow patterns. It is, therefore, useful to obtain more detailed flow information, escpecially with regard to the downstream distribution and the tidal variation of aortic flow. Studies using flow visualization techniques usually give only qualitative information and the major handicap of hot-film anemometry is the

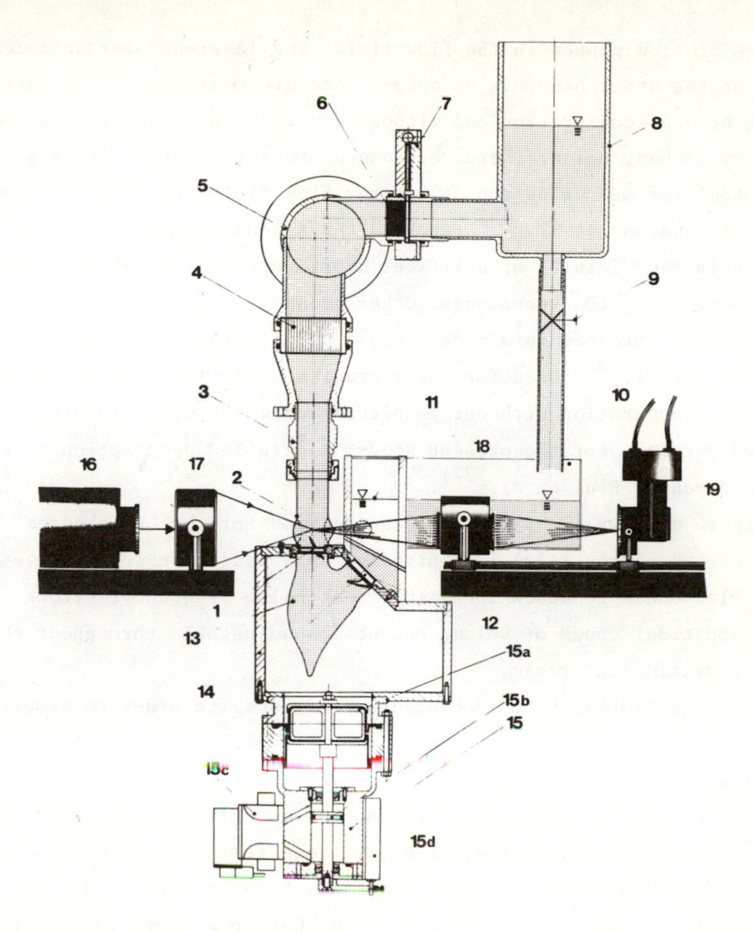

Figure 8: Schematic of Aachen pulsatile flow test rig as used e.g. for Laser-Doppler-Anemometry measurements: (1) aortic valve; (2) elastic aortic root; (3) electromagnetic flow probe; (4) characteristic resistance; (5) adjustable compliance; (6) peripheral resistance; (7) adjusting mechanism for (6); (8) fluid reservoir; (9) adjusting throttle; (10) atrial reservoir; (11) left atrium; the atrial housing has a central slit in order to provide a free laser beam path; (12) mitral tilting disc type valve; (13) elastic ventricular sac; (14) rigid Plexiglass housing; (15) hydraulic pump; (15a) low-pressure piston; (15b) high-pressure piston; (15c) electromagnetic servovalve; (15d) displacement transducer; (16) He-Ne laser; (17) transmitting optics; (18) receiving optics; (19) photomultiplier.

presence of flow probes in the flow field. The laser-doppler anemometer (LDA), on the other hand, is an opto-electronic measuring system without the need of calibration and without any influence on the flow field caused by probes, holders, etc. Because of these features, it is a unique tool for measuring the downstream flow field of heart valve prostheses. Yoganathan et al. [49] reported their measurements of velocity profiles in the vicinity of artifical heart valves under steady flow conditions, using LDA techniques. Other groups [1, 34] published the first results obtained in a mock circuit under pulsatile flow.

Hwang et al. [32] used for their studies a flow loop which was developed in cooperation with our Institute and which was also used, in a modified version, for the present study. A detailed description of this loop is given by Reul et al. [37].

Figure 8 shows our flow loop, as modified for the LDA studies. The motion of the piston (15a) is controlled by a hydraulic servo-valve and an electronic feedback loop, such that pulse frequency, stroke volume and tidal cours of volume output are adjustable throughout the entire physiological range.

In connection with tuned hydraulic elements (in order to simulate the physiological input impedance of the systemic circulation), quasi-physiological pressure and flow curves can be generated in the aortic region. The test fluid is a 36 per cent aqueous glycerol mixture with a dynamic viscosity of 3.6 m Pa s. These test conditions are suitable for comparative studies of different valve types and allow a better extrapolation to phenomena in vivo than do steady flow experiments.

The LDA system used in our experiments consists of a 15mW He-Ne laser, transmitting optics, with beam expander and frequency shifter, receiving optics, photomultiplier, and a tracker-type signal processor (TSI Model 1090). The system was operated in a dual beam forward scatter mode with a focal length of 120 mm. The dimensions of the ellipsoidal measuring volume, resulting from the intersection ot the two laser beams under the above conditions, were 0.3 mm (length) by 0.04 mm (width).

Figure 9 shows schematically the velocity fields of four investigated valves at the specific time instant of t = 180 ms after start of systole. The actual measured velocity profiles are indicated in each Fig..

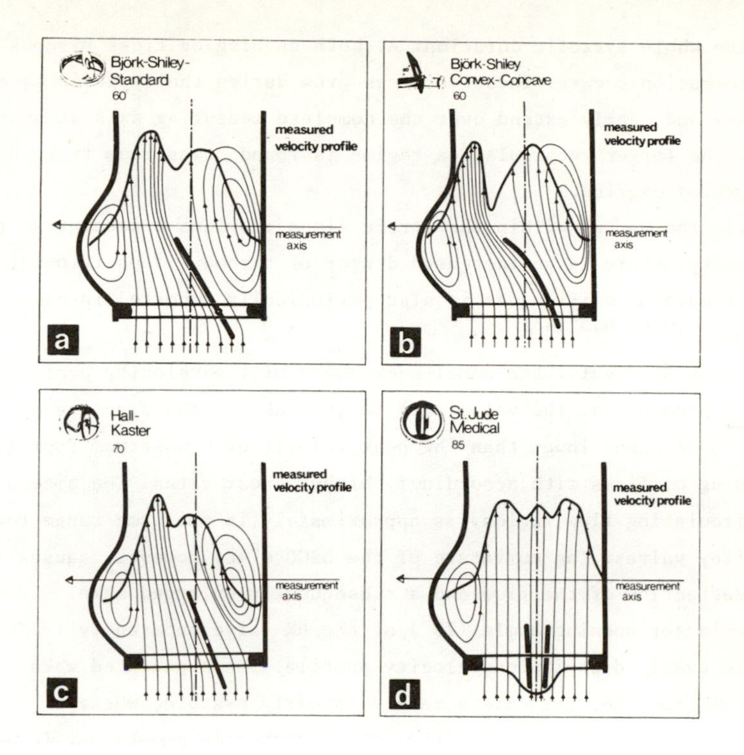

Figure 9: Schematic representation of the velocity fields at t = 180 ms after start of systole, based on the measured velocity profiles.

The LDA measurements in pulsatile flow facilitate a more detailed evaluation of the different valve types. Besides the absolute values of flow velocities, they also yield velocity distribution and the location of recirculation areas. The size and location of stagnate and recirculating flow regions can indicate where thrombus deposition and formation may occur. The distribution of the velocity field also yields the degree of efficient utilization of the available free cross-section area and facilitates the calculation of shear rates.

As can be seen from Figure 9 (for a specific time instant), the velocity field of the Björk-Shiley valve is characterized by two peaks downstream of the larger and minor opening orifice. Between these two peaks, in the wake of the disc, the axial velocity is markedly reduced during the whole systolic duration. At both opening orifices viscous vortex formation occurs. These vortices grow during the systolic ejection phase and nearly extend over the complete measuring axis at end-systole. The larger recirculation regime is found downstream from the minor opening orifice.

While the vortex within the aortic sinus provides a washout of the sinus cavity, there is an increased danger of thrombus deposition at the minor opening orifice, as is also periodically reported in clinical literature [4,5,6,16,31,48].

For the BSCC valve the camel-back shape of the velocity profile is even more pronounced. The velocities in the wake of the disc are on average 75 per cent lower than the peak velocities downstream from the two opening orifices with accordingly higher shear rates. The size of the recirculating flow regions is approximately in the same range for both Shiley valves. The curvature of the BSCC disc, however, causes a larger deflection of the flow and a subsequently enlarged wake.

The larger opening angle (70°) of the HK valve apparently leads to a more evenly distributed velocity profile. The pronounced wake downstream from the disc exists mainly in early systole, whereas during the late ejection period the velocity profile smoothens. In a similar manner to the Shiley valves, large vortices develop on both sides of the open disc; however, the vortex downstream from the minor ortifice is somewhat larger.

Because of its large free orifice area with leaflet opening angles of 85°, the SJM valve has the flattest velocity profile, with three small peaks and two dips downstream from the leaflets. The shear rates across the measuring axis are acordingly lower than for the other valves. It is of special interest that a recirculating flow region develops only within the sinus cavity, while in the opposite cross-section the flow velocities are almost entirely positive throughout systole. Only a transient end-systolic recirculation is measured. The viscous losses are lowest for this valve, al already expressed by the lowest pressure loss of all four valves.

Although the velocity measurements were carried out only for one measurement axis, this method gives a good insight into the structure of the velocity fields downstream from the valves. A more complete mapping of the flow downstream from the valves, together with measurements of laminar and turbulent shear stresses, may finally lead to complete characterization of the velocity fields and to specific clinical correlations.

Evaluation of the magnitude of the shear stresses in the blood in the vicinity of artificial valves has been used as a measure of the 'blood damaging' characteristics of the valve.

Determination of both turbulent shear stresses and wall shear stresses in the vicinity of artificial heart valves require specific measurements technique, such as hot-wire anemometry, Laser-Doppler-Anemometry and direct measurements of wall shear stresses by flush-mounted hot-film probes. The two first methods are very useful for the evaluation of bulk shear stresses and velocity profiles, however, have their limitations close to the wall because of probe size (hot wire) or uncertain measurement volume size (Laser). (Kreid [27]).

Measurements of Figliola et al. [17] and estimations by Yoganathan et al. [48] yielded turbulent shear stresses in the range of potential damage to blood components. We have concentrated on wall shear stresses in the valve ring and in the aortic root. This approach is justified by the fact that the combined effects of foreign wall materials contact and shear stresses result in potentially damaging shear stress levels about one order of magnitude lower than bulk shear stresses (Blackshear [9], Mohandas et al. [32]).

Hot-film anemometry was selected as a suitable measurement technique because of the good spatial and temporal resolution of the hot film sensors.

Miniature hot-film probes were flush-mounted within the valve rings of different valve designs.

The measurements-results for two mechanical prostheses with different closing mechanisms are shown in Figures 10 and 11.

The wall shear distribution at the valve ring is completely different between the B-S and the L-K valve, especially during diastole. During systole both valves show comparable wall shear stresses in the range below 10 N/m^2, depending on frequency. Whereas the shear stresses at the L-K valve ring remain low during diastole, extremely high values of up to 400 N/m^2 can be measured at the B-S valve. The striking difference of the wall shear distribution of the two disk valves during diastole is due to the different closing

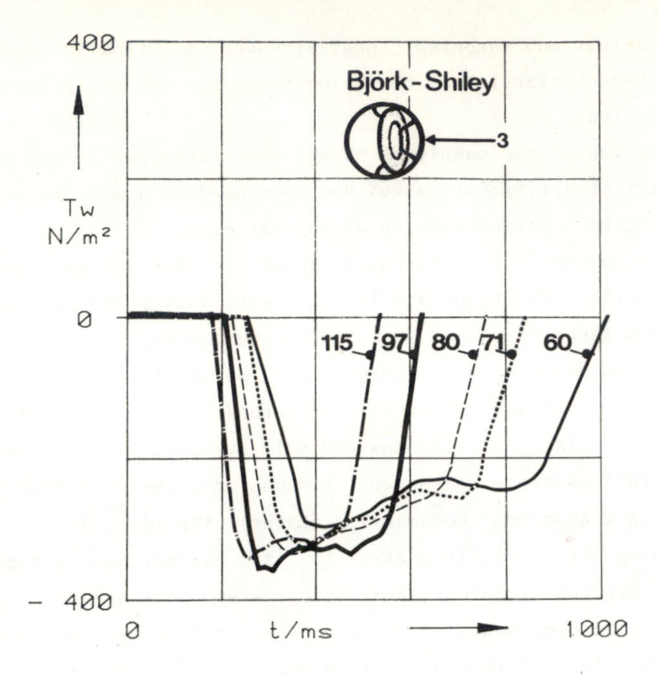

Figure 10: Wall shear-stress at the inner valve ring surface for frequencies from 60 to 115 beats/min (location 3) at a Björk-Shiley valve.

mechanisms. The disk of the Lillehei-Kaster prosthesis which rests on the valve ring prevents a diastolic regurgitation but, on the other hand, reduces the ratio of internal valve ring diameter to annulus diameter. The "full orifice" design of the Björk-Shiley prosthesis results in a large inner to outer diameter ration but also causes high shear loads in the gap between disk and ring during diastole, due to the pressure gradient across the valve and the resulting backflow rate.

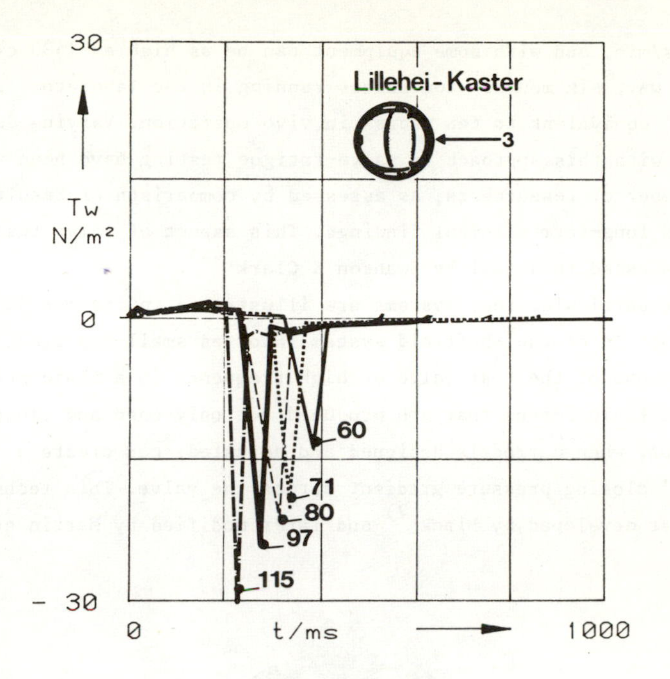

Figure 11: Wall shear-stress at the inner valve ring surface for frequencies from 60 to 115 beats/min (location 3) at a Lillehei-Kastor valve.

8. FATIGUE TESTING

In an earlier section reference was made to the concern of surgeons regarding long-term durability of tissue valves. This continues to be one of the major drawbacks ot these valves, and currently the only universally-accepted method of evaluating valve durability is by extensive routine clinical use. This has the inherent disadvantage that the assessment of any particular valve, or modified design, is not known for several years, by which time the reasons for failure may well be obscure. Although both mechanical and biological factors can influence valve failure, their individual effects often cannot be separated in the clinical implantation situation. These problems can be partially overcome by accelerated fatigue testing of valves in the laboratory. This technique can usually isolate the mechanical factors which might reduce the long-term durability of any prosthesis. Such test systems are designed to open and close the valve at a rate that is much greater than the average physiological value of around

70 beats/min, and with some equipment can be as high as 1400 cycles/min. In this way, six months' continuous running in the laboratory is 'mechanically' equivalent to ten years' in vivo operation. Varying degrees of success with this approach to valve-fatigue testing have been achieved by a number of researchers, as assessed by comparison of results with observed long-term clinical findings. This aspect of valve testing has been discussed in detail by Swanson & Clark [44].

Two particular test systems are illustrated in figures 12 and 13. The principle of the Sheffield system involves small-amplitude axial oscillations of the test valve at high frequency in a fluid medium. The inertial fluid forces that are produced not only open and close the valve but, when correctly designed and adjusted, can create a 'physiological' closing pressure gradient across the valve. This technique was first developed by Black [7] and later modified by Martin et al.[29].

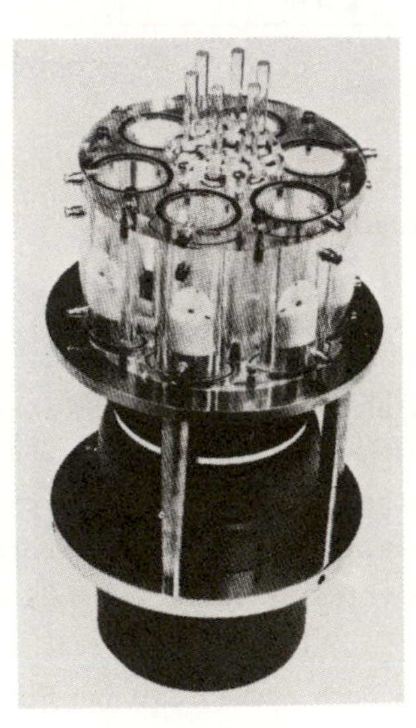

Figure 12: The Sheffield multivalve accelerated fatigue tester

The Aachen fatigue tester has only one rotating disk with an arrangement of ring-slits. During each rotation the valves are fully opend by the flow through the passing ring slits. In the closed state, the space below the valves is coupled to a ring channel, which again is

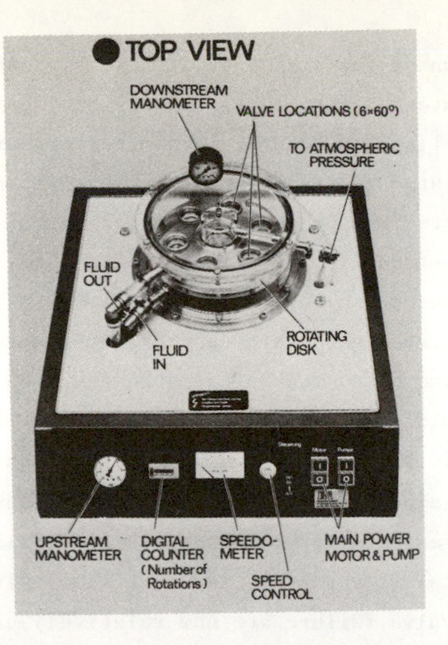

Figure 13: The Aachen multivalve accelerated fatigue tester

connected to atmospheric pressure. The loading pressure of the valves is fully adjustable as well as the rotational speed of the device.

Whilst in vitro accelerated fatigue testing can seperate some of the influences of mechanical and biological factors on valve durability, it does not distinguish easily amongst the effects of tissue selection and treatment, leaftet geometry, frame design and technique of construction. Some of these latter effects can be isolated by extensive laboratory evaluation of all the materials used in the construction of a valve. Carefully controlled tissue selection procedures and a thorough investigation of the mechanical properties of the chemically-treated material are essential prerequisites to the successful development of a tissue valve.

Glutaraldehyde, as noted by Woodroff [47)], has been used routinely to stabilize connective tissue for heart valve substitutes over the past twelve years. The thermal stability of treated connective tissue can be assessed in the laboratory by observing the change in length of a sample as the temperature of a surrounding aqueous solution is slowly increased. A dramatic shortening of sample length occurs at the critical 'heat shrink temperature'. High values of this critical temperature correspond to greater stability of the tissue.

Chemical modification of the tissue by different treatments will produce changes in its internal structure, usually by increasing the number of collagen cross-links. Since the mechanical properties of the tissue are intimately related to its structure, the effects of these treatments will be reflected in a change in the mechanical properties. Thus, measurement of the stress/strain response of the treated tissue will not only provide information on its mechanical strength, but will also give useful data concerning the efficacy of any particular treatment. In general terms, increased mechanical strength with minimum loss of flexibility, should enhance tissue leaflet durability.

Any material which is to be used in a heart valve substitute must have a good fatigue life since it will be required to withstand around forty million stress cycles per year for many years. This is equally as important for mechanical valves as it is for tissue valves. Although instances of mechanical valve failure are now relatively uncommon, they are not unheard of, and will almost always be catastrophic, as in the case of a strut fracture in a Björk-Shiley valve reported by Brubakk et al.[11].

In vitro testing of prosthetic valves is now a popular topic in many engineering laboratories. If the investigations are carried out in collaboration with experienced cardiac surgeons then there is every likelihood that the interpretation of the test results will be beneficial to future valve development. However, if there is no clinical input to the research, the work may be good in scientific terms but not immediately relevant to in vivo applications. This latter situation can also lead to the incorrect extrapolation of in vitro data to predict the in vivo performance of a heart valve substitute.

9. SUMMARY AND FUTURE DEVELOPMENTS

This paper has outlined the major events and developments which have occured in the field of heart valve replacement since its inception some thirty year ago. Clearly, it has not been possible to include references to all aspects of what has been a remarkably extensive field of endeavour involving many individuals and research centres. Neither has it been practicable to include details of all the currently-used heart valve substitutes. The omission of any particular prosthesis should not be taken to imply that it is either unworthy or unacceptable.

Initially, the subject was very much the province of the cardiac surgeon. However, the increasing complexity of valve development has resulted in the involvement of scientists from many different disciplines. It is unlikely that any new prosthesis could reach the stage of routine clinical use today without the support of a multidisciplinary research team.

In an earlier section, attention was drawn to the advantages and disadvantages of the two major categories of heart valve substitutes. It does not seem likely that there will be any substantial improvements to mechanical valves in the immediate future. Minor changes of design and materials of construction will produce only marginal gains in performance for these valves. However, in the case of tissue valves there is a much greater opportunity for further developments and improved designs. In the former category, a better understanding and analysis of the effects of tissue treatments should lead to greater long-term reliability. It is alos possible that researchers in the field of polymer science may yet develop a genuinely biocompatible man-made membrane of adequate strength, which could be used for the flexible leaflet prosthesis. From the point of view of design, newly developed techniques for analysing the stresses in tissue valve leaflets under load, should enable the development of geometrical configurations which result in minimum stress levels during function. This in turm will assist in increasing the long-term durability of the valve.

Even without improvements to the actual prostheses, there are many other facets to heart valve substitutes which require attention. At the present time there is no agreed standard of presentation of data for the various commercially-available valve models.

The differences in data can lead not only to confusion in the operating theatre, but also add to the difficulties encountered when making comparative in vitro evaluations of different valves. The only way to overcome these and other inconsistencies in the valve field is by the introduction of a 'Standard' which should include:

(1) a universally agreed terminologie;

(2) minimum requirements regarding properties of construction materials and standards of quality control;

(3) protocols for in vitro assessment. including upper and lower

limits for haemodynamic performance and fatigue testing;

(4) guidelines for the presentation of results of laboratory tests;

(5) protocols for minimum animal trial requirements; and

(6) acceptable methods of packaging and labelling, including consistent presentation of data.

Such data should include details of the sizes and other relevant descriptions regarding the materials and handling of the contents.

The advantages of such a 'Standard' would be to: (1) protect the consumer; and (2) assist the manufacturer to produce goods of adequate quality.

The ultimate consumer is, of course, the patient. However, initially the surgeon is the user, and such a 'Standard' would provide him with the knowledge that any approved heart valve substitute has met clearly-defined specifications of construction, durability and performance. From the manufacturers' point of view, the 'Standard' would provide a guide to the production of a good product.

As already noted, there is, as yet, no ideal or optium heart valve substitute. Consequently the search for improved designs will continue, as will the need for good clinical assessment through long-term follow up of individual patients. In this context the value of multicentre valve trials of the type referred to in this paper should be recognized.

R E F E R E N C E S

1 Affeld,K., Psolla,J., Lehmann,B. & Mohnhaupt,R. (1979). Proceedings of the 2nd Meeting of ISAO, 439 - 441

2 Ashraf, M., & Bloor, C.M., (1978) American Journal of Cardiology 41, 1185 - 1190

3 Bellhouse, B. & Bellhouse F., (1969) Circulation Research 25, 693 - 704

4 Ben-Zvi, J., Hildner, F.J., Chandraratna, P.A. & Samet, P., (1974) American Journal of Cardiology, 34, 538

5 Björk, V.O., (1978) Scandinavian Journal of Thoracic and Cardiovascular Surgery, 12, 18 - 84

6 Björk, V.O. & Henze, A., (1979) Journal of Thoracic and Cardiovascular Surgery, 78, 331 - 342

7 Black, M.M., (1973) In: Perspectives in Biomedical Engineering. Ed. R.M. Kenedy, Macmillan, London; pp 21 - 28

8 Black, M.M., Drury, P.J. & Tindale, W.B., (1982) Proceedings of the European Society for Artifical Organs, 9, 116 - 119

9 Blackshear, P.L., (1972) In: Chemistry of Biosurfaces (Edited by M.L. Hair, Vol. 2) 523, Marcel Dekker, New York

10 Broom, N.D., (1978) Journal of Thoracic and Cardiovascular Surgery, 76, 202 - 211

11 Brubakk, O., Simonsen, S., Kallman, L. & Fredriksen, A., (1981) Thoracic and Cardiovascular Surgeon, 29, 108 - 109

12 Bruss, K.-H., Reul, H., van Gilse, J., Knott, E., (1983) Life Support Systems 1, 3 - 22

13 Cartwright, R.S., Palick, W.E., Ford, W.B., Giacobine, J.W., Zubritsky, S.A. & Ratan, R.S., (1962) Journal of the American Medical Association, 180, 6 - 12

14 Emery, R.W. & Nicoloff, D.M., (1979) Journal of Thoracic and Cardiovascular Surgery, 78, 269 - 276

15 Emery, R.W., Anderson, R.W., Lindsay, W.G., Jorgensen, C.R., Wang, Y. & Nicoloff, D.M., (1979) Surgical-Forum, 30, 235 - 238

16 Fernandez, J., Samual, A., Yang, S.S., Sumathisena, Gooch, A., Varanhao, V., Lemole, G.M. & Goldberg, H., (1976) Chest, 70(1),12

17 Figliola, R.S. & Mueller, T.J., (1977) Journal of Biomechanical Engineering, 6, 173

18 Ghista, D.N., Reul, H., (1977) Journal of Biomechanics, 10(56), 313

19 Gibbon, J.H.jr., (1954)Minnesota Medicine, 37, 171 - 185

20 Housman, L.B., Pitt, W.A., Mazur, J.H., Litchford, B. & Gross,S.A., (1978) Journal of Thoracic and Cardiovascular Surgery,76, 212-213

21 Hufnagel, C.A., (1951) Bulletin of the Georgetown University Centre, 4, 128 - 130

22 Hwang, N.H.C., Lu, P.C., Sallam, A. & Reul, H., (1979) Proceedings of the 14th AAMI Annual Meeting, Las Vegas, 127

23 Ionescu, M.I. & Ross, D.N., (1969) Lancet ii, 335 - 338

24 Ionescu, M.I., Mary, D.A.S. & Abid, A., (1971) In: Late Results of Valvular Replacements and Coronary Surgery. Ed. G. Stalpaert et al. European Press, Ghent; p 56

25 Kaiser, G.A., Hancock, W.D., Lukban, S.B. & Litwak, R.S., (1969) Surgical Forum, 20, 137 - 138

26 Kramer, C., Gerhardt, H.J., Bleifeld, W. & Schwerin, H., (1975) Presented at the fifth over the water meeting of the Biological Engineering Society, Aachen (unpublished)

27 Kreid, D.,(1970) Ph.D. thesis, University of Minnesota

28 Martin, T.R.P., Palmer, J.A. & Black, M.M., (1978) Engineering in Medicine, 7, 229 - 230

29 Martin, T.R.P., Van Noort, R., Black, M.M. & Morgan, J., (1980) Proceedings of the European Society for Artificial Organs, 7, 315 - 319

30 Mc Millan, I.K.R., daley, R. & Mathews, M.B., (1952) British Heart Journal, 14, 42 - 46

31 Messmer, B.J., Okies, J.E., Hallmann, G.L. & Cooley, D.A., (1972) Journal of Cardiovascular Surgery, 13, 281

32 Mohandas, N., Hochmuth, R.M., Spaeth, E.E., (1974) Journal Biomed. Materials Res. 8, 119

33 Naumann, A. & Kramer, C., (1970) AGARD Conference Proceedings, 65, paper 4

34 Philips, W.M., Snyder, A., Alchas, P., Rosenberg, G. & Pierce,W.S., (1980) Transactions of the American Society for Artificial Internal Organs, 26, 43 - 49

35 Reece, I.J., Van Noort, R., Martin, T.R.P. & Black, M.M., (1982) Annals of Thoracic Surgery, 33, 480 - 485

36 Reis, R.L., Hancock, W.D., Yarborough, J.W., Glancy, D.L. & Morrow, A.G., (1971) Journal of Thoracic and Cardiovascular Surgery, 62, 683 - 689

37 Reul, H., Tesch, B., Schoenmackers, J. & Effert, S., (1974) Medical and Biological Engineering, 12, 431 - 436

38 Ross, D.N., (1962) Lancet ii, 487

39 Ross, D.N., (1967) Lancet ii, 956

40 Scotten, L.N., Walker, D.K. & Brownlee, R.T., (1979) Journal of Medical Engineering and Technology, 3, 11 - 18

41 Senning, A., (1966) Acta chirurgica Scandinavica Suppl. 356 B; pp 17 - 20

42 Senning, A., (1971) Thoraxchirurgie, 19, 304 - 308

43 Starr, A., (1960) Surgical Forum , 11, 258 - 260

44 Swanson, W.M. & Clark, R.E., (1981) Association for the Advancement of Medical Instrumentation, Arlington, USA, 16th Annual Meeting, Abstracts; p 96

45 Tindale, W.B., Black, M.M. & Martin, T.R.P., (1982) Clinical Physics and Physiological Measurement, 3, 115 - 130

46 Wain, E.H., Greco, R., Ignegeri, A., Bodnar, E. & Ross, D.N., (1980) International Journal of Artificial Organs, 3, 169 - 172

47 Woodroof, E.A., (1979) In: Tissue Heart Valves. Ed. M.I. Ionescu, Butterworths, London, p 349

48 Yoganathan, A.P., Corcoran, W.H., Harrison, J.R. & Carl, J.R., (1978) Circulation, 58(1), 70

49 Yoganathan, A.P., Corcoran, W.H., Harrison, E.C. & Carl, J.R., (1979) Medical and Biological Engineering and Computing, 17, 453 - 459

PROCEEDINGS OF THE II INTERNATIONAL CONFERENCE ON APPLICATIONS OF PHYSICS TO MEDICINE AND BIOLOGY
edited by Ž. Bajzer, P. Baxa & C. Franconi

ULTRASOUND TECHNIQUES FOR THE VASCULAR SYSTEM*

Tommaso D'Alessio - INFOCOM Dept. and Centro Ingegneria Biomedica
Universita di Roma

Abstract

Ultrasound diagnostic systems have undergone increasing widespread use, because they allow simple and non-invasive examinations to be made, thus providing valuable clinical information.

The simpler continuous wave doppler apparatuses are familiar to any physician dealing with vascular diseases, but the increasing diffusion of echographic and of pulsed doppler systems stresses the need for ever increasing mastering of these techniques.

In fact, even if from a technological point of view these systems can be considered as fairly developed, more work is needed in order to better present and exploit the information contained in the signals. We mention, for instance, the problems of image formation, presentation and processing, of extraction of suitable diagnostic parameters, or of defining "more objective" diagnostic methods, which could as far as possible be free from artifacts introduced by human operators. Moreover, neither safety and standardization requirements, nor the right place of ultrasound among other more or less new diagnostic techniques have been adequately defined until now.

With these problems in mind, in this talk I will try to focus on some crucial aspects, and present some points to think about, as a contribution for a frame of reference in this developing field.

1. Premise

When I began thinking of this speech, I discussed with some physicians I cooperate with and we agreed that standardization (of instruments and protocols) was, in our opinion, one of the pitfalls of ultrasound, which could seriously limit their usefulness and relevance.

In fact, I feel that, even if presently it is not possible to exactly define what will be the place of ultrasound among other diagnostic techniques, it is sure that the lack of standardization limits the possibility of obtaining significant results, of comparing on an objective basis the results obtained by different experimenters, and of evaluating the usefulness of this technique compared to the others. Moreover, in this context, human operator's skill and subjective intervention can more seriously affect the relevance and the interpretation of the signals provided by the instruments. In a provocative way, I will say that the presence of a human operator can even condition the processing of signals, because instrument outputs are often tailored to the operator and his limitations.

* This work has been partially supported by Ministero della Pubblica Instruzione - Roma.

Therefore, in the following, I will try to show that presently some not completely known pitfalls of ultrasound systems originate in the lack of standardization and in the subjective intervention of a human operator. To this purpose, and following my experience of a researcher involved in signal processing, I will look behind the signals provided by the instruments, and inside the "black boxes" of the instrument itself. This analysis will also provide guidelines for some problems the solution of which is nevertheless necessary before a more quantitative approach can be defined and the possibilities of these

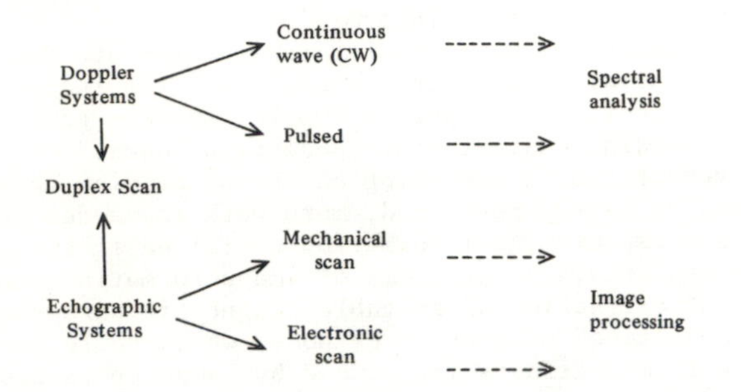

Tab. 1. Ultrasound techniques

instruments completely exploited. I will also briefly review some aspects of continuous wave (CW) and pulsed doppler flowmeters, and of echographic systems (see Table 1), but from the hypothesis that the fundamentals on these instruments are already known

2. CW doppler systems

The principle of doppler systems is well known: the ultrasonic wave radiated from a piezoelectric transducer is scattered by blood cells and received and processed by the receiver (Fig. 1). If the blood cells move with a velocity v, the received signal is shifted in frequency by a quantity f_D (positive or negative, depending on the verse of flow):

(1) $$f_d = \frac{2\ f_e \cdot v\cos\ \theta}{c}$$

where:
- f_e is the emission frequency

- c is the velocity of ultrasonic waves in the tissues (1500 m/s)
- θ is the angle formed by the wave direction and the velocity of the particles.

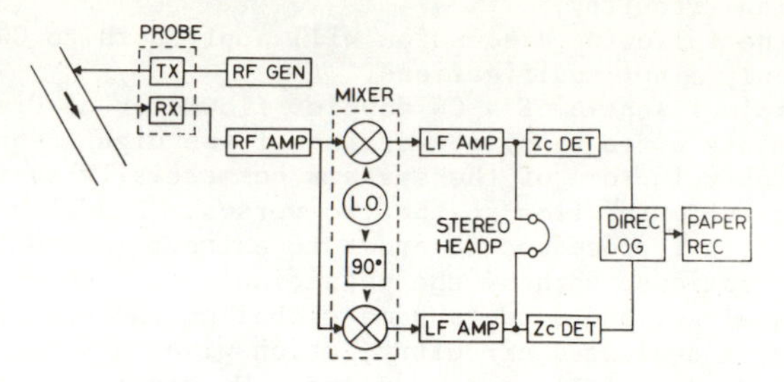

Fig. 1. Scheme of a CW doppler flowmeter.

This doppler shifted signal is detected and presented as an audio signal, to be subsequently processed. As velocity v typically falls in the range 0-100 cm/s and frequencies emitted range from 2 MHz to 10 MHz, the frequency shift falls in the audible band (0-15 kHz). Varying the emission frequency, the radiation pattern of the probe and the attenuation introduced by the tissues, which increases very much with frequency, will vary too. Therefore, on peripheral arteries, frequencies in the two ranges 4-5 MHz and 8-10 MHz (for the more superficial ones) are generally used.

In the case of blood, there is not a single velocity but a distribution of velocities in the vessel section, giving rise to a complex frequency spectrum. This means that the flow regimen determines the signal spectrum but, in a CW system, neither is it possible to biunivocally associate a frequency spectrum to a flow regimen nor is distance information provided.

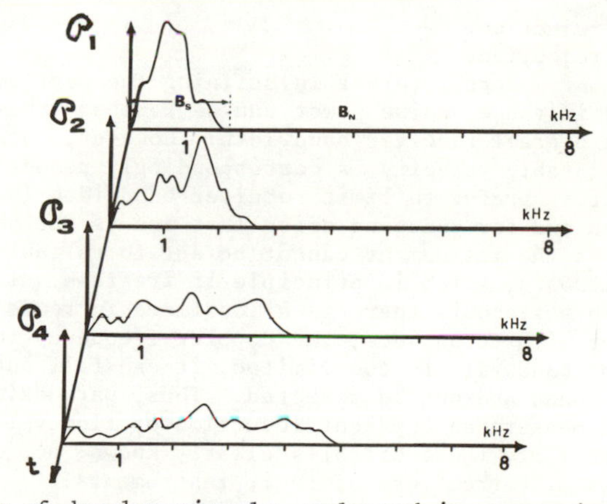

Fig. 2. Spectra of doppler signals evaluated in successive intervals during the cardiac cycle.

Moreover, we note that doppler signal is non-stationary because its spectrum varies as for amplitude, bandwidth and shape, following the cardiac action (Fig. 2). This fact imposes constraints on the signal processing circuitry.

Most of the following discussion will apply both to CW and pulsed systems, with only minor modifications.

In the typical scheme of a CW doppler flowmeter of Fig. 1, the receiver implements a frequency conversion of the high frequency signals, and separates (only in some of the systems commercially available) the two bands relative to the flow in the two verses. Then, the low frequency signals are processed in order to extract information connected to vascular alterations, both by the physician's ear (which acts roughly as a spectral analyser and detects some global characteristics of the spectrum) and by a dedicated circuitry, which gives a graphic output useful for recording or follow-up purposes. We mention that acoustic analysis (which has partially been overcome by spectral analysers) nevertheless gives some valuable information. The conventional processing is analog, and determines some integral characteristics of the spectrum (area, that is the power of the signal, or normalized first moment, that is mean velocity, or the normalized second moment, that is the zero-crossing density) the variations of which with time give information on the possible disease.

Obviously, the accuracy in the estimation of these quantities will depend on the characteristics of the flowmeters and on the algorithm used.

The most important parameters that influence flowmeter performances are:
- emission frequency
- emitted power (electrical-acoustic)
- probe characteristics
- receiver bandwidth
- Signal-to-Noise Ratio (SNR)
- dynamic range
- low frequency rejection
- unwanted band rejection.

All these parameters interact in defining the performances of the instrument. For instance, noise power can be reduced (thus improving SNR) by reducing overall receiver bandwidth. However, this means that the range of measurable velocity is correspondingly reduced. In fact, as the manufacturers prefer to limit receiver bandwidth (both for ease of implementation and for reducing noise that annoys the physician[a], it can happen that the instrument cannot be apt to reliably detect the presence of a stenosis, which in principle is its task. In fact, as in correspondence to a stenosis there is a localized increase in blood velocity together with turbulence, the doppler frequency increases but, if the instrument bandwidth is too limited, it can fall outside the useful frequency band and not be detected. Thus, paradoxically, the instrument could measure an apparent reduction of flow velocity. In my experience, neither are these pitfalls clearly known, nor do manufacturers report the limits of correct use of their instruments.

---

[a]In fact, it has been demonstrated[1)] that the "reliability" of the human operator diminishes with exposure to noise.

We mentioned that signal power spectrum is only an indirect indicator of the flow regimen occurring in the vessel. Moreover, the raw signal is in general processed in order to extract its zero-crossing density because the circuitry needed is simple. However, it is well known[2,3] that this quantity is related to the spectrum only in an integral way, is heavily affected by noise level and has an intrinsic low accuracy. Some other processors, such as quadratic and FFT processors, have therefore been devised in order to have more accurate estimates.

A quadratic processor (Fig. 3) consists of a (possible) shaping filter, a square law detector and an integrator (or some other kind of low pass filter), and can be used to obtain the power of the doppler signal (related to the radius of the vessel and thus to vessel elasticity) or the mean velocity of flow. These quantities are related in a more direct way than zero-crossing to physical parameters, and are less affected by measurement noise. As an example, in Fig. 4 you will see the results obtained by means of conventional zero-crossing detectors (curve A), and of quadratic

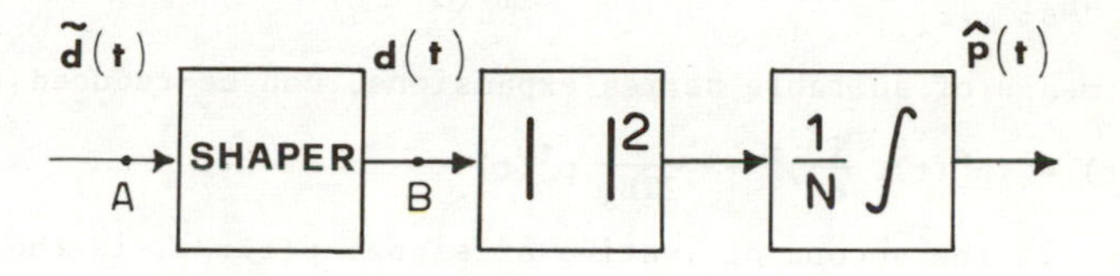

Fig. 3. Scheme of a quadratic processor. For power estimation, the shaping filter is not included.

processors (mean velocity, curve B) when used on altered vessels, with low signal-to-noise ratio, in the so-called supra-orbitary test.

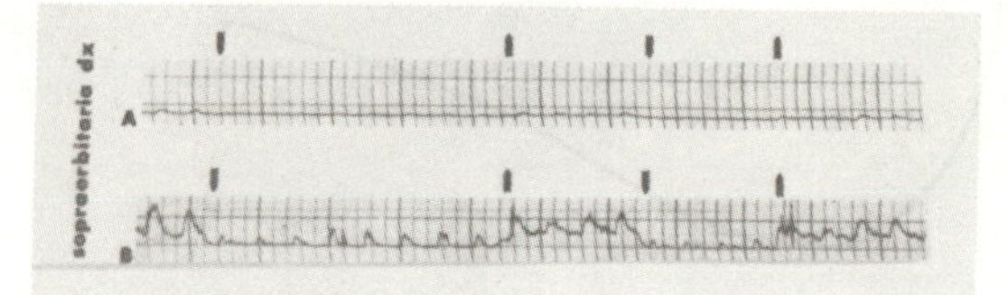

Fig. 4. In A) a conventional zero-crossing signal, in B) the mean velocity signal. Arrows indicate compression or release of the artery.

It also emerges that in some conditions the conventional processor (curve A) can neither provide a reliable estimate nor detect the presence of flow signals, and that mean velocity is a more interesting indicator of the state of an artery.

Anyhow, the characteristics of the processor can greatly affect estimation errors. Therefore, I will now discuss some not easily understood errors inherent in these processors. Consider, for instance, the estimation of signal power. A model for the doppler signal d(t) reasonably adequate in this case is the following:

$$d(t) = p^{1/2}(t).n(t) \tag{2}$$

where p(t) is the signal power (of which we look for an estimate p(t) from the raw signal d(t)) and n(t) is an ergodic Gaussian process. It can be demonstrated[4] that the m.s. estimation error has the following approximate expression (if we use a symmetric integrator in the quadratic processor):

$$\varepsilon^2(t) = E\{[p(t) - p(t)]^2\} = [\frac{1}{2T}\int_{-T}^{T} p(t-\alpha)d\alpha - p(t)]^2 + \frac{1}{4B_s T^2}\int_{-T}^{T} p(t-\alpha)p(t-\beta)\frac{\sin^2\omega_s(\alpha-\beta)}{\omega_s(\alpha-\beta)^2}\,d\alpha d\beta \tag{3}$$

which, by means of suitable series expansions, can be reduced to:

$$\varepsilon^2(t) \cong (p''(t).\frac{T^2}{6})^2 + \frac{1}{2TB_s}p^2(t) \tag{4}$$

where p''(t) is the second derivative of signal p(t). T is the integration time and B_s is the signal bandwidth.

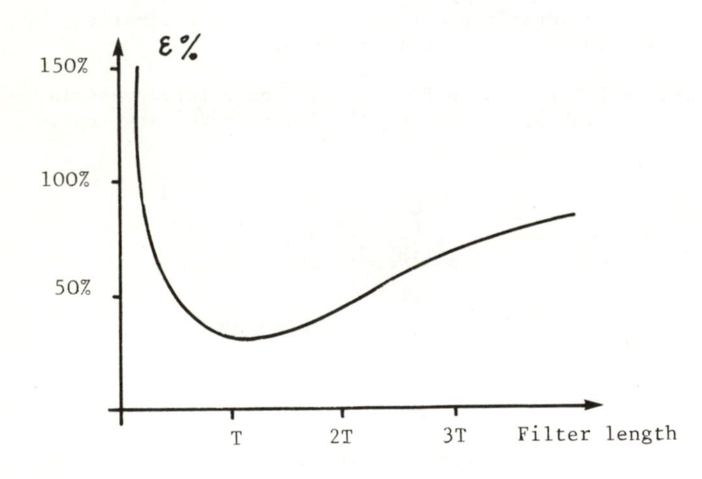

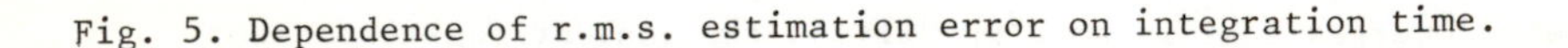

Fig. 5. Dependence of r.m.s. estimation error on integration time.

It is clearly emerges that the error varies with T, and has a minimum (Fig. 6) when:

$$(5) \qquad T \cong \left[\frac{9p^2(t)}{2B_s [p''(t)]^2} \right]^{1/5}$$

This also means that not necessarily the estimate with the "smoother look" is the one which gives the minimum r.m.s. error.

Similar results apply to the estimate of zero-crossing density or mean velocity. However, in general, physicians prefer to have a signal without any superimposed noise, even if due to the stochastic nature of the doppler signal. To this purpose, the manufacturer reduces the bandwidth of the processor. Sometimes, the physician himself can make a (subjective) adjustment of the filtering of signals but in so doing he alters both the amplitudes and the time relationship of signals. When he tries to make measurements on these signals with the aim of obtaining "more quantitative" data, he does not take into account the systematic errors introduced by this kind of processing.

In previous considerations, I did not consider the effect of doppler signal bandwidth variation during the cardiac cycle. As a general rule, we can say that estimation errors, in the presence of noise, increase when the signal bandwidth is narrower. In any case, quadratic processors are less affected than zero-crossing ones by measurement signal-to-noise ratio.

The results so far reported stress the need for a carefuly and "objective" optimization of doppler signal processors, by means of adaptive algorithms and more accurate estimation methods, and by the choice of more suitable parameters. This optimization is needed before the estimated signals could be used for diagnostic purposes or to build and check methods of extracting "more objective" parameters.

Moreover, it is necessary to be careful when some more sophisticated methods of signal analysis are introduced "downhill" from flowmeters. I mean, foe instance, spectral analysis which presently is implemented in real time with dedicated hardware connected both to CW and pulsed flowmeters. In general, a short-term spectrum (in 10-20 ms) is evaluated and the set of spectra during the cardiac cycle is presented on a screen by displaying time on the horizontal axis, frequency on the vertical axis while amplitudes are codified by a grey scale. However, if the signal bandwidth has previously been limited by the flowmeter, the spectrum analyser too will deal with distorted signals. Therefore, its results will be altered by this systematic error. Moreover, I note that spectral analysers *per se* are until now used in a fairly qualitative way, to determine the maximum frequency envelope of the signals during the cardiac cycle, or to obtain qualitative information from the whole shape of the short-term spectra.

However, I remember that spectrum characteristics are defined in a fairly fuzzy way, and with reference to visual analysis, just as a stenosis can be defined as mild or moderate or tight. This is a more general problem partly due to the lack of standardization of protocols, which makes difficult the achievement of an "objective" diagnosis and the exchange of results between different experimenters.

Sometimes, some indexes (i.e. ratio of maximum frequency in characteristic instants of time during the cycle) are evaluated, while it is claimed that it is possible to detect the presence and the level of stenosis from the possible spectral broadening, that is from the increase of maximum spectrum frequency in correspondence to a stenosis.

To this purpose, some comments are needed:

- spectral broadening depends not only on flow regimen but also on the divergence of ultrasonic beam and on the sample volume (this is in particular true for pulsed systems). Therefore, the contribution to spectral broadening due to the instrument has to be known before the measurements made on spectra could be properly interpreted;
- moreover, presently some rational procedures for the discrimination of the signals from the ever-present noise are lacking, and the decision on the presence or absence of signals is made on empirical grounds.

I think that this discrimination must be based only on a clear understanding of statistical characteristics of spectral estimates, and of their intrinsic limits. In fact, due to the non-stationarity of doppler signals, the spectral estimates that can be obtained are only a trade-off between accuracy in the estimation and observation time. Some studies[5] show that it is possible to define real-time algorithms that allow a rational discrimination (and therefore an objective maximum frequency estimation) to be made.

Finally, spectra are displayed on a screen by means of a number of grey levels (16 in general) which is suitable for a "visual" analysis, but limits the possibility of implementing filtering algorithms on a computer in order to improve diagnosis.

3. Pulsed Doppler systems

A CW doppler system does not allow us to get information on the distance of the vessel from the probe. However, if the emitted wave is modulated (in general by means of a short pulse) and a proper gating circuitry is provided, is possible to measure the distance of the various points insonated (Fig. 6). In fact, the doppler shift can be detected and the velocity profile on the vessel diameter

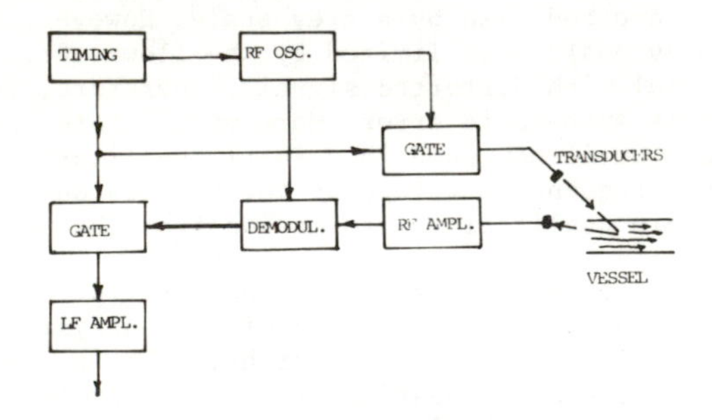

Fig. 6. Scheme of a pulsed doppler flowmeter.

reconstructed by opening a suitable gate in correspondence to the various points on the diameter, that is after the time the wavefront requires to travel the two-way path from the probe to the point. As the sample volume, that is the zone of the vessel insonated, is very small (a few cubic millimeters) only a restricted portion of the vessel is examined: therefore a resolution in distance of 0.7 - 1 mm can be achieved and the velocity profile in the section can be reconstructed.

The doppler spectrum is then processed in the very same way as CW signals, by means of zero-crossing detectors, or quadratic processors or FFT transformers. As the range of velocities in the sample volume is more limited than that of CW systems, the signal spectrum has a narrower bandwidth. This means that the errors in the estimation of mean velocity or zero crossing or maximum frequency envelope can be made smaller. However, also the range of velocities measurable by the instrument is more limited because of the existence of a "maximum non-ambiguous range". In fact, the following relation holds[6]:

$$f_{max} \cdot d_{max} \leq k \frac{c^2}{8f_e} \qquad (6)$$

where f_{max} is the maximum measurable frequency and d_{max} is the maximum admissible depth. Therefore, maximum depth and maximum frequency cannot be independently chosen. This also means that pulsed systems cannot be able to measure the high frequencies present in correspondence to a stenosis.

Therefore, even if, in principle, the measurement of the velocity profile is appealing because it can give more immediate information on the flow regimen, there are some drawbacks:

- the range of measurable velocities is more restricted than that of CW systems;
- the use of the pulsed doppler is more complicated and requires a greater skill for proper positioning of the probe than that required for CW systems;
- physicians have got good experience of CW systems, but are not yet able to adequately interpret the information on flow profiles.

To summarize the effect of human operators, we may conclude that:

- the subjective judgement of the operator has a great weight, so that large differences in interpretation of the same signals may occur. This also means that the repetition of the measurements is questionable and their variability according to the operator may be large.
- the operating instructions for the physician are insufficient as to the correct understanding of the use of the apparatus and are often restricted to a list of clinical cases. Therefore, the personal skill of the operator can also affect the attainment of significant results. Moreover, a scarce understanding of the limits and possibilities of apparatuses often leads one to attribute an "absolute" value to results whose validity the operator is not able to master. Finally,
- often the manufacturer designs the instrument to provide signals with a "smooth look", more agreeable to the physician. This procedure can lead to introduce limitations in the performances of systems.

4. Echographic systems

As another example not only of the influence of a human operator, but also of a greater flexibility of ultrasound techniques, I will discuss some figures relative to echographic (B-scan) systems (Fig. 7). Presently, these systems are widely used in different fields

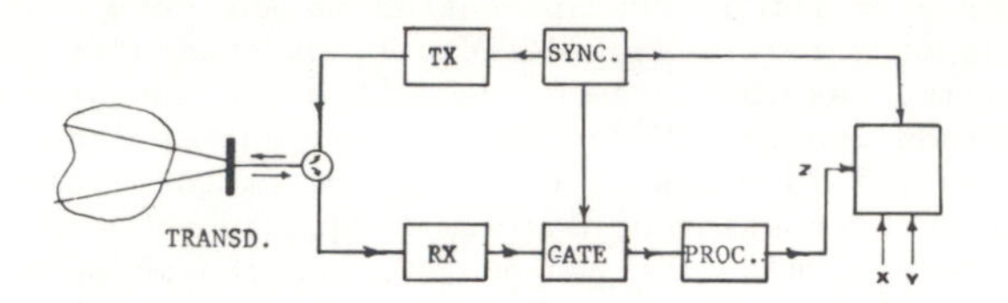

Fig. 7. Scheme of a B-Scan echographic system

(cardiovascular, neurological, gynecological, etc.). In the vascular field, there is a tendency toward a synergic use of echo systems (which give anatomo-morphological information) coupled with doppler systems (which give haemodynamic and functional information).

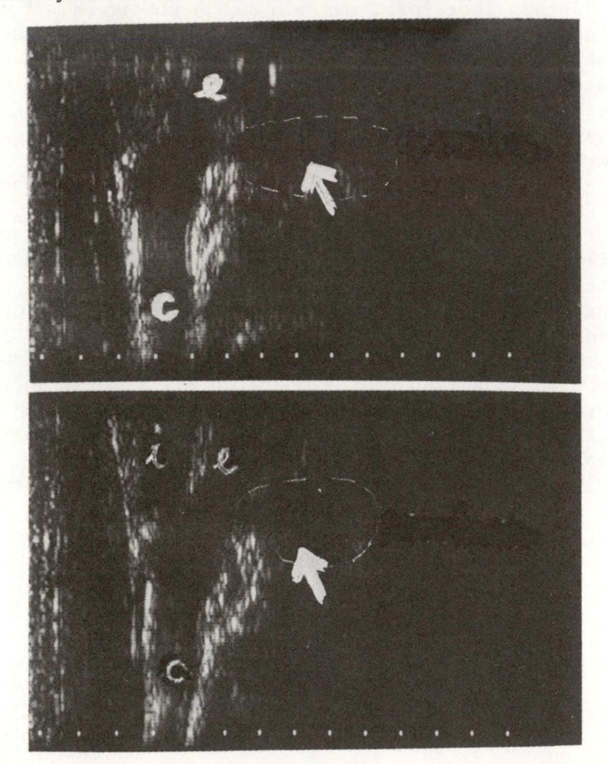

Fig. 8. B-Scan images from a calcific plaque. Arrows indicate the shadows (Courtesy of Prof. E Zanette, Rome).

In Fig. 8a, the shadow leads us to suspect the presence of a stenosis due to a (possibly) calcific plaque, the allocation of which is not completely clear. By choosing a different angle of view, the plaque can correctly be located as lying on the wall of the internal carotid (Fig. 8b). In fact, the subsequent anatomo-pathological examination enabled us to determine that the plaque with shadow was calcific. Should this conclusion be confirmed by more extensive experimentation, it would be possible to obtain information not only on morphology but also on the nature of plaque, which can be relevant to the evolution of the disease.

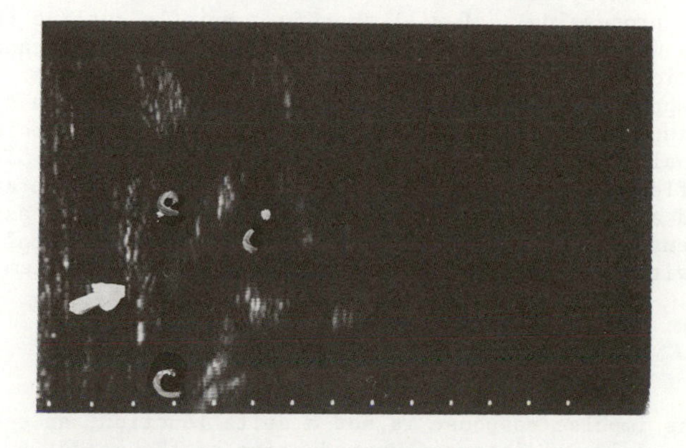

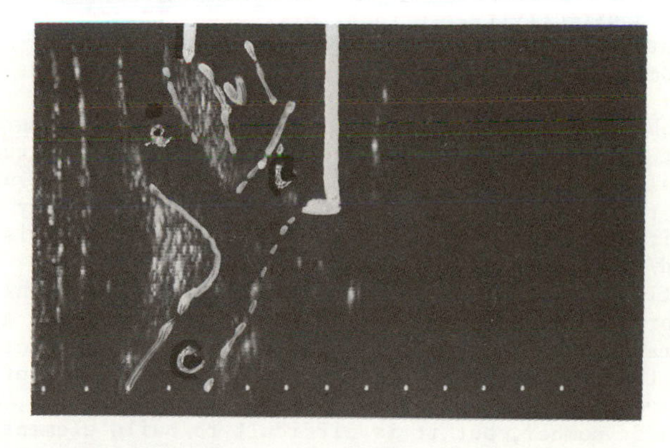

Fig. 9. B-Scan images from neck vessels (Courtesy of Prof. E. Zannette, Rome)

In Fig. 9a, the very low signal (marked with an arrow) happened to be originated by a vein, while the right image is that of Fig. 9b, in which the part in the white frame corresponds to the image of Fig. 9a.

These examples show both that a skilled physician is needed and that ultrasound systems provides good flexibility to resolve some doubtful cases can also give some functional information.

As has been shown, presently the interpretation of echographic images relies completely on the operator. Often, some facilities for a pre-processing of signals are provided, in order to facilitate the task of the human operator. In principle, a rational approach would require the joint optimization of the whole system (from the transducer to the image processing). I will therefore briefly outline the possible improvements which can be achieved by deepening our understanding of the image formation process.

The approach used and which seems to prove useful leads us to re-considering the whole image formation process from the point of view of signals which pass through some suitable networks. This means that the reflexion/scattering function s(x,y;d) from scatterers located at a depth d at the interior of the body may be considered as the results of a bi-dimensional convolution of the reflectivity function r(x,y) of the object with the impulse h(x,y;d) of the ecographic system that is[7,8,9]:

$$s(x,y;d) = h(x,y,d)**r(x,y). \tag{7}$$

As this impulse response is not a delta function, an echographic system will provide a distorted reproduction of the object examined.

This approach represents an interesting re-formulation of the concept of antenna directivity. In fact, ultrasound transducers can be considered as antennas, for which a directivity diagram can be defined (Fig. 9). From the typical radiation pattern reported in the figure, it emerges that:

- the main lobe has a non-zero width, typically of a few degrees. Therefore, it illuminates not a single point but a (restricted) zone;
- some lateral lobes appear, with an amplitude which is not negligible with respect to the main lobe. They emit and receive energy even from directions different from the main lobe, therefore giving rise to ambiguity and/or to masking phenomena.

Presently echo transducers are often built as electronically scanned linear arrays, and can therefore be considered as a group of elementary transducers which radiate an acoustic wave and interact to give rise to the overall radiation pattern. Some improvements are possible, by reducing the dimensions of individual elements, or increasing their number, but it is difficult to build elements with too reduced dimensions.

The approach through the impulse response is more fruitful than the more traditional one, because it also gives the possibility of reducing these distortions. In fact, besides the difficulties in estimating the impulse response of the echographic system, and implementing the algorithms, it is possible in principle to remove, by a suitable deconvolution, the distortion introduced.

5. Conclusions

What conclusions can be drawn about the possible applications and the right place of ultrasound techniques? Moreover, what strategies in the use of systems of different complexity (CW, pulsed echo) and what is the specific role of each technique? In fact, it is not possible to think of indiscriminately using apparatuses of great complexity, for all patients (asymptomatic or not), because of the cost of these exams. Therefore, what is the optimum trade-off between costs and benefits, in particular if we consider that no system can give a definite answer, but only eliminate a (limited) percentage of uncertainty?

In my opinion, ultrasound analysis should be used only after an exhaustive anamnesis and clinical evaluation which must provide clearly stated guidelines for the operator. Moreover, the results of ultrasound analysis must be integrated with all the "history" of the patient, because the ultrasound exam is only one of the possible tests useful in order to make a vascular diagnosis.

Ultrasound analysis can be used as a first screening test in a decision tree, in order to select patients who must undergo further more invasive analysis. As an example, I report a decision protocol used at the III Cattedra of Clinica Neurologica (Dept. of Scienze Neurologiche) in Rome and which has proved to be useful in order to

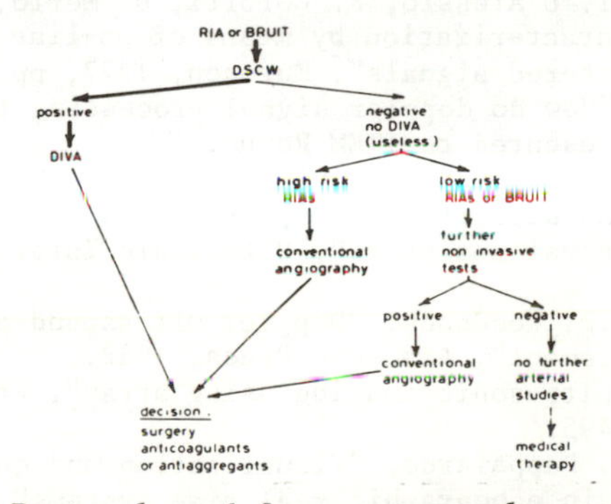

Fig. 10. Protocol used for neurovascular analysis.
(Courtesy of Prof. C. Fieschi, Rome)

optimize the diagnostic path (Fig. 10). It is obvious that by means of non-invasive techniques high-risk patients can be spared more invasive unnecessary tests.

Obviously, this requires that the operator master the technique and the instruments, knowing their exact limits and possibilities. In this context, the presence of a Biomedical Technical Staff in the Hospitals or diagnostic centres can be very helpful. As a very interesting example, I should like to mention the experience of the Clinical Engineering Service of Trieste Hospitals, which is involved in a work of selection, assessment and maintenance of biomedical equipment, and training of medical personnel.

Moreover, there is a need of internationally stated specifications and of Centres of evaluation of ultrasound apparatuses, in order to assess the quality and the suitability of the instruments. To this purpose, the EEC has formed a Special Working Group on Doppler Ultrasound, with the aim of defining technical specifications for the instruments, and protocols of analysis and of interpretation of doppler signals, to be experimented in the EEC countries and successively adopted to guarantee a uniform use of these techniques.

In conclusion, I think that even if ultrasound systems can be considered as fairly developed from a technological point of view, more work is needed in order to standardize instruments and protocols with the aim of more completely exploiting the possibilities of these instruments.

References

1) V.C. Roberts, A. Sainz, "Reduction of operator fatigue in doppler ultrasound blood flow investigations", J. Biomed. Eng., 3, 1981, pp. 140-142.
2) M. Lunt, "Accuracy and limitations of the ultrasonic doppler flow velocimeter and zero-crossing detector", Ultrasound in Med. and Biol., 2, 1975, pp. 1-10.
3) A. Cavallaro, T. D'Alessio, R. Loforti, U. Merlo, "Local circulatory system characterization by means of on-line processing of ultrasonic scattered signals", Eurocon, 1977, pp. 312-319.
4) T. D'Alessio, "How do doppler signal processing techniques affect diagnosis?", presented to ITEM Revue.
5) A Cavallaro, T. D'Alessio, C. Di Giuliomaria, R. Sacco, A. Catanzariti, "An algorithm to improve signal-to-noise discrimination in doppler spectral analysers", Ultrasonic International, Halifax, 1983.
6) P. Atkinson, J.P. Woodcock, "Doppler ultrasound and its use in clinical measurement", Academic Press, 1982.
7) A. Makovski, "Ultrasonic imaging using array", Proc. IEEE, 67, 4, 1979, pp. 484-495.
8) E. D'Ottavi, M. Pappalardo, "Visualization and quantification of acoustic beams in echographic real time systems", IEEE Trans. on SU, 1982, 11.
9) T. D'Alessio, "Interazione tra problemi tecnologici e di elaborazione di segnale nei sistemi ecografici", Automazione e Strumentazione, 9, 1983.

PROCEEDINGS OF THE II INTERNATIONAL CONFERENCE ON APPLICATIONS OF PHYSICS TO MEDICINE AND BIOLOGY
edited by Ž. Bajzer, P. Baxa & C. Franconi

NMR STUDIES OF METABOLISM IN VIVO

David G. Gadian
Department of Physics in relation to Surgery,
Royal College of Surgeons of England, 35-43
Lincoln's Inn Fields, London WC2A 3PN
ENGLAND

ABSTRACT

An introduction is given to the type of information that NMR studies of metabolism can provide. The scope and limitations of the technique are discussed, with particular reference to the spatial selectivity that can be obtained.

1. Introduction and historical perspective

Although it is only recently that NMR has received widespread attention from clinicians, the idea of using the technique to study living systems is by no means new. For example, almost 40 years ago, very soon after the first successful NMR experiments had been performed [1,2] Bloch apparently obtained a strong proton signal on placing his finger in the radiofrequency coil of his spectrometer. In 1955, Odeblad and Lindstrom [3] reported the observation of low-resolution ^{1}H signals from several mammalian preparations, and Singer [4], in 1959, described the use of ^{1}H NMR to measure blood flow in the tails of mice, suggesting the possibility of making similar measurements of blood flow in human beings. In the 1962 edition of his book 'Physical Chemistry', Moore posed a question (which was at that time hypothetical) about a biochemist who wanted to use ^{31}P NMR to study the transformation of ATP to ADP in muscle [5].

These examples illustrate that at least some of the potential of NMR in biology has been appreciated for many years, but unfortunately the early experiments were limited in scope by the relatively poor quality of the instrumentation that was available at that time. The development of high-field superconducting magnets in the late 1960's together with the emergence of Fourier transform NMR revolutionized the scope of NMR and made a rapid impact on the use of the technique in structural studies of purified proteins and other biological molecules. Perhaps surprisingly, it emerged rather more slowly that NMR might have extensive applications in the study of living systems.

Then, in 1973 Moon and Richards [6)] reported high-resolution ^{31}P NMR studies of intact red blood cells. They detected signals from 2,3-diphosphoglycerate, inorganic phosphate and ATP, and showed how the spectra could be used to determine intracellular pH. At about this time it was also shown that ^{13}C NMR could be used to follow the end products of metabolic pathways [7,8)]. Meanwhile, Damadian [9)] had shown that certain malignant tumours of rats differed from normal tissues in their ^{1}H NMR properties, and suggested that ^{1}H NMR might therefore have diagnostic value. Then Lauterbur described an NMR imaging method for studying the spatial distribution of molecules within a sample, coining the work 'zeugmatography' for this new method [10)]. In 1974 it was reported that ^{31}P NMR spectra could be observed from muscle freshly excised from the hind leg of the rat [11)]. Signals were observed from ATP, phosphocreatine, and inorganic phosphate, and sequential changes could be observed in the concentrations of these compounds and in the intracellular pH, reflecting the fact that in these early experiments the muscles were not being supplied with oxygen. Similar changes were observed by Burt _et al._ [12)] for other muscle types.

These studies established the basis for much of the subsequent research involving the use of NMR as a non-invasive method of studying living systems. Metabolic and imaging studies proceeded more or less independently of each other for several years, each evolving with its own characteristic technology. As the power of the two types of study became increasingly apparent, interest was generated in examining ways in which the two technologies could be interrelated.

This article provides a general introduction to the type of information that metabolic NMR studies can provide. A brief discussion is also given of the technology that is special to this type of NMR, and finally we consider the feasibility and difficulties of combining metabolic and imaging studies.

2. What can NMR tell us about metabolism?

The nuclei that have been most widely used for metabolic studies are ^{13}C and ^{31}P, but it seems likely that ^{1}H and perhaps ^{19}F NMR will also become increasingly important for studies of this type. In this section, we shall discuss in general terms the type of information that ^{31}P, ^{13}C and ^{1}H NMR can provide.

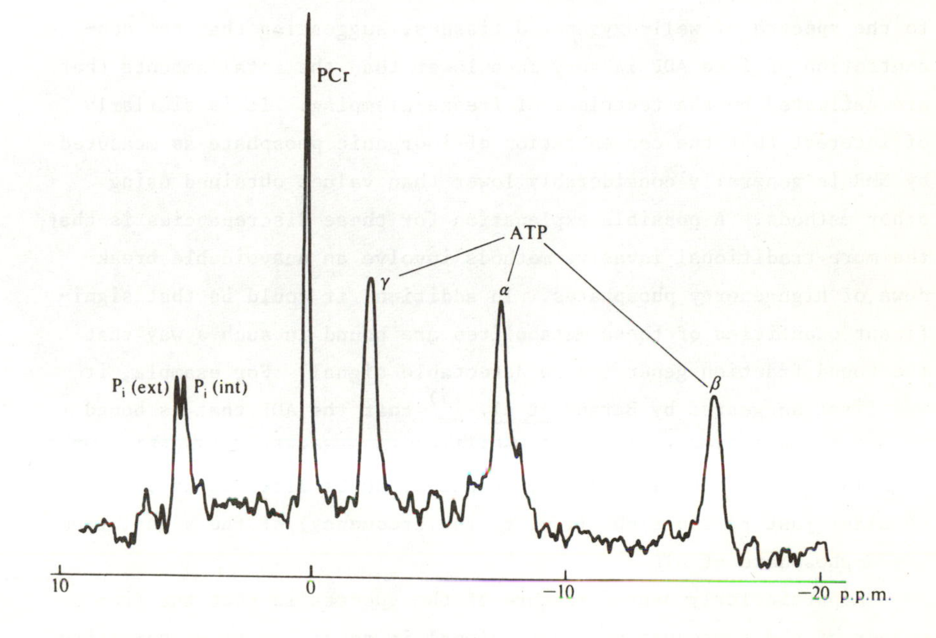

Fig.1. ^{31}P NMR spectrum obtained in 8 minutes at 73.8 MHz from a Langendorff-perfused rat heart. PCr refers to phosphocreatine, and P_i (int) and P_i (ext) to inorganic phosphate in the intracellular space and perfusion medium respectively. The chemical shifts are expressed relative to the PCr signal. (From Grove et al. 1980).

Figure 1 shows a ^{31}P NMR spectrum obtained from a perfused rat heart, which can be maintained in good physiological condition within the spectrometer for many hours (see ref. 13)). The spectrum contains signals that can readily be assigned to the β, α, and γ phosphates of ATP, phosphocreatine and inorganic phosphate. The simplicity of the spectrum reflects the fact that narrow signals are observed only from mobile phosphorus-containing compounds that are present at concentrations of above 0.2-0.5 mM; highly immobilized compounds such as membrane phospholipids produce very broad signals, which often show up in the spectrum as a sloping baseline, while compounds that are present at concentrations below 0.2 mM produce weak signals that may be lost in the noise. If ADP were present in mobile form at sufficiently high concentration, it would generate two signals overlapping with the signals from the γ and α phosphate of ATP. It is of considerable interest that ADP generally makes no detectable contribution to the spectra of well-oxygenated tissues, suggesting that the concentration of free ADP is very much lower than the total amounts that are estimated by the technique of freeze-clamping. It is similarly of interest that the concentration of inorganic phosphate as measured by NMR is generally considerably lower than values obtained using other methods. A possible explanation for these discrepancies is that the more traditional invasive methods involve an unavoidable breakdown of high-energy phosphates. In addition, it could be that significant quantities of these metabolites are bound in such a way that the bound fraction generates no detectable signal. For example, it was first suggested by Barany _et al._ [15] that the ADP that is bound to muscle myofilaments is too immobilized to generate detectable NMR signals. NAD produces signals that can often be detected as a shoulder just to the right (i.e. to low frequency) of the signal from the α-phosphate of ATP.

A particularly useful feature of the spectra is that the frequency of the inorganic phosphate signal is sensitive to pH variations in the normal physiological range. This signal therefore provides a monitor of intracellular pH. The sensitivity to pH arises because the state of ionization of inorganic phosphate changes in the physiological range (the compound has a pK_a of about 6.75), and this accounts for the observation that the signal in the spectrum of Figure 1 is split into two components; inorganic phosphate is present within both the intracellular space and the perfusion medium, and these two environments have pH values that differ by about 0.3 pH units.

The frequencies of the ATP signals are sensitive to the binding of divalent metal ions, as was first shown by Cohn and Hughes [16]. On the basis of titrations performed _in vitro_ (see, for example, ref. 17), it can be concluded from the spectrum of Figure 1 that the ATP in the heart is predominantly complexed to divalent metal ions, presumably Mg^{2+} ions, and similar conclusions regarding the state of ATP have been reached for other tissues. This conclusion is of interest because the state of ATP _in vivo_ has a considerable influence on its biological activity, and also because it enables information to be obtained about the concentration of free Mg^{2+} _in vivo_.

The concentrations of the various metabolites are, under certain conditions, proportional to the areas of their respective signals, and therefore metabolic processes can be followed simply by monitoring how the signal areas vary with time. However, it should be noted that careful controls are necessary in order to quantify the concentrations of metabolites, particularly in studies where there may be some uncertainty as to precisely where within the sample the signal is coming from (for example, when using surface coils). The heart spectrum of Figure 1 was obtained in 8 minutes, but adequate signal-to-noise ratios can usually be obtained in times much shorter than this; the time resolution for kinetic studies is often about one minute, and can sometimes be as low as 15 seconds. For some experiments, the time resolution can be greatly enhanced by synchronising the collection of data with physiological activity [18,19].

NMR can readily be used in this way to study the changes in metabolite levels and pH that are associated with muscular contraction, ischaemia etc. (see 'Further reading' at end of article). However, under most circumstances, the metabolic state of tissue will remain constant with time. What information can NMR provide about this steady state? In addition to providing a non-invasive method of monitoring metabolite levels and intracellular pH, NMR also enables us to measure the rates of certain reactions taking place under steady-state conditions. In one type of experiment, a magnetic label in the form of a suitable isotope can be added to the sample, and the metabolic fate of this label can be followed simply by monitoring the spectra as a function of time. This forms the basis of many ^{13}C NMR studies, as described below. In the second type of experiment, termed saturation transfer NMR, no exogenous label is introduced. Instead, a type of 'magnetic labelling' is achieved by using the NMR spectrometer to alter the magnetic properties of a given compound. Such an experiment is illustrated in Figure 2.

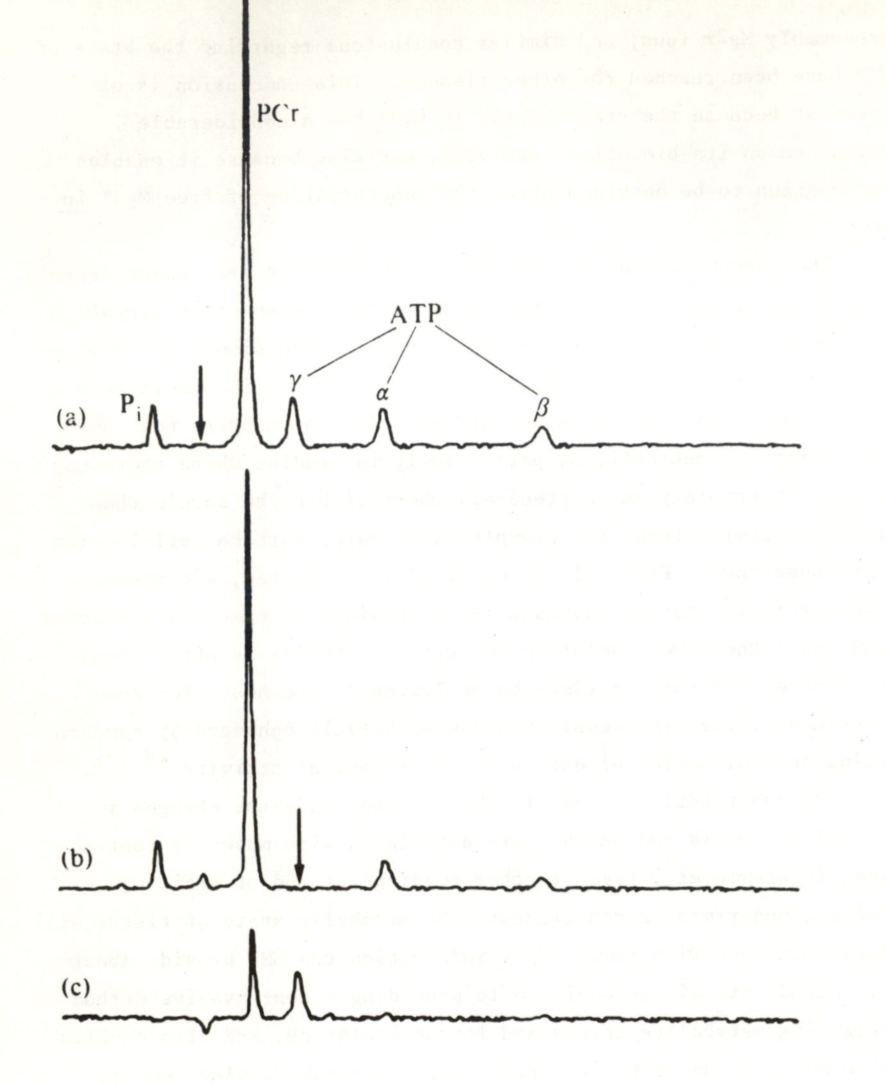

Fig.2. ^{31}P NMR spectra of frog gastrocnemius muscles obtained at 73.8 MHz, with selective irradiation applied at the frequencies indicated by the arrows. The irradiation applied in (a) acts as the control for (b), in which selective irradiation is applied to the signal from the γ-phosphate ATP. The difference between (a) and (b), given in (c), gives the extent of saturation transfer from the γ-phosphate of ATP to phosphocreatine. (Adapted from Gadian et al. 1981).

Figure 2(a) shows a normal spectrum of frog gastrocnemius muscle. In the spectrum of Figure 2(b) a selective radio-frequency pulse is applied to the peak from the γ-phosphate of ATP, causing it to disappear as a result of the process known as saturation. (Effectively, the γ-phosphate is being labelled with zero magnetization). If the γ-phosphate is exchanging with the phosphorus of phosphocreatine, then the lack of signal intensity will be transferred from ATP to phosphocreatine, resulting in a reduction in the intensity of the phosphocreatine signal. The extent of this reduction (see Figure 2(c)), together with the measurement of the spin-lattice relaxation time T_1, permits the rate of interconversion to be evaluated. This particular exchange process takes place by means of the reaction catalysed by the enzyme creatine kinase, and saturation transfer measurements have provided some very interesting results about the activity of this enzyme *in vivo*. By observing saturation transfer from ATP to inorganic phosphate, information can also be obtained about the steady-state synthesis (and hence also hydrolysis) of ATP, which is of particular interest as it can provide a measure of the rate at which chemical energy is being utilised *in vivo* [21,22).

Moving on to other nuclei, the abundant isotope of carbon, ^{12}C, has no magnetic properties and does not produce NMR signals. Therefore ^{13}C, which is only 1% abundant, is used. In order to obtain detectable signals, the compounds under observation must be highly concentrated (for example, glycogen in liver; see refs. 23, 24 or fats, see ref.25). Otherwise it is necessary to enrich the sample with ^{13}C-labelled material, and in this way it is possible to perform ^{13}C NMR studies that are analogous in many respects to radioactive tracer studies using ^{14}C. Unfortunately, a large amount of ^{13}C label must be used, and this may be very expensive.

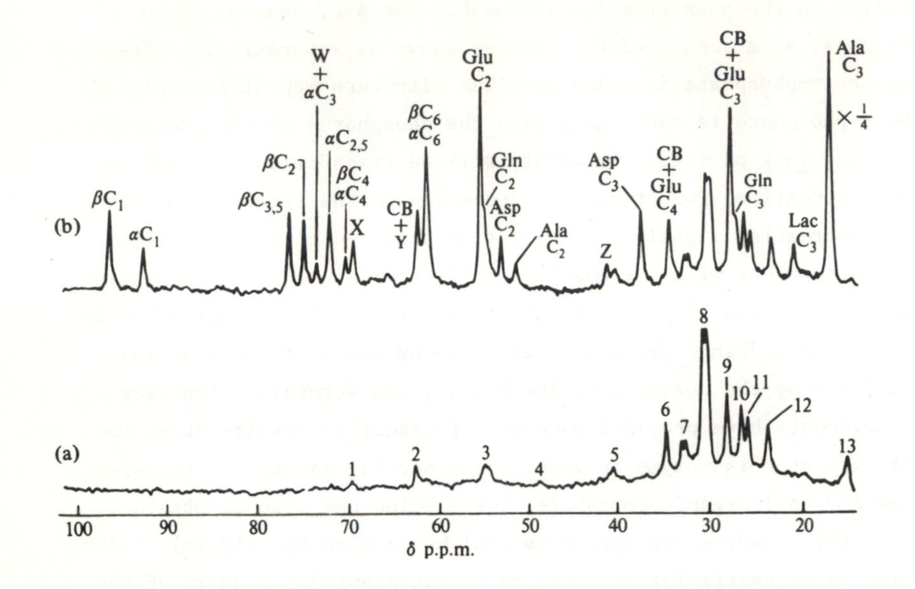

Fig.3. ^{13}C NMR spectra obtained at 90.5 MHz from a perfused mouse liver at 35°C. (a) ^{13}C natural abundance spectrum accumulated before the substrate was added. The peaks labelled 1, 2, 6, 7, 8, 9, 11, 12, and 13 have been assigned to the triglycerides of palmitic, oleic, and palmitoleic acids. 8 mM (3-^{13}C) alanine and 20 mM unlabelled ethanol were then added at 0 min and again at 120 min. (b) Spectrum obtained during the period 150-180 min. Peak assignments are as follows: $\beta^{C}1$, $\alpha^{C}1$, $\beta^{C}3,5$, $\beta^{C}2$, $\alpha^{C}3$, $\alpha^{C}2,5$, $\alpha^{C}4$, $\beta^{C}6$, $\alpha^{C}6$ are the carbons of the two glucose anomers; Glu $^{C}2$, glutamate; Gln $^{C}2$, glutamine; Asp $^{C}2$, aspartate; Ala $^{C}2$, alanine; Lac $^{C}3$, lactate; CB, cell background peak; W, X, Y, and Z, unknowns. (Adapted from Cohen et al. 1979).

Figure 3 shows ^{13}C NMR spectra that were obtained from a perfused mouse liver (a) before and (b) 150-180 minutes after the addition of ^{13}C-labelled alanine and unlabelled ethanol. The natural abundance spectrum accumulated before the addition of substrate shows signals that have been assigned to the triglycerides of palmitic, oleic and palmitoleic acid, which are present in the preparation at high concentrations. Following the addition of ^{13}C-labelled substrate, signals can be observed from a wide range of metabolites, including glucose, aspartate, glutamate, glutamine, alanine and lactate, and by observing the rate and extent of incorporation of label into individual carbons of these molecules, detailed information can be obtained about the relative rates of many of the reactions of gluconeogenesis. ^{13}C NMR clearly has an advantage over ^{31}P NMR that in principle a much larger range of molecules is amenable to study. However, the scope of the technique is at present limited by the availability and price of suitably labelled compounds.

^{1}H NMR is much more sensitive than ^{31}P or ^{13}C NMR, and therefore in favourable circumstances it can detect metabolites that are present at concentrations below 0.2 mM. In addition, because of the ubiquity of the hydrogen atom, very many compounds are in principle accessible to study. However, this ubiquity also leads to problems, firstly because the spectra contain so many overlapping signals that they defy interpretation, but also because water (and sometimes fats) produces an enormous signal that can mask the signals of interest. These problems have been overcome for studies of cellular suspensions (see, for example refs. 27,28), and at least to some extent have now been overcome for intact tissues [29,30]. Certainly the observation of metabolites such as lactate and a number of amino acids could produce some most interesting results, and ^{1}H NMR studies of metabolism could prove to be most fruitful.

Although the large ^{1}H signal from water is a nuisance in studies of metabolism, it can of course be used to great advantage in NMR imaging, as many of the other articles clearly demonstrate.

3. Studies of whole animals and humans

The studies described above were performed on isolated tissues. Extension to whole animals and humans relied on the development of (i) wide-bore magnets with the homogeneity and field strength required for metabolic studies, and (ii) methods of focusing or localizing on a particular region of interest within the animal or human. At the time of writing, suitable magnets that are large enough to accommodate human limbs have been available for almost three years, whereas whole body systems have only just become available. For this reason the majority of human studies so far described have been of the forearm. In particular, ^{31}P NMR spectra have been observed from forearm muscle at rest, during exercise and during recovery, both in control subjects and in patients with muscle disorders. There is not scope in this article to discuss the results that have been obtained; the interested reader is referred to refs. 31-34; for further references see the review by Gadian 1983 (see also ref. 35 for ^{31}P studies of cerebral metabolism in newborn infants). Here, we shall discuss the scope and limitations of the methods of localization that have been employed.

Two such methods have been successfully used for in vivo studies. The first of these involves the use of an unconventional type of radiofrequency coil, which has been termed a surface coil [36], which in its simplest form is a circular loop of wire, of one or more turns. If a surface coil is placed adjacent to an object, then under normal circumstances (but see below) most of the signal detected by the coil will originate from an approximately disk-shaped region of the object immediately in front of the coil, of radius and thickness approximately equal to the radius of the coil. A surface coil therefore provides a suitable, and remarkably simple method of localizing on a superficial region, and is well suited to examining the metabolic state of skeletal muscle and brain. A large number of such studies have been described, mainly using the ^{31}P nucleus, but also using ^{13}C and ^{1}H NMR (see 'Further reading'). However, it should be stressed that the volume that is observed is not precisely defined, and does not have sharp boundaries. Moreover, techniques of varying degrees of sophistication have been developed for altering the spatial selectivity of the coil [37,38]. For example, there is a simple procedure for nulling the signal generated by the surface region immediately in front of the coil [37]. The main incentive for developing such techniques is to enable surface coils to be used for focusing on internal organs. However, for the techniques that have been developed so far, the use of surface coils alone is not sufficient to provide satisfactory focusing for internal regions; additional localizing methods such as topical magnetic resonance (see below) have to be used. Moreover, it should be pointed out that even if satisfactory focusing can be achieved, the attainment of sufficient signal-to-noise becomes an acute problem if the region of interest is well separated (by about a coil radius or more) from the coil.

Surface coils can of course be used to study internal organs if coupled with surgery. In fact, studies have now been described in which coils have been chronically implanted in animals for kidney and heart studies in vivo [39], and it is even possible to use a catheter NMR probe [40]. However, there are obvious disadvantages to such an approach, and there is a clear need for developing techniques for localizing on internal regions without there being an requirement for surgery. One technique that has been successfully used has been termed topical magnetic resonance (TMR). This method employs a special type of field homogeneity coils, which profile the static magnetic field in such a way that the field is very homogeneous over a central, approximately spherical volume, but elsewhere is very inhomogeneous. As a result, narrow signals are observed only from the central region [41,42,43]. Topical magnetic resonance has been used in animal studies for monitoring liver metabolism [41], and in conjunction with the use of surface coils for studying the rat kidney [37].

An alternative approach to these methods of localization is to image the metabolic state throughout, say, the head, in an analogous manner to imaging proton density or relaxation times using ^{1}H NMR. In principle, the two technologies can be combined, as several workers have shown (see, for example, refs. 44,45,46). However, it should be noted that the metabolites that are detected have typical concentrations of about 1-5 mM, whereas the protons observed in images are present at concentrations of up to about 100 M. Furthermore, ^{13}C and ^{31}P NMR are inherently less sensitive than ^{1}H NMR, and for these reasons there are severe sensitivity problems associated with imaging ^{13}C- or ^{31}P-containing metabolites. Acceptable signal-to-noise ratios can only be obtained at the expense of a dramatic reduction in spatial resolution as indicated by the following arguments.

The concentration of a given phosphorus-containing metabolite is lower than that of protons in water by a factor of about 40,000. If we ignore the lower sensitivity of the ^{31}P nucleus (this might in any case be offset by working at a higher field than is most commonly used for ^{1}H imaging), this would mean that a given volume element would generate a ^{31}P signal that is 40,000 times smaller than its corresponding ^{1}H signal. In order to obtain sufficient signal, the volume would need to be increased by a factor of 40,000. The linear spatial resolution would be down by a factor of $(40,000)^{1/3}$, i.e. about 35, so that the spatial resolution would be several centimetres. Using an alternative argument, we could predict that the whole head might produce detectable signal in a single scan. A volume element comprising 1% of the head would produce 1% of that signal, and sufficiently high signal-to-noise could only be achieved by accumulating for 100^2 scans, i.e for about 3 hours if the scans are obtained at intervals of 1 second. (Signal increases according to the number of scans, noise according to the square root of the number of scans, and so the signal-to-noise ratio increases according to the square root of the number of scans). This argument leads to a similar estimate for the achievable spatial resolution. It therefore remains to be seen how valuable the imaging of metabolism will prove to be. However, it should be noted that for superficial regions the spatial resolution can be improved because it is possible to use relatively small surface coils. For example, excellent spectra have been obtained from forearm muscle using surface coils of diameter 2.5 cm [34], and even smaller coils could probably be used successfully for muscle studies.

In considering the various ways in which it is possible to combine imaging with metabolism, another type of approach is to perform ^{1}H imaging and localized spectroscopy on the same instrument. The feasibility of this has been demonstrated by Bottomley _et al._ [47], who have obtained ^{1}H images and localized ^{31}P and ^{13}C spectra of the human head.

4. Conclusions

NMR provides a powerful non-invasive method of studying the metabolism of living systems. Studies of appropriate animal models and of humans should considerably enhance our understanding of the metabolism of healthy and diseased tissue, and of the response to therapy. However, it must be appreciated that the spatial resolution of metabolic NMR is very much worse than for ^{1}H imaging, primarily because of the relatively low concentrations of the metabolites.

5. Acknowledgement

The author was formerly at the Department of Biochemistry University of Oxford, where many of the ^{31}P NMR experiments described here were performed. The author thanks the Rank Foundation and Picker International for support at the Royal College of Surgeons of England.

6. Further reading

Recent books and review articles

Gadian, D.G. (1982) Nuclear magnetic resonance and its applications to living systems. Oxford University Press, Oxford.

Gadian, D.G. (1983) Ann. Rev. Biophys. Bioeng. 12, 69-89.

Gadian, D.G. and Radda, G.K. (1981) Ann. Rev. Biochem. 50, 69-83.

Griffiths, J.R. Iles, R.A. and Stevens, A.N. (1982) Prog. Nucl. Magn. Reson. Spectrosc. 15, 49-200.

Roberts, J.K.M. and Jardetzky, O. (1981) Biochim. Biophys. Acta. 639, 53-76.

The book of abstracts from the 2nd Annual Meeting of the Society of Magnetic Resonance in Medicine, held in San Francisco in August 1983, contains a large number of abstracts describing recent studies. It is published by the Society of Magnetic Resonance in Medicine (P.O. Box 9750, Berkeley, California 94709).

7. References

1) Bloch, F., Hansen, W.W. and Packard, M. (1946) Phys. Rev. 70, 474.

2) Purcell, E.M., Torrey, H.C. and Pound, R.V. (1946) Phys. Rev. 69, 37.

3) Odeblad, E. and Lindstrom, G. (1955) Acta Radiol. 43, 469-476.

4) Singer, J.R. (1959) Science 130, 1652-1653.

5) Moore, W.J. (1962) Physical Chemistry. Problem No. 36, Chapter 14. Longmans Green, Harlow, Essex.

6) Moon, R.N. and Richards, J.H. (1973) J. Biol. Chem. 248, 7276-7278.

7) Eakin, R.T., Morgan, L.O., Gregg, C.T. and Matwiyoff, N.A. (1972) FEBS Lett. 28, 159-264.

8) Sequin, U. and Scott, A.I. (1974) Science. 186, 101-107.

9) Damadian, R. (1971) Science. 171, 1151-1153.

10) Lauterbur, P.C. (1973) Nature. 242, 190-191.

11) Hoult, D.I., Busby, S.J.W., Gadian, D.G., Radda, G.K., Richards, R.E. and Seeley, P.J. (1974) Nature. 252, 285-287.

12) Burt, C.T., Glonek, T. and Barany, M. (1976) J. Biol. Chem. 251, 3584-2591.

13) Garlick, P.B., Radda, G.K. and Seeley, P.J. (1979) Biochem. J. 184, 547-554.

14) Grove, T.H. Ackerman, J.J.H., Radda, G.K. and Bore, P.J. (1980) Proc. Natl. Acad. Sci. USA 77, 299-302.

15) Barany, M., Barany, K., Burt, C.T., Glonek, T. and Myers, T.C. (1975) J. Supramol. Struct. 3, 125-140.

16) Cohn, M. and Hughes, T.R. (1962) J. Biol. Chem. 237, 176-181.

17) Gadian, D.G., Radda, G.K., Richards, R.E. and Seeley, P.J. (1979) In Biological Application of Magnetic Resonance (ed. R.G. Shulman) pp. 463-535. Academic Press, New York.

18) Dawson, M.J., Gadian, D.G. and Wilkie, D.R. (1977) J. Physiol. 267, 703-735.

19) Fossel, E.T., Morgan, H.E. and Ingwall, J.S. (1980) Proc. Natl. Acad. Sci. USA 77, 3654-3658.

20) Gadian, D.G., Radda, G.K., Brown, T.R., Chance, E.M., Dawson, M.J. and Wilkie, D.R. (1981) Biochem. J. 194, 215-228.

21) Matthews, P.M., Bland, J.L., Gadian, D.G. and Radda, G.K. (1981) Biochem. Biophys. Res. Commun. 103, 1052-1059.

22) Shoubridge, E.A., Briggs, R.W. and Radda, G.K. (1982) FEBS Lett. 140, 288-292.

23) Stevens, A.M., Iles, R.A., Morris, P.G. and Griffiths, J.R. (1982) FEBS Lett. 150, 489-493.

24) Sillerud, L.O. and Shulman, R.G. (1983) Biochemistry. 22, 1087-1094.

25) Alger, J.R., Sillerud, L.O., Behar, K.L., Gillies, R.J., Shulman, R.G., Gordon, R.E., Shaw, D. and Hanley, P. (1981) Science. 214, 660-662.

26) Cohen, S.M., Shulman, R.G. and McLaughlin, A.C. (1979) Proc. Natl. Acad. Sci. USA 76, 4808-4812.

27) Brown F.F., Campbell, I.D., Kuchel, P.W. and Rabenstein, D.C. (1977) FEBS Lett. 82, 12-16.

28) Brindle, K.M., Brown, F.F., Campbell, I.D., Foxall, D.L. and Simpson, R.J. (1982) Biochem. J. 202, 589-602.

29) Yoshizaki, K., Seo, Y. and Nishikawa, H. (1981) Biochim. Biophys. Acta 678, 283-291.

30) Behar, K.L., den Hollander, J.A., Stromski, M.E., Ogino, T., Shulman, R.G., Petroff, O.A.C. and Prichard, J.W. (1983) Proc. Natl. Acad. Sci. USA 80, 4945-4948.

31) Ross, B.D., Radda, G.K., Gadian, D.G., Rocker, G., Esiri, M. and Falconer-Smith, J. (1981) New Eng. J. Med. 304, 1338-1342.

32) Edwards, R.H.T., Dawson, M.J., Wilkie, D.R., Gordon, R.E. and Shaw, D. (1982) Lancet. i, 725-731.

33) Radda, G.K., Bore, P.J., Gadian, D.G., Ross, B.D., Styles, P., Taylor, D.J. and Morgan-Hughes, J. (1982) Nature. 295, 608-609.

34) Taylor, D.J., Bore, P.J., Styles, P., Gadian, D.G. and Radda, G.K. (1983) Mol. Biol. Med. 1, 77-94.

35) Cady, E.B., Costello, A.M. de L., Dawson, M.J., Delpy, D.T., Hope, P.L., Reynolds, E.O.R., Tofts, P.S. and Wilkie, D.R. (1983) Lancet. i, 1059-1062.

36) Ackerman, J.J.H., Grove, T.H., Wong, G.G., Gadian, D.G. and Radda, G.K. (1980) Nature. 283, 167-170.

37) Balaban, R.S., Gadian, D.G. and Radda, G.K. (1981) Kidney Int. 20, 575-579.

38) Bendell, M.R. and Gordon, R.A. (1983) J. Magn. Reson. 53, 365-385.

39) Koretsky, A.P., Wang, S., Klein, M.P., James, T.L. and Weiner M.W. (1983) In book of abstracts from 2nd Annual Meeting of the Society of Magnetic Resonance in Medicine (see under 'Further Reading'). pp. 199-200.

40) Kantor, H.L., Balaban, R.S. and Briggs, R.W. (1983). In book of abstracts from 2nd Annual Meeting of the Society of Magnetic Resonance in Medicine (see under 'Further Reading'). p.192.

41) Gordon, R.E., Hanley, P., Shaw, D., Gadian, D.G., Radda, G.K., Styles, P., Bore, P.J. and Chan, L. (1980) Nature. 287, 736-738.

42) Hanley, P.E. and Gordon, R.E. (1981) J. Magn. Reson. 45, 520-524.

43) Gordon, R.E., Hanley, P.E. and Shaw, D. (1982) Prog. Nucl. Magn. Reson. Spectrosc. 15, 1-47.

44) Bendel, P., Lai, C. and Lauterbur, P.C. (1980) J. Magn. Reson. 38, 343-356.

45) Brown, T.R., Kincaid, B.M. and Ugurbil, K. (1982) Proc. Natl. Acad. Sci. USA 79, 3523-3526.

46) Haselgrove, J.C., Subramanian, V.H., Leigh, J.S. Jr., Gyulai, L. and Chance, B. (1983) Science. 220, 1170-1173.

47) Bottomley, P.A., Hart, H.R., Edelstein, W.A., Schenck, J.F., Smith, L.S., Leue, W.M., Mueller, O.M. and Redington, R.W. (1983) Lancet. ii, 273-274.

PROCEEDINGS OF THE II INTERNATIONAL CONFERENCE ON APPLICATIONS OF PHYSICS TO MEDICINE AND BIOLOGY
edited by Ž. Bajzer, P. Baxa & C. Franconi

CLINICAL CONSIDERATIONS IN THE DESIGN AND APPLICATION OF NMR IMAGING SYSTEMS

William R. Brody

Stanford University
Department of Radiology
Stanford, CA 94305
and
University of Trieste
Institute of Radiology
Trieste, Italy

ABSTRACT

NMR imaging promises to have a great impact on diagnostic methods in medicine. The potential advantages of NMR, including its supposed lower deleterious effects compared to x-radiation, the excellent sensitivity to soft-tissue abnormalities and the potential for imaging physiologic and biochemical parameters, must be offset by some of the difficulties facing the implementation of the technology and its integration into the hospital environment. This paper discusses the interface between the design and the application of NMR imaging systems.

INTRODUCTION

Since the introduction of NMR imaging early in the previous decade, there has evoloved an increasing awareness of the almost awesome potential of NMR as a relatively non-invasive diagnostic tool for medicine.

While NMR has applications for diagnosis based upon the creation of an image as well as in the measurement of bulk tissue parameters (e.g., flow, pH, etc.) the major clinical and commercial interest is focussed currently upon the use of NMR as an instrument to produce cross-sectional images of the body, and therefore this paper will be oriented primarily to those imaging applications of the technology. Strictly speaking, the proper terminology for this application is Magnetic Resonance Imaging (MRI) and I will endeavor to use the appropriate abbreviation where applicable.

This paper presents a discussion of the advantages and limitations of MRI with respect to the clinical applications. however, because most papers dealing with MRI are elucidating more clearly than I the applications of NMR, I will devote much more space to a discussion of the problems and limitations of the methodology. This is not to be construed to be a negative emphasis of MRI - quite to the contrary, as I share the excitement and enthusiasm for MRI - but rather to stimulate our colleagues in physics and engineering to advance the applications of MRI through the solution of these problems.

As a disclaimer, I must state at the outset that in the following comparison of MRI to existing imaging methods, most of the discussion centers around x-ray computed tomography. This is not intended to slight other imaging modalities. However, at this juncture, MRI images are closest in form to those of x-ray CT, and the competetion is clearly defined. Space does not permit me the luxury of an exhaustive listing of the advantages and limitations of all the various imaging modalities.

ADVANTAGES AND LIMITATIONS OF MRI

The advantages of nuclear magnetic resonance for body imaging include the following:

1. Less injurious than x-radiation
2. Improved sensitivity to soft-tissue differences
3. Imaging of other parameters and other elements
4. Isotropic imaging

The known hazards of nuclear magnetic resonance appear to be far less, at current exposure levels, than the risks of ionizing radiation. While we should always be concerned with potential hazards (known and unknown) and minimize the energy levels used for any diagnostic procedure, the prevailing opinion is that NMR has a high safety factor for routine clinical uses. While the same argument can be made for ultrasound, the limited tissue discrimination and lack of resolution and depth of penetration of ultrasound restrict its application far more than MRI. Hence the impact of MRI will be far greater than ultrasound, especially in radiation sensitive applications -- e.g., pediatrics, obstetrics, and in screening large populations for disease.

The most appealing factor in NMR images is the outstanding sensitivity to the different structures characterizing what radiologists have come to call the "soft-tissues" of the body. The major structures of interest in the body contain water, calcium salts, or fat. The different water density structures, e.g., the components of the brain (gray and white matter, blood vessels, and cerebrospinal fluid) have very subtle differences in their bulk attenuation to x-rays (the so-called tissue "contrast"). For example, the attenuation difference between gray and white matter is on the order of 0.5 percent, and the gray/white matter discrimination has been one of the tests of image quality on an x-ray CT scanner. The discrimination between body tissues based upon hydrogen density is not great; however the differences in T1 and T2 relaxation rates are large, and therefore MRI images based upon T1 and/or T2 differences show outstanding soft-tissue differentiation.

The initial hope was that MRI would not only provide excellent anatomic differentiation of normal structures, but would also allow differentiation among various pathologic alterations, e.g., the differentiation between a tumor and an abscess, or the separation of benign from malignant tissues. While this discrimination is sometimes achieved, MRI has not up to this point demonstrated substantial advantages over x-ray CT for tissue characterization using density, T1 and T2 parameters.

The possibility of creating images based upon additional NMR information such as chemical shift, blood flow, or even the imaging of elements other than protons continues to excite investigators in the field, and offers, perhaps, the opportunity for tissue characterization and discrimination far beyond that achievable with x-ray CT. If the developments in NMR spectroscopy performed in-vitro can in any way be reproduced in-vivo with spatial localization, MRI spectroscopy will become an important and exciting tool for clinical investigation with substantial consequences. Some months to years later the results of these studies will be translated to routine clinical application.

A word about blood flow: because flowing blood carries excited spins out of the plane of excitation, this information can be used to measure blood flow non-invasively. In addition, and perhaps even more important than the quantitative estimation of flow, is the observation that blood vessels, normally not clearly identified on x-ray CT or ordinary radiographs without the administration of iodinated contrast media, are clearly seen on MRI scans. Hence, MRI provides a completely non-invasive modality for evaluating the course of arteriosclerotic disease. In addition to detecting narrowings and other flow abnormalities in vessels, MRI can image the atherosclerotic plaque in the vessel, a unique and important contribution.

MRI provides isotropic imaging capability. In contrast to x-ray computed tomography in which spatial resolution in the axial direction is superior to that in other planes of orientation (coronal or saggital), MRI acquires data with no preferred direction or orientation. Regardless of whether planar or volumetric data acquisition are employed, one can specify any desired plane of orientation.

The limitations of MRI include the following:

1. Slow image acquisition

2. Limited spatial resolution

3. Skeletal tissues not imaged directly

4. Siting problems

5. Expense

6. Limited knowledge

One of the reasons that MRI met with early skepticism on the part of the medical community was the long imaging times required for image acquisition. From the experience with x-ray computed tomography where there were substantial improvements in image quality whenb the scan time was reduced first from three minutes to twenty seconds, and then from twenty seconds to three seconds, physicians extrapolated to MRI with the supposition that images acquired over five to fifteen minutes would be similar in quality to first generation CT scanners (and hence not competetive with current CT technology).

However, the extrapolation has proved invalid, and MRI images have very quickly become competetive with current generation CT scanners. Without dwelling on some of the explanations for this phenomenon, let me point out that the slow imaging time of MRI remains one of its major limits. First, critically ill patients and pediatric patients will be difficult to study without sedation. Second, motion effects are undoubtedly in current MRI images (even though they do not necessarily appear as streak artifacts as in CT) and contribute to degraded spatial resolution. Third, patient throughput, and hence the cost per exam are limited by image acquisition time. Although the use of volumetric acquisition - in which multiple planes can be acquired simultaneously -- helps to compensate for the slow acquisition time, it is not a substitute for rapid scanning, and introduces some problems of its own (computational requirements, image reconstruction time, and dynamic range).

It is a little unfair to discuss spatial resolution without considering contrast sensitivity, since the two

parameters of system performance. X-ray CT is vastly inferior to conventional film radiography in terms of its ability to resolve high contrast objects; however, CT excells in the visualization of low constrast phenomenon. Likewise, MRI is primarily a tool for enhancing the visualization of low contrast soft-tissue structures. MRI is inferior even to CT in its spatial resolving capabilities, but superior in many applications of low contrast detectability. What is probably most fair to say is that MRI, despite its rapid progress in improved image detail, could use further enhancement of small structure visibility, at least for some applications.

Another apparent limitation of MRI is its inability to image directly skeletal structures. I say apparent, because already information is appearing on the use of MRI for detection of abnormalities of the spine, including metastatic carcinoma and infection. While the mineral substance of bone is not imaged, the fat and associated tissue components mwithin the marrow spaces are imaged, and one can infer by indirect evidence of marrow involvement, the status of the cortical bone substance. However, some method of directly imaging cortical and cancellous bone would improve the utility of MRI even further, and I am spectulating that specific pulsing sequences will be devised to effect this goal.

Finally, there remain the problems of finding a home for NMR in the hospital or clinic environment, and the eventual task of paying for the acquisition and operation of this costly equipment. These problems go hand in hand because the trend to high field superconducting magnet technology for MRI leads to the requirement for an isolated enclosure for the system, often at considerable cost. In addition, the hardware for the supercon technology bring the system cost to nearly double that of

CT. Finally, the operational costs of superconducting NMR systems is also high. We shall return to these problems later.

The last major item is the problem of our lack of understanding of tissue pathology and its effect on the MRI image and on the NMR parameters (density, T1, and T2). In CT, considerable knowledge was gained "on the fly" so to speak, as clinical experience with the new CT technology grew. A similar phenomenon is occurring with MRI, with one important distinction. Whereas with CT the number of parameters controlling data acquisition was limited to a very few (kVp, slice thickness; slice orientation; use of contrast agents), the number of possibilities with MRI is extremely large (type of image data acquisition, pulsing sequence, pulse parameters, field strength, gradient strength). With CT we had an absolute numerical reference standard (in Hounsfield units), whereas investigators continually warn us that the T1 and T2 data presented are not necessarily relevant to another system.

CLINICAL APPLICATIONS

We will briefly discuss the potential role of MRI in clinical imaging, as artificially divided into the following categories:

1. Those that compete with existing diagnostic imaging modalities.

2. Those that represent new applications without existing competition from other methods.

The primary sphere of competition for MRI comes from x-ray computed tomography. With an installed base of roughly 5000 scanners, a large staff of physicians trained in the operation and interpretation of CT scans, the scope and sphere of CT diagnosis is both firmly established and increasing daily. Initially limited to neurological applications, CT expanded to the body and more recently to the heart (the latter through the development of CT devices for millisecond data acquisition). In addition, CT has become increasingly used for spine examinations, and has replaced the myelogram as the primary diagnostic modality in degenerative disk disease in many centers.

In comparison to CT, MRI provides better soft-tissue sensitivity, and possibly somewhat better soft-tissue discrimination than CT. Two important advantages of MRI -- the lack of ionizing radiation and the ability to see certain soft-tissue abnormalities without the intravascular infusion of iodinated contrast agents -- are offset by the increased cost and long scanning times of the modality. Already, however, MRI seems to established its superiority in applications where CT has difficulty: for example, in the posterior fossa of the cranium, where CT is plagued by streak artifacts, and in imaging the spinal cord. In addition, its use in the pediatric age group will be important, although for the largest application (neurologic studies in the newborn), MRI must compete with ultrasound, a simpler, less expensive and readily employed method.

The domain of abdominal imaging has not been as readily explored, and at the moment, MRI has made few inroads into the domain of body CT. This results from the fact that initial systems have been more specifically tailored to head studies, but perhaps there may also be problems in MRI body imaging resulting from patient motion, RF penetration and phase shift disturbances on the NMR image.

As mentioned above, the advantages of MRI in terms of its non-invasive character, its high soft-tissue sensitivity, and its ability to image flow, would dictate other major applications for MRI. Two that we have been investigating are the use of MRI for breast imaging, and MRI angiography.

Breast imaging has been employed using ultrasound, infrared, light transillumination, and x-ray methods (both conventional mammography and CT). For screening large populations of females at low risk for breast carcinoma, the x-ray methods have fallen into disrepute because of the radiation hazard. The experience with CT for breast imaging indicated that in order to have effective tumor detection, intravenous contrast agents had to be employed. MRI overcomes both of these limitations, although introducing other problems, including positioning of the breast optimally in the magnetic field, and the probable lack of detection of small calcifications within the breast (the latter indicative of a malignancy). Nonetheless, the relative advantages of MRI should dictate the extension of considerable research efforts to determine the applicability of this modality to this important problem.

One of the striking and consistent findings beginning with the earliest in-vivo NMR images has been the visualization of blood vessels. When spins are flipped via selective excitation for the purpose of producing a free induction decay signal, moving blood exhibits different T1 and T2 parameters from surrounding stationary tissues because the spins flipped by the 90 degree pulse, for example, are carried out of the imaging plane by the blood flow. Hence, in this particular example, moving blood generates no NMR signal. Thus, blood flow, and blood vessels can be directly visualized without the requirement for instilling an artifical opaque medium as in x-ray angiography. The contrast between the interior of blood vessels (moving blood) and the vessel wall is high, and forms the basis for a non-invasive method to determine the presence or absence of blood vessel narrowing.

In addition, MRI images have shown directly the presence of lipid laden atherosclerotic plaque, something that is difficult if not impossible to accomplish with other imaging modalities. On the surface, it would appear that MRI has a great future for detection and quantification of atherosclerotic vascular disease. The issue of improving spatial resolution as well as the development of specific algorithms for vessel imaging with NMR is an important research area.

The extension of chemical NMR techniques to in-vivo imaging may provide yet another frontier for clinical applications, and certainly will provide an important tool for research. Yet to be determined are methods for simultaneous imaging/spectroscopy, albeit with reduced spatial resolution over MRI imaging alone. Considerable effort is being expended in industr, and university research laboratories to try to establish targeted imaging/spectroscopy.

DISCUSSION OF PROBLEMS OF MRI FOR USE IN CLINICAL DIAGNOSIS

The limitations of MRI have been presented briefly. In the present section I would like to discuss some of these in further detail within the context of the integration of MRI into the patient care environment. The primary focus of this section is on the cost of the MRI technology, or rather the cost/benefit ratio. However, since the clinical applications of MRI are still in the early phase of evolution, we cannot be very specific about the benefits of MRI, since these will likely develop (and increase) rapidly over the next 3-5 years.

From the standpoint of cost, it is easiest to compare an MRI system with a CT system in both its form and function. Both share a number of common elements -- cross-sectional anatomic imaging, excellent soft-tissue contrast sensitivity, the use of digital data acquisition and computer reconstruction hardware and software. By way of comparison, the system components for MRI and CT differ in two respects:

1. Replacement of the x-ray generation system by a magnet, gradient and RF coils and associated magnet and RF power supplies.

2. Necessity for site preparation, replacing the lead-lined room with one that isolates the magnet from the surrounding environment.

On both of these points one finds that the trend toward high-field axial superconducting magnets is a costly one, in that the basic magnet and associated components increase the system cost (500-800K dollars for CT, versus 1.2-2.5 million dollars for MRI) as well as the operating cost (for liquid helium and nitrogen), plus the installation often requires the construction of a separate structure away from the hospital radiology department.

Thus, given the present trends to large bore

superconducting magnets for MRI, the costs of an MRI scan promise to be a factor of two or greater over that of a CT scan. Is the benefit from MRI likely to be that much increased to justify the additional expense, or are there alternatives to reduce the purchase, installation and operating costs? Let us consider the possibilities of reducing the cost of an NMR imaging system.

To reduce cost, we must simplify the magnet and/or provide shielding to make the installation less costly. First, we are confronted with the problem that in order to decide which magnet technology to explore, we need to know the desired field strength for operating the MRI system. The answer to this critical question is unknown. From the literature we can find isolated data points accumulating at 0.15, 0.35, 0.5, 1.0, and 1.5 Tesla for head images, and somewhat less experience for abdominal imaging. For applications not involving spectroscopy, the educated guess seems to indicate 0.5-0.6T as optimal, but this is really speculative, with no scientific reports yet in the literature. Other factors, such as shielding of the RF, gradient coil design and gradient strength, eddy currents, pulsing sequence and reconstruction method all enter into the quality of the reconstructed image. Hence comparison from one system at 0.3 to another at 0.6 as regards the influence of field strength is risky.

Notwithstanding the issue of field strength, let us consider some possibilities for magnet technology:

1. Superconducting

2. Resistive

3. Permanent

Superconducting magnets have produced the highest quality MRI images to date. Whether that is due to better field uniformity, better stability or to higher field strength is not apparent from the literature. Acquisition, installation and operating costs all rise as field strength increases, and the problems of gradient coil power supply design, noise resulting from pulsing the gradients, and cryogen consumption all must be factored into when opting for higher fields.

Proposed solutions for decreasing the installation and operating costs of the supercons include the fabrication of ferromagnetic enclosures to shield the stray fields around the magnet (requiring between 10-20 tons of iron at 1.5T), and the use of liquid helium refrigerator and helium recycling to reduce cryogen consumption. Both proposals are under investigation in industrial laboratories.

Resistive magnets have been fabricated using axial and transverse field geometries at field strengths of up to 2.2T. Transverse fields would appear to be advantageous because the RF pickup coil can be helical rather than saddle geometry, thereby providing a factor of 2 increase in SNR. AT higher fields, power consumtion and magnet cooling become a problem. Shielded configurations for both geometries have been developed, having the additional advantage of reducing the power consumtion for a given field strength in addition to solving the shielding and installation problems. The fact remains, however, that to date, resistive systems have not produced images competetive with superconducting magnet systems. Whether this is due to uniformity, field strength, stability, or some other aspect is unknown.

Permanent magnets provide the possibility of reduced

operating costs and self-shielding (analogous to shielded resistive systems). If rare earth materials are employed to attain high fields (up to 0.5T), cost of the magnet becomes a problem. Ferromagnets require considerable bulk to achieve adequate fields. One commercial concern offers a permanent magnet system with field strength of 0.5T and weight of 100 Tons.

There is need for considerable forethought into the role of MRI versus the cost of these exams. While the medical community is moving rapidly to acquire and utilize these systems, there are few countries, including the United States, that can afford the large acquisition of this high cost-technology. After the past decade in which the rapid dissemination of CT technology occurred, we now have, in the US, a large installed base of CT technology. Is MRI to replace or augment this installed base? If it replaces it, does the increased benefit justify the increased exam costs, and what happens to the installed base of CT systems in the meantime? If MRI is synergistic, will the employement of MRI lead to a cost reduction or cost increase? Somehow, the anwers to these questions should be forthcoming.

Finally, there remains the problem of educating radiologists in the use of MRI instumentation, as well as the issue of informing the general physician population as to the indications and the limitations of MRI for diagnostic evaluation.

REFERENCES

1. Mansfield, P and Morris, PG: NMR Imaging in Biomedicine. New York: Academic Press, 1982.

2. Wherli, FW, MacFall, JR and Newton, TH: Parameters determining the appearance of NMR Images. Ch5 in Advanced Imaging Techniques, Volume Two in Modern Neuroradiology, TH Newton and DG Potts, Eds. San Francisco: Clavadel Press, 1983.

3. Crooks, LE, Mills, CM, Davis, PM, et al: Visualization of cerebral and vascular abnormalities by NMR imaging. The effects of imaging parameters on contrast. Radiology 144: 843-852, 1982.

4. NMR: Nulcear Magnetic Resonance. Guideline Report. AHA Hospital Technology Series, Vol 2, No 8, 1983.

PROCEEDINGS OF THE II INTERNATIONAL CONFERENCE ON APPLICATIONS OF PHYSICS TO MEDICINE AND BIOLOGY
edited by Ž. Bajzer, P. Baxa & C. Franconi

DETERMINANTS OF WATER PROTON RELAXATION RATES IN TISSUE

Seymour H. Koenig and Rodney D. Brown, III
IBM T. J. Watson Research Center
Yorktown Heights, New York 10598
U.S.A.

ABSTRACT

It is well established that the longitudinal magnetic relaxation rate of solvent protons, $1/T_1$, increases markedly in homogeneous protein solutions as the magnetic field is reduced well below the traditional NMR range. For a 5% solution of protein of 10^5 Daltons, for example, $1/T_1$ increases from about 50% above the pure water rate at 20 MHz to five times the water rate at 0.01 MHz. For tissue, including blood, the behavior is similar. Data for blood show that extracellular water has ready access to the hemoglobin inside red cells, which causes the enhanced relaxation. The extent to which cell water can sample the spatial structure of solid tissue, and how this structure influences relaxation rates, is as yet unknown. Nonetheless, the relaxation data for tissue can be accommodated within the conceptual framework developed previously for analyzing homogeneous solutions of diamagnetic proteins. The variation of $1/T_1$ with field differs among tissues, and its magnitude at a given field can vary by more than a factor of three, far more than does the water content of the tissues.

Solute paramagnetic ions also increase the relaxation rates of solvent protons. These ions can be introduced intravenously, are known to accumulate in specific organs, and therefore have potential utility as contrast-enhancing agents in NMR imaging. Mn^{2+} and Gd^{3+}, for example, produce characteristic dependencies of $1/T_1$ on magnetic field that vary with the chemical state of the agent. The possibility exists, therefore, for following the biochemistry of these agents.

Introduction

NMR imaging differs in a fundamental way from competitive modalities, X-ray tomography, for example, in that a major determinant of image contrast is the dynamic, rather than the static, properties of tissue. Whereas contrast in X-ray images is determined by the absorption coefficient of tissue, which in turn can be deduced from knowledge of the density of electrons in the X-ray path, contrast in NMR proton images is multiparametric: it depends on the proton concentration (which varies little from one tissue to another), but it also depends on the rate at which the magnetization of an ensemble of protons returns to its equilibrium state after a perturbation. This rate is a tensor quantity, generally expressible by the two parameters $1/T_1$ and $1/T_2$, where T_1 is the longitudinal and T_2 the transverse relaxation time. T_1 is the time constant for the component of magnetization parallel to H_o to approach its thermal equilibrium value, once perturbed, and T_2, in liquids and tissue, is the time constant for decay of the transverse component to its equilibrium value of zero. NMR imaging is practical only because T_1 and T_2 vary significantly among different types of tissue, and may also differ substantially from each other in a given tissue (or region thereof).

The relaxation rates $1/T_1$ and $1/T_2$ are functions of H_o, the static magnetic field, and this variation depends on the microscopic structure of the particular tissue in a way that remains to be understood.[1-3] Moreover, the intrinsic relaxation rates of tissue can be altered by the introduction of paramagnetic agents;[4] these contrast agents not only alter $1/T_1$ and $1/T_2$, but also drastically change their functional dependence on H_o.[5] Thus, the question of how to optimize contrast in a particular situation is complex, and an understanding at the molecular level of proton relaxation behavior should contribute to the ultimate clinical utility of NMR imaging.

We have made extensive studies of the magnetic field dependence of relaxation processes of solvent nuclei in solutions of homogeneous protein,[1,2,5-9] using specialized instrumentation developed for the purpose.[7] The studies have involved both diamagnetic proteins[1,2] and metalloproteins containing paramagnetic ions.[5-9] More recently, in collaboration with other investigators, we have extended these studies to a number of human tissues, normal and abnormal,[10] observing the dependence of $1/T_1$ on H_o for both diamagnetic samples and those containing paramagnetic contrast agents.[11]

In what follows, we present a survey of the phenomenology plus a limited discussion of its implications. The data are presented as the variation of $1/T_1$ with H_o, the latter expressed both in units of Tesla and as the Larmor precession frequency of protons in the field H_o. We call the variation of $1/T_1$ with H_o a Nuclear Magnetic Relaxation Dispersion (NMRD) profile. Measuring the analogous $1/T_2$ profile is quite arduous,[12] but rather unnecessary once the $1/T_1$ profile is known, since the two are related by theory and the $1/T_2$ profile can be predicted fairly well within our present understanding.

Results

Diamagnetic Protein Solutions. Fig. 1 shows the $1/T_1$ NMRD profile for a "typical" solution of diamagnetic protein,[2] here a 6.5 wt. % solution of yeast alcohol dehydrogenase, 160,000 Daltons. The background rate of protein-free solvent is also shown. Each data point results from a least squares fit of a single exponential to 23 pairs of measurements of magnetization and time. The rms deviations are generally less than ±1% throughout this work, and the absolute uncertainty of the results is of this order as well, as may be inferred from the small scatter of the points about the fitted curve through the data. The latter derives from a four-parameter heuristic expression that provides a convenient way of characterizing and cataloging the data.[2] The four parameters are A, the amplitude in the limit $H_o \rightarrow 0$ of a term that is highly dispersive, and that inflects at a "correlation frequency" indicated by ν_c; D, a term that is field-independent in the range indicated, and which remains after the dispersive term goes to zero at high field; and a parameter (not shown) that characterizes the deviation of the observed dispersion from a Lorentzian form, shown dashed, which is the variation expected on the basis of a simple model of two-site exchange of water molecules between bulk solvent and a short-lived hydration shell of the protein.[1]

Relaxation of protons in pure water comes about by fluctuations in the magnetic dipolar interactions of neighboring proton pairs. Addition of solute protein influences the relaxation rates by altering the dynamics of the solvent molecules. From what is known about the solvent-solute interactions in protein solutions,[1,2,13] the A-term arises from the influence of the rotational Brownian motion of the protein molecules on the motion of solvent; the experimental observations are that information about this thermal motion is conveyed (by a mechanism not fully established) to the solvent water molecules. It has been rather thoroughly demonstrated that the values of A and ν_c vary with temperature, protein molecular weight and shape, and solvent viscosity as would be expected from the known behavior of the rotational Brownian motion.[1,2] The D-term, which remains at high fields (frequencies), at which the proteins can be regarded as stationary, arises from alteration of the dynamics of solvent water molecules at the protein-solvent interface.[2,13] This term might be expected to be rather larger in tissue because of the extensive area of its microscopic structure.

From what is known of the theory, and the limited $1/T_2$ NMRD profiles available, the A and D contributions as $H_o \rightarrow 0$ are the same for $1/T_2$ and $1/T_1$. However, at the higher fields, the D-term contribution to $1/T_2$ remains constant whereas the A-term disperses only to about ~30% of its low field value. Thus at high fields, $1/T_1$ is dominated by D and $1/T_2$ by A and, as a consequence, at intermediate fields, say 5-30 MHz, where NMR imaging is being done, relaxation in protein solutions (and undoubtedly in tissues as well) involves contributions from two distinct molecular mechanisms that contribute to $1/T_1$ and $1/T_2$ in significantly different proportions.

Diamagnetic Animal Tissue. Fig. 2 shows NMRD profiles for a sample of rabbit blood at normal hematocrit, at two temperatures, both fresh and congealed, the latter because the sample had been kept at 4 °C for somewhat over two days. The profiles are very much like those for a cell-free protein solution and, indeed, have been shown to differ very little from those for cell-free solutions of hemoglobin at concentrations found within erythrocytes.[9] This is surprising at first since, at normal hematocrit, ~25% of the water is intracellular and the remainder extracellular, whereas the relaxation data that underly each point in Fig. 2 indicate[9] that all waters are equivalent. The conclusion is that, on the time scale of these experiments, i.e., for times of the order of T_1, intra- and extracellular water molecules interchange locations many times; the cell membrane in this instance serves as a sac to confine the protein, but not the water.[9] The temperature dependence is that expected from the change in viscosity of the solvent with temperature, as alluded to above.[1,2]

The fact that the NMRD profiles are little altered by the state of the blood—congealed compared with liquid—says that the gross morphological features of tissue (which blood is) are not necessarily good guides to the relaxation behavior of water in that tissue. The one immediate inference is that it will not be straightforward to predict the nature of the changes in relaxation, and therefore in image contrast, knowing only the morphological changes induced by a particular abnormal state of tissue. These results are similar to earlier results for solutions of sickle hemoglobin studied through the gelatin transition,[9,14] and the explanation is presumably similar: congealing of the rabbit blood under our conditions means polymerization of a small amount of plasma protein, which in turn entrains the remaining blood in interstices too large to alter the thermal motion of solvent water in any substantive fashion.

Fig. 3 shows NMRD profiles for several tissues from rat.[3] The lines through the data points result from the four-parameter fits used in the earlier figures.[2] Two major points should be noted: (1) the behavior of fat tissue is qualitatively different from the others, and (2) these are qualitatively like the data of Figs. 1 and 2, though the magnitudes of the relaxation rates vary about 2-fold from liver to spleen. The latter is not due to differences in water content, which are minor, nor can the different relaxation rates be ascribed in a straightforward fashion to variations in average protein size from one tissue to another, else the profiles with the greatest rates at low field would have the lowest rates at high field. What determines these NMRD profiles, quantitatively, is an open question, though it is clear that (in all but fat) we are observing mobile water that interacts with the tissue structure at the macromolecular level. For fat, the intensity of the signal indicates that we must be observing protons from lipids as well as from water, even though the decay of the magnetization can still be fit to a single exponential with an uncertainty, in these data, of about $\pm 1.2\%$. For all the tissues in Fig. 3, the relative extent to which protons either from water in different compartments, or from water and fat together, contribute to the observed NMRD profiles remains an open question. The data for liver tends to scatter more about the smooth curve through the data than do the results for the other tissues. This is a real effect and results from changes in the properties of the liver during the approximately 30 minutes required to collect the data. (The data are not taken in order of field, but scattered so as to show up time-dependent effects.)

The data in Fig. 4 address the question, implicit in Fig. 3, regarding the range of variability of the NMRD profiles among individuals. The data here show the variability among species (rat, dog, and rabbit) for two tissues, liver and spleen. Though the liver profiles differ from each other, as a group they are quite distinct from the three profiles for spleen, which are almost identical (particularly so when account is taken of the lower temperature of the rat spleen.)

A close look at Fig. 3 shows that the data at 3 MHz for all tissues but fat lie somewhat above the smooth curves through the data. In Fig. 5, this region is shown enlarged for two samples of rat heart, one perfused with D_2O until about half the rapidly exchangeable protons had been displaced. Two peaks are seen in the data, contributing about 10% to the relaxation rate, that have previously been identified[15] as cross relaxation of protons by the ^{14}N nuclei of protein when the (magnetic field dependent) energy of the protons becomes equal to the energy of one of the electric quadrupolar transitions of the ^{14}N nuclei of tissue protein. The decreased peak amplitude for the partially deuterated sample indicates that the coupling is not direct, but involves transport of magnetic energy (spin diffusion) through at least one intervening exchangeable proton. This built-in "contrast-agent" suggests the possibility that diamagnetic agents containing nuclei with large quadrupolar splittings might be developed for contrast enhancement in NMR imaging.

Human Tissue, Normal and Abnormal. Of major importance to the utility of NMR imaging as a clinical modality is the possibility of significant and predictable differences between the relaxation rates of diseased tissue and "normal" tissue, the latter meaning both tissue from individuals free of disease, and uninvolved tissue adjacent to abnormal regions. In principle, all three may differ in their relaxation properties. Early workers have claimed that relaxation rates decrease in diseased, particularly tumorous, tissue.[16] The data in Fig. 6 are not consistent with this view.

Fig. 6 shows NMRD profiles of tissue resected from two elderly patients (each over 80); the tissue was diagnosed histologically as containing invasive adenocarcinomas of the colon in the tumorous region, and negative at the margins. The profiles of the relaxation *times* rather than *rates* are shown, to emphasize the distinctions in the field region at which imaging may be done. It is seen that the two non-involved specimens (controls) have essentially identical NMRD profiles, whereas the two diseased samples deviate in opposite directions from the controls, by amounts that exceed the spread between the two controls. These data, of course, are from only two patients, and the tissue specimens had been frozen for one month between surgery and our measurements. Nonetheless, the remarkable similarity of the NMRD profiles of the two controls suggests that the altered profiles of the diseased tissue are real, and not artifacts related to the post-operative history of the tissue.

It is apparent that, for colon specimens, the situation will be found to be rather complex. The situation with other tissues may well be simpler, as seen below.

Fig. 7 shows NMRD profiles of breast tissue from a 61-year old female patient; the histological diagnosis of the solid tumor was of invasive duct carcinoma. The distinction of the two profiles is that the normal tissue has a profile much like that of fat (cf. Fig. 3), whereas the diseased tissue has a dispersion profile like that of non-fat tissues, the latter being consistent with the view that the tumor cells derive from non-fat tissue and retain characteristics of their source. This point has been considered previously, where it was noted that contrast in images of diseased breast tissue is due to the fat-cell content of uninvolved tissue, particularly in older patients.[16,17] The data in Fig. 8, which includes those of Fig. 7 plus profiles of (male) peritoneal fat handled in two different ways, reinforces the interpretation of the data of Fig. 7. One conclusion immediately apparent from the results in Fig. 7 is that the contrast between normal and tumorous breast tissue in T_1-weighted images will be greater at higher fields.

Paramagnetic Contrast Agents. There is often a need to enhance image contrast beyond that obtainable from native tissue, even under optimized conditions. This has been the practice for many forms of X-ray imaging, including CT, and it is possible in NMR imaging as well. Relaxation rates of solvent protons are enhanced by paramagnetic ions, particularly if tissue water has access to the inner coordination sphere of these ions. Ions like Mn^{2+} and Gd^{3+} are particularly good relaxing agents, but questions relating to toxicity, organ specificity, and the biochemical fate of these ions *in vivo* require that these ions be introduced in chelated forms.

Unlike the situation with heavy metal ions as used in X-ray image-enhancement, where the increase in contrast depends only on the increased electron density, the influence of paramagnetic ions on the relaxation rates of water depends on the accessibility of solvent to these ions, and on the chemical environment of the ions. For ions in solution, i.e., hydrated aquoions, the relaxation effects are well understood, there are ample data, and the theory is well developed.[5] In solutions of these ions complexed with protein, the situation is more complex.[6-9,12] Each ion-protein solution appears to be a special case; most data can be understood retrospectively, but reliable prediction is as yet very difficult. The variability of the data is illustrated in the next two figures.

Figs. 9, 10 show the NMRD profiles of solutions of Gd^{3+} and Mn^{2+} ions, each in three different chemical environments: hydrated aquoions; chelated with EDTA (ethylenediaminetetraacetic acid); and bound to (different sites of) the protein concanavalin A. The NMRD profiles clearly depend on the chemical environment of the ions; this behavior is well understood, and has been discussed in depth elsewhere.[5,6] The major question is the extent to which this understanding can be transferred to *in vivo* studies, and to predicting contrast enhancement in NMR imaging. For example, if chelated ions are introduced *in vivo*, will they remain so, and will the interactions of chelate with tissue influence the relaxation rates and the associated NMRD profiles? The next figure shows results of a preliminary study.

Fig. 11 shows the NMRD profiles of liver from a white rabbit control, and livers from two rabbits, ~2 kg, sacrificed 15 minutes after intravenous injection of 0.6 ml of a 0.12 M solution of Mn^{2+}-chelate complex (a diaminopropanoltetraacetic acid). Also shown is the NMRD profile of a 2.3 mM dilution of the Mn^{2+}-chelate complex *in vitro*. The shapes of the NMRD profiles of the two Mn^{2+}-containing liver samples, compared to the control, indicate (cf. Fig. 10) that Mn^{2+} ions have been taken up by the liver (as anticipated from previous experiments) and immobilized by binding to a rather large macromolecular structure. Moreover, the magnitude of the enhancement over background, which corresponds to an enhanced relaxation rate, per Mn^{2+} ion available, that is far greater than for the solution of Mn^{2+} chelate, indicates that the Mn^{2+} ions have been removed from the chelate and, though immobilized, are accessible to tissue water.

Discussion

We have presented the phenomenology of the NMRD profiles of protons in tissues: normal, abnormal, and with paramagnetic contrast agents, as well as the profiles of homogeneous solutions of protein and ions that model the tissue samples. For the longer run, we would like to obtain sufficient data, and develop adequate insight, to be able to predict, and modify, contrast in NMR images. But beyond this, we want to understand the underlying physical mechanisms, at the molecular level, that determine the observed longitudinal and transverse relaxation rates in tissue. The problem is a difficult one, however, in part because of the question of compartmentalization of water in tissue.

In discussing the NMRD profiles of blood, a simple tissue, Fig. 2, two compartments were considered: intra- and extracellular water. Nonetheless, all water molecules in blood have access to all cells as well as to all the extracellular solvent, so that any macroscopic, though very small, portion of the sample would display identical relaxation behavior. More complex tissue may well be different; one presumes that water in solid tissue may be restricted to compartments with exchange of water among the compartments slow on the time scale of T_1, so that what one measures is a superposition of proton signals from various compartments of the sample, each possibly with a different value of $1/T_1$. The problem, as will be demonstrated, is that data from such an intrinsically heterogeneous sample will appear indistinguishable from that of a homogeneous sample under typical experimental circumstances. The next figure illustrates this point.

Fig. 12 shows the NMRD profiles of dog liver, spleen, and a mixed sample comprised of half of each of the first two. (The data were taken at 5 °C to minimize degradation of the samples by enzymatic or bacterial decay.) What is apparent from the figure is that the mixed sample gives relaxation rates that are close to the average of the rates of each contributing component. What is not apparent from the data shown, and what is of particular note, is that the data underlying each point on the NMRD profile of the mixed sample could be represented by a single exponential with an rms uncertainty of about $\pm 1.2\%$ in each case, compared with $\pm 1\%$ for the homogeneous samples. This (perhaps surprising) fact, that two exponentials can appear as one, is demonstrated explicitly in the next figure.

Fig. 13 shows two exponential decays, with time constants that differ by a factor of two, and the fit of a single exponential to a composite curve that is the equally weighted sum of the two. The intent is to model the data of Fig. 12. The time range represented is in accord with experimental procedures, in that noise obscures the data when more than three time constants have elapsed. For these circumstances, the composite curve can be fit by a single exponential with an rms uncertainty of $\pm 0.4\%$. For two exponentials that differ by a factor three in their decay constants, the uncertainty increases to $\pm 2.7\%$, still small for most experimental situations.

The foregoing suggests that a synthetic, rather than an analytic, approach will have to be taken to understand relaxation in tissue. That is, the contribution to the NMRD profiles of the different substructures of tissue will have to be understood first, and these then summed to synthesize the overall tissue profiles. It probably will not be possible to do the converse: to analyze an NMRD profile of a particular tissue and resolve it into its constituent contributions. Nonetheless, once the basic contributions are understood, and information regarding how these contributions are altered, for example, by disease, it should be possible to predict changes in NMRD profiles of human tissue for a variety of states, both normal and abnormal.

Acknowledgments

The unpublished data in Figs. 7, 8 are from a collaboration with Professor Y. Wang of Thomas Jefferson Medical School; those of Fig. 11 from a collaboration with Dr. G. Wolf of the VA Medical Center, Philadelphia, PA; and those in Fig. 12 are from a collaboration with Professor D. Adams of Brigham and Women's Hospital, Harvard Medical School.

Figure Captions

Figure 1. NMRD profile of solvent protons in a solution of yeast alcohol dehydrogenase (160,000 Daltons), as a function of magnetic field strength (indicated both in Tesla and the proton Larmor frequency at that field). The sample is a 65 mg/ml solution, near neutral pH. The data, taken at 6 °C, are typical of solutions of globular proteins. There are two different protein contributions to the data: one, with a magnitude A in the low field limit, decreases monotonically to zero at high fields, and inflects at ν_c; a second, with magnitude D, is smaller and constant in the range of field shown here. The solvent contribution is indicated by $1/T_{1w}$. The solid curve is a fit of the data to the Cole-Cole expression, known to represent the data very well, whereas the dashed curve is the Lorentzian form anticipated on the basis of a simple two-site model of relaxation. (After reference 2.)

Figure 2. NMRD profile of solvent protons in fresh (liquid) and aged (congealed) rabbit blood, measured at 5 °C and 35 °C. The curves through the data points result from fits of the Cole-Cole expression to the data. The background contributions of water is indicated. (After reference 3.)

Figure 3. NMRD profile of protons in several rat tissues at 30 °C. The curves through the data points result from a fit of the Cole-Cole expression to the data. The background contribution of water is indicated. (After reference 3.)

Figure 4. Comparison of NMRD profiles of protons of liver and spleen from three mammalian species: rat, dog, and rabbit. The data for rat are at a somewhat lower temperature than those for dog and rabbit. The correction for this is small and, when made, would make the spleen profiles for the three species indistinguishable. (After reference 3.)

Figure 5. NMRD profiles of protons of two rat hearts, measured at 10 °C, one perfused with D_2O to replace about half of the protons by deuterons. The 1-4 MHz region is expanded to demonstrate peaks in the NMRD profiles. (After reference 3.)

Figure 6. NMRD profiles of resected tissue from two patients with adenocarcinoma of the colon. Samples in each case were from the tumors and from the uninvolved marginal areas.

Figure 7. NMRD profiles of resected tissue from a patient with invasive duct adenocarcinoma of the breast. Samples were from the tumor and the neighboring uninvolved tissue.

Figure 8. The NMRD profiles of Fig. 7 plus two samples of peritoneal fat, each handled differently, to indicate the similarity of the profiles of fat and of normal breast tissue (cf. Fig. 3).

Figure 9. NMRD profiles of solutions of Gd^{3+} ions in three chemical environments: the (hydrated) aquoion; chelated with excess EDTA; and complexed with excess Zn^{2+}-Ca^{2+}-concanavalin A, a diamagnetic protein that binds Gd^{3+} ions well. (After reference 6.)

Figure 10. NMRD profiles of solutions of Mn^{2+} ions in three chemical environments: the (hydrated) aquoion; chelated with excess EDTA; and complexed with excess demetalized concanavalin A at the major binding site for Mn^{2+} ions. The dispersion of the aquoion data at low fields is unique to Mn^{2+} and relates to a contact interaction that makes the $1/T_2$ NMRD profile of Mn^{2+} aquoions unique as well (cf. reference 6).

Figure 11. NMRD profiles of livers from normal rabbit (control), and from two rabbits sacrificed 15 minutes after intravenous injection of 0.3 ml/kg of a 0.12 M solution of Mn^{2+} chelate solution (see text). The NMRD profile of a 2.3 mM dilution of the chelate solution is also shown. The data indicate that the Mn^{2+} has been taken up by the liver, removed from the chelate, and bound to macromolecules.

Figure 12. NMRD profiles of samples of dog liver and spleen and of a sample composed of approximately equal volumes of liver and spleen. The average of the first two profiles is also shown for comparison with the measured profile for the mixed tissue.

Figure 13. Exponential decays with time constants that differ by a factor of two, and a comparison of the average of the two exponential curves with a fit of a single exponential to the average, composite decay.

<u>References</u>

1. Koenig, S.H., & Schillinger, W.E. (1969),. J. Biol. Chem., **244**:3283-3289.

2. Hallenga, S.H., & Koenig, S.H. (1976) Biochemistry **15**:4255-4263.

3. Koenig, S.H., Brown, R.D., III, Adams, D., Emerson, D., & Harrison, C.G. (1984) Invest. Radiol. **19** (In press.)

4. Brasch, R.C. (1983) Radiology **147**:781-788.

5. cf. Koenig, S.H., & Brown, R.D., III (to be published) for additional background.

6. Koenig, S.H., Baglin, C., Brown, R.D., III, & Brewer, C.F. (To be published.)

7. Brown, R.D., III, Brewer, C.F., & Koenig, S.H. (1977) Biochemistry **16**:3883-3896.

8. Koenig, S.H., Brown, R.D., III, Bertini, I., & Luchinat, C. (1983) Biophysical J. **41**:179-187.

9. Lindstrom, T.R., & Koenig, S.H. (1974) J. Mag. Resonance **15**:344-353.

10. Brown, R.D., III, Koenig, S.H., & Wang, Y. (Manuscript in preparation.)

11. Koenig, S.H., Brown, R.D., III, Cooperman, H., & Wolf, G. (Manuscript in preparation.)

12. Koenig, S.H., Brown, R.D., III, & Studebaker, J. (1971) Cold Spring Harbor Symp. Quant. Biol. **36**:551-559.

13. Koenig, S.H. (1980) In: Rowland SP, ed. ACS Symposium Series, No. 127, Water in Polymers, 157-176.

14. Lindstrom, T.R., Koenig, S.H., Boussios, T., & Bertles, J.F., (1976) Biophysical J. **16**:679-689.

15. Winter, F., & Kimmich, R. (1982) Biochem. Biophys. Acta **719**:292-298.

16. cf. Mansfield, P., & Morris, P.G. (1982) NMR Imaging in Biomedicine: New York, Academic Press.

17. Mansfield, P., Morris, P.G., Ordidge, R.J., Pykett, I.L., Bangert, V., & Coupland, R. E. (1980) Phil. Trans. R. Soc. Lond. B **289**:503-511.

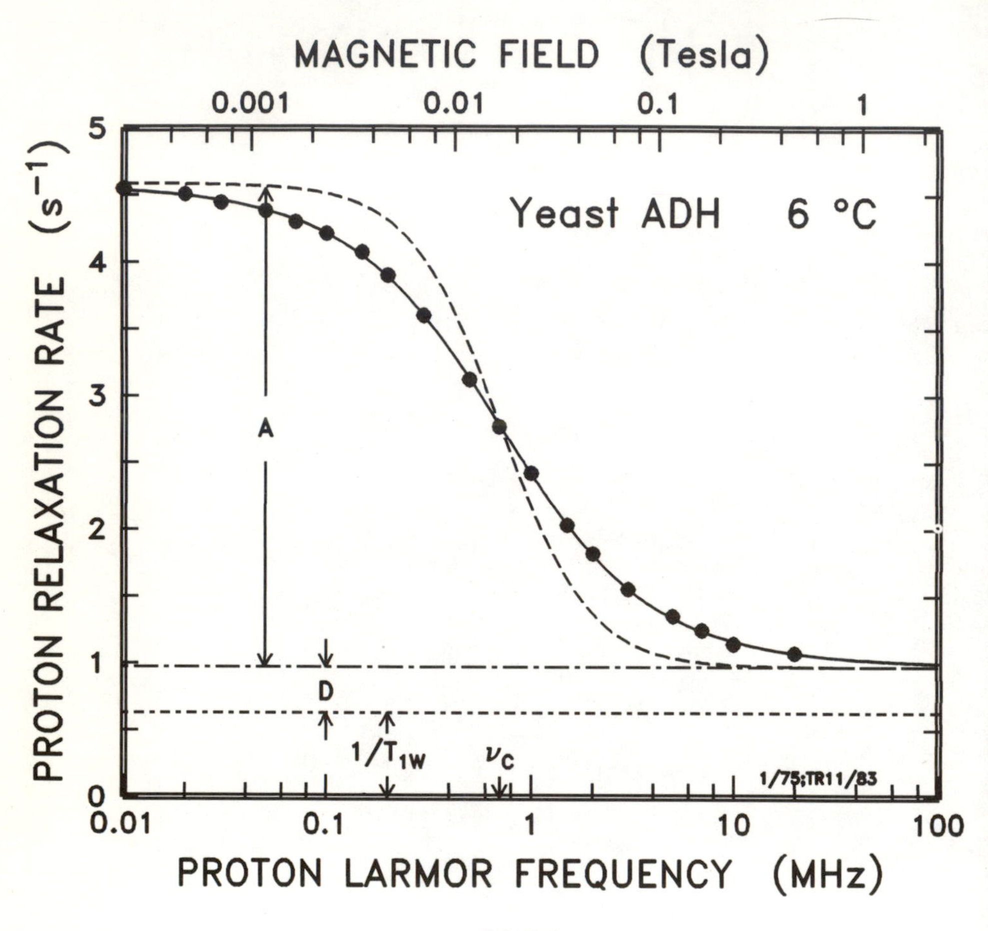

MAGNETIC FIELD (Tesla)
0.001
0.01
0.1
1
Yeast ADH 6 °C
PROTON RELAXATION RATE (s⁻¹)
5
4
3
2
1
0
A
D
1/T1W
νC
1/75;TR11/83
0.01
0.1
1
10
100
PROTON LARMOR FREQUENCY (MHz)

FIGURE 1

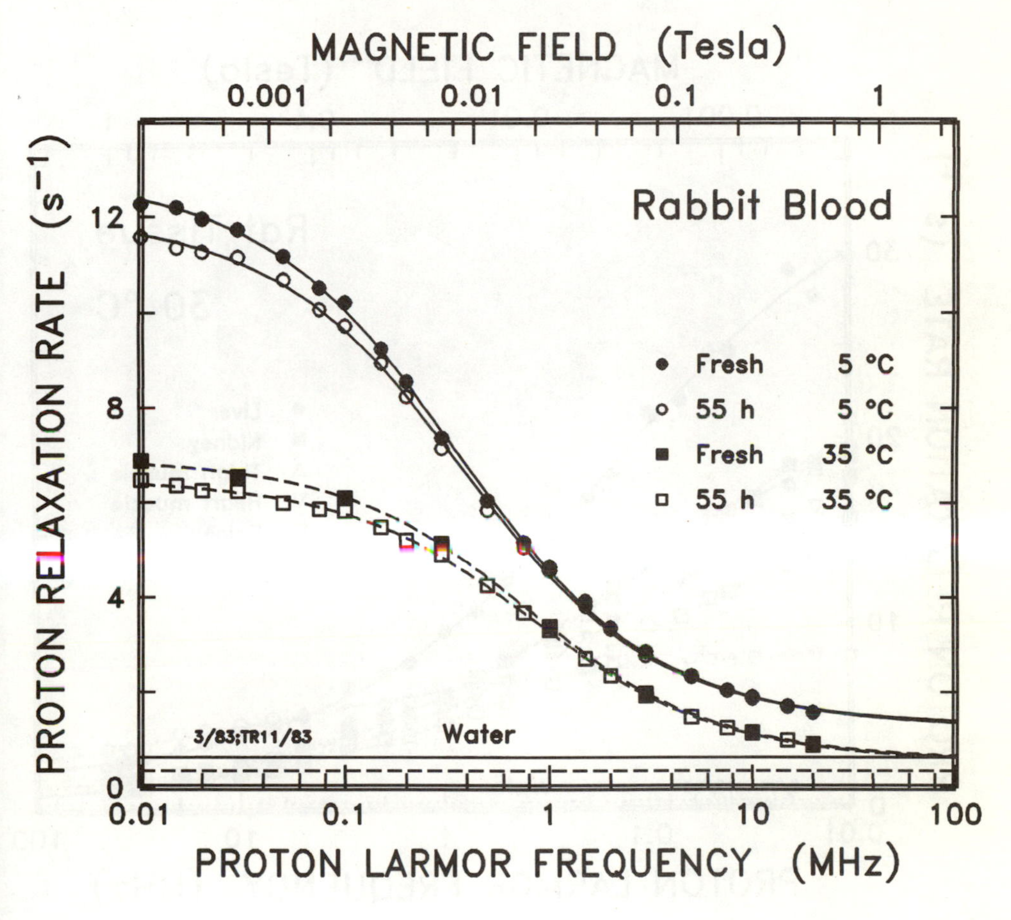
MAGNETIC FIELD (Tesla)
0.001
0.01
0.1
1
Rabbit Blood
PROTON RELAXATION RATE (s^{-1})
12
8
4
0
Fresh 5 °C
55 h 5 °C
Fresh 35 °C
55 h 35 °C
3/83;TR11/83
Water
0.01
0.1
1
10
100
PROTON LARMOR FREQUENCY (MHz)

FIGURE 2

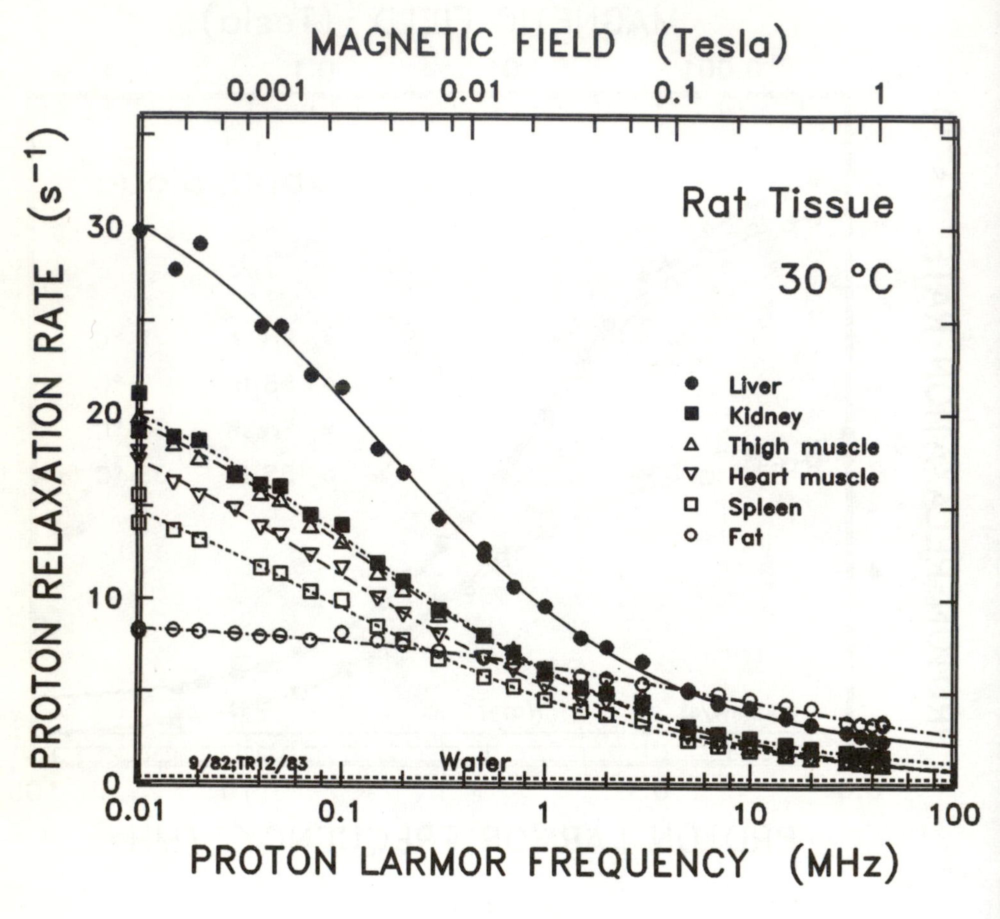
MAGNETIC FIELD (Tesla)
0.001
0.01
0.1
1
PROTON RELAXATION RATE (s⁻¹)
30
20
10
0
Rat Tissue
30 °C
Liver
Kidney
Thigh muscle
Heart muscle
Spleen
Fat
9/82;TR12/83
Water
0.01
0.1
1
10
100
PROTON LARMOR FREQUENCY (MHz)

FIGURE 3

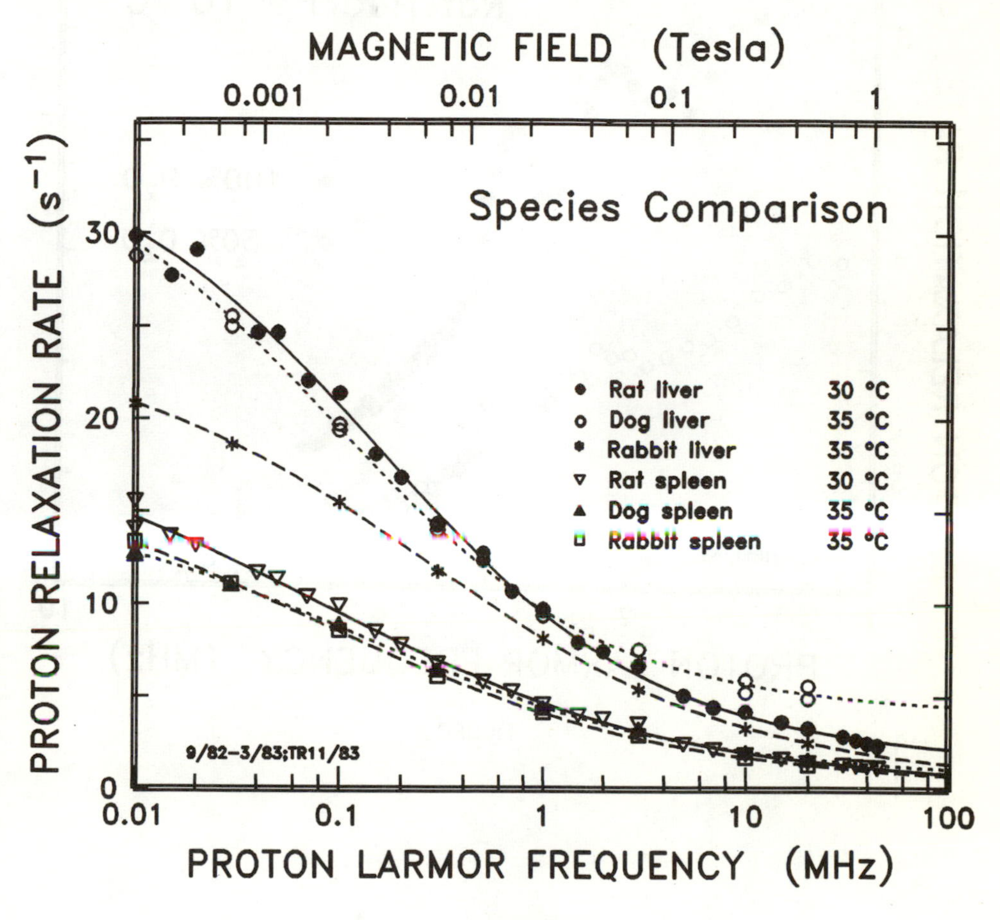
MAGNETIC FIELD (Tesla)
0.001
0.01
0.1
1
Species Comparison
PROTON RELAXATION RATE (s^{-1})
30
20
10
0
Rat liver 30 °C
Dog liver 35 °C
Rabbit liver 35 °C
Rat spleen 30 °C
Dog spleen 35 °C
Rabbit spleen 35 °C
9/82-3/83;TR11/83
0.01
0.1
1
10
100
PROTON LARMOR FREQUENCY (MHz)

FIGURE 4

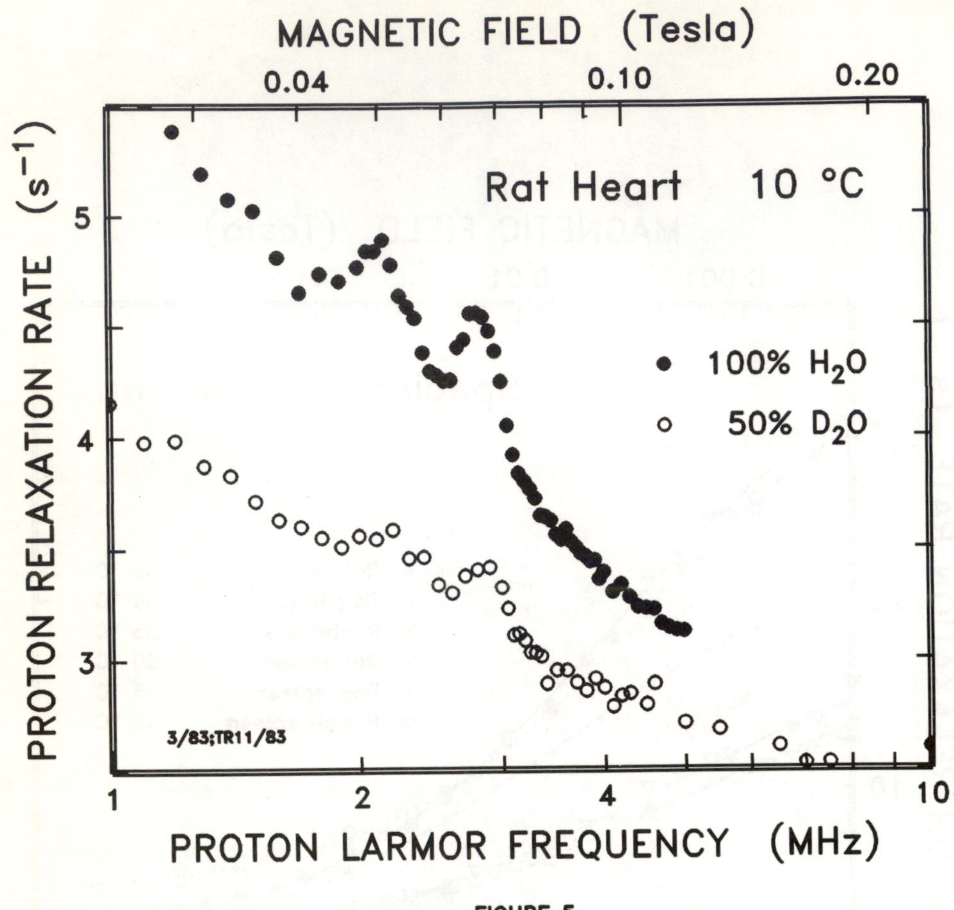

MAGNETIC FIELD (Tesla)
0.04
0.10
0.20
Rat Heart 10 °C
100% H_2O
50% D_2O
PROTON RELAXATION RATE (s^{-1})
5
4
3
3/83;TR11/83
1
2
4
10
PROTON LARMOR FREQUENCY (MHz)

FIGURE 5

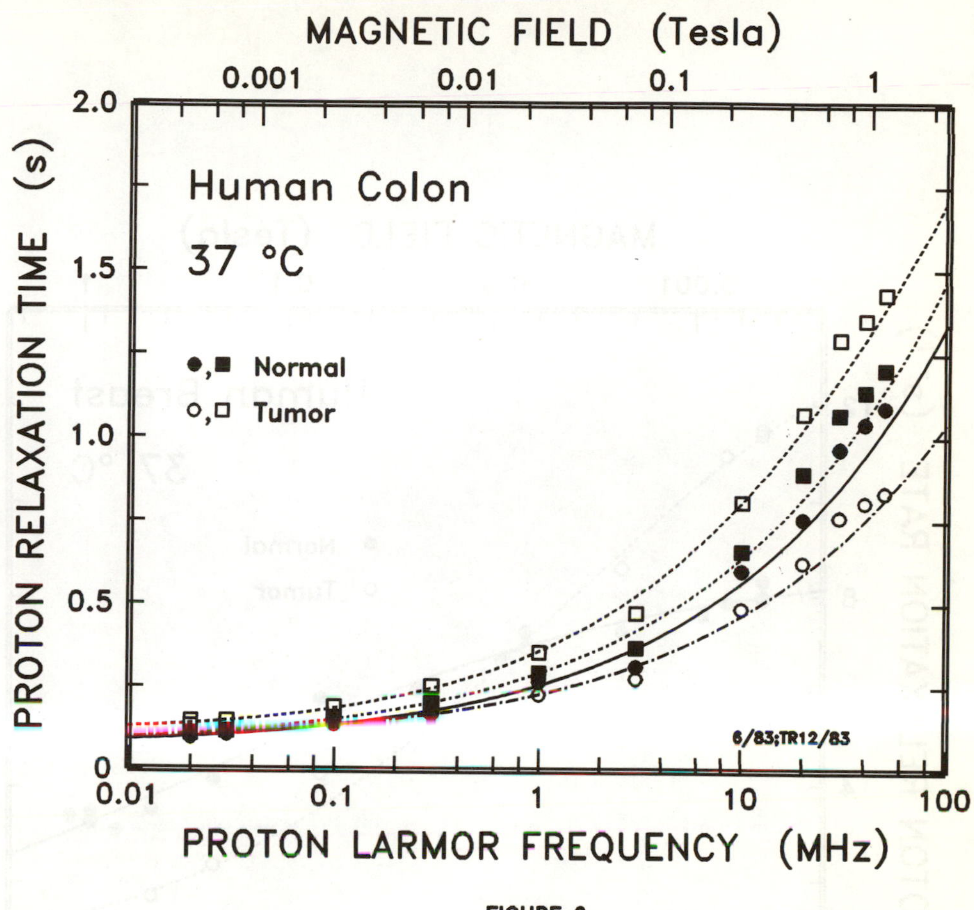
MAGNETIC FIELD (Tesla)
0.001
0.01
0.1
1
2.0
1.5
1.0
0.5
0
PROTON RELAXATION TIME (s)
Human Colon
37 °C
●,■ Normal
○,□ Tumor
6/83;TR12/83
0.01
0.1
1
10
100
PROTON LARMOR FREQUENCY (MHz)

FIGURE 6

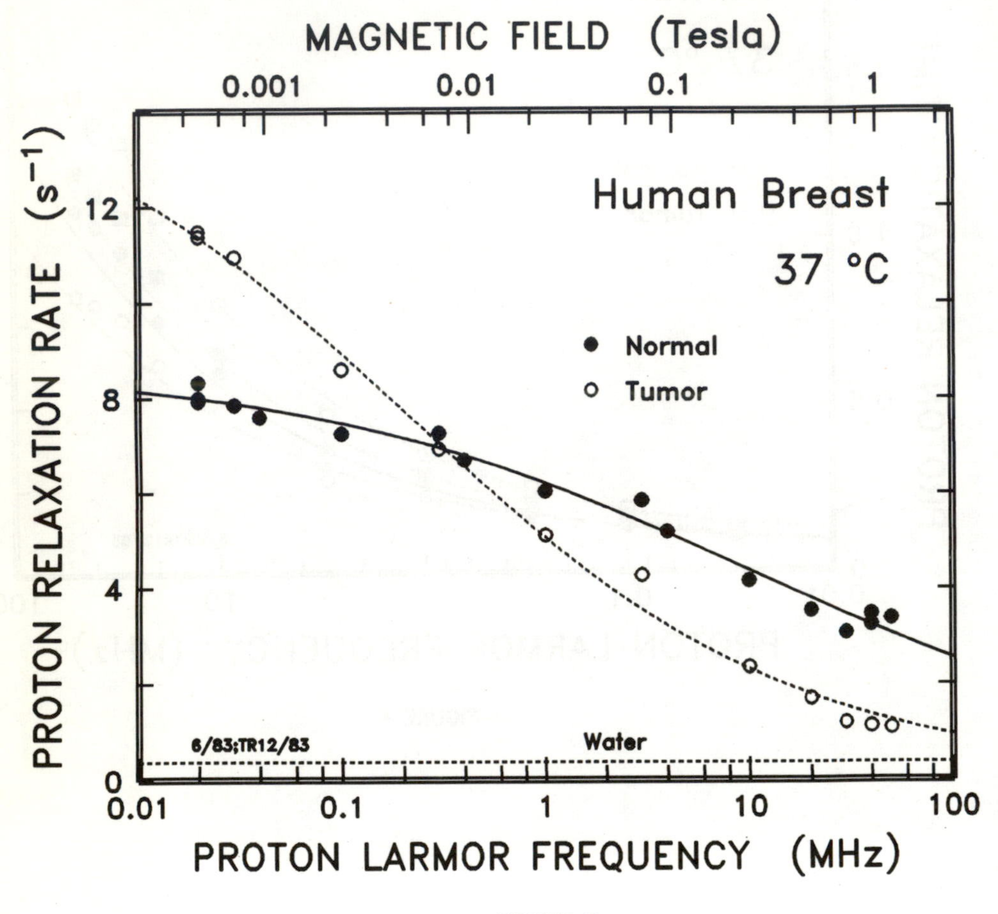
MAGNETIC FIELD (Tesla)
0.001
0.01
0.1
1
Human Breast
37 °C
Normal
Tumor
PROTON RELAXATION RATE (s^{-1})
12
8
4
0
6/83;TR12/83
Water
0.01
0.1
1
10
100
PROTON LARMOR FREQUENCY (MHz)

FIGURE 7

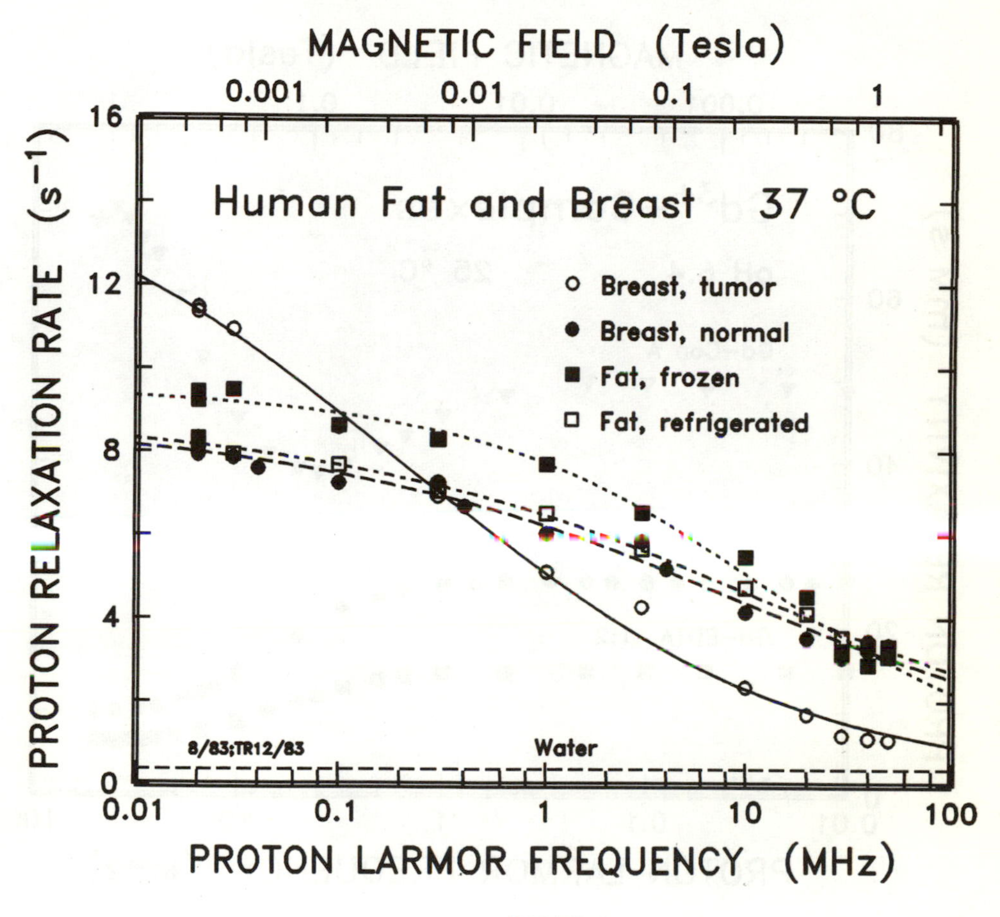
MAGNETIC FIELD (Tesla)
0.001
0.01
0.1
1
16
12
8
4
0
Human Fat and Breast 37 °C
Breast, tumor
Breast, normal
Fat, frozen
Fat, refrigerated
PROTON RELAXATION RATE (s^{-1})
8/83;TR12/83
Water
0.01
0.1
1
10
100
PROTON LARMOR FREQUENCY (MHz)

FIGURE 8

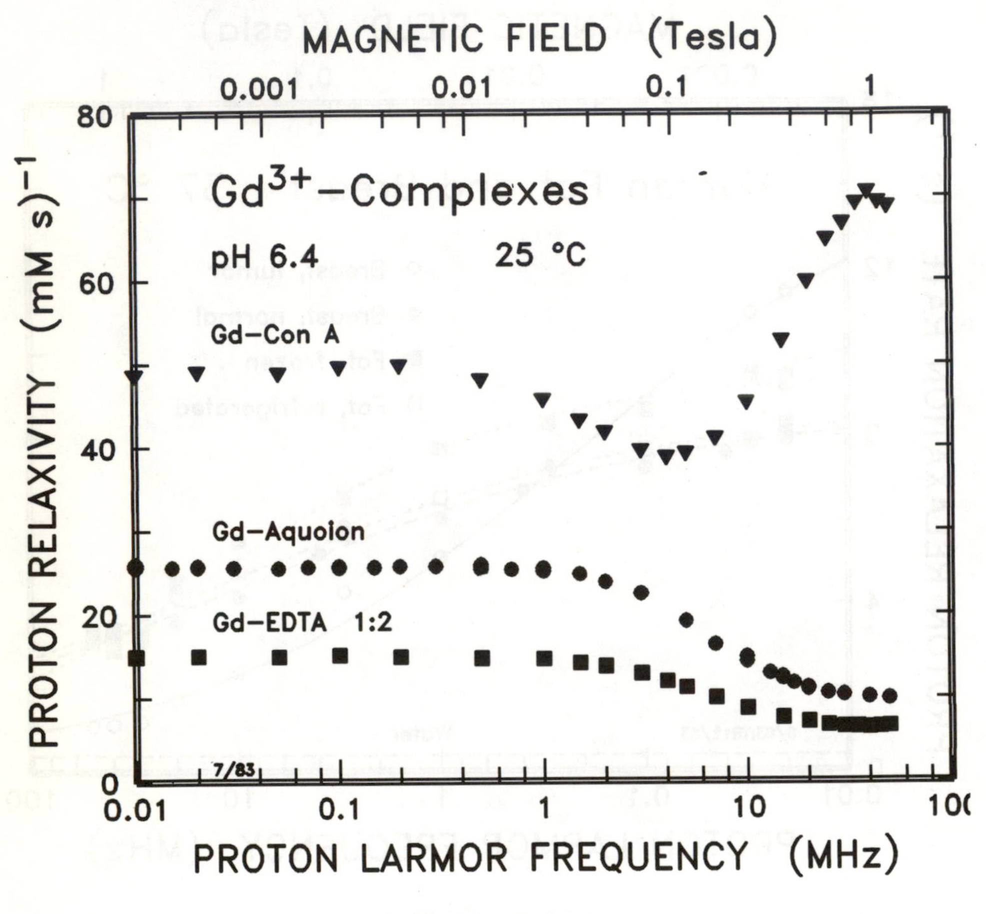
MAGNETIC FIELD (Tesla)
0.001
0.01
0.1
1
80
60
40
20
0
Gd3+-Complexes
pH 6.4
25 °C
Gd-Con A
Gd-Aquoion
Gd-EDTA 1:2
7/83
PROTON RELAXIVITY (mM s)-1
0.01
0.1
1
10
PROTON LARMOR FREQUENCY (MHz)

FIGURE 9

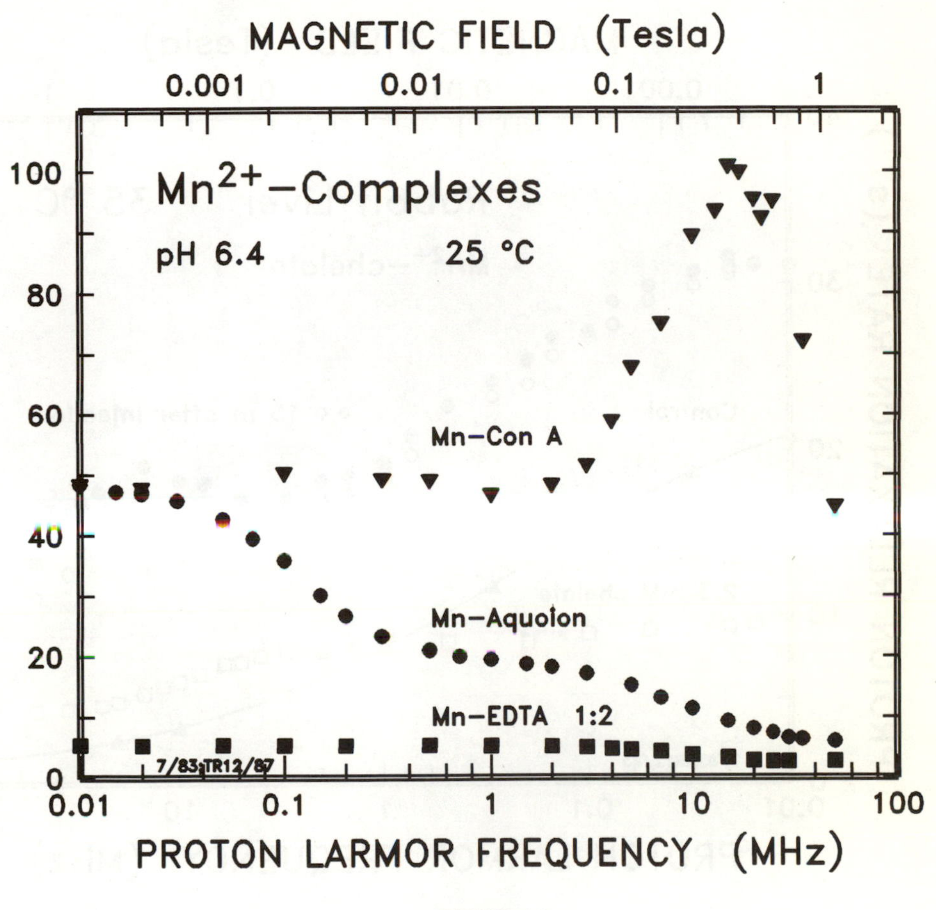
MAGNETIC FIELD (Tesla)
0.001
0.01
0.1
1
100
80
60
40
20
0
Mn^{2+}–Complexes
pH 6.4
25 °C
Mn–Con A
Mn–Aquoion
Mn–EDTA 1:2
7/83;TR12/87
0.01
0.1
1
10
100
PROTON LARMOR FREQUENCY (MHz)

FIGURE 10

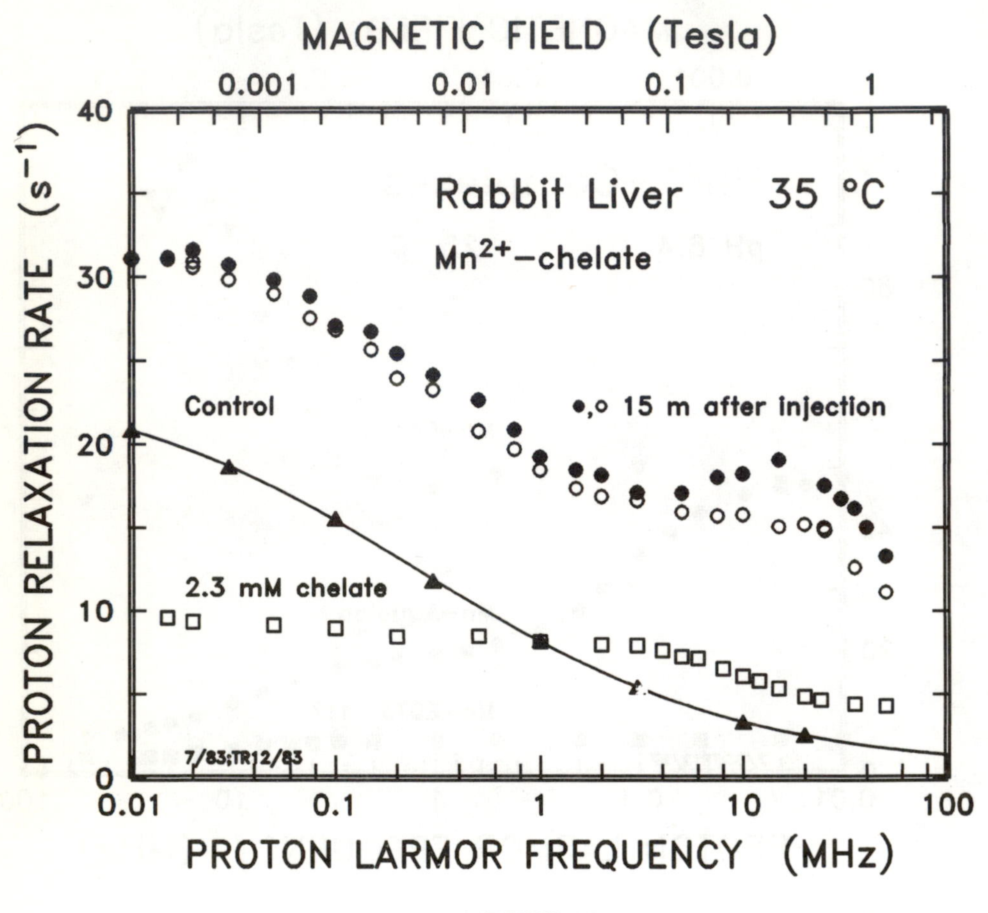
MAGNETIC FIELD (Tesla)
0.001
0.01
0.1
1
40
30
20
10
0
PROTON RELAXATION RATE (s^{-1})
Rabbit Liver 35 °C
Mn^{2+}–chelate
Control
●,○ 15 m after injection
2.3 mM chelate
7/83;TR12/83
0.01
0.1
1
10
100
PROTON LARMOR FREQUENCY (MHz)

FIGURE 11

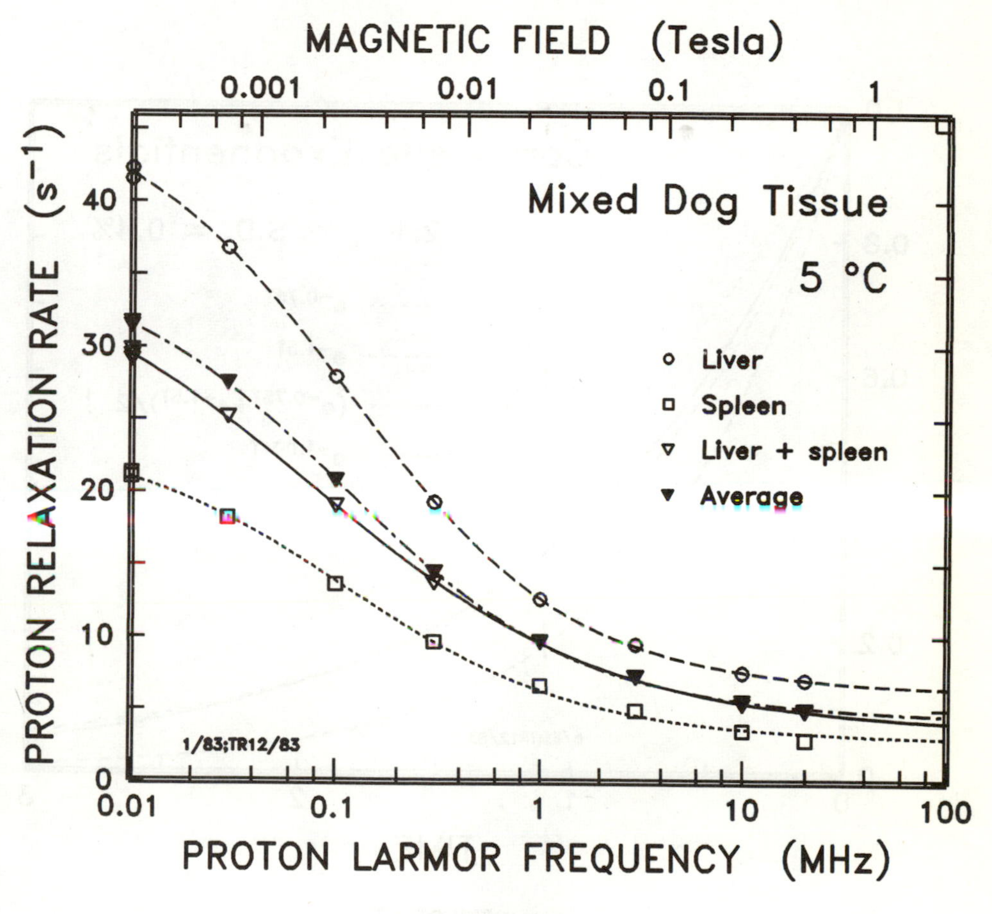
MAGNETIC FIELD (Tesla)
0.001
0.01
0.1
1
PROTON RELAXATION RATE (s^{-1})
40
30
20
10
0
Mixed Dog Tissue
5 °C
Liver
Spleen
Liver + spleen
Average
1/83;TR12/83
0.01
0.1
1
10
100
PROTON LARMOR FREQUENCY (MHz)

FIGURE 12

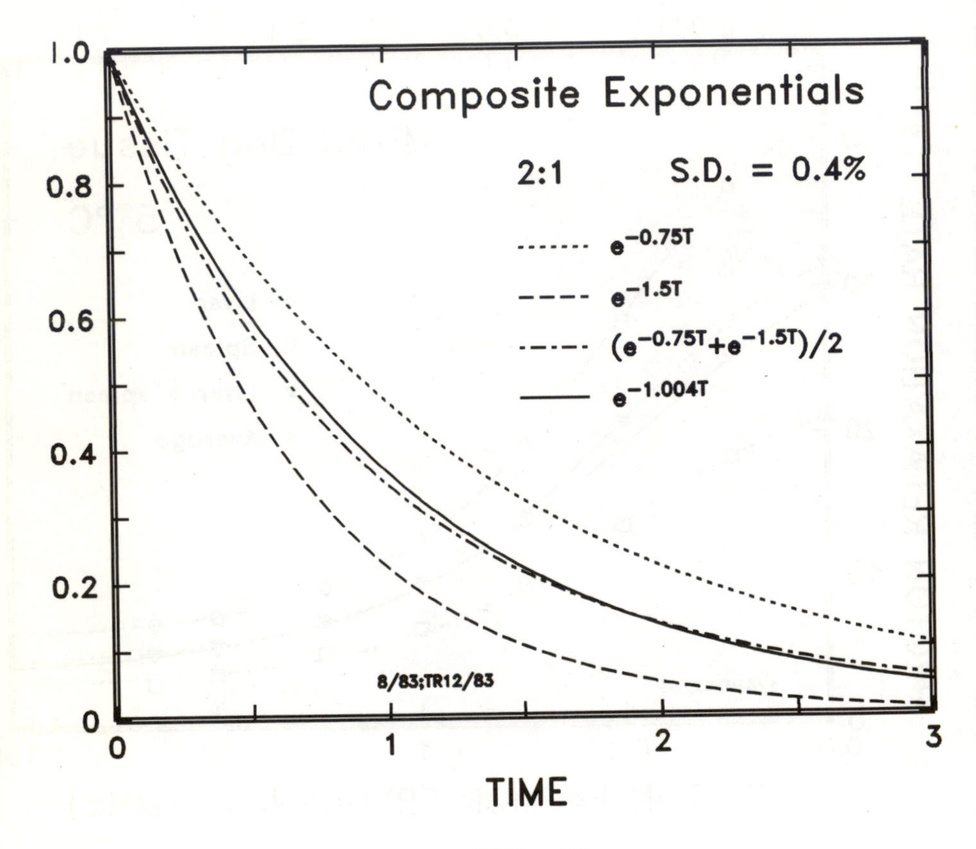

Composite Exponentials
2:1 S.D. = 0.4%
$e^{-0.75T}$
$e^{-1.5T}$
$(e^{-0.75T}+e^{-1.5T})/2$
$e^{-1.004T}$
8/83;TR12/83
1.0
0.8
0.6
0.4
0.2
0
0
1
2
3
TIME

FIGURE 13

PROCEEDINGS OF THE II INTERNATIONAL CONFERENCE ON APPLICATIONS OF PHYSICS TO MEDICINE AND BIOLOGY
edited by Ž. Bajzer, P. Baxa & C. Franconi

SINGLE PHOTON EMISSION COMPUTED TOMOGRAPHY (SPECT) AND POSITRON EMISSION TOMOGRAPHY (PET)

M.M. Ter-Pogossian
Division of Radiation Sciences
Washington University School of Medicine
St. Louis, Missouri 63110
USA

The important developments in nuclear medicine imaging included single photon emission computed tomography (SPECT) and positron emission tomography (PET). In both modalities, the distribution of a systemically administered radionuclide is reconstructed, as transverse tomographic images, from multiple measurements taken around the subject. The method of image reconstruction in SPECT and PET is that utilized in transmission computed tomography. In SPECT the image forming variable is the distribution of a gamma-emitting radionuclide; in PET it is the distribution of a positron-emitting radionuclide detected through the annihilation radiation generated when positrons undergo annihilation in matter while in CT this variable is the distribution of x-ray attenuation properties in the tissues imaged.

The methods of image reconstruction in SPECT and PET are identical, but there are differences between these two modalities. Most of the imaging devices utilized in SPECT are scintillation cameras of conventional design which have been fitted with a mechanical system designed to provide them with motions necessary to acquire information at different projections around the patient and with a suitable computer system. A few SPECT devices have been specifically designed for that purpose. The radiopharmaceuticals used in SPECT are, with few exceptions, those developed for conventional nuclear medicine. SPECT exhibits a practical advantage in nuclear medicine imaging. The instrumentation is directly adapted from the nuclear medicine armamentarium and so are the radiopharmaceuticals. The images obtained by SPECT have a much higher contrast resolution than those obtained conventionally. This higher contrast not only provides better images but it leads to the possible use of radiopharmaceuticals marginally suitable for conventional imaging. SPECT has also weaknesses. The conventional collimation of the gamma radiation yields a field of view which varies with depth, a condition unfavorable for tomographic reconstruction. The accurate correction for the attenuation of the gamma radiation in tissues is either difficult or practically impossible. Thus, SPECT measurements are seldom quantitatively accurate. The necessity of taking a number of measurements at different angles with adequate counting statistics make SPECT measurements long, typically of the order of 10 to 20 minutes. Finally, the physiological limitations of many of the radio-pharmaceuticals used in conventional nuclear medicine largely remain in SPECT studies.

The PET instrumentation is specifically designed for that purpose. Most modern PET devices include rings of scintillation detectors surrounding the patient. The detection of the annihilation radiation by coincidence yields sensitivity highly uniform isocount in depth and permits the accurate attenuation correction. Thus the PET images may be highly quantitative. The electronic collimation of the annihilation radiation provides a high sensitivity than SPECT systems for the patient's radiation exposure. This higher sensitivity can be translated in either higher spatial contrast or temporal resolution. PET exhibits the advantage that carbon-11, oxygen-15, nitrogen-13, and fluorine-18 provide excellent labels for a very large number of radio-pharmaceuticals potentially useful in physiological examinations. The main disadvantages of PET derive from the fact that the short-lived radionuclides most useful for PET studies require the use of a costly and complex cyclotron and the labeling of radiopharmaceuticals with the short-lived radionuclides calls for on site labeling capabilities which are considerably more extensive than in conventional nuclear medicine.

A general comparison of the merits SPECT and PET allows the following comments. SPECT provides, in a limited number of nuclear medicine studies, higher contrast images than obtained by conventional methods. However, the spatial resolution in SPECT, under favorable conditions about 1.5 cm FWHM, and the temporal resolution, approximately 20 min, are either comparable or poorer than in conventional nuclear medicine. In PET, the spatial contrast and temporal resolutions are vastly superior to those observed in conventional nuclear medicine. Indeed, a spatial resolution of better than a centimeter (about 8 mm) with a contrast resolution of a few percent and a temporal resolution of better than 10 seconds are achievable. The variety of radiopharmaceuticals useful in physiological studies have already permitted, with PET, the study of variables such as organ blood flow, organ metabolism, distribution of receptor sites, etc. which are impractical by any other method.

It appears at this time that while SPECT is going to remain a modestly useful adjunct to conventional nuclear medicine, while PET provides a far reaching tool in the understanding of normal physiology, of pathophysiology, and which has already provided new physiological information.

PROCEEDINGS OF THE II INTERNATIONAL CONFERENCE ON APPLICATIONS OF PHYSICS TO MEDICINE AND BIOLOGY
edited by Ž. Bajzer, P. Baxa & C. Franconi

FUNCTIONAL IMAGING IN NUCLEAR MEDICINE

M. Della Corte - M.R. Voegelin
Ist. di Medicina Nucleare e Fisica Medica
Università di Firenze - V.le Morgagni,85
FIRENZE - ITALY

ABSTRACT

Problems involved in the obtainement of quantitative data from dynamical series of γ - camera images are discussed. The meaning of parametric and functional images is considered and the most commonly used methods to obtain such images are reviewed.
Examples of the procedures are given.

In the development of Nuclear Medicine the two most important improvements have been the use of the labelling isotope 99mTc and the coupling of the γ - camera with a computer.

The lowering of the patient'dose and the consequent possibility of supplying higher activities of the tracer, from one side, and the possibility of a mathematical treatment of large quantities of data in very short times, from the other, have been fundamental achievements in the study of physiopathological functions "in vivo".

Such studies are carried out by taking a series of - camera images integrated in successive temporal intervals, namely a dynamical series.

---

*) This work was supported by the CNR Progr.Fin.Tecn.Biom.

The temporal distribution of a tracer in a structure such as an organ can be followed and diagnostic information obtained. A very important progress in this direction is the possibility of obtaining quantitative data in addition to morphological ones and, at the same time, the development of procedures for the improvement of image quality.

The major difficulties in obtaining quantitative data from dynamical series of γ - camera images are :

1) The identification of the regions of interest (ROI) for which the activity-time function must be established.
2) The choice of a sampling time interval in order to avoid temporal resolution loss and to preserve the statistical meaning of the sample.
3) The evaluation of the instrumental errors (field disuniformity, loss for dead time, etc.).
4) The evaluation of the background and of the overlapping effect in the ROI.

Furthermore, other more "conceptual" difficulties are present :

1) The choice of the mathematical model for a physio-pathological interpretation of the time functions.
2) The search of sub-regions with a different behaviour from the hypotised mathematical model.
3) The problems arising in the obtainement of concentration-time functions from counts-time functions.
4) The uncertainty in defining regions of low activity.

Some of these difficulties can be overcome by the introduction of parametric and functional images. Nowadays this is possible by the use of modern powerful calculations means.

Because the contents of each pixel of the image are a function of the spatial coordinates and time, the series of the images corresponds to a set of N x N signals. Each of these signals is a function of time that can be represented by a set of parameters. The spatial distribution of the parameter's values gives the parametric maps.

The knowledge of parametric maps allows us to identify spatial regions with the same temporal behaviour and the true functional images can be obtained from a model of the physiopathological fenomena.

In the process leading to functional images the logical procedure is totally reversed as compared to the standard method of extraction of data from dynamic series. In fact, while in the standard procedure the ROI are identified at first and afterwards the time relationship is established, in the functional images the temporal variation in each pixel is first identified followed by the delimitation of the spatial regions having the same temporal behaviour.

This procedure has been introduced by Mc Intyre et al. about 15 years ago, however, the amount of time needed for calculations has prevented its clinical application.

As already pointed out the content of each pixel in successive time intervals is obtained from the dynamical series of images.
Thus a temporal histogram is obtained :

$$C_{xy}(t) = C_{xy}(t_1), C_{xy}(t_2), \ldots\ldots, C_{xy}(t_M)$$

where x, y are the spatial coordinates of the pixel.

If we represent each temporal histogram by K parameters ($K \leq M$), under the hypothesis that all histograms can be represented by an equal number of K parameters, we get K matrices each representing the spatial distribution of a parameter, this is a parametric map.

In order to attribute a physiopathological meaning to

a parametric map the following conditions must be fulfilled even if with some degree of approximation.

1) The temporal histogram must represent the counts coming from a well defined element of volume. This condition is only approximately fulfilled taking into account the attenuation of radiation and the spatial dispersion caused by the revelation system.
2) The volume element corresponding to a temporal histogram must not change its position during the temporal series.

This second condition cannot be fulfilled in structures such as heart and lungs for which the use of a synchronizing device is necessary.

A further limitation comes from the high number of parameters required for highly variable histograms. In spite of these limitations parametric maps do offer some advantages:

1) Temporal series can be summarised in a few maps.
2) Informations on "local" functionality can be obtained.
3) Quantitative data about relative volumetric variations and distrectual kinetic patterns can be obtained.
4) The background noise can be accurately evaluated.

When collecting data with the purpose of obtaining parametric maps, it is preferable to avoid any data pre-treatment such as smoothing procedures. This because a correlation between neighbouring elements can be introduced.

Parametrical images can be obtained by the following three methods:

A) <u>Numerical methods</u>

These methods don't require too much calculation time and the clinical interpretation of images can be easily carried out. Parameters are chosen from the characteristical features of the histograms (maximum or minimum times and values, points or maximum derivative and so on).

Maps are very noisy and a pretreatment of data is unavoidable. Generally this is performed on each histogram by a mobile-mean smoothing procedure. A parabolic smoothing on L points is commonly used.

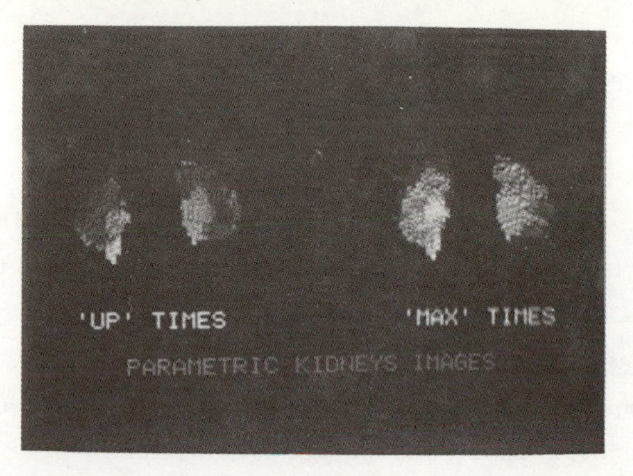

Fig.1 - Functional images obtained by numerical methods (kidneys)

B) Analytical methods

These methods need a previous knowledge of the mathematical model of the studied function in order to represent the temporal histogram by the hypothetical function of k parameters

$$C(t) = f(t, a_1, a_2, \dots\dots, a_k)$$

Each histogram is considered as a statistical sample from which values of the parameters can be estimated by a minimum squares procedure.

The most utilized functions are:

$A e^{-kt}$ (wash out) $\qquad A e^{-kt} t^{b}$ (transit)

$A e^{-kt} + B e^{-ht}$ (cumulation) $\qquad A_i e^{-k_i t}$ (Comp. Syst.)

$A \frac{(1-kt)}{ht} / t^{3/2}$ (transit) $\qquad a + bt$

Analytical methods do have some drawbacks which can be hardly overcome.

1) The hypotized mathematical model is not applicable to all regions of the image. Generally a segmentation procedure is needed with the consequential higher incidence of errors.
2) An histogram can be the result of two or more overlapping phenomena, if so, the choice of an analitical function becomes difficult.
3) Parameters are often correlated one another, this produces a loss of meaning.

The time required for calculations is remarkable, furthermore, the occupation of memory grows up with the square of the number of parameters.

C) <u>Factorial methods</u> [2][3][4]

Each temporal histogram is represented by the weighed-mean of a set of k basic independent functions

$$C(t) = p_1 f_1(t) + p_2 f_2(t) + \dots\dots\dots + p_K f_K(t)$$

The weights p_i can be obtained by the relation:

$$p_i = \sum c(t) f_i(t)$$

The set of parameters or factors p_i represent the histogram without error. These parameters contain informations about the shape of the histogram and the experimental noise.

The first problem to solve is the choice of the type and the number of basic functions. The solution depends on the shape of the histograms and on the noise/signal ratio.

A well known set of basic ortonormal functions is the Fourier transform

$$C(t) = \bar{C} + \sum a_i \,\mathrm{sen}(w_0 i) + \sum b_i \cos(w_0 i)$$

$$a_i = \sum c(i) \,\mathrm{sen}(w_0 i) \quad ; \quad b_i = \sum c(i) \cos(w_0 i)$$

$$A_i = \sqrt{a_i^2 + b_i^2} \qquad \phi_i = \mathrm{tang}^{-1} \frac{a_i}{b_i} .$$

In this case the parameters are the coefficients a_i and b_i or, alternatively, the magnitudes A_i and phases ϕ_i of the components.[5]

If the histogram has a rapid temporal variation, the number of parameters needed goes up with a consequent increase in the noise/signal ratio.

It is normally very difficult to attribute some physio-pathological meaning to these parameters, however, this is possible in the case of periodical phenomena like those encountered in cardiac (blood pool gated) and lung studies.

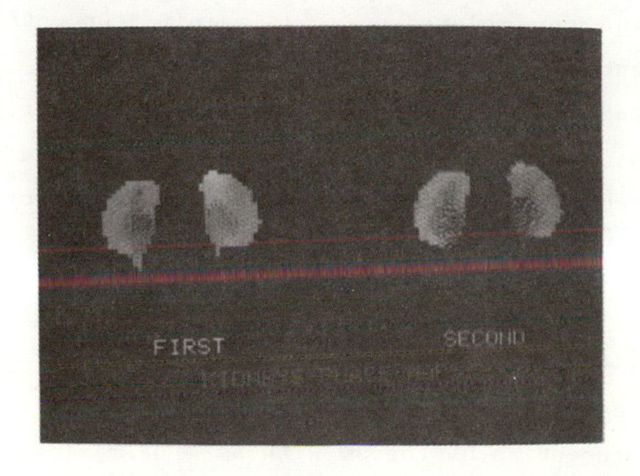

Fig. 2 - Phase maps of kidneys. First and second component (Fourier)

A second set of basic functions comes from the Hotelling transform (method of principal components). A number of N histograms of k points can be considered as a random sample of vectors in a k - dimensional space. The eingenvectors A_i corresponding to eingenvalues . of the variance - covariance matrix are linearly independent and can be taken as a set of k basic functions.

An histogram is then represented as

$$c_i - \bar{c}_i = \sum_{1}^{k} p_{ij} A_{ij} \quad ; \qquad C = AP$$

where $\bar{c}_i$ is the mean value of the histogram, A is the matrix of eingenvectors and P is the vector of the parameters. The vector P is given by

$$P = A^T C \ .$$

The transform $C = AP$ is a rotation of axis around the origin in such a way that the variance corresponding to each coordinate is equal to the corresponding eigenvalue.

If we put the eingenvalues in a decreasing order, the variance is "concentrated" in the first components. The variance due to uncorrelated noise is mainly "concentrated" in the farthest components.

We can then use only the first m components to represent the histogram, negletting the last $k-m$ ones. This gives a substantial reduction of the noise with a little error in the representation.

It may be shown that in this case the error is given by

$$\varepsilon^2 = \sum_{m+1}^{k} \lambda_i \ .$$

In most cases no more than 5 or 6 components are required to represent an histogram with a 5% of maximum error, which is the error made in representing each component with his mean value

As an example of application of this method, the procedure to obtain functional images of the cardiac region is shown step by step.[8] This procedure with some modifications can be applied to several other dynamic series (renal, epatic, cerebral, etc.).[9]

The γ -camera was a Selo KR6 on line with a computer HP 21 MX with 128 K words and a disk storage of 15M Byte.

Each subject, premedicated 20 min. earlier with 25 mg i. v. of PYP and 100 ml of a saturated solution of $KClO_4$ per os. was injected i. v. with a bolus of 0.3 mCi/kg of 99mTc ($Na^+TcO_4^-$). A first pass analysis was normally carried out. A labelling of red cells "in vivo" took place in the subject.[1]

Collection of data began 30 min. after the injection of the tracer. Images were taken in LAO 45° position. Acquisition was made in list - mode during 3 min. together with a trigger signal (R wave of ECG).

Data from 60 subjects distributed in equal number between normal, infractions and anginous have been collected.

In each case the distribution of the heart cycle periods T has been obtained. Data with periods differing more than 5% from the mode of the distribution T_o were rejected.

The data have been summed up in phase for intervals of 10 ms. A total number of $\frac{T_o}{0.01}$ 64 x 64 images was obtained for the cardiac cycle of each subject.

A number of 32/cycle-images were generated from the list mode data.

The baricenter of the image resulting from all the 32 images was determined and a mask of 32 x 32 pixel around this point was obtained. This region contained the whole heart during the cycle.

64 temporal histograms each of 32 points were calculated. Each histogram corresponded to the temporal series of the content of 4x4 pixels.

Each histogram was normalized to unit area.

The mean histogram for all the subjects was then determined

Calculation of the histograms $C_i - \overline{C}_i$ and construction of the variance - covariance matrix was carried out.

Determination of the eingenvalues λ_i and eingenvectors A_i of such matrix was performed. (Fig. 3 and table 1).

The pool of histograms could be divided in three sets corresponding to the two functional regions, atria and ventricules, and to the background. The shape of the histograms belonging to each region must be different. This suggests that the parameters representing the histograms have a different distribution. If it is so an identification of the functional regions can be obtained from parametric maps.

A mathematical model of the cardiac function is not available, thus the shape of the histograms corresponding to atrial and ventricular regions was obtained from the selection of a number of histograms from the region images.

Regional histograms were very different. They were characterized by the values

$$E\% \doteq \frac{max - min}{max} \cdot 100 \text{ , } T_{max} \text{ and } T_{min}$$

Each histogram belonging to a subject was classified as

Background	if	$E\% \leq 5\%$	
Atria	if	$E\% \geq 7\%$	$10 \leq T_{max} \leq 18$
			$T_{min} \leq 10$ or $T_{min} \geq 22$
Ventricles	if	$E\% \geq 10\%$	$7 \leq T_{min} \leq 18$
			$T_{max} \leq 7$ or $T_{max} \geq 18$

About 25% of all the histograms did not fulfil any of the previous criteria, these were taken out from the analysis.

Parameters have been calculated for the histograms

TABLE I

EIGENVALUES (x 10^4)

(in decreasing order)

5.627	1.060	0.305	0.281
0.190	0.146	0.115	0.100
0.080	0.068	0.056	0.047
0.040	0.030	0.023	0.016
0.012	0.009	0.006	0.004
0.003	0.002	0.001	0.000
0.000	0.000	0.000	0.000
0.000	0.000	0.000	0.000

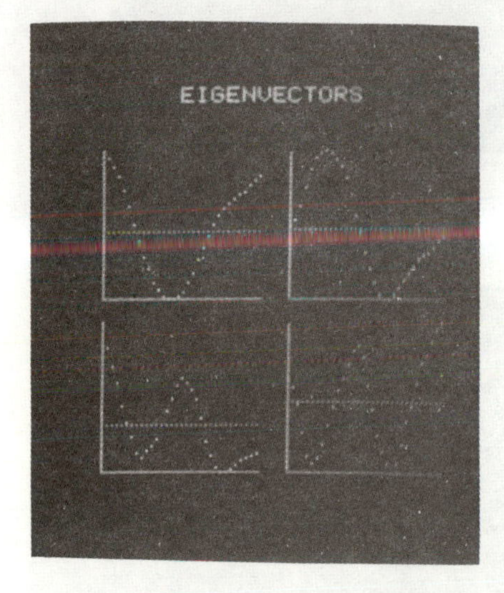

Fig. 3 - Eigenvectors corresponding to eigenvalues = 5.627 , 1.060 , 0.305 , 0.281

of each class together with mean values and variance.

The distribution of the first parameter is shown in fig. 4.

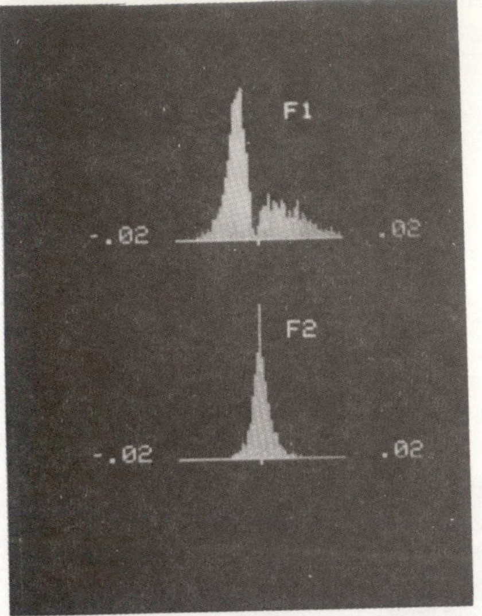

Fig.4 - Distribution of the first and second parameters. Left atria, right ventricles.

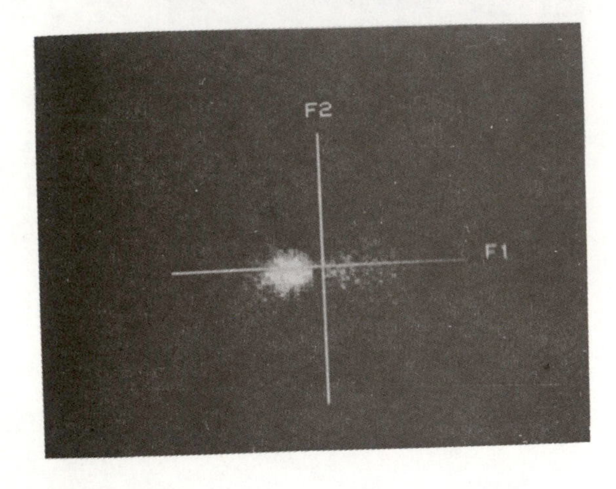

Fig. 4bis - Distribution of the second parameter as a function of the first.

As we can see its value is sufficient to discriminate between the three populations. Thresholds values for the separation between populations were calculated

Threshold background - atria - 0.0045

Threshold background - ventricules 0.0012

Following the observation that in all cases:

$$\varepsilon^2\% = \sum_{6}^{32} \lambda_i \Big/ \sum_{1}^{32} \lambda_i \leq 5\%$$

The number of factors used in the representation of histograms was 5. However, when considering each separate case, the number of factors to minimize the noise/signal ratio can be different.

When the number of parameters is too low, the error given by

$$\varepsilon^2 = \sum_{m+1}^{K} \lambda_i$$

may be too high. On the other hand, in the presence of a high noise, even a too high number of parameters may increase the noise/signal ratio.

Following a simulation procedure, an empirical formula giving the number of factors necessary to minimize the noise/signal ratio has been obtained:

$$NF = 0.48\ 10^{-4}\ p_1^{1.03}\ CP^{1 - 0.3 \log p_1}$$

where p_1 is the value of the first parameter and CP the total number of counts in the histogram.

The procedure described above, carried out on the pool of 60 subjects has given the following data:

1) The set of eigenvectors.
2) The criteria for the attribution of each histogram to atria, ventricles or background.
3) The criteria for the choice of the number of parameters necessary to the representation of the histograms.

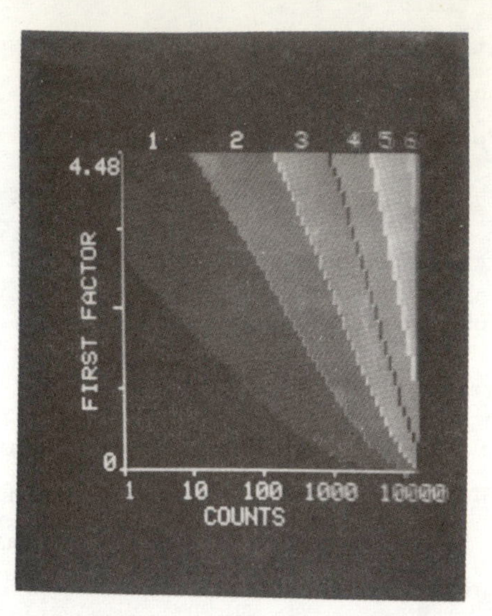

Fig. 5 - Optimal number of factors for a given value of the first parameter and the total number of counts.

Although this procedure is cumbersome requiring a long calculation time, (about 15 minutes) it is done once for ever.

The dynamical series belonging to a new subject is then treated as follows:

1) As previously described, a set of 32 x 32 histograms is calculated one for each pixel of the mask.
2) Using the set of eigenvectors previously obtained, the first 5 parameters are calculated.
3) Maps of the 5 parameters and of the total counts are generated.
4) From the value of the first parameter and of the total number of counts in each histogram the parameter's number necessary for the reconstruction is obtained.
5) Histograms are reconstructed with a number of parameters as in 4).

6) Separation of the atrial and ventricular histograms is carried out and corresponding maps are obtained.

7) With a method of evaluation of the minimum (identification of the septus) the left and right ventricles are identified.

8) Curves of the total counts for each ventricle and atria are obtained.

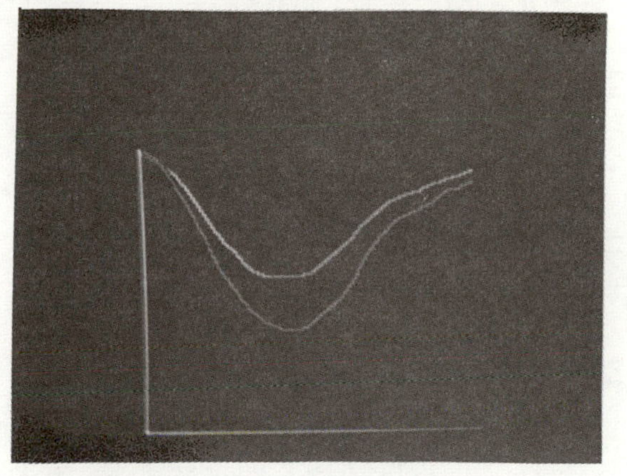

Fig. 6 - Curves of the total counts for left and right ventricles.

9) Constants of relevant physiopathological interest on each histogram (times and values of the minimum, times of maximum derivative, etc.) and their mean values on each region are obtained.

10) Functional maps are generated and data of interest are printed.

The time required for this procedure does not exceed two minutes.

This method gives us the possibility of further achievements: the possibility of correcting for the heart movement during the cycle. The possibility by using the information contained in other parameters not takent into account in the procedure described above to discriminate normals from pathological subjects.

VENTRICULAR FUNCTIONAL DATA

MEAN VALUES FROM HISTOGRAMS

	Dx	Sx	
End-systolic time	0.36 ± 0.12	0.40 ± 0.07	sec
Ejection Fraction	49 ± 31	62 ± 23	%
Time to peak ejection rate	0.13 ± 0.04	0.16 ± 0.03	sec
Peak ejection rate	2.3 ± 1.4	2.7 ± 1.2	sec^{-1}
Time to peak filling rate	0.58 ± 0.04	0.59 ± 0.04	sec
Peak filling rate	1.6 ± 1.0	1.90 ± 0.8	sec^{-1}

VALUES FROM THE MEAN HISTOGRAM CALCULATED IN THE REGION

	Dx	Sx	
End-systolic time	0.40	0.41	sec
Ejection Fraction	47	65	%
Time to peak ejection rate	0.16	0.18	sec
Peak ejection rate	2.3	3.6	sec^{-1}
Time to peak filling rate	0.59	0.61	sec
Peak filling rate	1.7	2.7	sec^{-1}

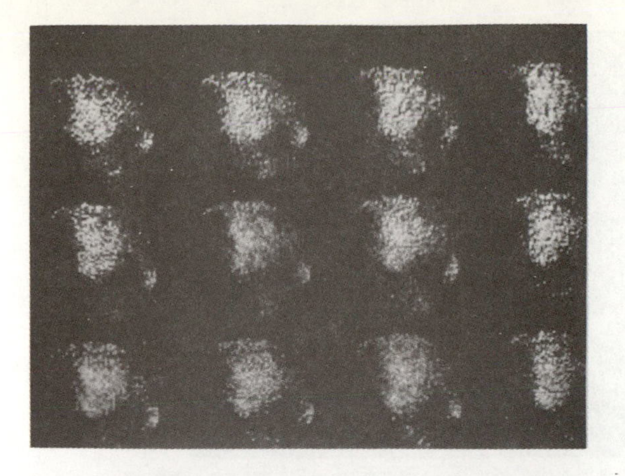

Fig. 7a - Original images (dynamical series)

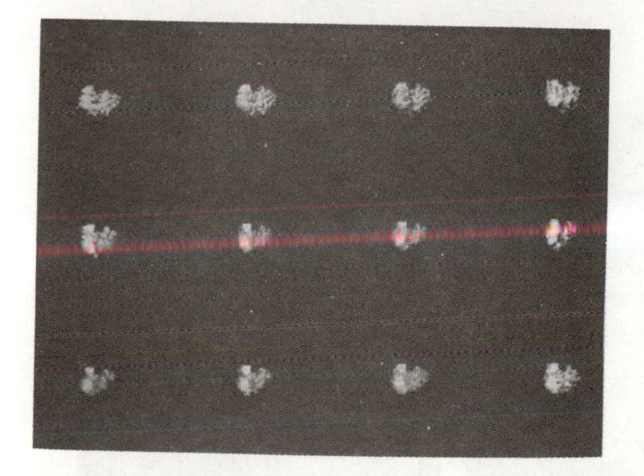

Fig. 7b - Reconstructed ventricular images

A research project aimed to obtain an increase in diagnostical information from the identification of sub-classes of histograms related to the parameters is now in progress in our laboratory.

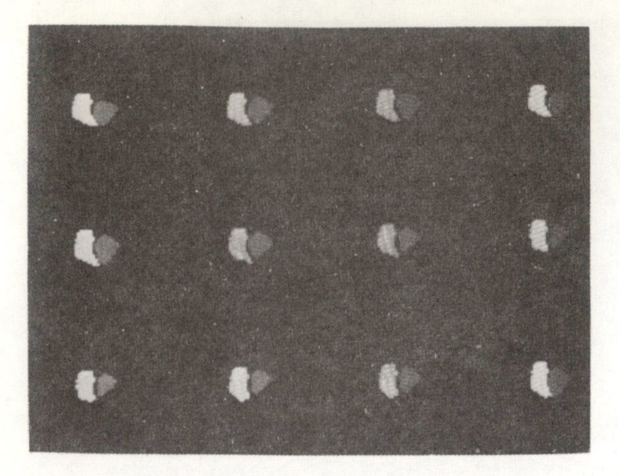

Fig. 7c) - Ventricular masks

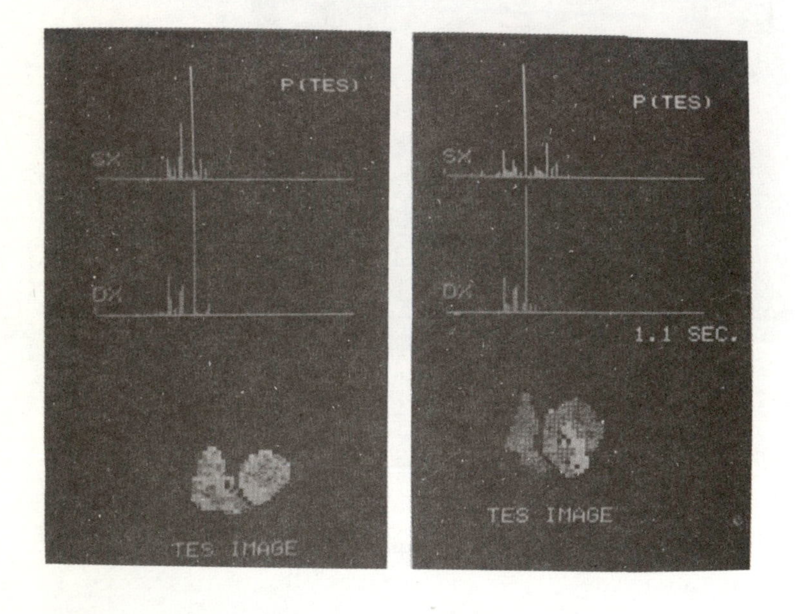

a) b)

Fig.8 - Functional images of TES (End - Systolic Time) and its distribution in left and right ventricles. a) Normal subject, b) Infarction of left ventricles.

<u>BIBLIOGRAPHY</u>

1) Pavel D.G., Zimmer A.M., Peterson V.M. - In vivo labelling of red blood cells with 99mTc: a new approach to blood pool visualisation.
J. Nuclear Med. 18 - 305 - 1977

2) Oppenhim B.C., Appledorn C.R. - Functional renal imaging through factor analysis.
J. Nucl. Med. 22 - 417 - 1981

3) Schmidlin P. - Quantitative evaluation and imaging of functions using pattern recognition methods.
Phys. Med. Biol. 24 - 385 - 1979

4) Bazin J.P., Di Paola R. - Advances in Factor Analysis application in Dynamic function studies.
Proc. III World Conf. Nucl. Med. Biol. - Pergamon Press 1981 - 82

5) Malda H., Takeda K., Nakagawa N., Toguchi M., Konisci T., Hamada M. - Investigation of new cardiac functional imaging using Fourier Analysis of Gated Blood Pool study.
Proc.III World Conf.Nucl.Med.Biol.- Pergamon Press 1981-82

6) Herman H.H. - Modern Factor Analysis.
Univ. of Chicago Press - Chicago 1970

7) Bartoli c., Cappellini V., Voegelin M.R. - A digital technique for processing dynamic heart scintigraphic images by using Hotelling Transform.
Proc.Int.Workshop on Time varing process - Florence 1981-73

8) Voegelin M.R., Pupi A. - Automatic filtering and histogram reconstruction of renal sequences with Hotelling Transform.
J.Nucl.Med. 24 - P.94 - 1983

9) Voegelin M.R., Bisi G., Bartoli C. - Proceso automatico de angiografia en equilibrio utilizando analisi factorial.
Rev. Espan. de Cardiol. 36 - Supp. 1 - 1 - 1983

PROCEEDINGS OF THE II INTERNATIONAL CONFERENCE ON APPLICATIONS OF PHYSICS TO MEDICINE AND BIOLOGY
edited by Ž. Bajzer, P. Baxa & C. Franconi

NUCLEAR ANALYTICAL TECHNIQUES IN MEDICINE

Roberto Cesareo
Centro per l'Ingegneria Biomedica, Università di Roma "La Sapienza"
Corso Vittorio Emanuele II,244, 00186 Roma
ITALY

ABSTRACT

There are different ways for deducing analytical information about a sample from the interaction of low-energy radiation with it. In this paper the following nuclear analytical techniques are described,developed in the "Centro per l'Ingegneria Biomedica" of the University of Rome "La Sapienza":

Transmission of radiation
Scattering of radiation
X-Ray Fluorescence

Transmission measurements are very useful for resolving "in vitro" or "in vivo" "two components" problems.Typical MDL values are 100 ppm for an element in solution.
Scattering of radiation,in particular the evaluation of the elastic to inelastic scattered peaks ratio,can be very useful for "in vivo" analysis. A typical MDL value is 0.5 mg of element in low atomic number matrix.
Energy dispersive X-Ray Fluorescence is very suited for "in vitro" analysis of trace elements in biological samples. MDL vary between about 1 ng to 1 μg, for elements ranging from Z = 20 to Z = 92.

1. Introduction and theoretical background

When monoenergetic X - or gamma rays, of energy between about 10 keV and 150 keV interact with matter, different effects occurr,which contribute to the absorption,scattering or transmission of the incident radiation. These effects are: a) photoelectric effect,which gives rise to the absorption of an incident photon and to the contemponaneous emission of X-rays characterising the chemical elements constituting the sample; b) Compton effect,which gives rise to the scattering of a photon reduced in energy (inelastic scattering); c) Rayleigh effect,which gives rise to a scattered photon of the same energy (elastic scattering).

Therefore the irradiated sample not only absorbs the incident radiation,it also scatters the radiation both at the original energy and at reduced energy. The spectrum which can be obtained by collecting the radiation emited by the sample consists therefore not only of the characteristics lines of the elements we are interested in, but also of characteristic lines and background from the source of exciting radiation.

Consequently, when a sample is irradiated by a flux of N_0 photons of energy E_0 , the following can occur (Figure 1):

1a.Transmission of radiation [1)]

Part of the incident radiation can be transmitted by the sample and detected. The ratio of incident to transmitted photons, N_0/N, is given by the Equation:

$$\ln (N_0/N) = \frac{\mu(E_0)\,\delta\, x}{\delta} \qquad (1)$$

where:

$\mu(E_0)$ ($cm^2 \cdot g^{-1}$) is the total mass attenuation coefficient of the specimen at the energy E_0;

δ ($g \cdot cm^{-3}$) is the physical density of the specimen:

x (cm) is the thickness of the specimen.

The total mass attenuation coefficient of a material containing a mixture of elements can be approximately evaluated from the coefficients for the constituents elements according to the weighted sum:

$$\frac{\mu(E_0)}{\delta} = \sum_{1\,i}^{n} c_i \frac{\mu_i(E_0)}{\delta_i} \qquad (2)$$

where c_i is the concentration by weight of the i-th constituent.

The mixture rule is thought to be valid for $E_0 < 10$ keV and more than 1 keV from a photoelectric absorption edge.

Transmission measurements of N_0 and N at various energies E_1, E_2..... allows, through Eq.(1) to determine the corresponding attenuation coefficients μ_1, μ_2 .... and therefore,by employing Eq.(2) to determine the concentration values c_1, c_2....of the elements constituting the sample,if previously known.

It can be observed that,in order to determine the sample composition (i.e. the n terms c_1, c_2 ....c_n) , at least (n-1) transmission measurements are required. In fact, an additional condition is obviously given by:

$$c_1 + c_2 + \ldots\ldots c_n = 1$$

It is important to note, following Eq.(1), that the energy E_o of incident radiation should be modulated as a function of the thickness of the sample and its approximate composition, in order to have an optimal transmission. A general approximate rule which connects the attenuation coefficient of the sample at energy E and the optimal thickness is:

$$x_{opt.} = \frac{2}{\mu} \quad g \cdot cm^{-2} \qquad (3)$$

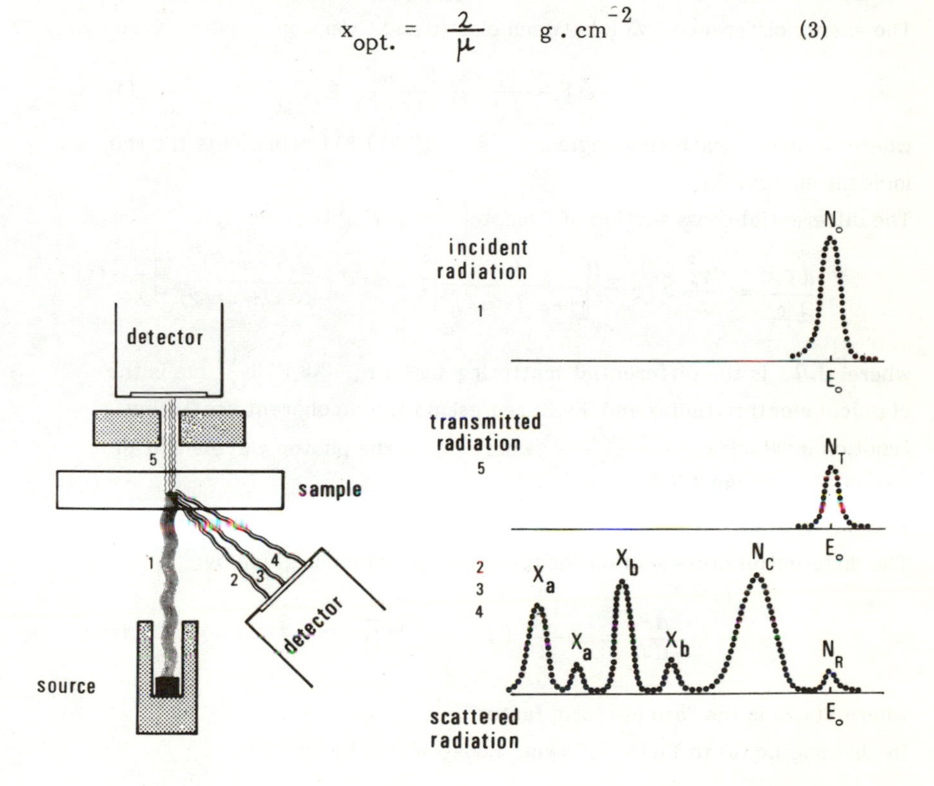

Fig.1 - Transmission, scattering and X-Ray Fluorescence of low-energy monoenergetic X - or gamma rays.

1b.Scattering of radiation [2]

As pointed out above incident radiation can be scattered by the sample in all directions elastically (mainly by Rayleigh effect) or inelastically (by Compton effect). This last effect will produce in the diffused X-ray spectrum a peak of energy E_C, lower than the energy E_0 of the elastic peak.
The energy difference ΔE between elastic and Compton peaks is given by:

$$\Delta E = \frac{(1-\cos\vartheta)\,\alpha}{1 + (1-\cos\vartheta)}\, E_0 \qquad (4)$$

where ϑ is the scattering angle and $\alpha = E_0(\mathrm{keV})/511$ represents the reduced incident energy.
The differential cross section of Compton scattering is given by:

$$\frac{d\sigma_c}{d\Omega} = \frac{r_0^2}{2} S(y,Z)\left\{\frac{1}{[1+\alpha(1-\cos\vartheta)]^2}\left[1+\cos^2\vartheta + \frac{\alpha^2(1-\cos\vartheta)^2}{1+\alpha(1-\cos\vartheta)}\right]\right\} \qquad (5)$$

where $d\Omega$ is the differential scattering angle, $r_0 = 2.8 \cdot 10^{-13}$ cm is the classical electron radius and S(y,Z) represents the incoherent scattering function in which $y = \frac{\mathrm{sen}\,\vartheta/2}{\lambda}$,where λ is the photon wavelenght in angströms = 12398/E(eV).
The Rayleigh effect gives rise in the diffused spectrum to a peak of energy E_0.
The differential cross section for the Rayleigh effect is given by:

$$\frac{d\sigma_R}{d\Omega} = \frac{r_0}{2}\,(1+\cos^2\vartheta)\,[F(y,Z)]^2 \qquad (6)$$

where F(y,Z) is the "atomic form factor".
By dividing Eq.(6) to Eq.(5) following Equation can be written:

$$\frac{d\sigma_R}{d\sigma_c} = \frac{(1+\cos^2\vartheta)[1+\alpha(1-\cos\vartheta)]^2}{\left[1+\cos^2\vartheta + \frac{\alpha^2(1-\cos\vartheta)^2}{1+\alpha(1-\cos\vartheta)}\right]}\;\frac{[F(y,Z)]^2}{S(y,Z)}. \qquad (7)$$

When $\alpha << 1$, Eq.(7) can be written in the form:

$$\frac{d\sigma_R}{d\sigma_C} \simeq \left[1 + \alpha(1-\cos\vartheta)\right]^2 \frac{\left[F(y,Z)\right]^2}{S(y,Z)} \qquad (7')$$

For backscattered radiation, Eq.(7) and (7') can be written in the form:

$$\frac{d\sigma_R}{d\sigma_C} = \frac{(1+2\alpha)^3}{(1+2\alpha)+2\alpha^2} \frac{\left[F(y,Z)\right]^2}{S(y,Z)} \qquad (8)$$

and

$$\frac{d\sigma_R}{d\sigma_C} \simeq \frac{\left[F(y,Z)\right]^2}{S(y,Z)} \qquad (9)$$

respectively.

When calculations are carried out, taking into account the factors F(y,Z) and S(y,Z) the following can be deduced:

- the ratio N_R/N_C between elastically and inelastically scattered photons (proportional to the ratios given in Eqs. (7) and (7') can be approximated, at a given angle and incident energy, by:

$$N_R/N_C = k\,Z^n \qquad (10)$$

where k is a coefficient of proportionality, Z is the atomic number of the element constituting the sample or the mean atomic number of the sample and n a factor which depends on Z, E_o and ϑ.

For $\vartheta = 180°$, $E_o = 60$ keV, the ratio N_R/N_C versus Z is shown in Figure 2.

In Figure 3 is also shown the ratio N_R/N_C versus Z, for various angles and for E_o = 30, 60 and 120 keV.

The elastic scattered counts decrease rapidly as the scattering angle increases. Therefore a small angle favours detection of elastic scattered photons. However the smaller the scattering angle, the smaller the separation between the Compton and coherent peaks in the pulse height spectrum.

1c. X-Ray Fluorescence [3)]

In rapid temporal succession with the above cited photoelectric effect, the atoms of the irradiated sample emit characteristic X-Rays: the energy of these X-rays is characteristic of the constituents elements.

A sample containing several elements will therefore emit as many X-rays as there are elements,and the intensity of the X-rays is proportional to the elemental composition.

The energy of the emitted X-rays is in any case lower than the energy of the incident radiation. The technique based on the above cited phenomena is called X-Ray Fluorescence (XRF) and is well suited for qualitative and quantitative analysis.

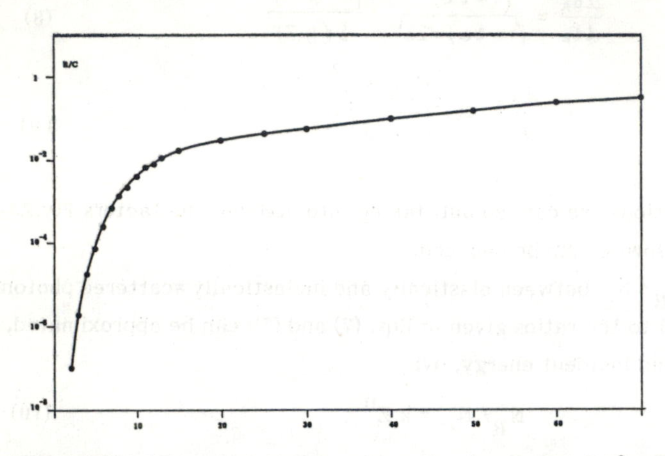

Fig.2 - Rayleigh to Compton counts versus atomic number Z, for E_o = 60 keV and scattering angle ϑ = 180°.

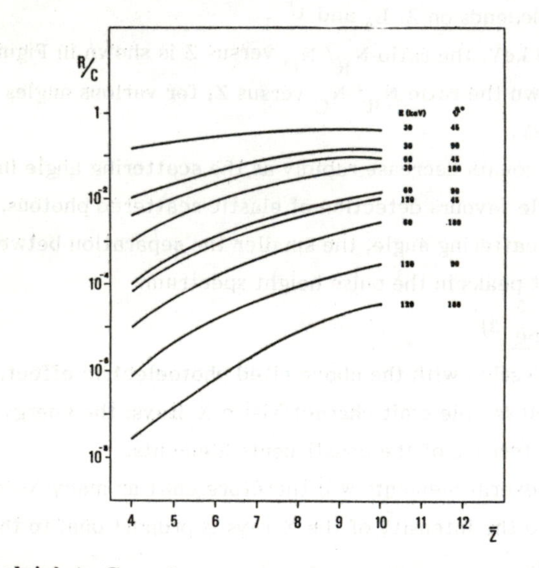

Fig.3 - Rayleigh to Compton counts versus atomic number Z (in the biological atomic number region) for E_o = 30,60 and 120 keV and scattering angles of 45°,90° and 180°.

X-Ray Fluorescence is,strictly speaking,an atomic and not a nuclear method, but it involves exactly the same instrumentation and detectors than nuclear analytical techniques.

The analytical information of the XRF technique, as will be shown in Par.2, is relative to a very low depth of the sample,and decreases with the decreasing of the atomic number Z of the emitting element.

By considering an element a having concentration c_a (% in weight) in a matrix with total attenuation coefficient at incident and fluorescent radiation energies E_0 and E_a, $\mu_t(E_0)$ and $\mu_t(E_a)$ respectively,and by irradiating the sample with monoenergetic radiation of energy E_0, the fluorescent flux of X-rays of element a is given by:

$$N_a = \frac{N_0 K \omega_a \mu_{ph.a} c_a}{\mu_t(E_0) + \mu_t(E_a)} \left[1 - e^{-[\mu_t(E_0) + \mu_t(E_a)]m} \right] \tag{11}$$

where:

K is the overall geometrical and detector efficiency at the fluorescent energy E_a;

ω_a is the fluorescent yield of element a;

$\mu_{ph.a}$ and μ_t are the photoelectric absorption coefficient for the element a and the total attenuation coefficients of the sample at energies E_0 and E_a.

m is the mass per unit area of the sample.

Two extreme cases can be considered for the sample:

a) infinitely thick sample. This is the case of liquids,alloys,minerals and so on, when the thickness of the sample is larger than a few tenths of mm. For infinitely thick samples following condition must be satisfied:

$$m\left[\mu_t(E_0) + \mu_t(E_a) \right] \gg 1 \; . \tag{12}$$

In this case, Eq.(11) can be written in the form:

$$N_a = \frac{N_0 K \omega_a \mu_{ph.a} c_a}{\mu_t(E_0) + \mu_t(E_a)} \tag{13}$$

showing a non-linear dependence of the fluorescent counts N_a versus the elemental concentration c_a (Figure 4).

b) infinitely thin samples. For practical purposes, thin samples are obtained by depositing and drying liquids on thin filters, by vacuum deposition, by applying paints, by electrodeposition and so on.

For infinitely thin samples following condition must be satisfied:

$$m\left[\mu_t(E_o)+\mu_t(E_a)\right] << 1 \quad . \tag{14}$$

In this hypothesis Eq.(11) can be written in the form:

$$N_a = N_o\, K\, \omega_a\, \mu_{ph.a}\, m\, c_a \tag{15}$$

showing a linear relationship between the fluorescent counts N_a and c_a (Figure 4).

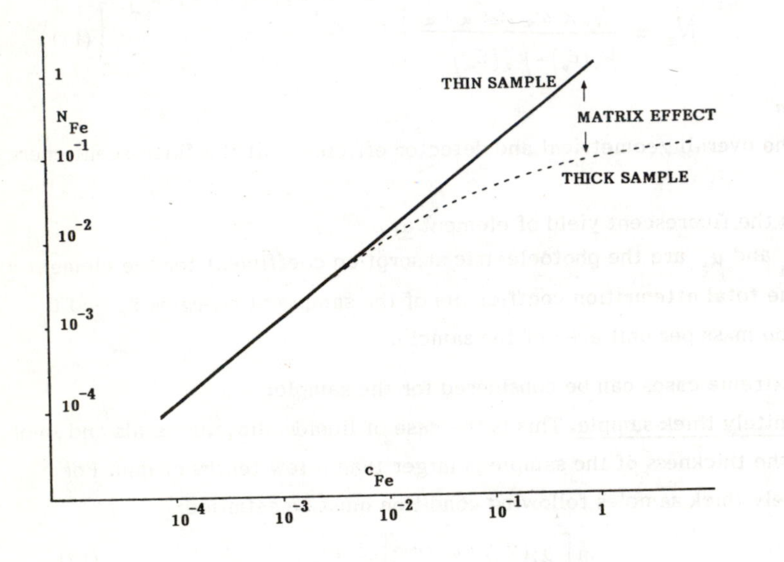

Fig.4 - Linear relationship between Fe-fluorescent counts and Fe-concentration for "infinitely thin" samples. For 'thick samples" the dependence is non-linear.

It is only in the case of very thin samples that the mass or the concentration of an element is proportional to peak intensity. As the sample thickness increases the intensity increases non-linearly reaching an effective maximum at a thickness which depends on the energy of the radiation and the composition of the sample.

In Fig.5 are shown the Fe-counts versus thickness,for solutions containing 1% and 10% Fe respectively.The thin sample condition is the best one for XRF analysis.

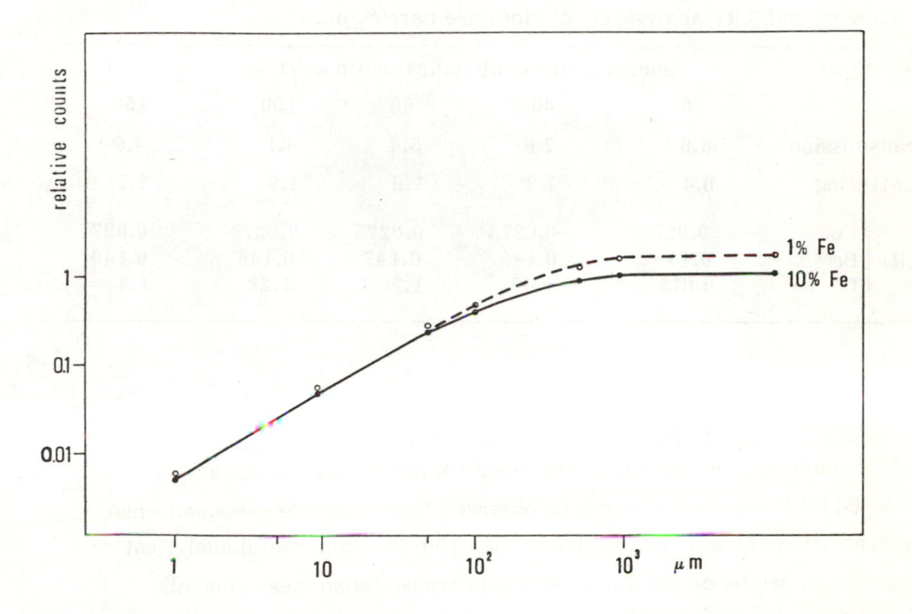

Fig. 5 - Fe-counts for XRF analysis versus thickness of Fe-solution.

2. Penetration of the radiation

The processes above cited,i.e. transmission,scattering and X-Ray Fluorescence due to X or gamma incident radiation can be usefully employed for deducing information about the irradiated sample. While transmission measurements which involve traversing the sample give rise to information of the whole specimen (volume analysis), scattered radiation and still more X-Ray Fluorescence,involve a definite thickness,connected with the penetration of incident radiation inside the sample,with the scattering or X-Ray Fluorescence at a certain depth inside the sample and with the "return" of secondary radiation.

All these effects are connected with the energy of incident and secondary radiation, and with the composition and density of the sample.
When radiation of 20,40,60,100 and 150 keV is employed,in Table I are shown the thicknesses of a typical biological matrix (water) containing Fe,Br and I, for which the detected "output" radiation is reduced to one half (radiation is assumed to be ortogonal to the sample).

Table I: Typical thicknesses (in cm) of sample "explored" when transmission, scattering and XRF analysis of solutions are carried out

technique		energy of incident radiation (in keV)				
		20	40	60	100	150
transmission		0.9	2.6	3.4	4.1	4.6
scattering		0.4	1.2	1.6	1.9	2.2
XRF	Fe	0.027	0.0274	0.0275	0.0276	0.0277
	Br	0.13	0.145	0.147	0.148	0.149
	I	0.015	1.09	1.2	1.28	1.3

3. Applications to Medicine

3 a. Transmission measurements (experimental set-up and results)

1. Transmission of a single monoeneregetic X or gamma ray beam

When Eq.(2) is considered, it can be observed that in simple cases,and when concentration values c_i are not lower than 100-500 ppm, useful analytical information can be deduced from accurate transmission measurements.
If, for example, a solution is considered containing an element a with concentration c_a, then the attenuation coefficient at incident energy E_o, following Eq.(2) can be written as a function of c_a:

$$\mu_t(E_o) = \mu_{H_2O}(E_o) + \mu_a(E_o)\, c_a \quad . \qquad (16)$$

If $c_a << 1$ and condition given by Eq.(3) are considered, then, from Eqs.(1) and (16) the transmitted photons N can be calculated for different values of concentration c_a in solution. A minimum detectable limit (MDL) of the transmission technique can be calculated in this hypothesis. It is defined as the element mass or concentration which would give a peak intensity above local background equivalent to 3 times the error in the local background. The following Eq. can be therefore written:

$$(MDL)_a = \frac{3}{2\sqrt{N_o e^{-2}}\left(\mu_a/\mu_{H_2O}\right)} \quad (17)$$

In Table II are shown some values of the MDL for various elements in solution,energies and values of N_o.

Table II - Minimum detectable limit for various elements in solution,by employing transmission measurements

Element	secondary target and its energy (in keV)		MDL (ppm) $N_0 = 10^6$	MDL (ppm) $N_0 = 10^7$
Fe	Ni	7.45	100	25
Zn	Ge	9.9	80	25
Br	Sr	14.2	46	14
Rb	Zr	15.7	42	13
Ag	Sb	26.3	33	10
I	Pr	35.8	35	11
Ba	Sm	39.8	43	13
Pb(L)	Zr	15.7	42	13

Figure 6 shows a schematic diagram of the experimental arrangement. The radioactive sources which have been employed and their characteristics are listed in Table III. Alternatively, an X-Ray tube was employed, with different secondary targets, for energies between about 4 and about 20 keV. The hightly collimated photons were detected by a NaI(Tl) scintillator coupled to a pulse height analyser, or by a HPGe detector,having an energy resolution of 250 eV at 6.4 keV.

Because of the high degree of accuracy required, particular care was taken in lowering possible correction factors,such as the small angle scattering and the statistical fluctuations. It was demonstrated that,in the hypothesis of small angle scattering, the ratio of the number of scattered photons S to the number of transmitted photons T reaching the detector is given by:

$$S/T = N_e r_o^2 \vartheta^2 d^2 \quad (18)$$

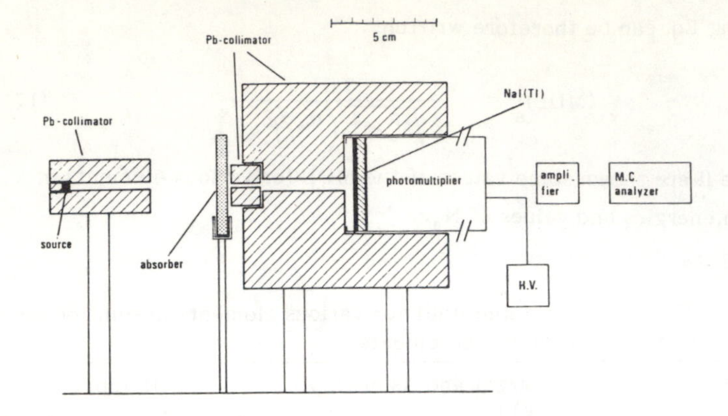

Fig.6 - Experimental arrangement for transmission measurements.

Table III - Radioactive sources employed for transmission measurements and their characteristics

source	energy (in keV)	intensity (photons/s)
Cd^{109}	22.1	$2.5 \cdot 10^{6}$
	25.2	$4.2 \cdot 10^{5}$
Am^{241}	59.6	$1.8 \cdot 10^{7}$
Cd^{109}	88.0	$9 \cdot 10^{4}$
Co^{57}	122.1	$3 \cdot 10^{6}$
	136.5	$3.9 \cdot 10^{5}$

where N_e is the number of electrons per unit volume of the absorber, r_o is the classical electron radius, ϑ the scattering angle and d the lenght of the absorber.

When attenuation coefficients have to be measured with an accuracy better than 0.5%, the number of detected photons should be always greater than 10^5. In our experiments the source-to-absorber and the absorber-to-detector solid angles were chosen so that the ratio S/T was less than 10^{-3}.

Experimental measurements on I-solutions at 23 keV,40 keV and 60 keV are shown in Figures 7 and 8.

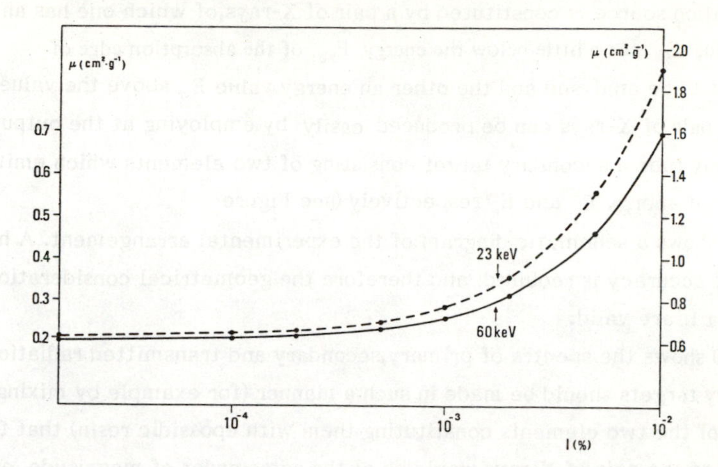

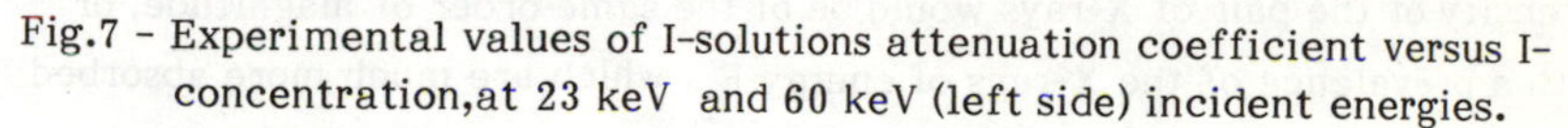
Fig.7 - Experimental values of I-solutions attenuation coefficient versus I-concentration,at 23 keV and 60 keV (left side) incident energies.

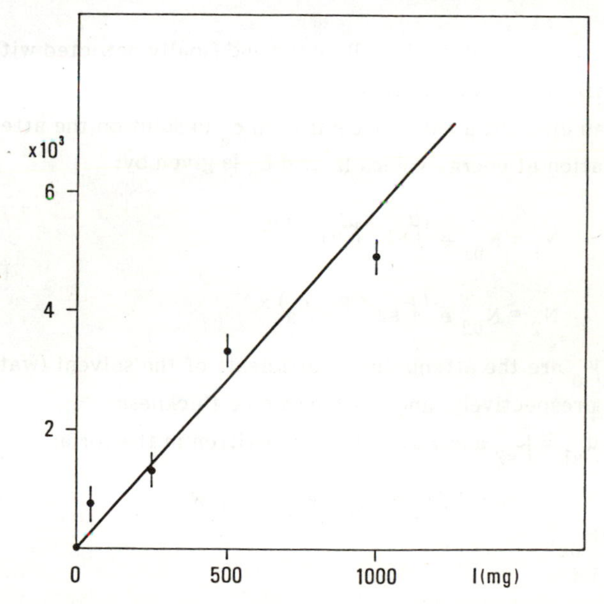

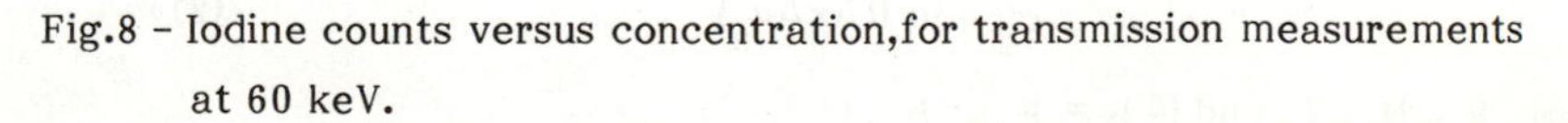
Fig.8 - Iodine counts versus concentration,for transmission measurements at 60 keV.

2. Differential attenuation of X-rays

In this method,the photoelectric absorption edge of the element a to be analysed is employed.

The radiation source is constituted by a pair of X-rays,of which one has an energy value E_1 just a little below the energy E_{ab} of the absorption edge of element a to be analysed and the other an energy value E_2 above the value E_{ab}. The pair of X-rays can be produced easily by employing at the output of an X-ray tube a secondary target consisting of two elements which emit K-X-rays of energy E_1 and E_2 respectively (see Figure

Figure 9 shows a schematic diagram of the experimental arrangement. A high degree of accuracy is required, and therefore the geometrical considerations of Par.3 a 1 are valid.

Figure 10 shows the spectra of primary,secondary and transmitted radiation. Secondary targets should be made in such a manner (for example by mixing powders of the two elements constituting them with epossidic resin) that the intensity of the pair of X-rays would be of the same order of magnitude, or with a prevalence of the X-rays of energy E_2, which are much more absorbed by the solution.

Transmitted radiation are strongly collimated and finally detected with a high resolution HPGe semiconductor detector.

By considering an element a with concentration c_a in solution,the attenuation of incident radiation at energy values E_1 and E_2 is given by:

$$N_1 = N_{01}\, e^{-(\mu_{s1} + \mu_{a1} c_a)\, x}$$
$$N_2 = N_{02}\, e^{-(\mu_{s2} + \mu_{a2} c_a)\, x} \qquad (19)$$

where μ_s and μ_a are the attenuation coefficients of the solvent (water) and of the element a respectively, and x is the sample thickness.

Since $E_1 \simeq E_2$, $\mu_{s1} \cong \mu_{s2}$ and Eqs.(18) can be written in the form:

$$\ln (N_1/N_2) = c_a x\, (\mu_{a2} - \mu_{a1}) w \qquad (20)$$

where $w = N_{01}/N_{02}$.

From Eq.(20) it has:

$$c_a = \ln R/(x \Delta \mu_a) \qquad (21)$$

where: $R = N_1/N_2$ and $(\Delta \mu_a = \mu_{a2} - \mu_{a1})$

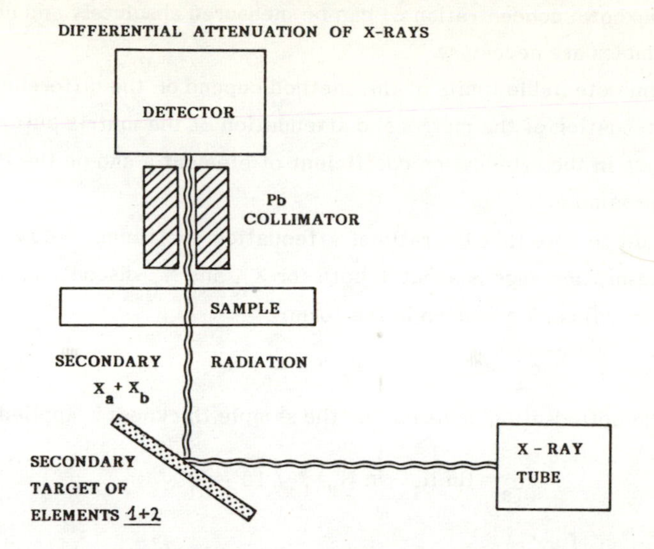

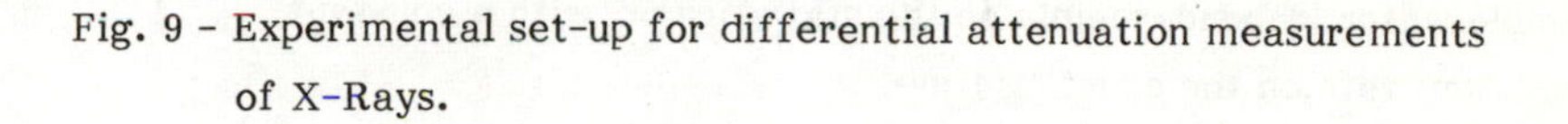
Fig. 9 - Experimental set-up for differential attenuation measurements of X-Rays.

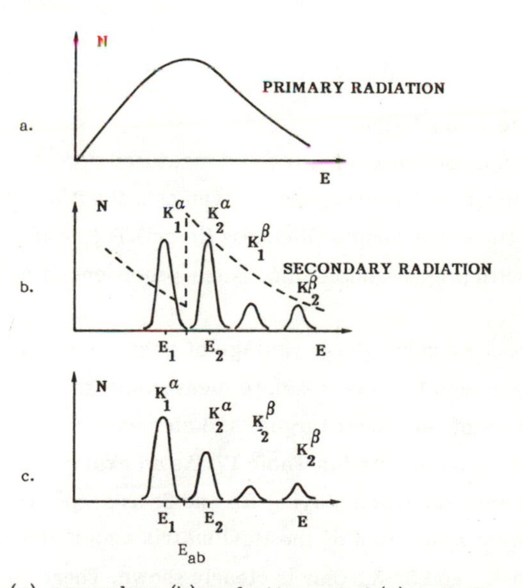

Fig.10 - Primary (a), secondary (b) and attenuated (c) radiation. The sample containing an element a, is characterized by a photoelectric jump (dotted curve) with an energy value E_{ab} between E_1 and E_2.

Therefore,unknown concentration c_a can be measured absolutely,and no standard samples are necessary.

The minimum detectable limits of this method depend on the difference between attenuation of the matrix and attenuation of the matrix plus element a, on the jump in the attenuation coefficient of element a and on the thickness x of the sample.

It is important to note that the ratio of attenuation coefficient below and above the absorption edge is about 6, both for X_K and X_L discontinuity. Therefore Eq.(20) can be written in the form:

$$c_a = (\ln N_1 - \ln N_2) / 5 \mu_{a1} x \qquad (21')$$

and when the optimisation criterion for the sample thickness is applied,then:

$$c_a = (\ln N_1 - \ln N_2) \mu_s / 10 \mu_{a1}$$

where $\mu_s = \mu_{s1} \cong \mu_{s2}$.

The accuracy in determining the unknown concentration c_a, for a given matrix, for example water, depends mainly on the statistics or, with a constant detector count rate,on the counting time.

A relatively simple expression can be derived for the expected precision:

$$\frac{\Delta c_a}{c_a} = \frac{2}{(Nt)^{\frac{1}{2}} \Delta\mu_a c_a x} (4+2e^{\Delta\mu_a c_a x}+2e^{-\Delta\mu_a c_a x})^{\frac{1}{2}}$$

where Nt= count rate x count time.

Fig. 11 shows a comparison of calculated and measured precision for the case of I-solutions analysis. The precision is given as a function of iodine concentration (in g/g), times the sample thickness (in cm).The result clearly demonstrates that with proper sample dimensions precisions of better than 3% are achieved.

The described method,which has the advantage of being sensitive only to the element to be analysed and to allow absolute measurements,can be applied to different combinations of secondary targets and elements in solution. Employable combinations are listed in Table IV. As an example,bromine solutions have been analysed by employing Rb and Sr as secondary targets.In Figure 12 the different absorption of the H_2O matrix and of the solution containing 0.8% Br,for Rb-K_α and Sr-K_α pair is clearly shown. There can also be observed the different attenuation of Rb-K_β and Sr-K_β rays. Figure 13 shows the calibration curve of the ratio $R = N_{Rb}/N_{Sr}$ versus Br concentration.

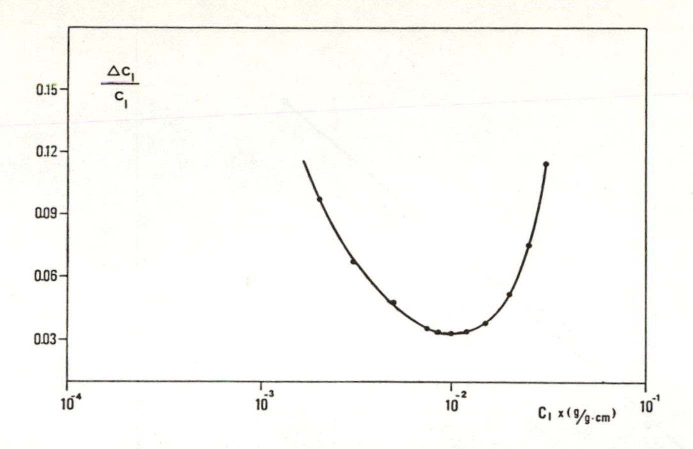

Fig. 11 - Calculated precision for I-solution in transmission measurements

Table IV - X-ray tube - secondary targets combinations for differential attenuation measurements

Element to be analysed	targets	E_1	E_2 (KeV)	E_{ab}
Br	Rb-Sr	13.4	14.1	13.5
Rb	Y -Zr	14.9	15.7	15.2
Ag	Sn-Sb	25.2	26.2	25.5
I	Ba-La	32.0	33.2	33.2

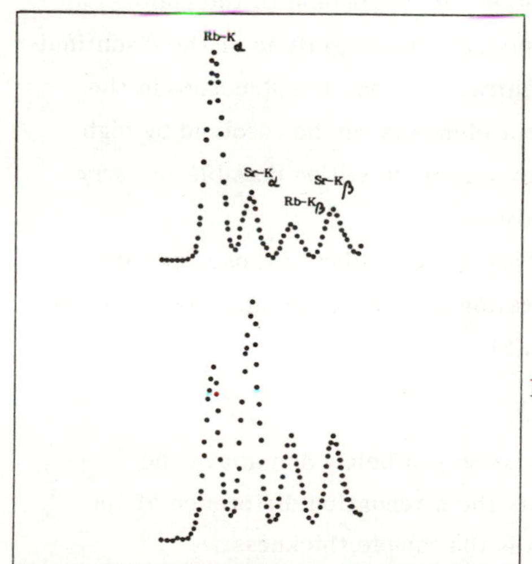

Fig.12 - Different attenuation of H_2O and H_2O+ 0.8% Br, for Rb-Kα and Sr-Kα pair of incident radiation.

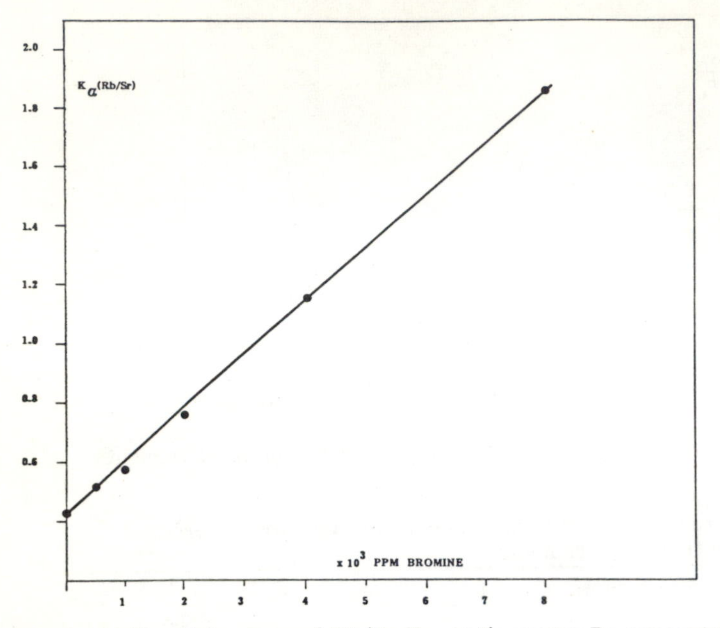

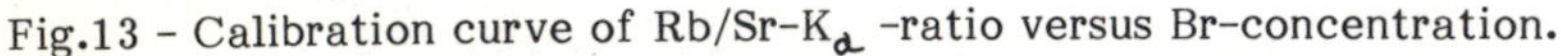

Fig.13 - Calibration curve of Rb/Sr-K_α -ratio versus Br-concentration.

3. <u>Absorption edge densitometry</u>

Solution assay by absorption edge densitometry provides the total concentration of an element <u>a</u> in a sample of well-defined geometry.

The measured transmission of a continuum photon source through a sample displays discontinuities at discrete energies corresponding to the photo-electric jumps of the elements in the sample. The magnitude of the disontinuity is a function of the elemental concentration. When the structure in the transmission spectrum due to component elements can be resolved by high resolution energy dispersive spectrometers,then it will be possible to carry out simultaneous multiple element analysis.

The theoretical aspects of the method are quite similar to those described in the previous Par.3 a 2. The concentration c_a of an element <u>a</u> in solution is in a similar manner as described in Eq.(21):

$$c_a = \ln R/\Delta\mu\, x$$

where R is the ratio of the two transmission just below and above the absorption edge,$\Delta\mu = \mu_2 - \mu_1$ represents the attenuation difference at the energies E_2 and E_1 respectively, and x is the sample thickness.

Figure 14 shows the experimental set-up. Incident radiation is strongly

collimated at the input and output of the sample,in order to avoid that small angle scattered photons in the sample will reach the detector.
Figure 15 shows the spectra of primary and transmitted radiation. The selective absorption of primary radiation at the energy of absorption edge of element a is evident.
Primary radiation which corresponds to the direct output of an X-ray tube is filtered through an Al-filter,in order to reduce high- energy X-rays. Radiation transmitted through the sample is collimated and detected with a HPGe semiconductor detector.

3 b. Elastic and inelastic scattering

The experimental set-up for elastic to inelastic scattered radiation ratio is shown in Figure 16. The radiation is strongly collimated at the output of the source and at the input of the detector.
As pointed out in Par.1 b, part of the radiation inciding the sample is elastically (by Rayleigh scattering) or inelastically (by Compton scattering) diffused by the sample. The diffused spectrum consists therefore,besides the well known fluorescence X-rays peaks,on the elastic and inelastic scattered peaks. In the backscattering geometry, these peaks are generally well resolved with a semiconductor detector (Figure 17). The ratio of the elastic to inelastic peak (R/C ratio),which is independent of the geometrical arrangement,depends,as can be observed by considering Eqs.(8) and (9),on the scattering angle and on the mean atomic number $\bar{Z}$ of the sample. The knowledge of this ratio can therefore give information about the composition of the sample.
Figure 18 shows for example the ratio N_R/N_C for E_o = 59.6 keV (Am^{241}) and scattering angle ϑ= 155°, for pure elements,compounds and mixtures having atomic number or mean atomic number $\bar{Z}$ ranging from 5 to 17.
Figure 19 further shows the Rayleigh peak,keeping constant the Compton peak,for water solutions containing 1% Cu and 1% Pb.

Several authors have employed the R/C ratio for analytical purposes in the medical field,particularly for the determination of bone mineral density. In general small angles are preferred,because elastic scattering increases with the decreasing of the scattered angle,despite the reduced separation between

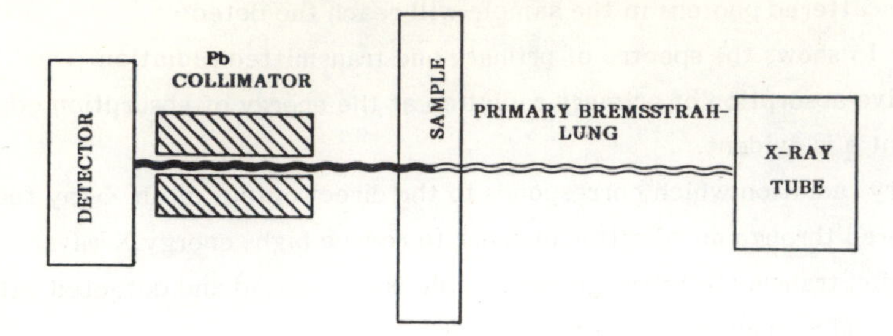

Fig. 14 - Set-up for absorption edge densitometry

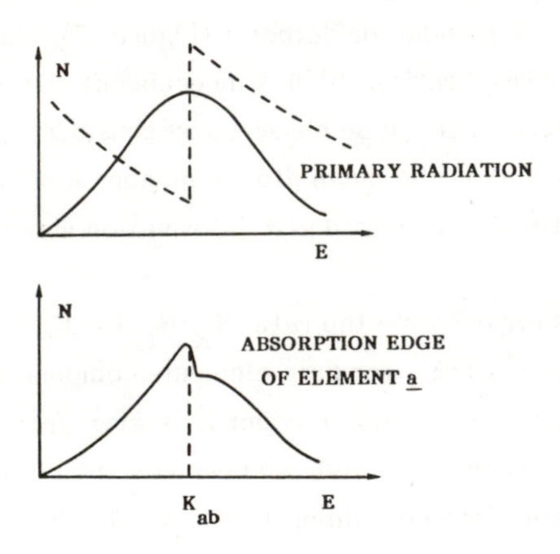

Fig.15 - Spectra of primary and transmitted radiation,by employing the absorption edge densitometry.

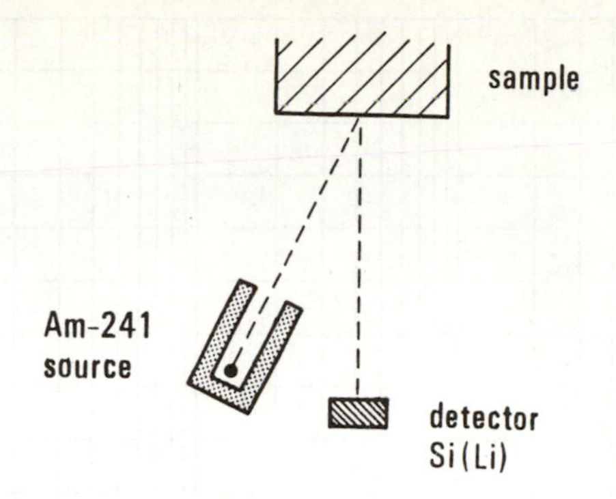

Fig. 16 - Experimental set-up for Rayleigh to Compton ratio measurements.

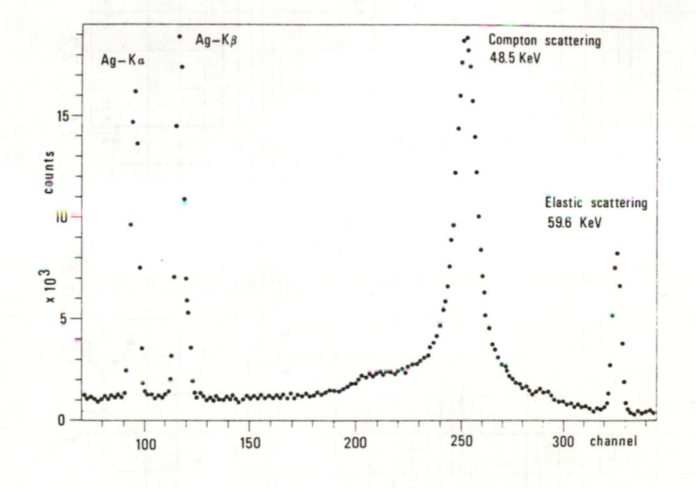

Fig.17 - Typical spectrum of radiation scattered by a silver sample.

the Compton and Rayleigh peaks. Therefore it should be noted that the "modulation" of R/C ratio versus Z is very much higher for increasing angles, in the low $\overline{Z}$ biological region (see Figure 3).

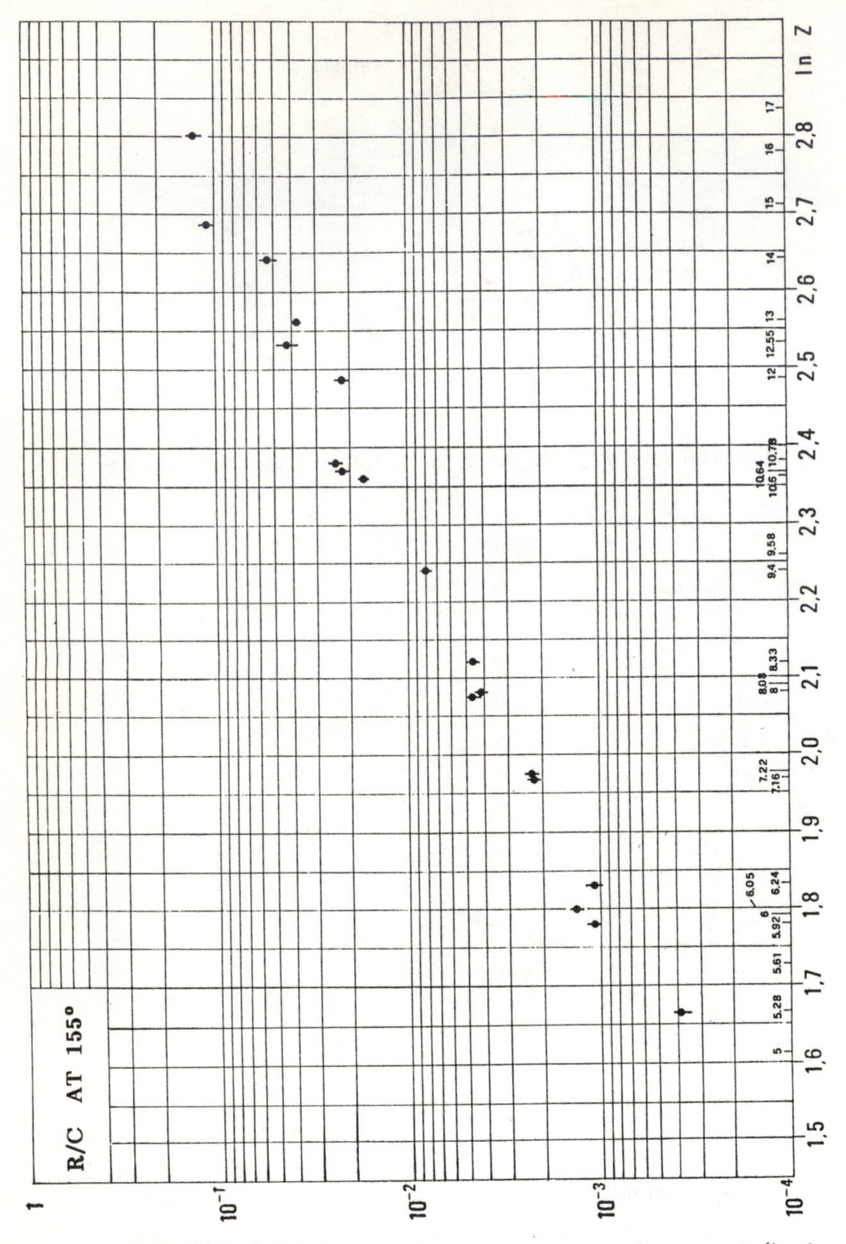

Fig.18 - Experimental values of Rayleigh to Compton ratio versus Z (in the biological region of atomic number) for E_0 = 59.6 keV and ϑ = 155°.

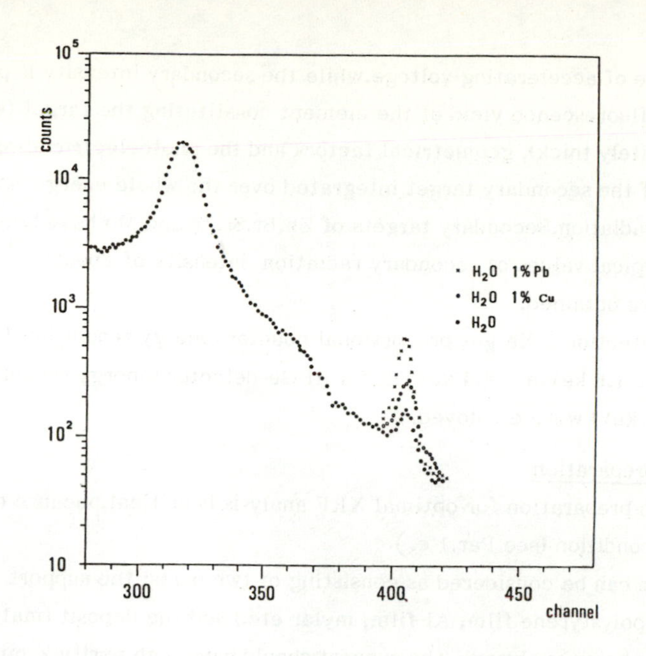

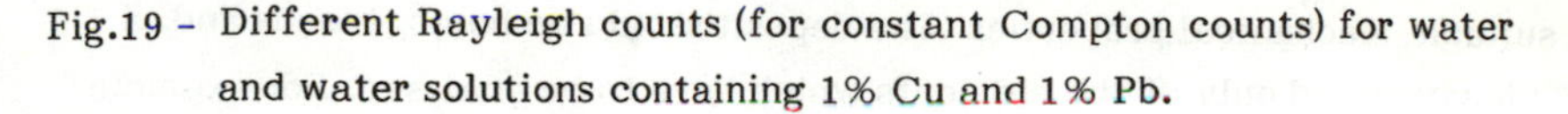

Fig.19 - Different Rayleigh counts (for constant Compton counts) for water and water solutions containing 1% Cu and 1% Pb.

3 c. Energy dispersive X-Ray Fluorescence (XRF)

Different kinds of equipment have been developed for the analysis of biological samples by means of energy dispersive X-Ray Fluorescence. The demand for higher sensitivities has become pressing because the development of research into the physiopathology of a trace element is generally restricted by the minimum detection limit for that element.

That has created the need for a new X-ray source such as an X-ray tube with secondary targets, that, delivering more flux and being capable of being flux and energy tuned,can give rise to better MDL.

1. Experimental set-up

Figure 20 shows a schematic representation of the experimental apparatus. An AEG X-ray tube (W anode) with external interchangeable secondary targets was used to obtain radiation of various energies and high intensity. The primary Bremstrahlung intensity is proportional to the beam current across the tube and the square of accelerating voltege,while the secondary intensity is proportional to the fluorescence yield of the element constituting the target (which must be infinitely thick), geometrical factors and the photoelectric absorption coefficient of the secondary target integrated over the whole energy range of the primary radiation.Secondary targets of Zn,Br,Sr,Zr and Mo have been employed. Typical values of secondary radiation intensity of about 10^9 photons/s were obtained.

For X-Ray detection,a Xe-gas proportional counter (energy resolution 1.2 keV at 6.4 keV and 1.6 keV at 14.4 keV) and a HPGe-detector (energy resolution 250 eV at 6.4 keV) were employed.

2.Specimen preparation

The specimen preparation for optimal XRF analysis is critical,because of the thin sample condition (see Par.1 c.).

The specimen can be considered as consisting of two parts: the support (filter paper,polystyrene film, Al-film, mylar etc.) and the deposit (matrix plus elements being analysed). The support should have high purity,$\bar{Z}$ minimum, and suitable mechanical properties. The deposit should be homogeneous and, if possible,composed only of the elements which have to be analysed. For example biological material is generally composed of an organic matrix that increases the scattered radiation and therefore the background level in the energy range of interest. The presence of the matrix contributes therefore largely to the saturation of the detector and the electronic chain.

In order to reduce matrix effects,preconcentration techniques can conveniently be used. In this work the biological liquids are preconcentrated by reducing them to ash. Thus most of the organic matrix evaporates,leaving an extract of trace elements.

3. Results

A.Analysis of serum samples

Different combinations of X-ray tube and secondary target were employed, for the detection of trace elements in human blood serum.A Zn-target was employed for the analysis of elements from Ca to Ni and a Sr-target for the analysis of higher Z elements. 0.5 ml of serum were ashed,deposited on a thin substrate and directly analysed.

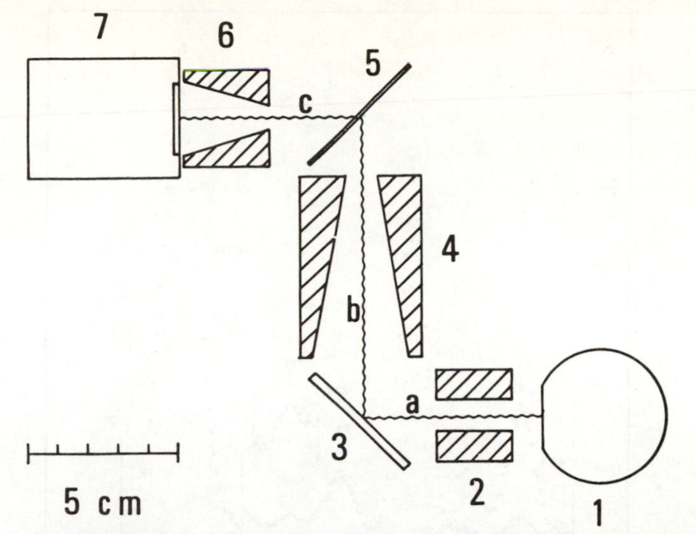

Fig.20 - Schematic set-up of the XRF spectrometer employing an X-ray tube with secondary target. 1) X-Ray tube 2) collimator 3) secondary target of element a 4)conic collimator covered by element a.5) sample 6) conic collimator 7) detector a)primary Bremsstrahlung radiation. b)secondary radiation (X-rays of element a). c)tertiary radiation

Calibration curves for standard Fe and Rb samples were carries out. The MDL at 3 S.D. confidence was found to be 1.6 ng/cm^2 and 1.5 ng/cm^2 respectively, in a measuring time of 10^3 s.

XRF spectra of a typical serum sample,excited by Zn and Sr-X-rays,and analysed by the HPGe detector,are shown in Figures 21 and 22.

The measured concentration values,compared with the mean normal values are shown in Table V.

Table V - Trace element concentration (in μg/ml)of a typical human blood serum sample,by XRF analysis.

Element	XRF (μg/ml)	normal concentration range
Cr	0.15	0.002 - 0.02
Mn	0.022	0.06
Fe	1.82	0.8 - 1.9
Ni	0.02	0.008-0.06
Cu	0.2	0.9 - 1.7
Zn	0.65	0.6 - 1.9
Pb	0.09	0.016-0.13
Se	0.09	0.1 - 0.3
Br	1.18	1 - 7

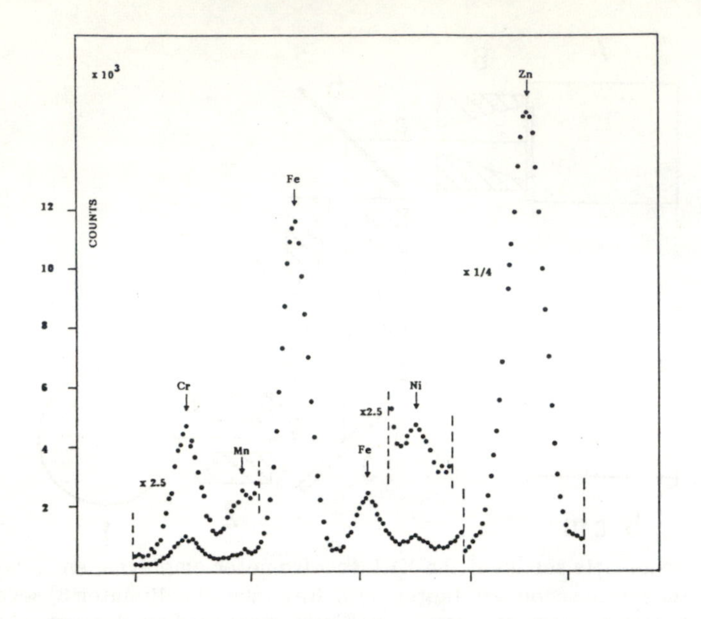

Fig.21 - XRF spectrum of 0.5 ml serum excited by Zn-secondary target-X-rays. Measuring time: 500 s.

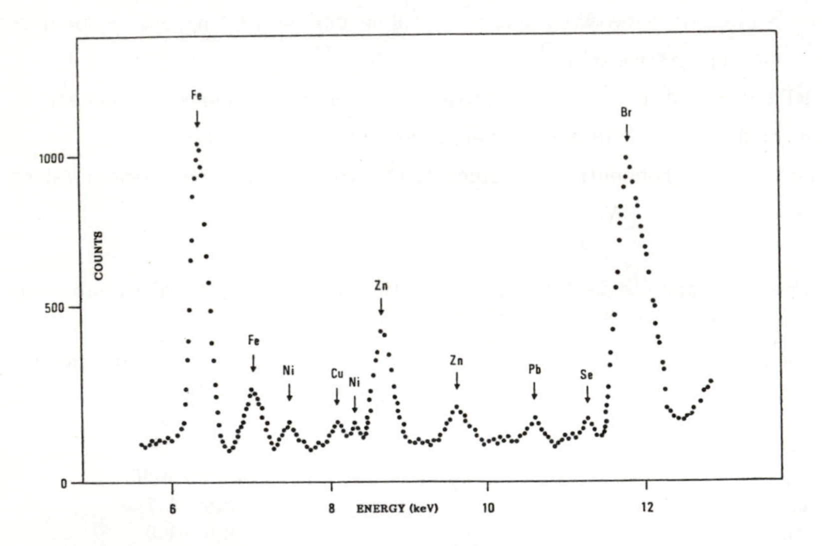

Fig.22 - XRF spectrum of 0.5 ml serum sample excited by a Sr-secondary target-X-rays. Measuring time : 500 s.

B. XRF - analysis applications in criminalistics

A considerable interest has arisen in applications of XRF-analysis to crime investigation. The areas covered are: 1)toxicology (applications involving trace element analysis in biological samples or in foodstuffs); 2)applications involving detection of small residues of transferred material (for ex.: analysis of ballistic residues); 3) applications involving multielement analysis in samples such as human hair,drugs,glasses,paints,cosmetics and so on.

-analysis of ballistic residues

A consolidated method for the detection of gunpowder residues on the hand of a person who had discharged a weapon is based on the neutron activation analysis method. Submicrogram quantities of Sb and Ba were detected,contained in cartridge primer and removed from the hand of a suspect. These elements are detectable with high efficiency with NAA; therefore higher quantities of Pb are generally contained in cartridge primers,and Cu,Fe and Zn can be found in ballistic residues,due to the metallic barrel of the gun and of the bullet jacket. Energy dispersive XRF analysis is very well suited for the analysis of these last elements contained in gunshot residues.Following measurements were carried out:

-direct analysis of primer components and gunpowder residues;

-analysis of gunshot residues collected at various distances on different targets; strips were cut out from the target and directly analysed;

-analysis of ballistic residues removed from the hand of the person who had discharged the weapon by cleaning the hand with a filter paper inhibited with a proper solution.

Two representative XRF spectra showing the analysis of ballistic residues of Cu-jacketed bullets are shown in Figure 23 . Elements Fe,Cu,Zn and Pb are detected,while Sb and Ba,if present,are not detectable owing to the low energy of the L-X-rays.

XRF- spectra of ballistic residues at various distances from the firearm are shown in Figure 24 .The distribution profile of Pb for various firing distances from weapon's tip to target is shown in Figure 25. Revolver calibers are 38 Special and 7.65 Parabellum,and targets were made on cotton, wool and synthe-

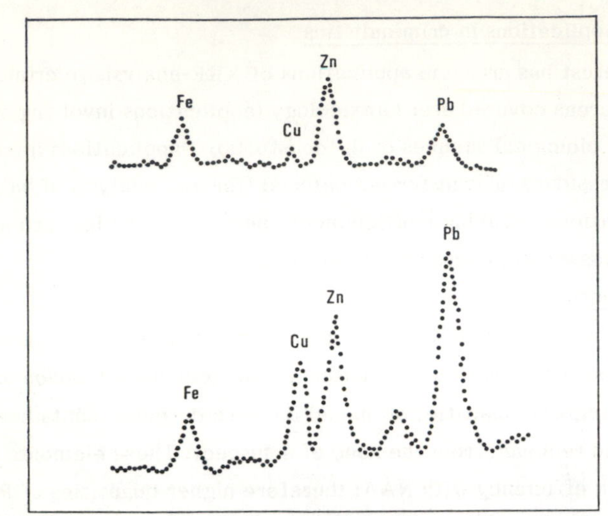

Fig.23 - XRF spectra of ballistic residues collected from the hand of the person who had discharged the weapon. Upper curve:Cal.7.65; Lower curve: Cal. 38 Spec.

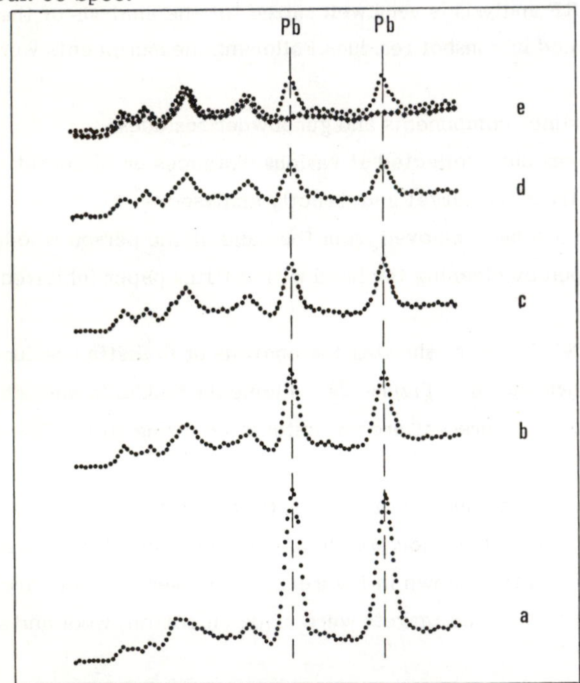

Fig.24 - XRF spectra of ballistic residues collected at various distances from the firearm. The trend of the Pb-peak versus distance is evident.

thic tissue. The variation in Pb concentration (c_{Pb}) versus distance d from the firearm is given by the Equation:

$$c_{Pb} = A\, e^{-bd}$$

where A and b are parameters for each weapon,ammunition type and tissue. In Table VI are reported the Pb-content values collected on filters at different distances from the firearm,the content removed from the hand of the person who had discharged the weapon,and the content in gunpowder and cartridge primers.

C. Analysis of urinary calculi

Because of the frequent occurrence of kidney and bladder stones and the resultant morbility,more information regarding the pathogenesis of the disease is required. Study of the stone composition and structure is of prime importance from the therapeutic and prophylactic viewpoint. Furthermore the composition of the stone's nucleus must bear a fundamental relationship to causation and perhaps the nucleating process may be caused by trace amounts of foreign bodies.

Urinary stones,having a mass of about 100 mg to 50 g were grated in different regions with a carborundum file until about 5-20 mg powder were obtained. The powder,mixed with a neutral glue is deposited on a thin substrate and directly irradiated with the X-ray tube with a Mo-secondary target.

The X-rays were detected by means of a HPGe detector.

Figure 26 shows an XRF-spectrum of a typical kidney stone. The presence of trace elements such as Cr,Fe and Pb is remarkable.

The trace element variation across the stone sections is an argument which requires attention,particularly with reference to the centre of the stone. Great variations were found for the trace elements profile in a stone section.

D. Trace elements in bone

Concentrations of the most important trace elements in bone (in μg/g) are reported in Table VII . A great dispersion in the values can be observed.

The most important trace elements in bone are probably Pb and Sr.

Lead poisoning has been recognized for centuries.Increasing use and consequent dissemination of lead in the environment has increased the incidence of chronic low-level lead-toxicity.

More than 90% of the total body burden of Pb in the adult is stored in the bone, of which 70% is located in the cortical bone.

Table VI - Lead concentration in gunpowder, primer residues, at various distances from weapon's tip and in the shooter's hand

N.	gunshot mark	gunshot calibre	gunshot type	gunpowder	primer residues mg powder	primer residues Pb (μg)	Pb at different distances from weapon's tip (μg) 20 cm	40 cm	Pb (μg) in gunshot residues from shooter's hand filter	ion-exchange resins
1	FIOCCHI	7.65 mm	FMC	traces of Sb,Ba,Pb	10	19	8	1.2	0.15	3.3
2	GECO	7.65 mm	FMC	traces of Fe	5.5	65	33	3.5	1.2	0.8
3	WINCHESTER	7.65 mm	FMC		3	40	5	5	0.3	1.1
4	HIRTENBERG	7.65 mm	FMC		11	120	5.7	0.6	0.22	1.0
5	WINCHESTER	7.65 mm	SHP	traces of Pb	4	50	20	-	3.6	3.0
6	SPEER	38 Special	LRN	traces of Pb	11	170	43	10	0.57	1.0
7	SPEER	38 Special	SWC	Pb (~6 μg), Ba	7	90	80	20	1.2	3.0
8	WINCHESTER	38 Special	LRN	-	8	110	30	5	0.36	1.8
9	SPEER	38 Special	JHP	traces of Pb	12	120	25	-	2.1	2.3
10	WINCHESTER	38 Special	LHP	traces of Pb	8	100	55	-	2.4	5.9
11	LAPUA	38 Special	WC	Pb (~12 μg)	7.5	100	60	4	0.8	3.4
12	NORMA	38 Special	LRN	-	15	200	60	10	1.0	2.8
13	LAPUA	38 Special	LRN	traces of Pb	15	210	77	8	0.8	2.3
14	WINCHESTER	44 Magnum	SWC	-	27	220	10	-	0.8	8.0
15	NORMA	44 Magnum	JSP	-	11	120	25	-	1.2	2.2
16	FEDERAL	44 Magnum	JHP	-	14	120	12	9	0.3	3.6

§ Shots were fired with following weapons: a. half-automatic Beretta 7.65 mm Mod. 70 pistol; b. Smith & Wesson cal. 38 Special, Mod.36 revolver; c. Smith & Wesson cal. 44 magnum, Mod.29 revolver.

Symbols: FMC: full metal case; SHP: silvertip hollow point; LRN: lead round nose; SWC: semi wadcutter; JHP: jacketed hollow point; LHP: lead hollow point; WC: wadcutter; JSP: jacketed soft point .

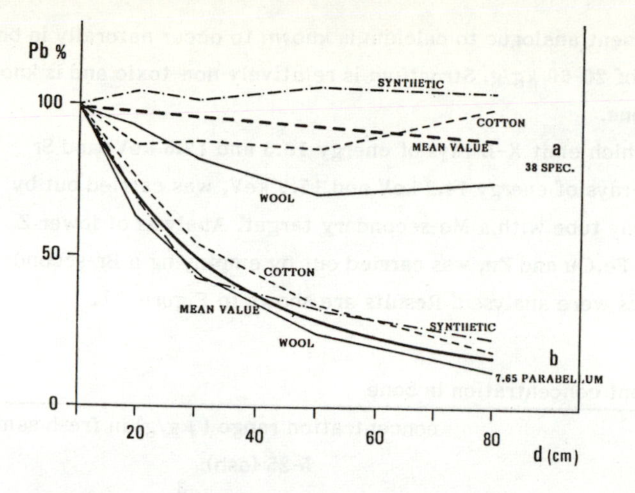

Fig.25 - Distribution profile of Pb versus distance for various targets and for 38 Spec. and 7.65 Parabellum ammunitions.

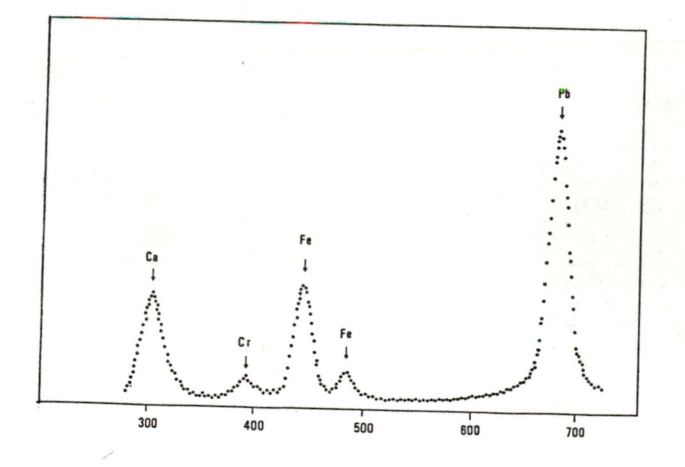

Fig.26 - XRF spectrum of a typical kidney stone.

Strontium,an element analogue to calcium,is known to occur naturally in bone in concentration of 20-60 μg/g. Strontium is relatively non-toxic and is known to turn over in bone.

Analysis of Pb (which emit X-L-rays of energy 10.5 and 12.6 keV) and Sr (which emit K-X-rays of energy 14.2 keV and 15.9 keV, was carried out by employing an X-ray tube with a Mo-secondary target. Analysis of lower Z elements,such as Fe,Cu and Zn, was carried out by employing a Br-secondary target. Bone slices were analysed. Results are shown in Figure 27.

Table VII - Element concentration in bone

Element	concentration range (μg/g) in fresh samples
Ba	5-25 (ash)
Ca	$(100 - 180) . 10^3$
Cu	1 - 20
Fe	5 - 500
Pb	10 - 40
Sr	45
	(100 - 160) ash
Zn	50 - 120

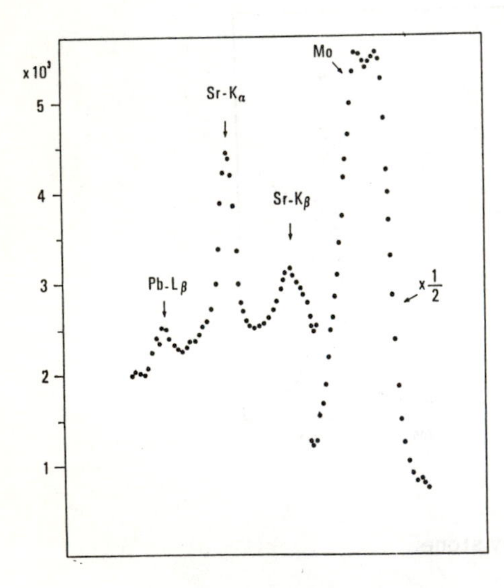

Fig.27 - XRF spectrum of a thin bone slice.

ACKNOWLEDGMENTS

The author thanks Drs.E.Alexiu and M. Sgarbazzini for helpful assistance in the analysis of forensic samples.
The author thanks also Dr.M. Gallucci for assistance in the analysis of kidney stones.
The X-ray fluorescence analysis of Dr.Ing. G. Viezzoli and the technical assistance of Mr. D. Tarsitano are gratefully acknowledged.

REFERENCES

1) R.Cesareo,M.Giannini, Elemental analysis by means of X-Ray attenuation measurements, Nucl. Instr. & Methods, 169, 551-555 (1980).

K.Kouris,N.M.Spirou,D.F.Jackson, Minimum detectable quantities of elements and compounds in a biological matrix, Nucl. Instr. & Methods, 187, 539-545 (1981)

2) P. Puumalainen,A. Uimarihuhta,H. Olkkonen,E.M. Alhava, A coherent/ Compton scattering method employing an X-ray tube for measurement of trabecular bone mineral content, Phys. Med. Biol. 27, 425-429 (1982).

S.S. Ling et al., The measurement of trabecular bone mineral density using coherent and Compton scattered photons in vitro, Med. Phys. 9, 208-216 (1982).

3) R. Cesareo, X-Ray Fluorescence analysis of thin biological samples, in X-Ray Fluorescence in Medicine, Field Educational Italia, 1982, 1-42.

PROCEEDINGS OF THE II INTERNATIONAL CONFERENCE ON APPLICATIONS OF PHYSICS TO MEDICINE AND BIOLOGY
edited by Ž. Bajzer, P. Baxa & C. Franconi

ADVANCES IN RADIATION DOSIMETRY

S. Mascarenhas (Inst. of Phys. and Chem., S. Carlos, S.P. Brasil 13560

and

J.R. Cameron (Dept. of Med. Phys., Univ. of Wisconsin, Madison, WI, 53706, USA)

This paper discusses some recent advances in radiation dosimetry in the areas of: Electron Spin Resonance Dosimetry (ESRD); Electret Radiation Dosimetry (ERD); Photoacoustical Radiation Dosimetry (PARD); Piezoelectric Radiation Dosimetry (PERD); and Thermoluminescent Dosimetry (TLD).

ELECTRON SPIN RESONANCE DOSIMETRY (ESRD)

This section is based on a paper on ESR DOSIMETRIC PROPERTIES OF BONE by Ignez Caracelli, M.C. Terrile and S. Mascarenhas which has been submitted to "Health Physics".

In 1972 Panepucci et al (Pa72) demonstrated the use of ESR in bones and teeth for radiation dosimetry. Ionizing radiation creates paramagnetic centers, the density of such centers produced in the inorganic bone component (hydroxyapatite) is linear with radiation dose over a wide range. In 1978, Ikeya (Ik78) calculated that the centers are stable for up to 10^6 years at room temperature. ESR was used by Mascarenhas et al (Ma73) to estimate the radiation dose to victims of the Hiroshima A-bomb. In 1974 Koberle et al (Ko73) showed that the ESR signal in bone is a reliable measure of the dose for irradiation in vivo. Mascarenhas et al (Ma82) used the technique to determine the age of fossil bones from the ESR signal induced by natural radiation.

Cylindrical samples of bovine medial tibia bone 1 cm long and 0.3 cm diameter were cut parallel to the longitudinal axis of the bone. As the high water content of the bone reduces the quality factor of the cavity, the samples were vacuum dried at 10^{-2} Torr at room temperature. To eliminate the ESR signal due to radicals formed in the organic matrix, the samples were heated for 10 minutes at 70°C.

A small ESR signal was detected in all non-irradiated samples corresponding to a density on the order of 10^{10} centers/mg. This signal was always present, independent of the techniques or tools used for cutting the bone. Its ESR signal is near the radiation induced signal. It is not confused with the radiation induced signals above 10^{10} spins/mg (Fig. 1). However, it affects the minimum detectable dose. A synthetic ruby inside the cavity was used as a secondary standard for intensity measurements.

The ESR spectrum consists of two radiation induced lines, a wide one with $g = 2.005$ and $\Delta H = 20$ G and a narrow one with $g_{//} = 1.9993$ and $g_{\perp} = 1.9997$ $\Delta H = 3.52$. The wide line is due to organic radicals formed in collagen (Pa72). It disappears after heating the sample at 70°C for a few minutes.

The small ESR signal in the unirradiated samples has $g = 2.0018$ and $\Delta H = 7.52$ G. This signal saturates with 10 mW of microwave power and does not increase with dose. The minimum detectable dose is limited by this signal to 0.5 Gy for ^{60}Co γ-rays, with a signal to noise ratio of 6:1. This minimum detectable dose corresponds to a density of about 10^{10} paramagnetic centers/mg., or, 3×10^9 paramagnetic centers/mg Ca. The number of centers increased linearly with dose up to 30 Gy.

Although all of the samples used in this work were from the same bone, Panepucci et al (Pa72) showed that various samples from human, bovine and canine femurs gave the same response within +/-7% over the range 2 Gy to 200 Gy. The stability of the signal is excellent and it is not subject to fading up to temperatures of 200° C. The minimum measureable dose is satisfactory for accident dosimetry. Since the technique works for teeth (Pa72), a tooth sample can be used for accident dosimetry. Measurements are possible with only a 10 mg sample. Suitable ESR equipment for this dosimetry technique is generally available at major laboratories but it may be desirable to design a special purpose ESR unit to be used for accident dosimetry. Such a device could be kept in qualified scientific institutions near nuclear power plants. Further work needs to be done to evaluate the ESR signal from fast neutron doses, which is important for nuclear accident dosimetry.

ELECTRET RADIATION DOSIMETRY (ERD)

The measurement of ionization by the collection of the ions produced in air is one of the oldest methods for measuring radiation. It will continue to be used because of the simplicity of the technique. The electric field to collect the ions is usually provided by a power supply or by means of a charged capacitor as in pen dosimeters. The use of an electret for this purpose was first suggested in 1954. The development of better electret materials, such as teflon has made this technique more useful.

An electret is an insulator that carries a semi-permanent electrical polarization. The electret charge produces a strong electrostatic field capable of collecting ions of opposite sign. If an electret surrounded by air is subjected to ionizing radiations, the charges created in the air will be attracted by the electret and be injected into it. Therefore the apparent electret charge will decrease with increasing radiation absorbed dose. This is the principle of the Electret Radiation Dosemeter (ERD).

Electret dosimetry has been reviewed by Gross (Gr80). A cylindrical electret ionization chamber type dosimeter has been studied for x- and gamma-rays (Ma79) (Ca80) and for neutrons (Ma82). The initial and final charges of the electret are measured by an induction method with an electrometer. The absorbed dose D is shown experimentally to be given by:

$$Q_f - Q_i = kD$$

where k is a constant determined by calibration.

The above equation applies to conditions of complete ion collection in the chamber. If these conditions are fulfilled, the calibration curve Q versus absorbed dose D will be linear.

The fact that the reading procedure does not erase the information, as happens with thermoluminescent dosimeters, and that the electret detector is reusable, which is not true for photographic dosimeters, make the ERD very useful for radiation dosimetry.

Electret ionization chambers can be designed with one or more electrets and in various shapes. Fig. 2 shows a simple design for an electret dosimeter similar to a cylindrical ionization chamber, with a sensitive volume of 3.5 cm^3, using a teflon electret.

A metal needle was used to support the cylindrical teflon electret (E). This needle is connected to the outer conductor (M), allowing the electret polarization to induce an electric field in the cylindrical region between the teflon and the outer jacket (W). The cap (C) on the extremity of the detector is removed during the electret charge measurement. Wall-materials for the outer jacket have a pronounced effect on the energy dependance of the ERD. The best energy dependance is obtained with a low Z wall. If plastic is used, the internal wall should be covered with a thin layer of graphite. Using the ERD without the jacket gives the ERD a very high sensitivity but it does not have a well defined sensitive volume.

The teflon electrets were produced by corona discharge in the air surrounding them. For the electret charge measurement a co-axial insulated metal chamber is connected to an electrometer (for example, Keithley model 610C). By measuring the charge before and after a known exposure the calibration curve can be obtained.

PHOTOACOUSTIC RADIATION DOSIMETRY (PARD)

This section is based on a paper by S. Mascarenhas, H. Vargas and C.L. Cesar (Ma84).

The basic phenomenon of photoacoustics (PA) was discovered by A.G. Bell in 1880. PA is the production of sound by a chopped light beam impinging on a cell containing a gas, liquid or solid. Photoacoustical spectroscopy (PAS) has been used for the investigation of many phenomena from infrared spectroscopy to microwaves. Although PA can be applied to any electromagnetic radiation, it had not previously been applied to the detection or dosimetry of ionizing radiation. The photoacoustical radiation dosimeter (PARD) for X-rays has a linear response with radiation intensity at a fixed radiation quality. It is a true absorbed dose meter, i.e. it measures energy absorbed in the sample being irradiated. The PARD is simple to construct and its use as a relative detector or quantitative dosimeter for all types of ionizing radiation appears possible.

The components of the PARD are shown schematically in Fig. 3. The PARD used a non-resonant photoacoustic cell with a condenser microphone as a detector. The x-ray beam was chopped by a rotating lead covered sectored disc. The acoustic signal was measured with a lock-in amplifier using a reference frequency generated by a light source (L) and detector (D) on either side of the chopper. The chopping frequency was varied from 10 to 200 Hz. The cell chamber was ~ 20 mm dia. and 2 mm deep. X-rays entered the cell through a 0.1 mm Be window and were absorbed in a 0.2 mm thick Pb detector. A duct 1 mm dia. led from the detection chamber to the microphone chamber. Measurements were made using 50 to 100 kVp x-rays with exposure rates at the detector of 2.6 x 10^{-6} C/kg/sec (10 mR/sec) to 5 x 10^{-5} C/kg/sec (200 mR/sec). The minimum measureable exposure rate at 90 kVp was 5 x 10^{-7} C/kg/sec (2 mR/sec).

Under the experimental conditions used, essentially all of the energy from the radiation beam is absorbed in the Pb disc and the PA signal should be proportional to intensity for a given x-ray spectrum and inversely proportional to the chopping frequency. This was found to be the case within the experimental error of about 2%. For x-ray beams of different qualities, the PARD signal should be proportional to the energy fluence rate in the beam, not to the exposure-rate. The energy fluence rate in the beam can be calculated from the exposure rate and the effective keV of the beam. The equivalent photon energy was determined by measuring the half-value layer (hvl) of the beam. Fig. 4 shows the measured response of the PARD as a function of equivalent photon energy. The points represent measured values and the line is the predicted theoretical response. Since the PARD was not calibrated absolutely, the experimental results were normalized at 27 keV to the theoretical curve. It can be seen that the PARD signal is proportional to the energy fluence of the beam. This is, of course, a relative response but in principle, the PARD can be calibrated absolutely using electrical energy.

The Photoacoustical Radiation Dosimeter may have important applications in dosimetry in general and for specific applications such as calorimetry. For diagnostic x-ray beams the response of the PARD is very sensitive to the spectrum for a given exposure rate. Thus the PARD may be useful to evaluate beam quality and may be useful in quality assurance measurements. Because of the 1/f response the PARD may be used to evaluate penetration depth of radiation being absorbed as has been demonstrated for light spectroscopy. Other possible applications of the PARD include dosimetry of other types of radiation such as gamma-rays, electrons, protons and neutrons. Measurement of the ionization produced in the gas volume may permit combined measurements of the exposure rate and the dose rate.

THE PIEZOELECTRIC RADIATION DETECTOR (PERD)

This section is based on a paper by M.R. de Paula, A.A. Carvalho, S. Mascarenhas and R.L. Zimmerman submitted to "Health Physics".

In this section we describe a new dosimetry method using the electrical signal produced in a piezoelectric material (PZT) when it is irradiated with a chopped x-ray beam in the diagnostic range. The piezoelectric crystal and related instrumentation constitutes the Piezoelectric Radiation Dosimeter (PERD). The piezoelectric element absorbs the radiation and provides a voltage proportional to the energy absorbed.

The electrical signal is a consequence of charges induced by thermal reorientation of electric dipoles in the material. This re-orientation arises from a combination of two effects: the temperature increase induces polarization charges proportional to the pyroelectric coefficient of the material. In addition, the thermal expansion causes additional polarization charges proportional to the piezoelectric stress constant.

The PERD (Fig. 5) consists of a lead zirconium titanate (PZT) crystal 17 mm x 17 mm and 3 mm thick, on a nylon support. The crystal has silver electrodes on opposite faces wired to a BNC connector. The detector is placed inside a chamber which is capable of being evacuated. A thin (25 μm) mylar window allows the entrance of the x-ray beams. The thickness of the detector absorbed more than 99.7% of the x-ray beam used in the studies.

The x-ray beam cross-section was collimated to be slightly smaller than the piezoelectric crystal area. The beam was chopped with a lead sectored disk on a variable speed motor.

With moderate x-ray intensities produced above 70 kV the signal can be obtained with the detector connected directly to a X-Y recorder.

The results described below were obtained by connecting the piezo-electric detector to a lock-in amplifier which measures the signal synchronized to the chopper frequency. The PERD voltage as a function of energy fluence rate of the x-ray beam at 4.8 Hz of chopping frequency is shown in Fig. 6.

The piezoelectric radiation dosimeter has the following characteristics for diagnostic x-ray beams:

1) It responds linearly to the radiation intensity for a given radiation spectrum;

2) It has an inverse frequency dependence in the 1.9 to 54 Hz chopping frequency band;

3) It responds linearly to the energy fluence rate of the radiation;

4) It has excellent stability;

5) It is simple to construct, inexpensive and rugged.

The PERD may have important applications in x-ray dosimetry. Other possible applications of PERD include dosimetry of gamma-rays, electrons, protons and neutrons. Since it is basically a microcalorimetric detector it may open new ways to absolute measurement of radiation.

A HIGH SENSITIVITY LiF THERMOLUMINESCENT DOSIMETER-LiF(Mg,Cu,P) (TLD)

This section is based on a paper by Wu Da-ke, Sun Fu-yin and Dai Hong-chen to be published in "Health Physics" May 1984. (Wu84).

At present there are two types of thermoluminescent dosimeters: one has good tissue-equivalence with poor sensitivity, such as Lif (Mg,Ti); the other type has high sensitivity with poor tissue-equivalency, such as $CaSO_4$(Tm or Dy). It would be desirable to have a high sensitivity TLD with good tissue equivalency. This section describes such a phosphor. In 1978, Nakajima et al. reported on their high sensitivity thermoluminescence dosimeter LiF(Mg,Cu,P) made with special grade LiF (Na78). Unfortunately, this material lost its high sensitivity after its first use. Recently, Wu Da-ke et al developed a high sensitivity LiF(Mg,Cu,P) phosphor which retains its sensitivity during reuse. LiF(Mg,Cu,P) has an effective atomic number close to that of human tissue, good energy response, and negligible fading. It is chemically stable and it is not sensitive to room light. In addition, LiF(Mg,Cu,P) has a low background signal without having to use nitrogen gas flow during readout.

A linear heating rate of 15°C/sec was used for producing the glow curves on a X-Y recorder. The heating method for the integration measurements used a preheat to about 130°C before integrating the signal. The maximum reading temperature was about 230°C. Fig. 7

shows the glow curve of LiF(Mg,Cu,P) exposed to 0.6 R from a ^{60}Co source. The curve was obtained with a heating rate of 15°C/s. The sample had been annealed under our standard conditions of 250°C for 15 minutes followed by 100°C for 2 hrs. LiF(Mg,Cu,P) has four glow peaks in the range from 0 to 410°C: the main peak is at about 210°C, and the other peaks are at about 125°, 270° and 360° C. The ratio of the height of the main peak to that of the lower temperature peak is about 70. LiF(Mg,Ti) as TLD-100* has ten peaks in the range of 0 to 400°C. The sensitivity of LiF(Mg,Cu,P) powder was compared to that of LiF (TLD-100). Phosphors were subjected to their standard annealing treatment before irradiation. The standard annealing of LiF (TLD-100) is 400°C for 1 hr, followed by 80°C for 24 hrs. The samples were given 500 mR of ^{226}Ra radiation at room temperature. TL measurements were made 24 hrs. later. The sensitivity of LiF(Mg,Cu,P) was 23 or 34 times higher than that of TLD-100 depending on which filter was used in the reader. Fig. 8 gives the response (TL/R) of LiF(Mg,Cu,P) over the range 0.3 mR to 10^4 R. It can be seen that LiF(Mg,Cu,P) has a linear response over the range from 0.3 mR to 1000 R. Above 1000 R, the response shows saturation. LiF (TLD-100) dosimeters show supralinearity at exposures above 1000 R, which is later followed by saturation at $\sim 10^5$ R.

The sensitivity of a TL dosimeter is largely determined by the radiation equivalence of the no-dose or background signal. This signal is not caused by ionizing radiation and cannot be eliminated by heat treatment. Its fluctuations seriously limit the ability to make low dose measurements. The background signal of LiF TLD-100 is equivalent to about 5 mR if N_2 gas flow is used, otherwise it is 10 mR or more. The background signal of LiF(Mg,Cu,P) averaged about 0.1 mR per mg. This result was obtained without N_2 gas flow. The decrease in the sensitivity after eight repeated uses was no more than 5%. The reusability of LiF(Mg,Cu,P) can be considered relatively good.

Light may affect TLDs by excitation or extinction of TL. That is, the light may cause an unirradiated TLD to produce light-induced TL, i.e. excitation. On the other hand, the light may reduce the intensity of the TL induced by ionizing radiation, i.e. extinction.

---

*Harshaw Chemical Co. Solon OH

There was no detectable excitation or extinction of TL under fluorescent or incandescent lamps. Both sunlight and ultraviolet light induced TL equivalent to about 1 mR and each caused about a 7% extinction of the stored TL signal.

Fading of the TL signal was studied in the following way. LiF(Mg,Cu,P) powder was given the standard annealing process and then irradiated with 10 R of ^{60}Co gamma-rays. The powder was stored at various temperatures for different periods of time and the TL responses were measured. The results obtained were compared with the reading obtained immediately after the irradiation. The TL responses of dosimeters stored one month at 40°C and two months at room temperature were within experimental error. After 30 days of storage at 70°C, the TL signal was about 73% of the initial value.

Tribo-thermoluminescence—TL due to mechanical vibration was not observed after vibrating the powder.

LiF(Mg,Cu,P) was used to measure the background in a non-radioactive laboratory. The TL was measured at various times. Fig.9 shows the glow curves obtained from 0 to 25 days. The areas under the curves as well as the peak heights of the curves increased linearly with time. The slope was 0.37 mR/day.

Some of the work reported in this article was supported by CNEN, Pro-Nuclear, CNP_q and FAPESP (Brazil).

REFERENCES

Ca68 Cameron J.R., Suntharalingam N. and Kenney G.N., 1968, Thermoluminescent Dosimetry, p 32 (The University of Wisconsin Press).

Ca80 Cameron J. and Mascarenhas S., 1980, Report of the 6th Int. Conference on Solid State Dosimetry, Toulouse, France.

Gr80 Gross B., 1980, in Topics in Applied Physics, Vol. 33 "Electrets" (Edited by G.M. Sessler, Springer Verlag).

Ik78 Ikeya M., 1978, "Electron Spin Resonance a Method of Dating", Archaeometry, 20, 147-158.

Ko73 Koberle G., Terrile M.C., Panepucci H.C. and Mascarenhas S., 1973 "On Paramagnetism of Bone Irradiated in vivo", An. Acad. Bras. Cien. 45, 157-160.

Ma73 Mascarenhas S., Hasegawa A. and Takeshita K., 1973, "ESR Dosimetry of Bones from Hiroshima A-Bomb Site", Bull. Amer. Phys. Soc., 18, 579, ref BK-8.

Ma79 Mascarenhas S. and Zimmerman R.L., 1979, Report of the 10th Anniversary Conference, Brazilian Association of Physicists in Medicine (Edited by S. Watanabe), Sao Paulo, Brazil.

Ma82 Mascarenhas S., Baffa O. and Ikeya M., 1982, "ESR dating of Human Bones from Brazilian Shell-Mounds (Sambaquis)", Am. Journ. Phys. Anthropology, 59, 415.

Ma84 Mascarenhas S., Vargas H. and Cesar C.L., 1984, "A Photoacoustical Radiation Dosimeter", Med. Phys. 11, 73-74.

Na78 Nakijima T., Murayama Y., Matsuzawa T. and Koyano A., 1978, "Development of a New Highly Sensitive LiF Thermoluminescence Dosimeter and its Application", Nuclear Instruments and Methods, 157, 155-162.

Pa72 Panepucci H., Mascarenhas S. and Terrile M.C., 1972, "Bone as a Dosimeter", Proceed. 1st. Latin Am. Conference on Med. Phys. and Rad. Protection, S. Paulo.

Pa77 Panepucci H. and Farach H., 1977, "ESR Spectra of quasirandomly oriented centers: Applications to radiation damage centers in bone" Med. Phys. 4.

Ro80 Rosencwaig A. Photoacoustics and Photoacoustical Spectroscopy, John Wiley and Sons, NY, USA (1980).

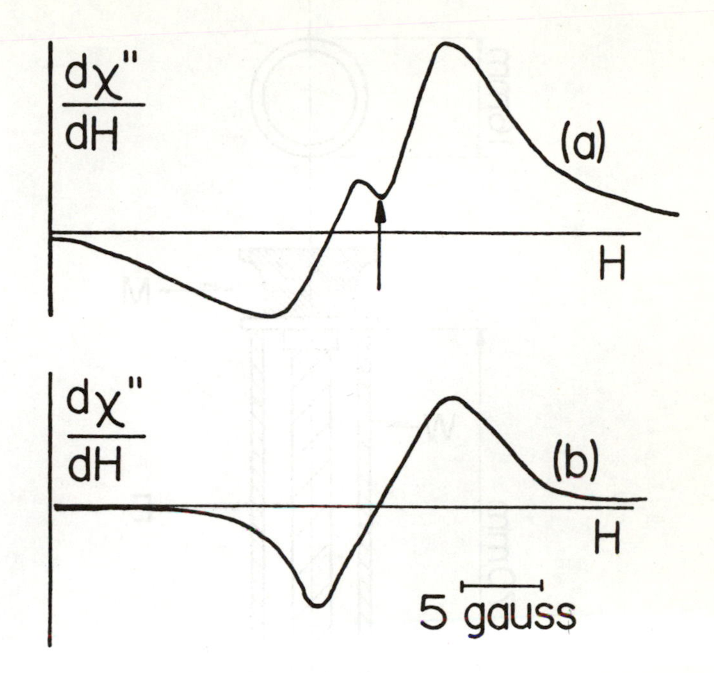

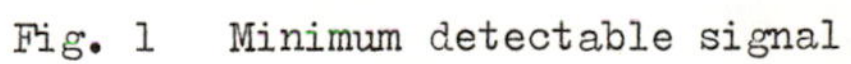

Fig. 1 Minimum detectable signal

a) ESR signal from non-irradiated bone;

b) ESR signal from bone given 0.5 gy

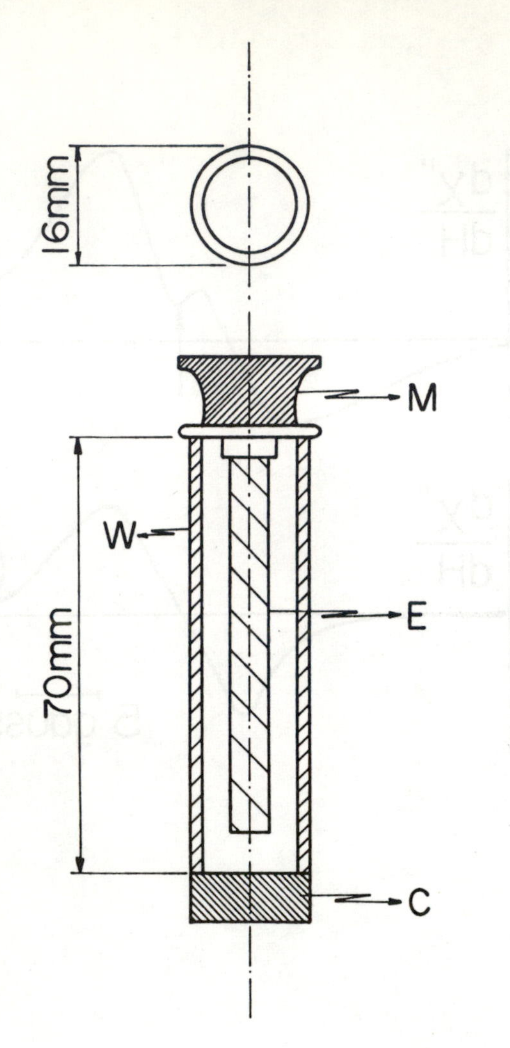

Fig. 2 A cylindrical electret radiation dosimeter: E: Teflon Electret, M: Outer jacket, and C: Aluminum cap

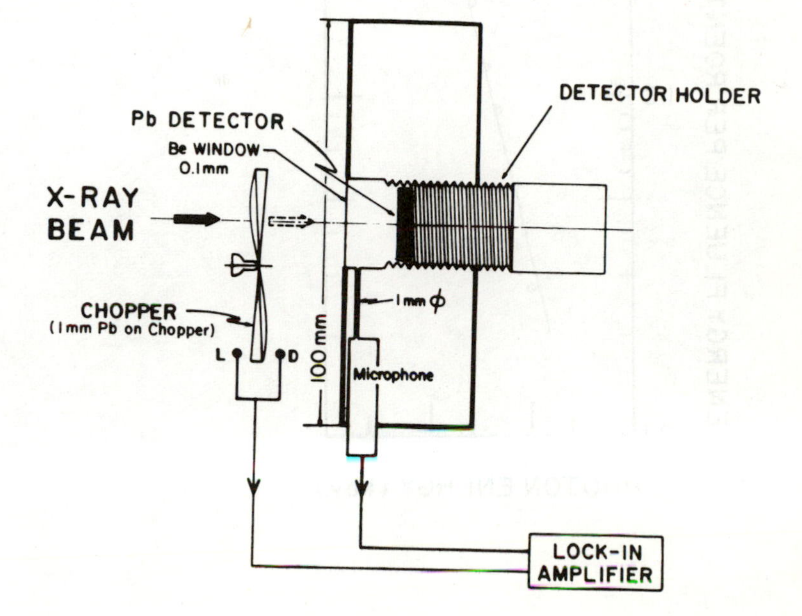

Fig. 3 Schematic diagram of the photoacoustic radiation dosimeter

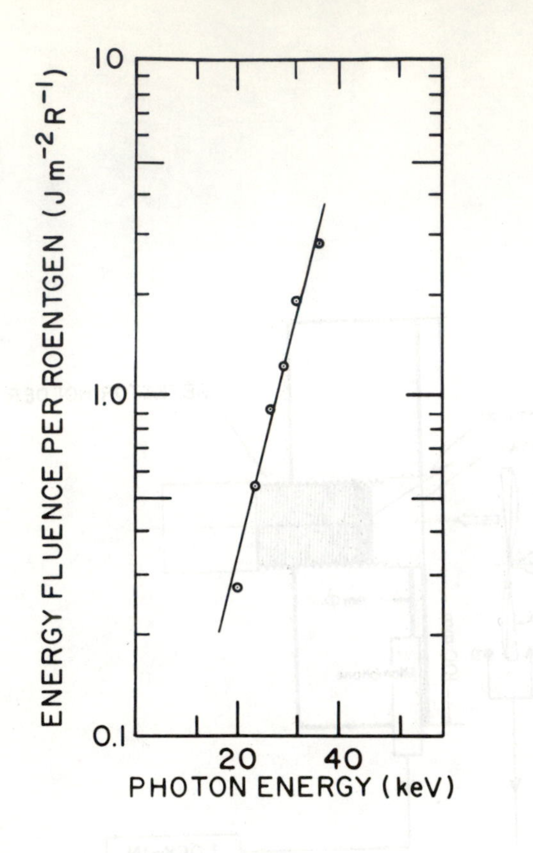

Fig. 4 PARD signal (normalized at 27 keV) for energy fluence per roentgen (dots) compared to the predicted response (solid line), versus equivalent photon energy.

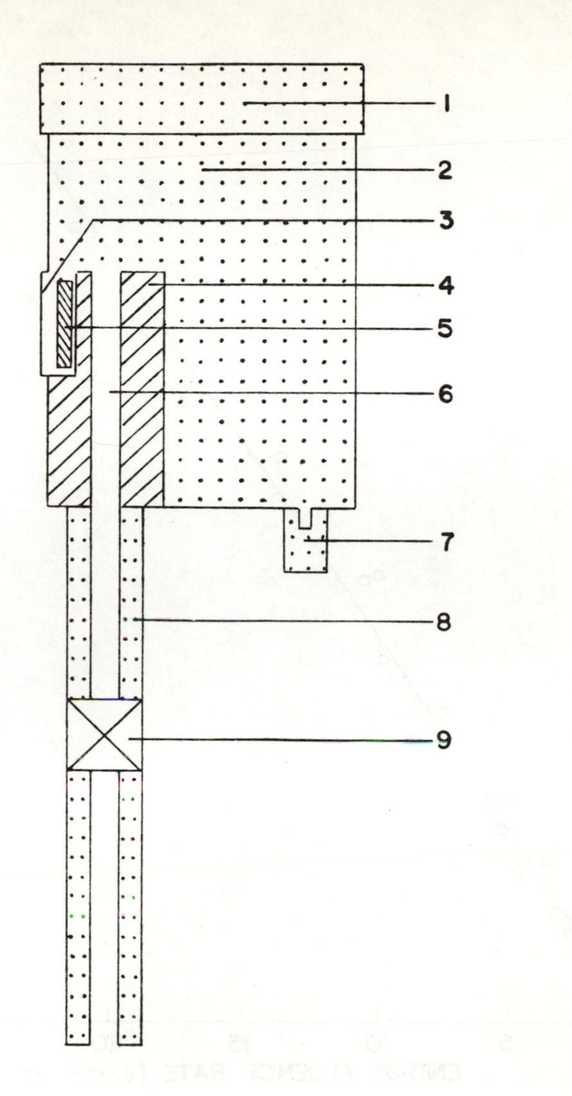

Fig. 5 Schematic drawing of the Piezoelectric Radiation Dosimeter (PERD). 1 - Cover, 2 - PERD casing, 3 - Mylar Window, 4 - Nylon support, 5 - Piezoelectric Crystal, 6 - Vacuum line, 7 - BNC connector, 8 - Casing support, 9 - Valve

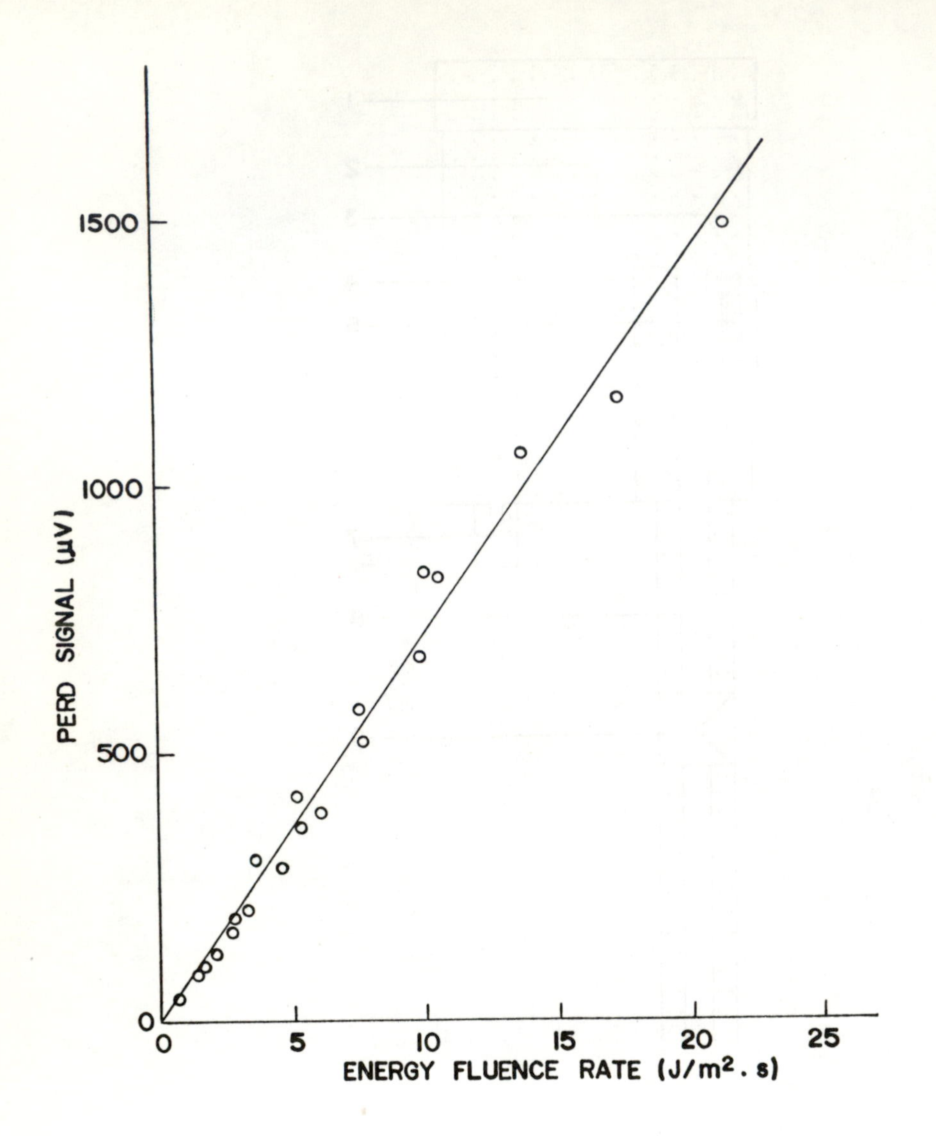

Fig. 6 PERD voltage as a function of energy fluence rate. Experimental conditions: Chopping Frequency: 4.8 Hz, Distance from the target to the detector: 7 cm, Irradiated Area: 1 cm^2, Total Filtration: 2 mm Al.

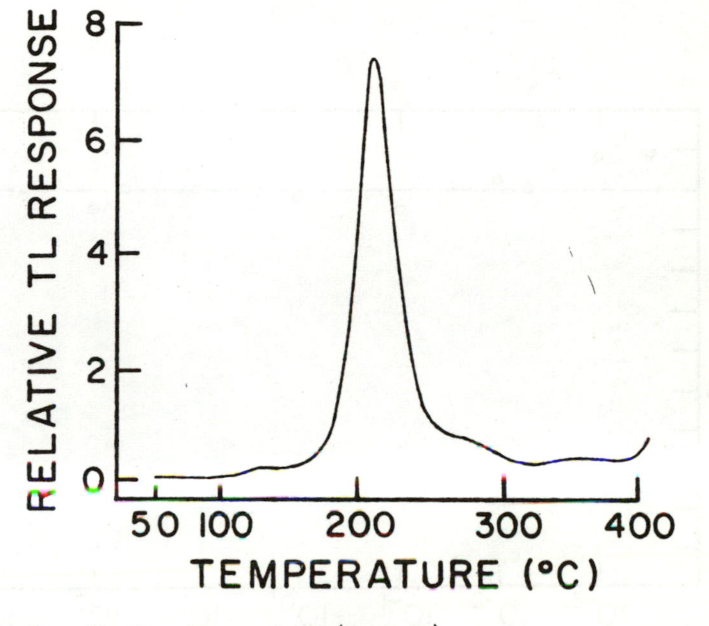

Fig. 7 Glow Curve of LiF(Mg,Cu,P)

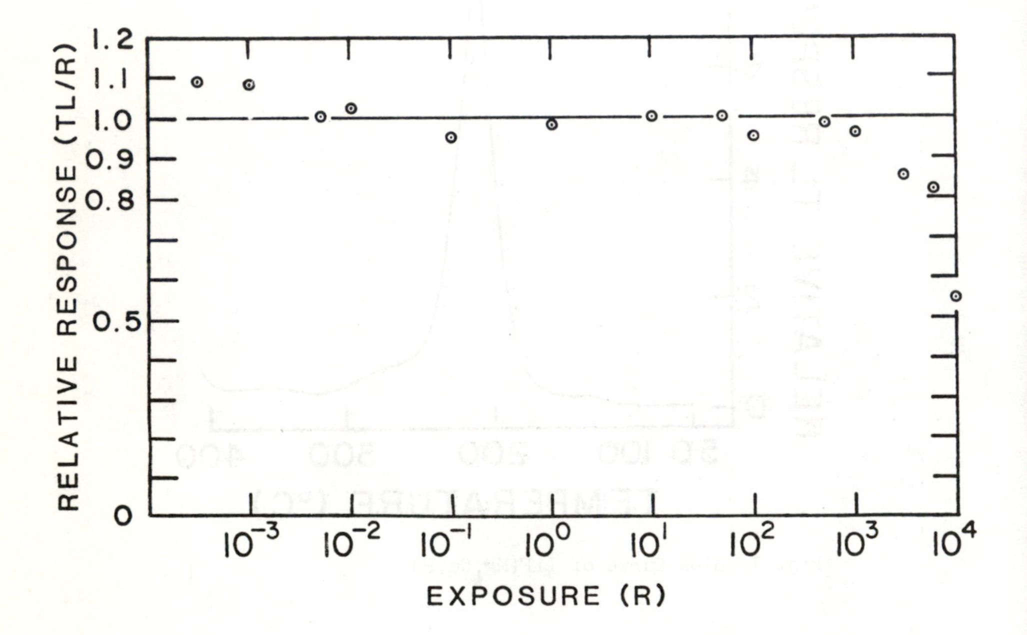

Fig. 8 TL/R for LiF(Mg,Cu,P) from 0.3 mR to 10^4 R

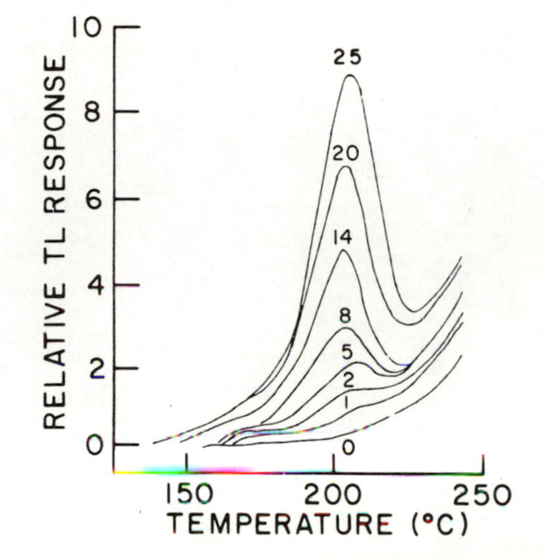

Fig. 9 Glow curves of LiF(Mg,Cu,P) stored in a laboratory for various number of days

PROCEEDINGS OF THE II INTERNATIONAL CONFERENCE ON APPLICATIONS OF PHYSICS TO MEDICINE AND BIOLOGY
edited by Ž. Bajzer, P. Baxa & C. Franconi

ADVANCED RADIATION DETECTORS

Fabio SAULI
CERN
CH-1211 Geneva 23
SWITZERLAND

1. Fast detectors in particle physics

The experimental methods in elementary particle physics have been greatly improved about fifteen years ago by the introduction, due to Charpak and collaborators, of the Multiwire Proportional Chambers, MWPC[1)]. With various improvements and modifications, this kind of electronic detector has replaced most of the old visual techniques; linked to a computer, a system of MWPC allows fast data collection and analysis. The accurate, bi-dimensional localization capabilities of the device in detecting ionizing radiation can be exploited in other fields of applied research; I will in these notes briefly summarize the main operational characteristics of gas detectors, and give several examples of application in biomedical research.

A MWPC consists in a plane of parallel, thin anodic wires stretched across an insulating support frame between two cathodes realized either with thick wires or foils. When a negative potential is applied to the cathodes, the anodes being grounded, an electric field structure develops as shown in fig. 1; typically, the anode wire diameter and spacing are 20 microns and one mmm and the anode to cathode gap 5 mm. The structure is flushed with a suitable gas mixture, as for example argon-methane. Electrons released within the gaps by an ionizing radiation drift towards the anodes and experience avalanche multiplication in the increasing electric field around the thin wires; proportional gains of 10^6 can be attained, thus allowing detection of even tiny amounts of ionization with inexpensive electronics. A simple amplifier followed by a discriminator provides a pulse at the location of the wire or wires geometrically facing the ionizing encounter.

As an illustration, fig. 2 shows a large surface MWPC ready to be installed; a typical setup used in high energy physics experiments may include several tens of detection planes with various geometries. The computer display of a complex event is shown in fig. 3 (from the UA1 detector built at the prton-antiproton collider at CERN). A large multiplicity of energetic charged particles are generated in the collision; from the curvature of the reconstructed tracks (the whole detector is installed in a large magnet) and other information the topology and kinematics of even complex events can be reconstructed.

Bi-dimensionality, essential for imaging neutral radiation (having a single ionizing encounter) can be achieved exploiting the induced signals on the two cathode planes, see fig. 4. The avalanche growing around the anode wire is seen by the cathodes as a charge distribution whose center-of-gravity represent the coordinates of the event. Realizing the cathode planes with wires or groups of wires, perpendicular to each other, and recording the distribution one can compute the x and y coordinates with good accuracy. As an example, fig. 5 shows an image obtained irradiating a bi-dimensional cathode readout MWPC with 1.5 keV x-rays through a mask with letters cut-out: the size of the writing is 4 x 2 mm^2, and the corresponding localization accuracy 40 microns rms[2).

Although opening up interesting perspectives in soft x-ray micrography and structure analysis, the readout method used to obtain the quoted result, that requires one amplifier and analog-to-digital converter on each cathode strip, is rather complex for laboratory use. Several simplified readout methods however have been realized, for example using electromagnetic delay lines coupled to the cathode planes[3)]. At the expense of a reduced localization accuracy (around 500 microns or so) and rate capability, such systems proved to be cheap and reliable to operate and are finding applications in various fields[4-6)].

For a more detailed description of MWPC operation and readout schemes see for example my review article[7)].

2. Nuclear scattering radiography

A medical diagnostics method that closely resembles an

elementary particle experiment has been developed[8,9]), see fig. 5. A beam of protons, of momentum around one GeV/c, hits a target and undergoes nuclear scattering. For thick targets, a large fraction of the interactions result in one or more nuclear fragments ejected at angles with the incoming beam directions. A set of MWPC, before and after the target, measure the space coordinates of the crossing tracks; simple geometry allows then to reconstruct the vertex of the interaction. Since the scattering probability is density-dependent, mapping of the vertex position provides a fully three dimensional measurement of the density distribution in the target. The result obtained exposing a preserved human head is shown in fig. 6: simple mathematical algorithms allow to display on the computer screen slices of selected thickness and orientations.

Obviously, the necessity of using a high energy beam seriously limits the use of such a system. There are however various attempts to use charged particle beams of similar characteristics for deep tumor irradiation; if this proves to be competitive or complementary to other radiation therapies, the described diagnostic system would allow to obtain an on-line distribution of intensity of irradiation within the patient with the possiblitiy of correcting the dose distribution during exposure.

3. Imaging of beta emitters

Many biological materials can be labelled with radioactive tracers such as ^{3}H, ^{14}C, ^{32}P, ^{35}S, ^{131}I that are beta-electron emitters, with energies from a few keV to a few MeV. Being heavily ionizing, electrons in this energy range are easy to detect with a MWPC. Bi-dimensional activity distributions can be obtained quickly and with good efficiency; since each electron is individually detected and recorded, such a detector does not suffer the well known linearity and over-exposure limitations of photographic plates. Sub-millimeter localization accuracies can be obtained mapping thin layers of biological material labelled with tritium, that emits low energy (short range) electrons; to preserve efficiency, either very thin windows or windolwless chambers have to be used. Fig. 7[10]) shows the activity distribution in a monolayer of mammalian cell colonies labelled with ^{3}H; the picture itself (obtained from a CRT screen) is saturated but isocount contours and intensity profiles can easily be obtained and displayed as shown. The method is used in the study of defective repair of mutant cells, and has the advantage of preserving the integrity of colonies.

A problem appears however when imaging labellers emitting high energy electrons, because of their extended range in the gas that spoils localization; for example, ^{14}C has an electron spectrum extending to 150 keV, with a range in a gas at atmospheric pressure of about one cm. A solution has been found using a device called multistep proportional chamber[11)]; it consists in a two-stages gaseous detector, having as first element of amplification a parallel plate avalanche chamber (fig. 8). In the strong and uniform electric field of the parallel plate, only charges released close to the upper grid receive full amplification, while the gain exponentially decreases for deeper penetrations. The lower amplification element, a MWPC as described before, localizes therefore preferentially the beginning of the ionization trail of electrons emitted from a support laying over the upper cathode. Fig. 9 shows an example of distributions measured with a multistep chamber in the detection of ^{14}C samples on a substrate; the frame size is 3 x 3 cm^2 [12)]. From the activity projections, shown in the picture, one can deduce a resolution of about 500 microns on the positioning. A detector of this kind is very suited for bi-dimensional radiochromatography.

4. Detection of photons

Photons can be detected in a MWPC if they are energetic enough to penetrate without too much absorption the thin window separating the detector from the emitter; this sets a practical lower limit to about 2-3 keV. On the other hand, at energies above 10-20 keV the gas itself begins to be transparent to the photons, and efficiency is quickly loss unless some special construction systems, to be described below, are adopted.

An example of the efficiency that can be obtained with a 6 cm layer of Xenon as a function of photon energy is shown in fig. 10. clearly, to extend the useful region of a detector one could increase its thickness: however for any application where the radiation is not perpendicular to the plane of detection, the fact that the conversion point is not determined introduces a prohibitive parallax error (fig. 11a). Use of a collimator allows to avoid the problem (fig. 11b) but has the obvious consequence of strongly reducing geometrical acceptance.

In the particular case of a point source of radiation, as for example in crystal diffraction studies, an elegant solution to the above-mentioned problem has been found with the so-called spherical drift chamber[13,14], see fig. 12. A drift volume, thick enough to have the required conversion efficiency, has a radial field structure defined by suitable grids and guard electrodes. The field lines bend towards the MWPC where amplification and detection is performed; as all conversions at a given angle occur along a field line, there is no parallax error. This device is being used in several laboratories for the analysis of complex proteins, suitably crystalized; fig. 13 (provided by the authors of Ref. 14) shows the operational system in some details.

Another way to increase the efficiency of a MWPC for hard x-rays is to interleave the gas detector layers with solid converters of suitable characteristics (depending on the x-ray energy, the converter thickness has to be adjusted in order to maximize the combined probability of converting a photon and of letting the corresponding photoelectron into the gas volume). A multilayer chamber of this kind has been used to perform in vivo bone densitometry through an absorption measurement using as a source ^{153}Gd 42 keV x-rays[15]; fig. 14 shows an image obtained with such a method. Since the clinical interest of such a measurement resides in very small differences in the absorption due to long-term de- calcification, a fully digital method providing detailed values at precisely identified locations can be unvaluable.

The same group has extended the use of such a detector to te 511 keV gamma rays resulting from positron decays[16]; fig. 15 illustrates the MWPC positron camera developed at the Rutherford laboratory and presently undergoing clinical tests in british hospitals. The two collinear gamma rays emmitted by the positron annihilation are converted in the thick cathode planes of the two sets and detected as described. With typical efficiencies of around 6% and spatial resolutions of 6 mm FWHM, a MWPC based positron camera is very competitive with commercial multi-crystal arrays. Fig. 16, from the same reference, shows the activity distribution in the thyroid of a patient injected with ^{124}I after different time intervals (1,6 and 24 hrs after injection). Time-dependent countour maps can be constructed to show the progress of activity distribution in vivo.

An alternative approach to the hard x-ray detection problem has been suggested independently by Charpak[17)] and Perez-Mendez[18)] and is illustrated in fig. 17. The drift volume of an otherwise conventional MWPC is filled with a matrix of multilayer foils electrically insulated; rows of holes, vertically aligned, allow the drift and collection of charged generated in the gas channels by photoelectrons created in the solid layers. Alternatively, the converter is realized with a technique similar to the one used to build microchannel plate amplifiers: a thin plate built with a stack of glass tubes constitues the converter, a slight electrical conductivity in the channels being provided by suitable processing. Both designs of detector are being tested in a clinical environment[19,20)]. A positron camera of the described conception is presently receiving clinical tests at the Geneva's Hospital (fig. 18). As an example fig. 19, from Ref. 20, shows the activity distribution in the spinal chord of a rabbit, injected with 100 microcuries of the positron-emitter ^{68}Ga (the original, displayed in false colours, shows remarkably the activity due to different intakes).

5. Conclusions and summary

The purpose of these notes is to provide an inevitably superficial overview of various attempts in using the techniques developed in high energy physics to the biomedical field. Because of their good space accuracies, resolution times and rate capabilities, fully electronics detectors based on the MWPC concept appear very promising. Being however still far from any form of commercialization, this family of fast detectors is still in the hands of a few, enthousiastic users that build and use the instruments themselves. Only the eventual interest of the biomedical community will help making the devices largely available to experimenters, as it happened to similar technologies.

REFERENCES

1) G. Charpak, Ann. Rev. Nucl. Sci. 20 (1970) 195.

2) G. Charpak, G. Petersen, A. Policarpo and F. Sauli, Nucl. Instrum. Methods 148 (1978) 471.

3) R. Grove, K. Lee, V. Perez-Mendez and J. Sperinde, Nucl. Instrum. Methods 83 (1970) 257.

4) S.N. Kaplan, L. Kaufman V. Perez-Mendez and K. Valentine, Nucl. Instrum. Methods 106 (1973) 397.

5) V. Perez-Mendez, IEEE Trans. Nucl. Sci. NS-23 (1976) 1334.

6) G. Charpak, Nucl. Instrum. Methods 156 (1978) 1

7) F. Sauli, Principles of operation of Multiwier proportional and drift chambers, CERN 77-09 (1977).

8) J. Saudinos, G. Charpak, F. Sauli, D. Townsend and J. Vinciarelli, Phys. Med. Biol. 20 (1975) 830.

9) G. Charpak and J. Saudinos, Progr. Nucl. Med. 7 (1981) 164.

10) A. Abbondandolo, S. Bonatti, R. Bellazzini, G. Betti, A. Del Guerra, M.M. Massai, M. Ragadini, G. Spandre and G. Tonelli, Radiat. Environ, Biophys. 21 (1982) 109.

11) G. Peterson, G. Charpak, G. Melchart and F. Sauli, Nucl. Instrum. Methods 176 (1980) 67.

12) A. Cattai, Nucl. Instrum. Methods 215 (1983) 3.

13) G. Charpak, C. Demierre, R. Kahn, J.C. Santiard and F. Sauli, Nucl. Instrum. Methods 141 (1977) 449.

14) R. Kahn, R. Fourme, R. Bosshard, B. Caudon, J.C. Santiard and G. Charpak, Nucl. Instrum. Methods 201 (1982) 203.

15) J.E. Bateman, M.W. Waters and R.E. Jones, RL 75-140 (1975).

16) J.E. Bateman, A.C. Flesher, J.F. Connolly, R. Stephenson, D.S. Fairweather, A.R. Bradwell and R. Wilkinson, RL 82-036 (1982).

17) A.P. Jeavons, G. Charpak and R.J. Stubbs, Nucl. Instrum. Methods 124 (1975) 491.

18) D. Chu, K.C. Tam. V. Perez-Mendez, C.B. Linn, D. Lambert, S.N. Kaplan, IEEE Trans. Nucl. Sci. NS-23 (1976) 634.

19) R.S. Hattner, C.B. Lim, S.J. Swann, L. Kaufman, V. Perez-Mendez, D. Chu, J.P. Huberty, D.C. Price and C.B. Wilson, IEEE Trans. Nucl. Sci. NS-23 (1976) 523.

20) A. Jeavons, K. Kull, B. Lindberg, G. Lee, D. Townsend, P. Frey and A. Donath, Nucl. Instrum. Methods 176 (1980).

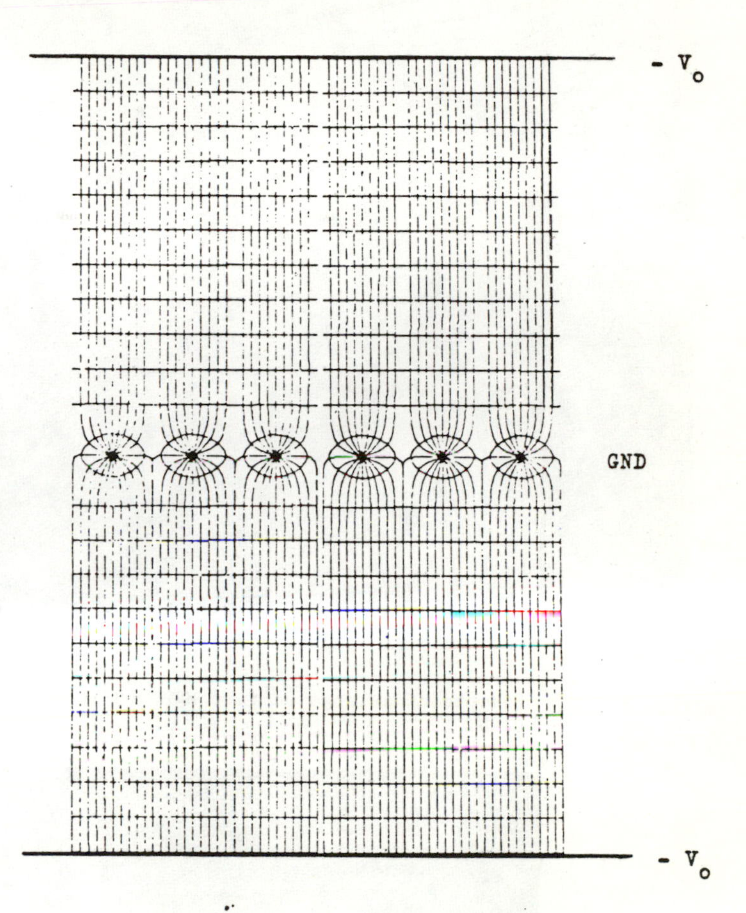

<u>Fig. 1</u> : Electric field and equipotential lines in a Multiwire Proportional Chamber.

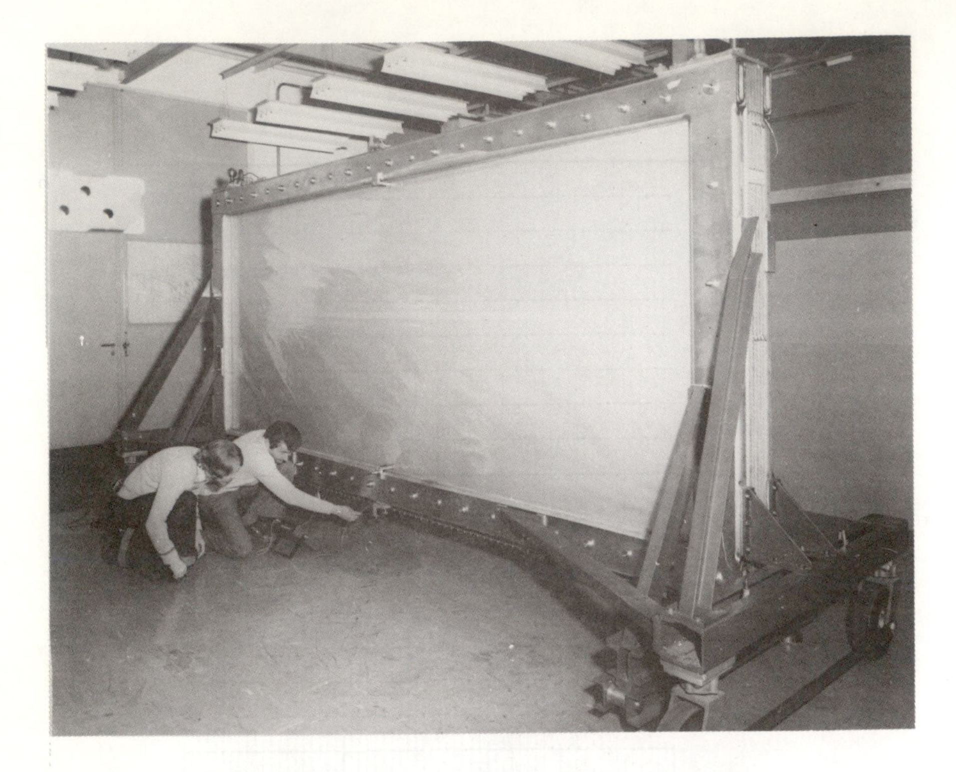

Fig. 2 : Example of a large surface MWPC built and operated at CERN.

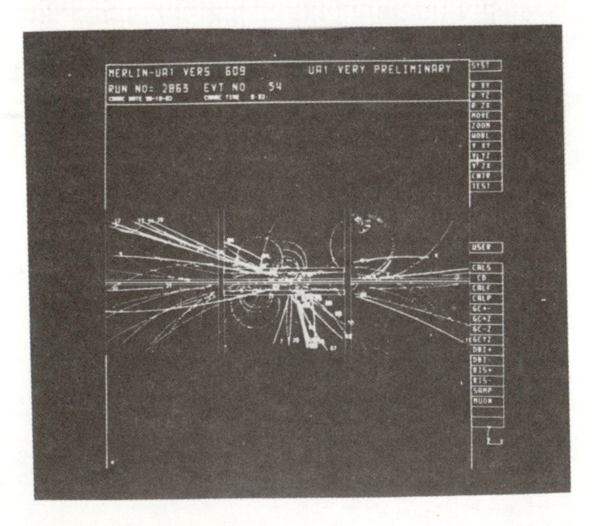

Fig. 3 : Computer display of a complex event obtained with a system of MWPC.

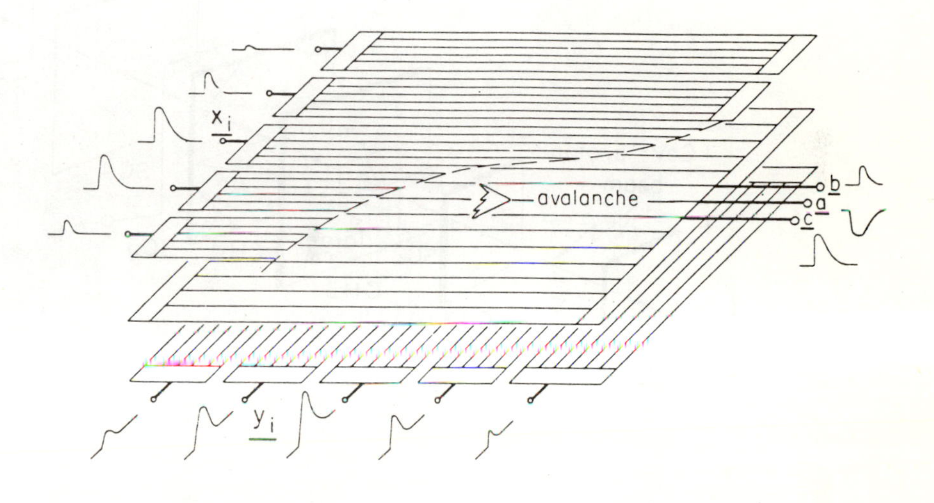

Fig. 4 : Principle of the localization method from the induced charges on the cathode planes of a MWPC.

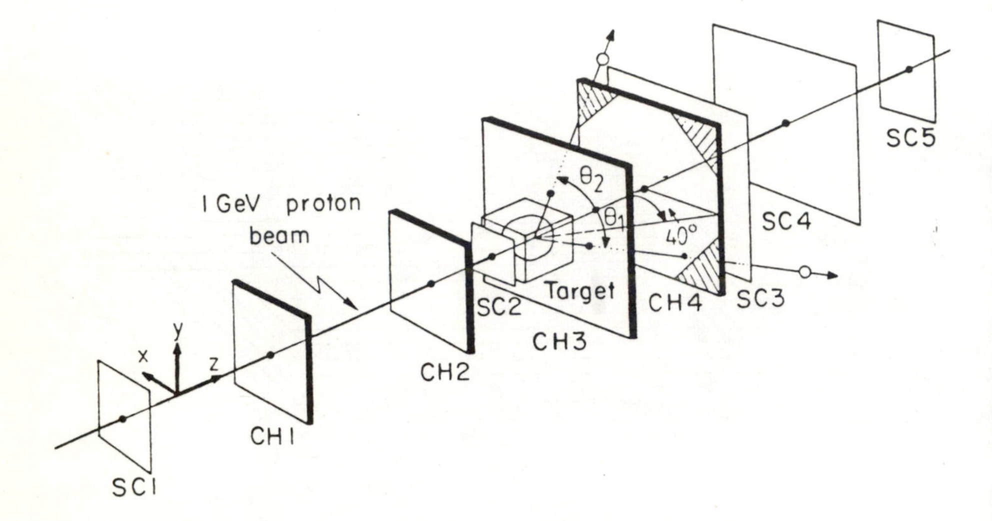

Fig. 5 : The proton scattering radiography setup.

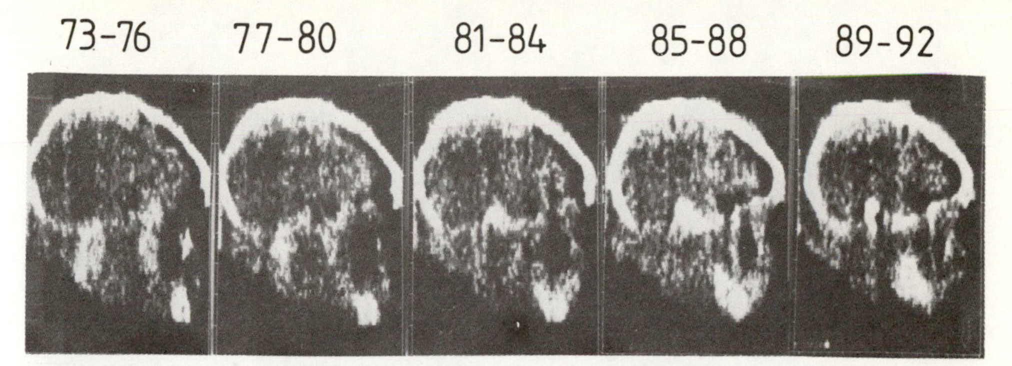

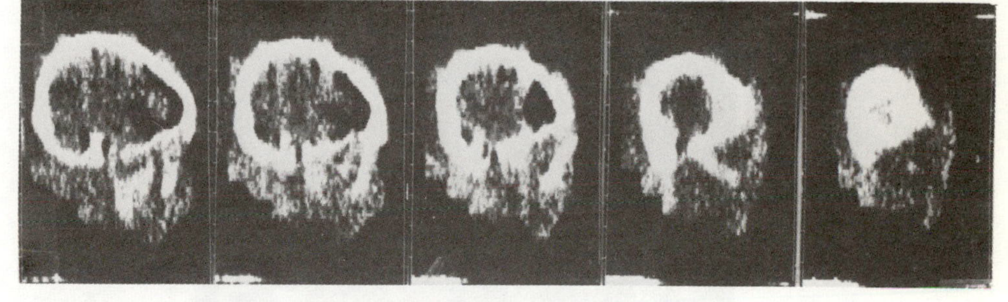

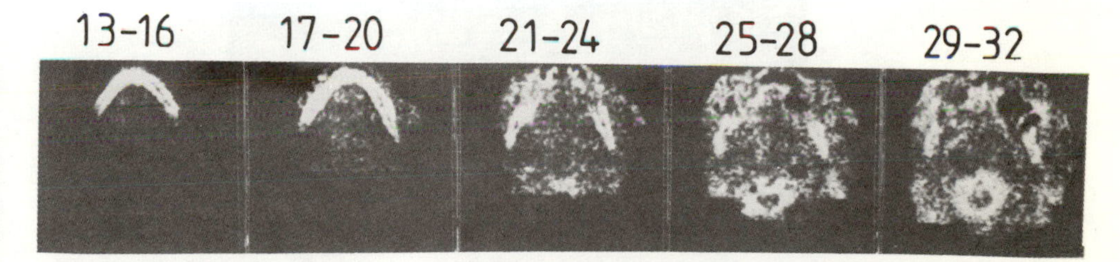

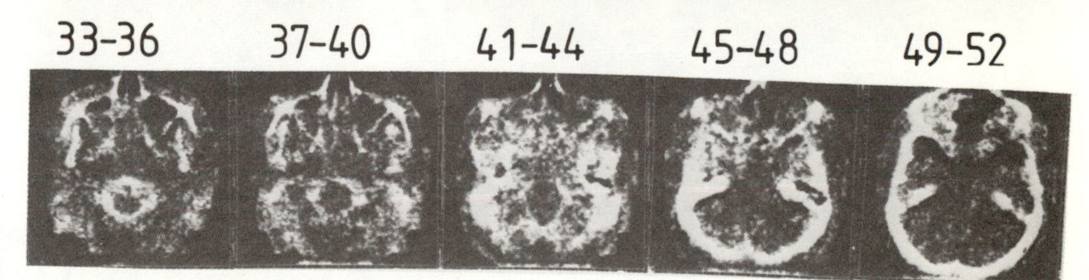

Fig. 6 : Computer display of a proton scattering tomography of the head along two different planes.

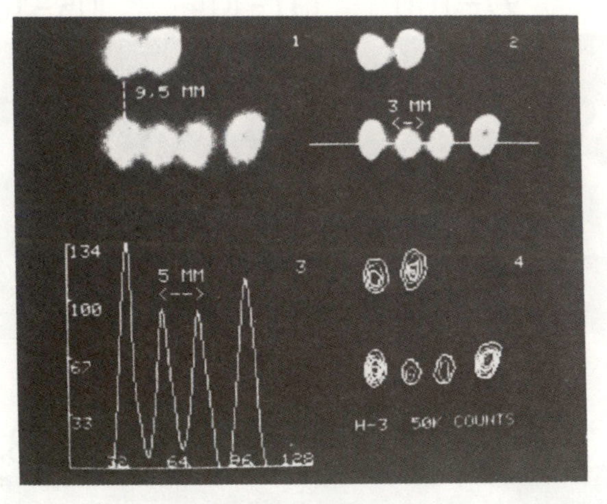

<u>Fig. 7</u> : Activity distribution of mammalian cell colonies labelled with tritium, measured with a MWPC.

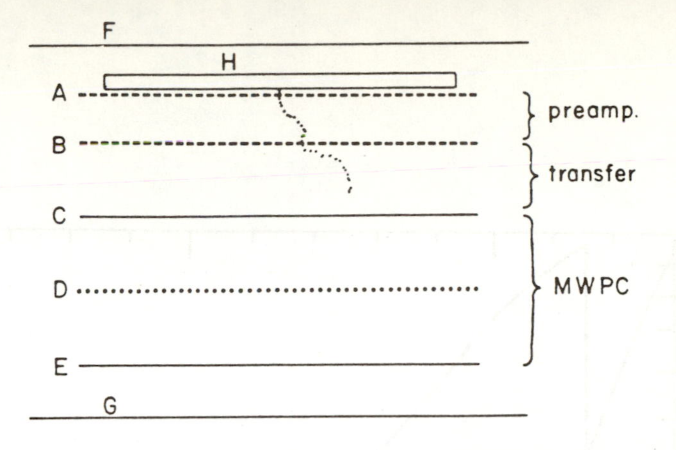

Fig. 8 : Use of a multistep proportional chamber to localize the emission point of long range electrons.

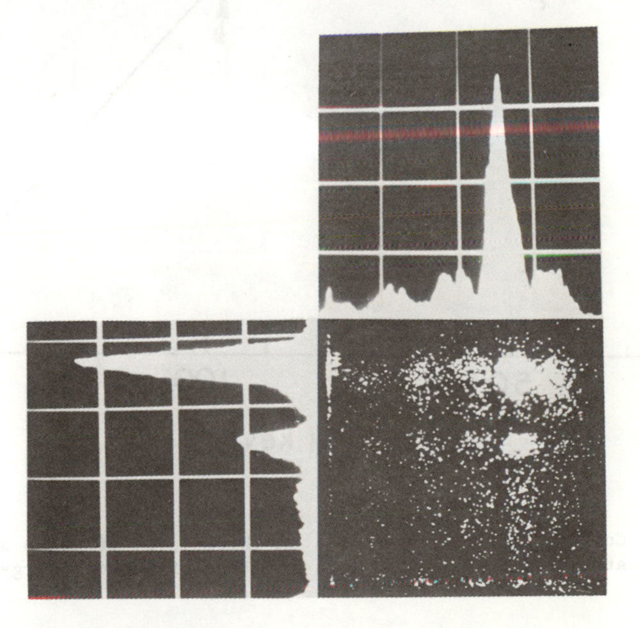

Fig. 9 : Bi-dimensional radiochromatography on ^{14}C obtained with a multistep chamber. Activity projections on the coordinate axes are also shown.

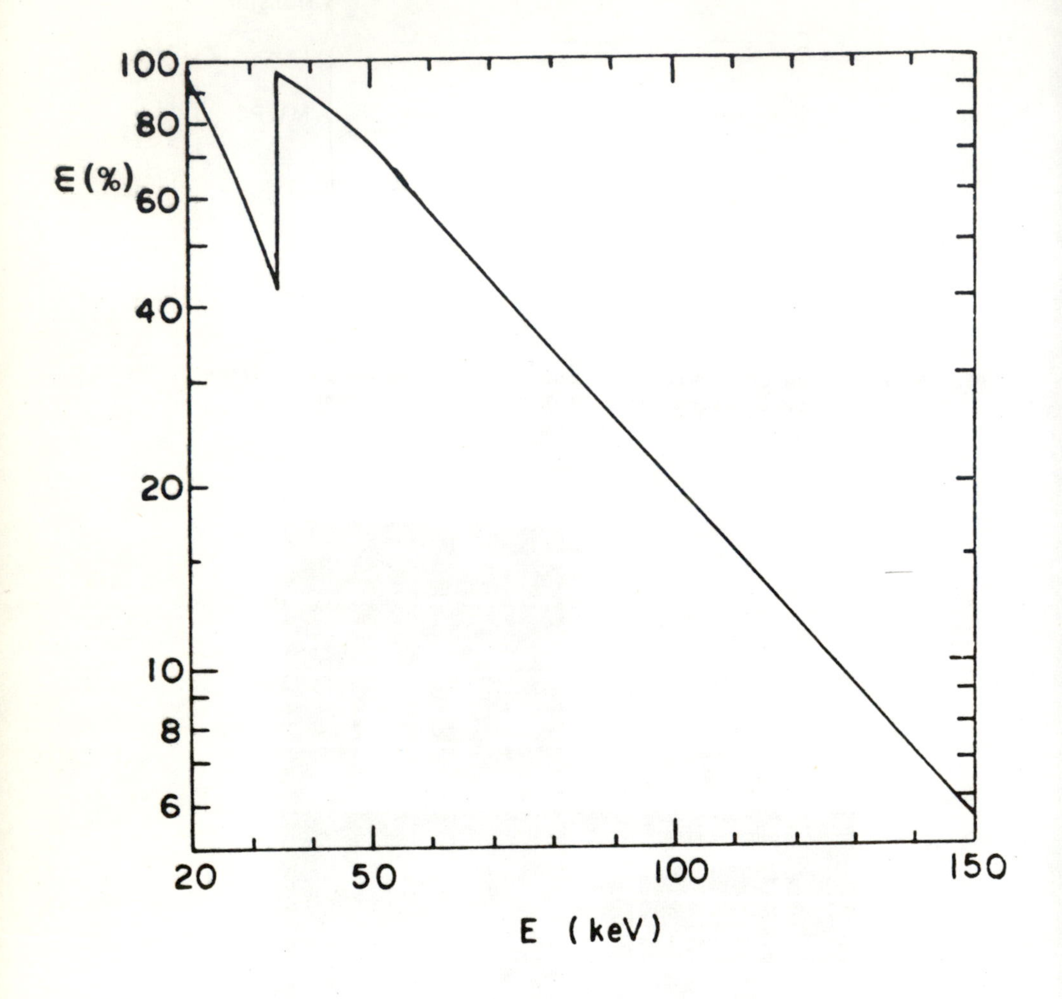

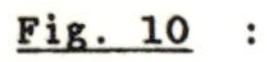

Fig. 10 : Computed efficiency of a 6 cm thick layer of xenon at atmospheric pressure, as a function of photon energy.

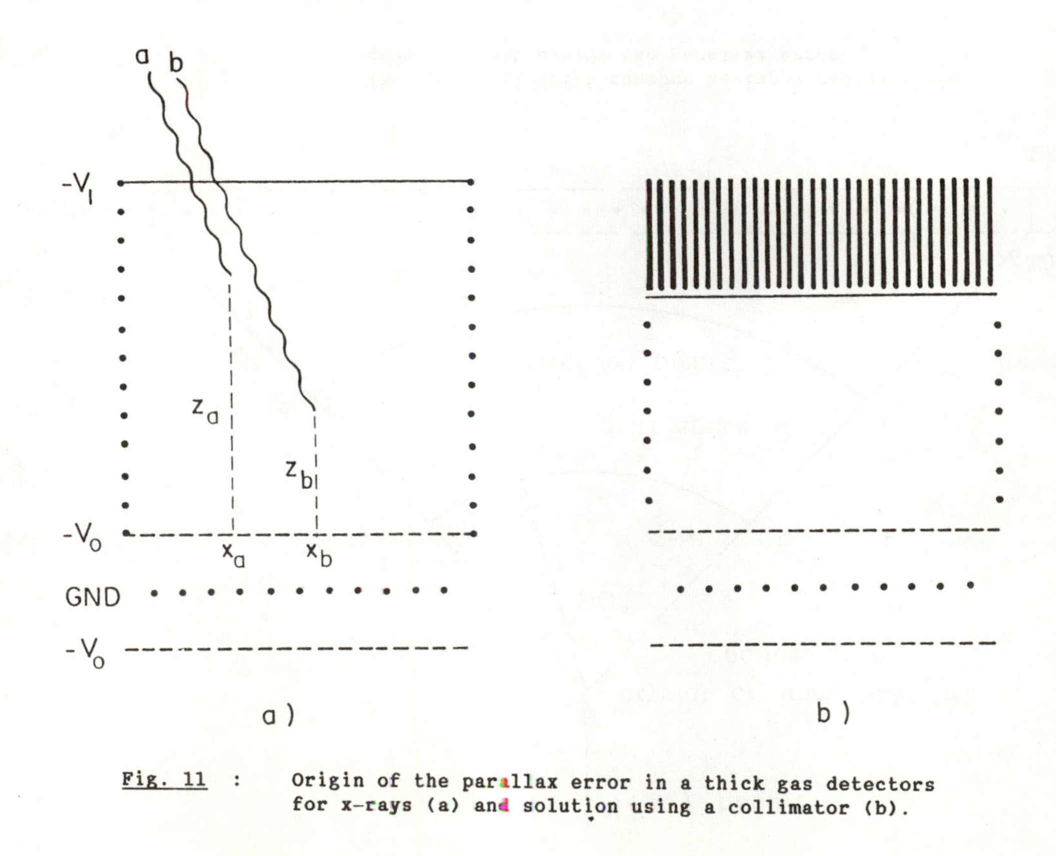

Fig. 11 : Origin of the parallax error in a thick gas detectors for x-rays (a) and solution using a collimator (b).

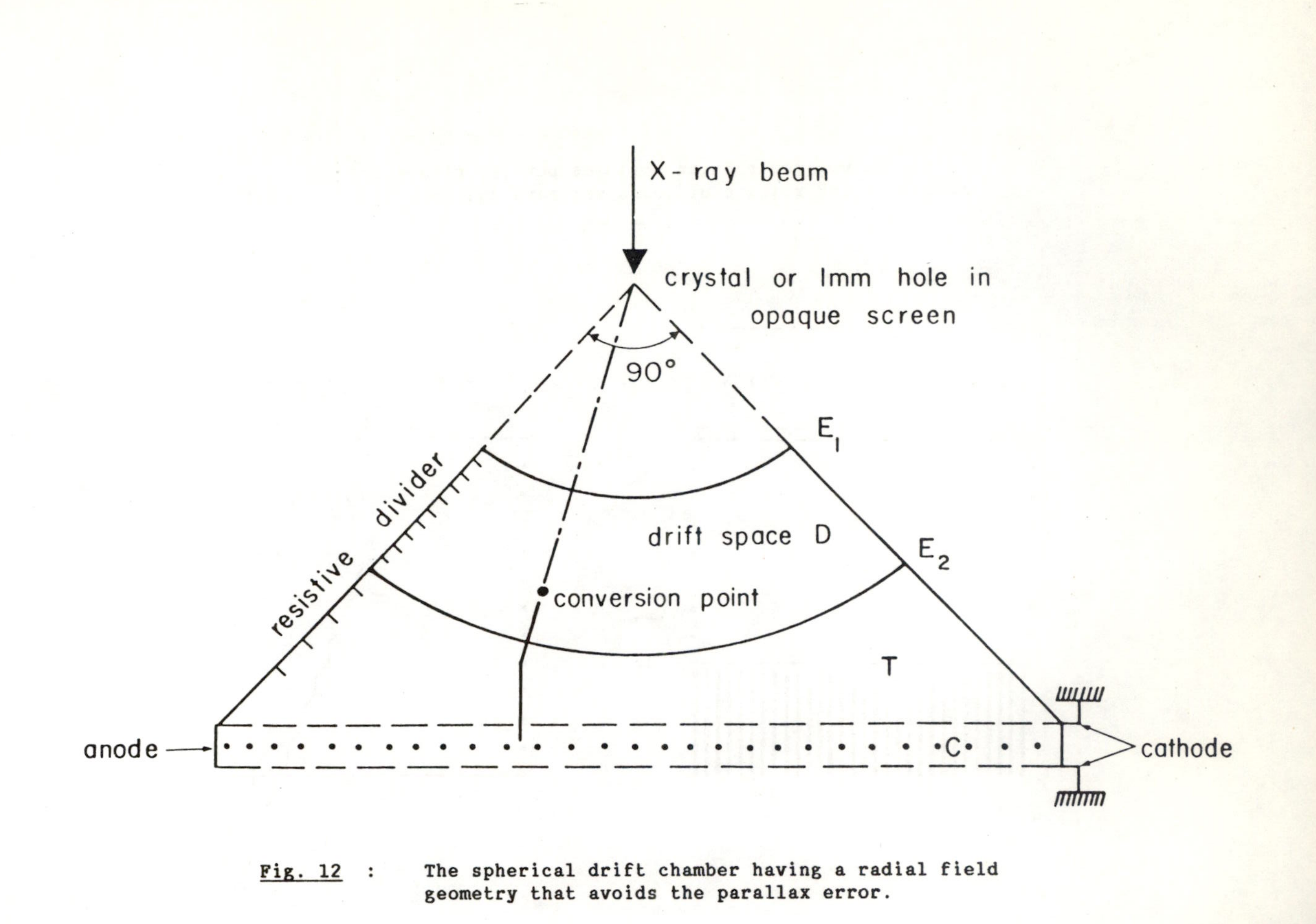

Fig. 12 : The spherical drift chamber having a radial field geometry that avoids the parallax error.

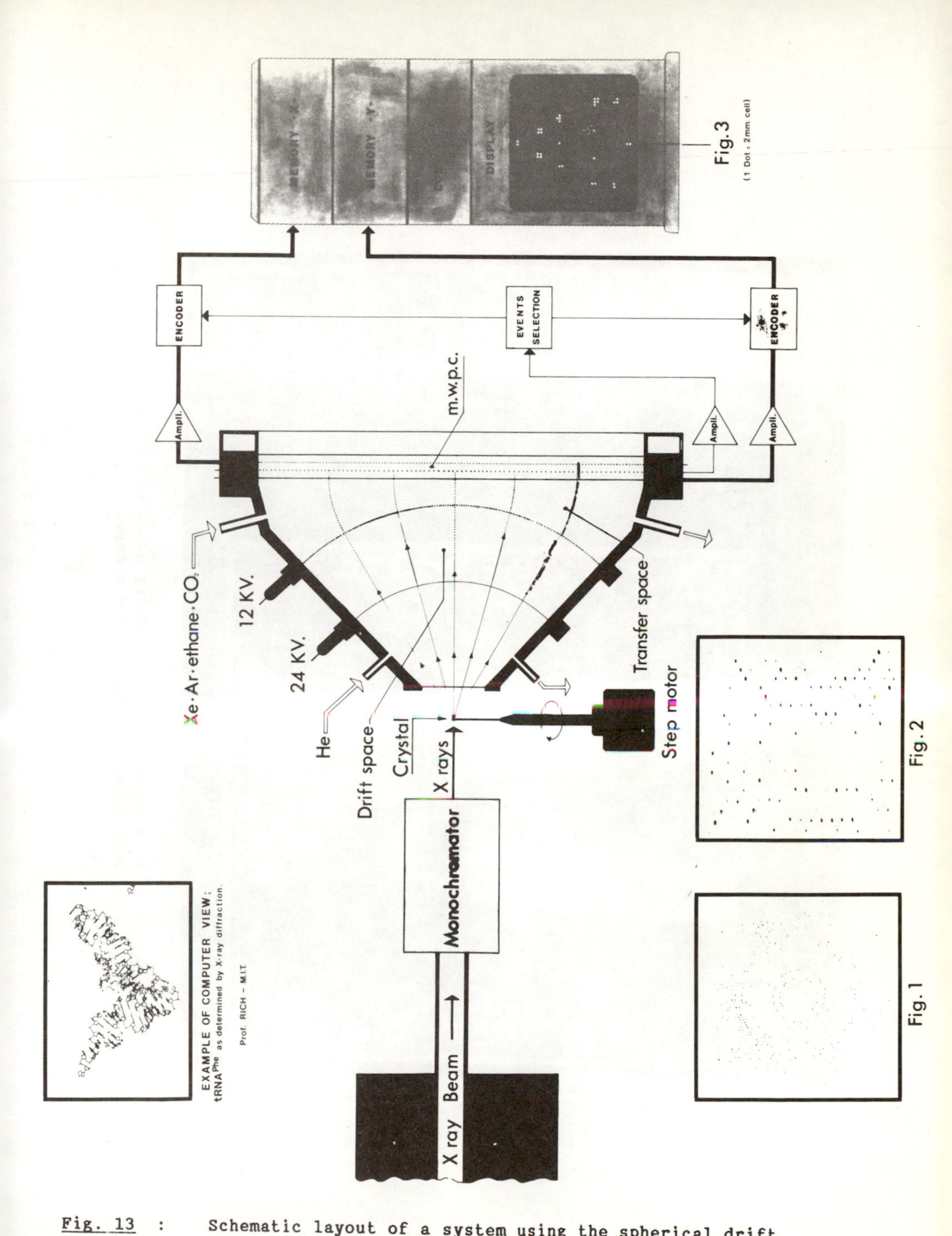

Fig. 13 : Schematic layout of a system using the spherical drift chamber for protein diffraction crystallography.

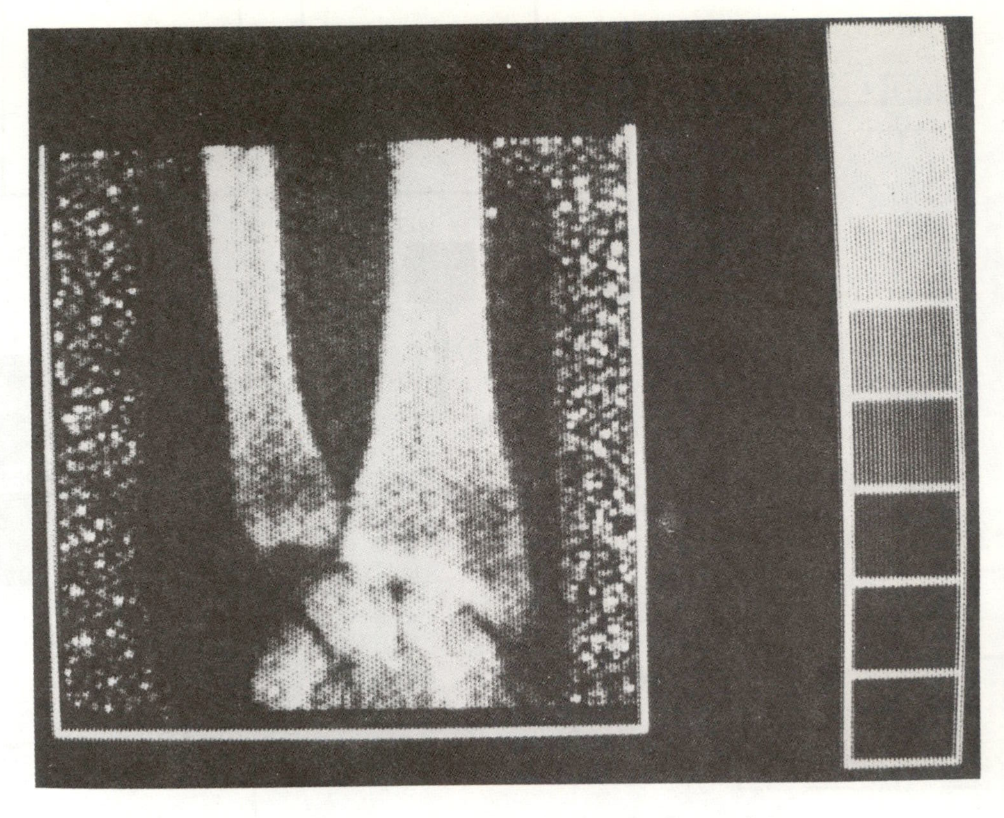

Fig. 14 : Bone absorption densitometry obtained with a multilayer MWPC with ^{153}Gd x-rays.

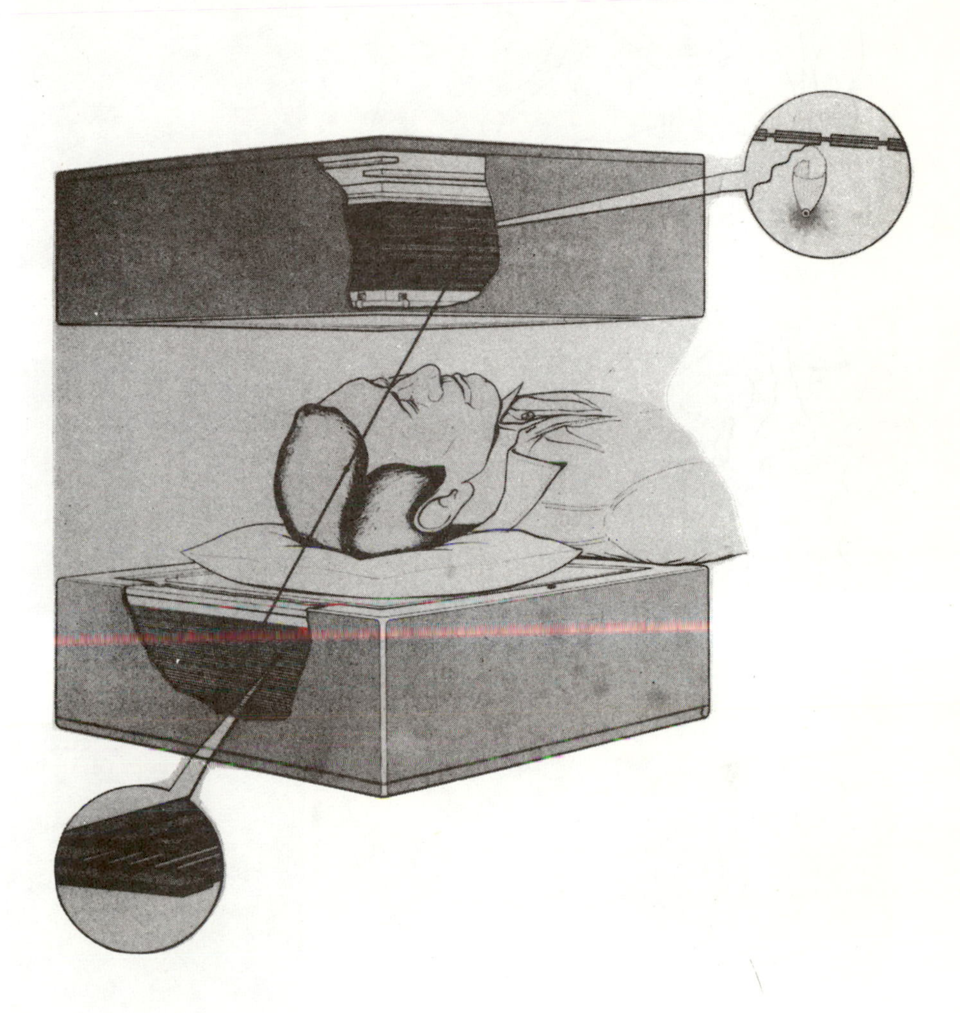

Fig. 15 : Schematic construction of the double, multilayer MWPC used as positron camera at Rutherford Lab.

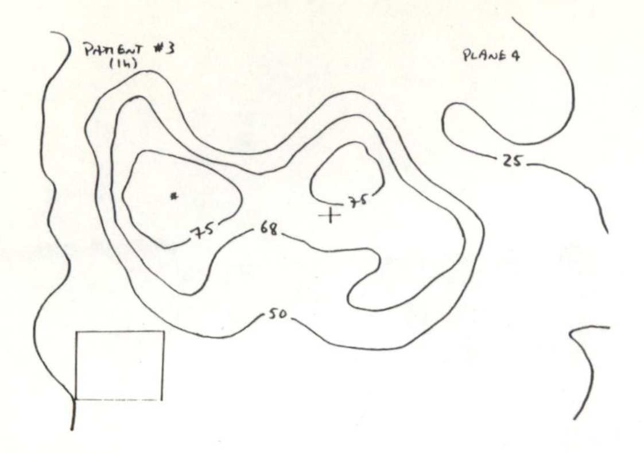

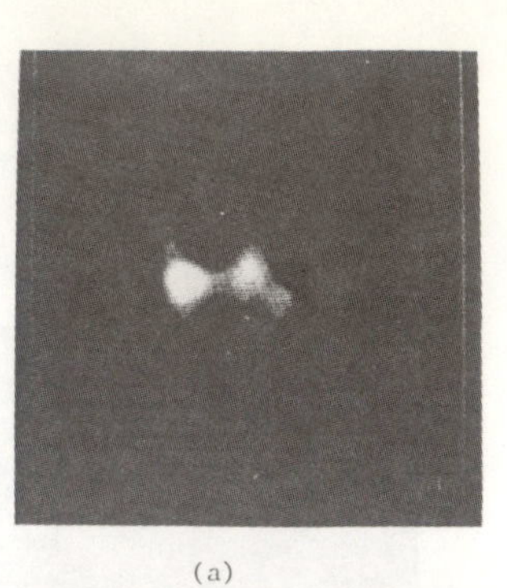

(a)

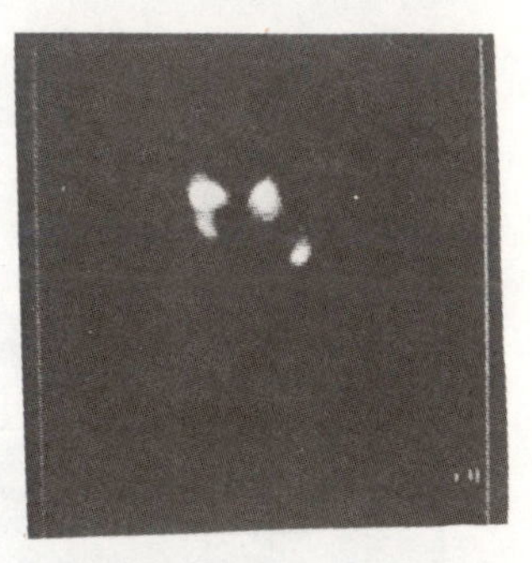

(b)

(c)

<u>Fig. 16</u> : Tomography and isocount distributions measured with a positron camera in a thyroid injected with ^{124}I after different time intervals.

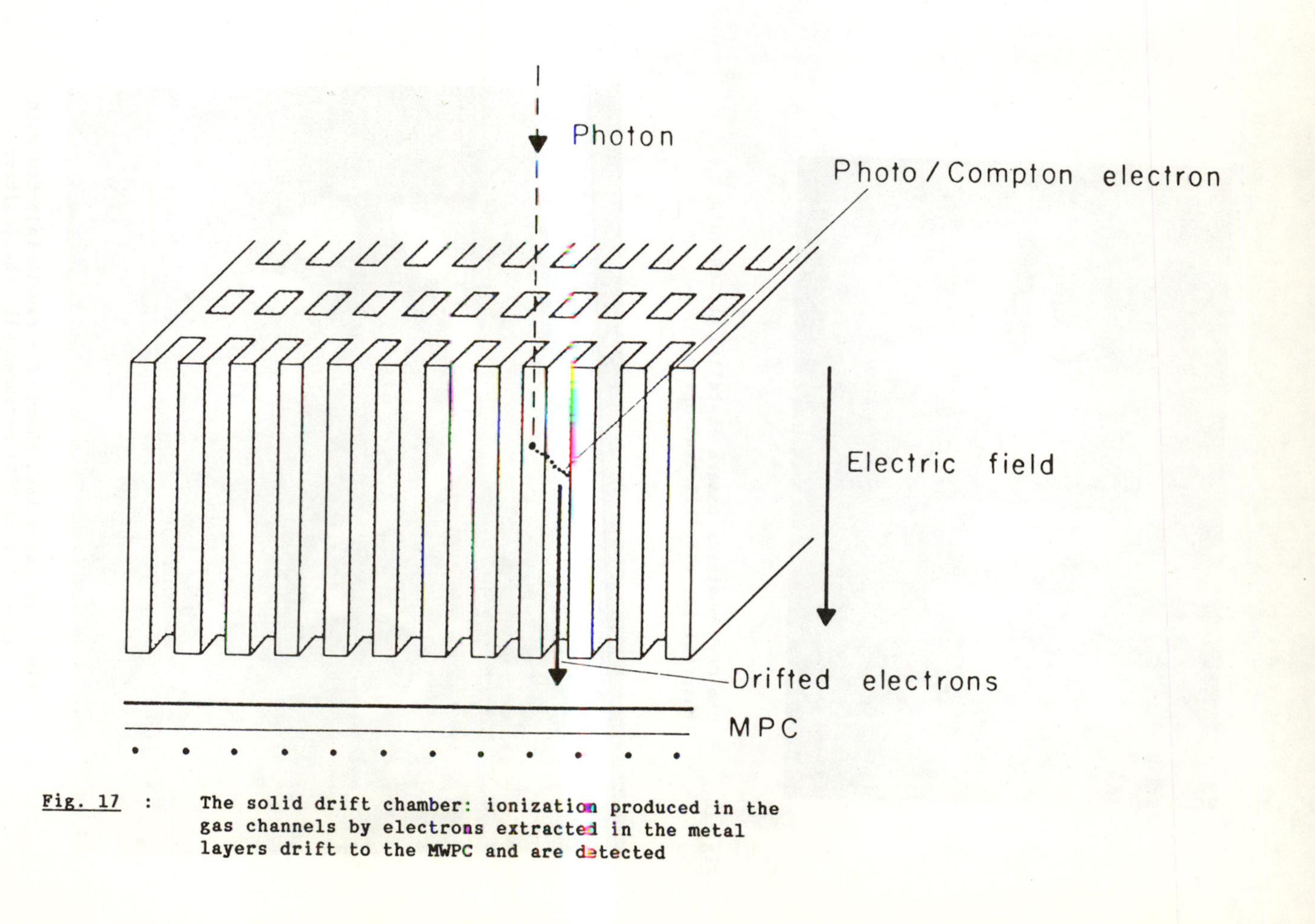

Fig. 17 : The solid drift chamber: ionization produced in the gas channels by electrons extracted in the metal layers drift to the MWPC and are detected

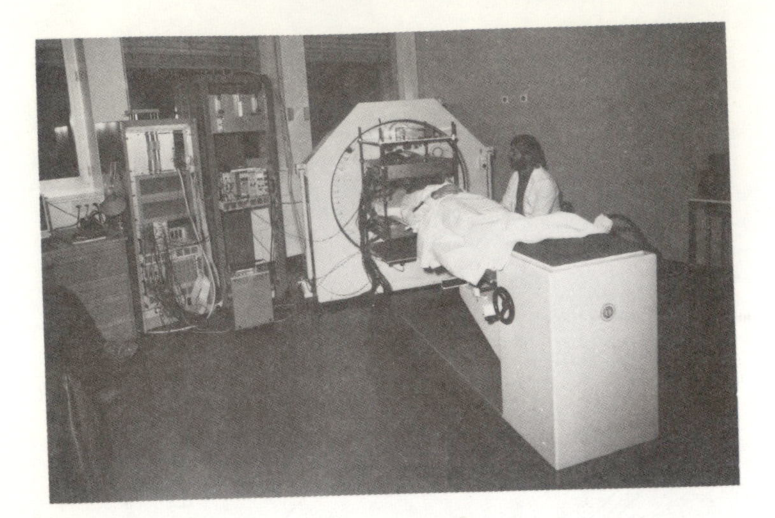

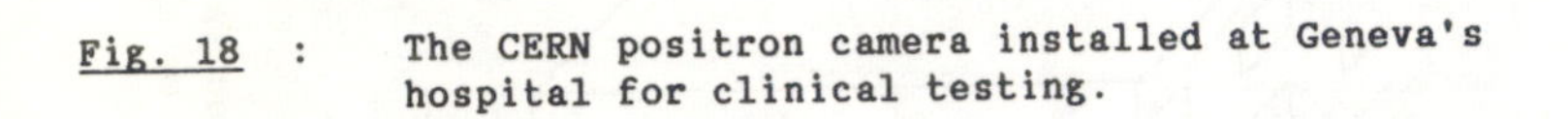

Fig. 18 : The CERN positron camera installed at Geneva's hospital for clinical testing.

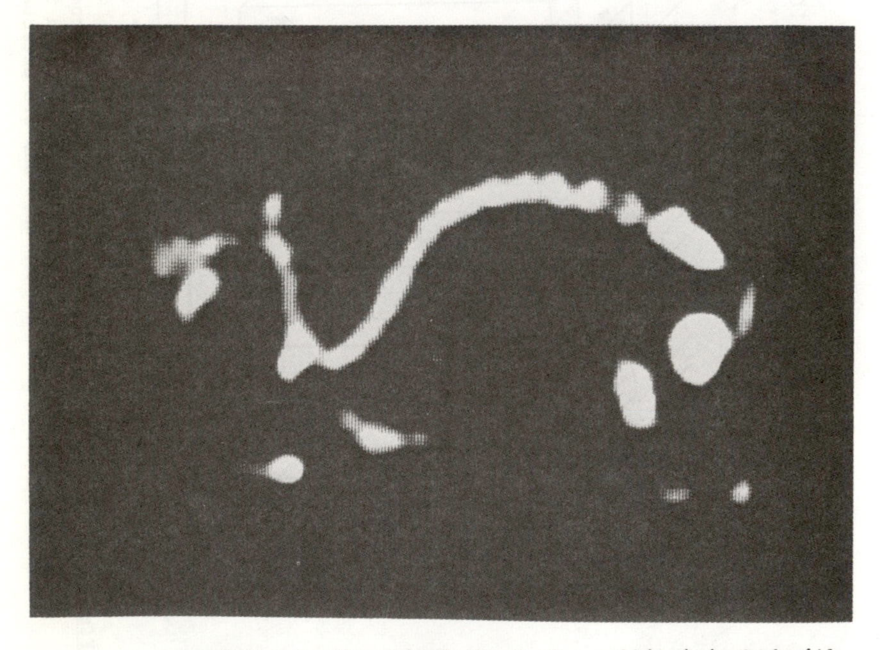

Fig. 19 : Activity in the spinal chord of a rabbit injected with 100 microcuries of ^{68}Ga measured with the positron camera.

PROCEEDINGS OF THE II INTERNATIONAL CONFERENCE ON APPLICATIONS OF PHYSICS TO MEDICINE AND BIOLOGY
edited by Ž. Bajzer, P. Baxa & C. Franconi

BIOLOGICAL BASIS OF THERMOSENSITIVITY OF TUMOR CELLS *

Bruno Mondovì and Pierluigi Riccio
Institute of Applied Biochemistry and CNR Center of Molecular Biology
University of Rome, Rome, Italy

The earliest clinical report concerning the effects of heat on cancer appears to be that of Bush in 1866[1] who described the complete disappearance of a histologically proven sarcoma of the face after two attacks of erysipelas.

Bruns reported in 1884[2] a terminal patient with multiple recurrent melanomas, who had an attack of erysipelas with a temperature of over 40°C for several days. All the tumors regressed completely and 8 years later the patient was still alive.

In 1893 Coley[3] discussed the cases of 38 patients with histologically diagnosed advanced cancer who had contracted accidental or deliberate infections of erysipelas with concomitant high fevers. The sarcomas regressed completely in the two patients who had the highest fevers, and their survival times were 27 and 7 years.

Nauts et al. reported[4] 30 selected patients with advanced cancer, treated with Coley's toxin (filtered extracts of Streptococcus and Bacillus Prodigiosus), 25 of which were alive and free of diseases over 10 years after the time of the treatment. It is interesting to note that all of these patients were treated with early batches of the toxin which were highly pyrogenic. Comparable results were not obtained when patients were inoculated with less pyrogenic batches: for some unexplained reason the authors[4] did not attribute the beneficial results to the high fever.

The first demonstration of a direct effect of heat on tumors come in 1967[5]: in 22 patients with cancer of the limbs, the temperature was raised to 41.5-43.5°C for several hours in 25 regional perfusions with prewarmed blood. Some of these patients are still alive today.

---

* In part supported by the Italian National Research Council - Special Project on the "Control of Neoplastic Growth", contract No.82.00360.96.

As the first results revealed more positive than expected, it became primarily important to clarify the nature of the biological factor(s) that make tumor cells selectively sensitive to high temperatures.

It is helpful to discuss some biological problems related to the employment of heat treatments both in the cancer therapy "in vivo" and in the experiments "in vitro". First, the terms "hyperthermia", "supranormal temperature" and "heat treatment" refer, unless specifically stated otherwise, to temperature which are above the normal temperature of the animal body (or tissue or cell culture) being studied. The majority of the researchers utilize a temperature range between 40 and 43°C. Temperatures above 43°C damage normal structures so fast that it becomes impractical to use them to selectively kill tumor cells.

Thermosensitivity appears to be a general property of the neoplastic cell acquired with the malignant transformation, as demonstrated by Giovanella[6)] on fibrosarcoma and normal fibroblast cells (fig. 1).

A selective and irreversible damage to the respiration of Novikoff hepatoma cells, produced by incubation at 42-44°C, has been demonstrated[7)]; in the same conditions normal and regenerating rat liver cells are not affected (see fig. 2). Respiration is also depressed in human osteosarcoma and human intestine carcinoma, but a longer incubation time at high temperature is required.[8)]

The inhibition of DNA, RNA and protein synthesis appears to be particularly evident after heat treatments in tumor cells as compared with normal ones. In fact, the incorporation of radioactive precursors of DNA in Novikoff hepatoma shows an inhibition of about 95% after 2 hours of incubation at 42°C or 30 minutes at 43°C, while in the control cells no inhibition is observed[9)].

Table 1 summarizes the main results obtained in our laboratory on several human and experimental tumors, including minimal deviation hepatoma.

These results confirm the hypothesis that all tumors are, more or less, sensitive to supranormal temperatures.

Cells of the same tumor type may vary in their sensitivity to heat, as Rofstad and Brustad[10)] demonstrated on human melanoma cells incubated at 42.5°C for various times both "in vivo", as a solid tumor growing in athimic nude mice, and "in vitro". The response to heat treatment was different in all the five melanomas studied. This finding may explain

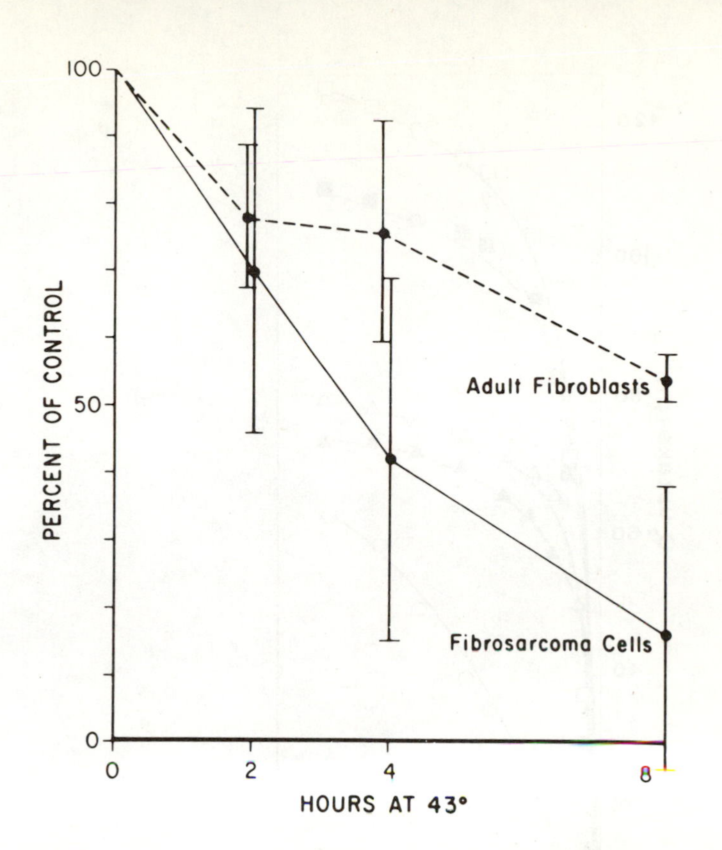

Fig. 1. The percentage of surviving fibrosarcoma cells and normal adult fibroblasts as a function of the duration of exposure to 43°C (from Ref. 6).

the different therapeutic results obtained for the same tumor type in different patients.

Another problem which has been studied is represented by the thermotolerance. During the heat treatment of patients with cancer, tumor

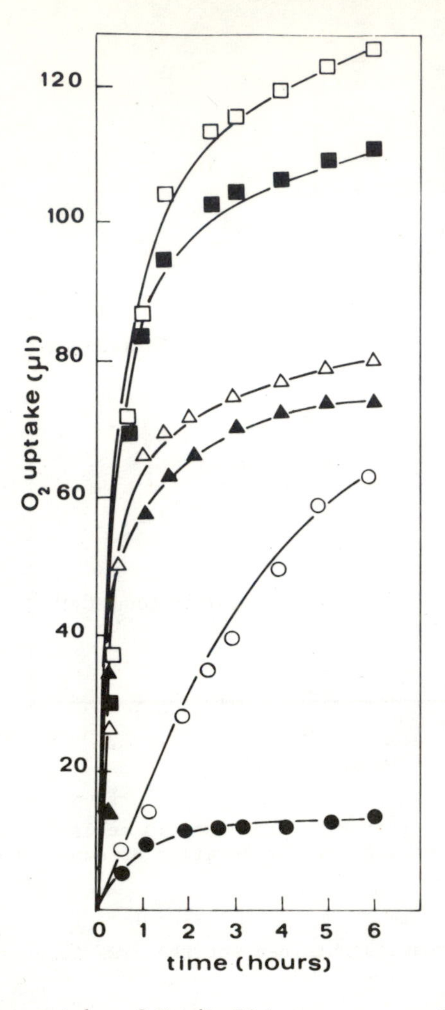

Fig. 2. Oxygen uptake of Novikoff hepatoma, rat liver and regenerating rat liver cells, incubated at 38°C and 42°C in the presence of 0.015 M D-Glucose and 0.013 M succinate. ●, Novikoff hepatoma, 42°C; O Novikoff hepatoma, 38°C; ▲, rat liver, 38°C; △ rat liver, 42°C; ■ regenerating rat liver, 42°C; □, regenerating rat liver, 38°C (from Ref.8).

cells become resistant to subsequent heat treatments. The extent of this phenomenon depends upon the time interval between one application and

TABLE I

Effect of "high temperature" on oxygen uptake and DNA synthesis in normal and neoplastic cells.

Values are expressed as % of the values obtained in equal samples preincubated for equal time at 38°C.

		Time of incubation at 43°				
		30 min	60 min	90 min	120 min	210 min
Human osteosarcoma cells	O_2 uptake	100	55.4	52.6	52.6	34.9
	DNA synthesis	—	—	—	—	37.8
Human rectum adenocarcinoma cells	O_2 uptake	100	77.6	47.3	37.7	30.0
	DNA synthesis	—	—	—	—	—
Rat Novikoff hepatoma cells	O_2 uptake	100	90	60.8	20.0	7.6
	DNA synthesis	20.2	—	—	11.8	—
Rat Morris 5123 "minimal deviation" hepatoma cells	O_2 uptake	—	100	—	90.2	61.4
	DNA synthesis	—	—	—	36.8	—
Rat regenerating liver cells	O_2 uptake	—	100	—	100	100
	DNA synthesis	110	—	—	112	—
Cell-free homogenate from Novikoff hepatoma cells.	O_2 uptake	—	—	—	102	—
	DNA synthesis	—	—	—	110	—

Table 1 (from Ref. 8).

another. In C3H breast carcinoma of the rat, thermotolerance increases with the time interval increase between one application and another to a maximum of 16 hours: it then decreases and disappears after 120 hours[11].

The "in vitro" experimenting has revealed very useful in the cancer research because of the possibility to easily manipulate the incubation medium and therefore to study the response of cells under different conditions.

Actually the composition of the incubation medium plays a very important role in modulating the effects that high temperature produces on the cells: in fact, a low pH enhances their sensitivity to heat[12].

When cells are cultured in the presence of D_2O they become more resistant to high temperatures. D_2O shifts the thermal denaturation curves of both DNA and proteins towards higher temperatures and increases protein aggregation, which results in increased thermostability[13]; the temperature threshold for cell killing is progressively shifted towards higher temperatures in the presence of increasing concentration of D_2O.

The effect of heat on tumor cells can be enhanced by the addition of some naturally occurring metabolites to the medium, such as polyamines or their oxidation products. Fig. 3 shows the enhancement of the thermal effect (Ben Hur et al.[14]) observed when spermidine is added to the culture medium.

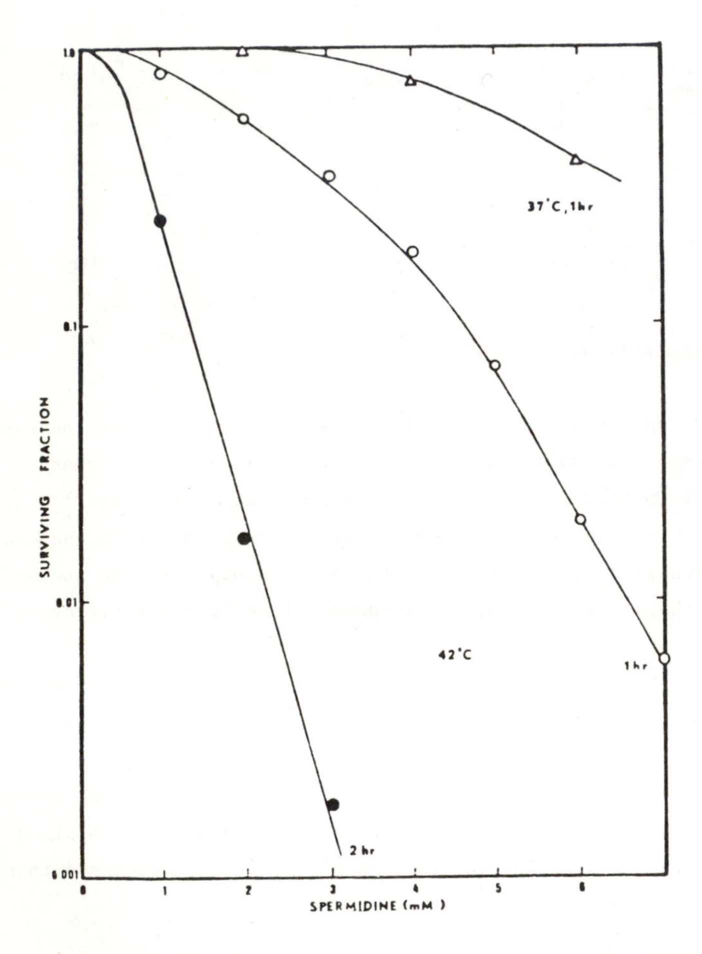

Fig. 3. Survival of Chinese hamster cells exposed to graded spermidine concentrations at 42°C and 37°C for 1 or 2 h, as indicated (from Ref.14).

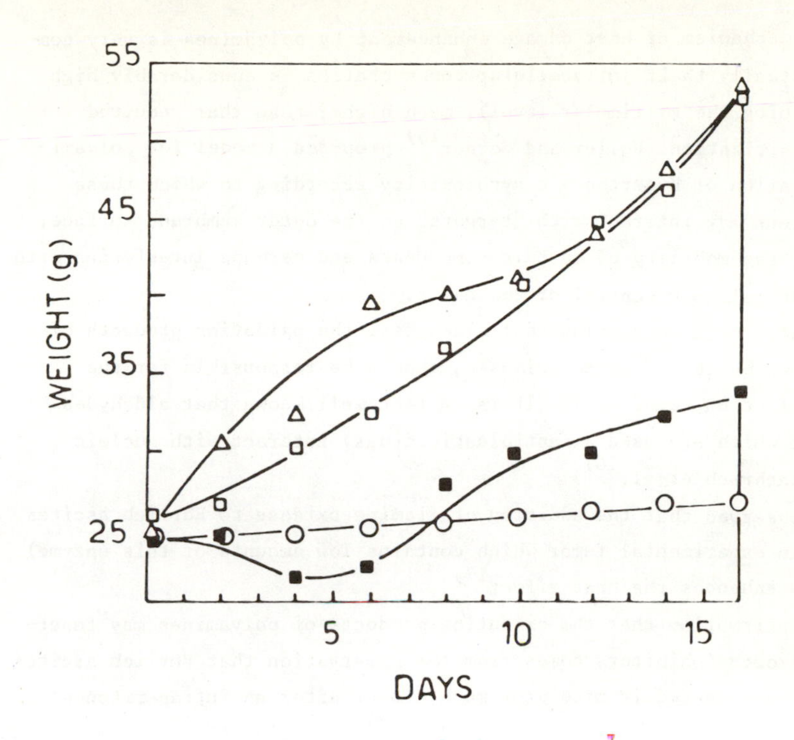

Fig. 4. (△) Tumor growth after inoculation of 3x10^7 viable tumor cells; (■) 24 h after the inoculation of 3x10^7 viable tumor cells, 0.3 units of immobilized DAO were injected intraperitoneally; (O) controls (mice without tumors); (□) 24 h after the inoculation of 3x10^7 viable tumor cells, 0.5 ml of ConA Sepharose, corresponding to the same amount used as a carrier of immobilized DAO, were injected into the peritoneal cavity. Groups of 7 mice were used for each experiment (from Ref. 18).

Radiation, and also some antibiotics used as antiblastic drugs, act in part indirectly by means of the production of radicals derived from oxygen, like superoxide (O_2) and its dismutation derivatives. The possible involvement of radicals and of the enzymes devoted to their elimination in the mechanism of hyperthermic killing have been studied. The cellular content of superoxide dismutase, catalase, glutathione reductase and glutathione peroxidase has recently been investigated[20] in tumor cells showing different sensitivity to heat, but no relationship between the activity of these enzymes and thermal sensitivity was found.

The mechanism of heat damage enhancement by polyamines is very complex; actually their intracellular concentration is considerably high (approaching the millimolar level), much higher than that required for heat sensitization. Fuller and Gerner[15] proposed a model for polyamine modulation of hyperthermic cytotoxicity according to which these polycations may interact with "targets" on the outer membrane surface, reducing the mobility of certain components and perhaps interfering with the osmoregulatory control of the cells.

We have data suggesting that aldehydes, the oxidation products of polyamines by way of amine oxidases, should be responsible for the toxic effect on tumor cells. It is in fact well known that aldehydes (some of which are used as antiblastic drugs) interact with nucleic acids (Bachrach et al.[16]).

We observed that the addition of diamine oxidase to Ehrlich ascites cells (an experimental tumor which contains low amounts of this enzyme) strongly enhances the heat effect[17].

A confirmation that the oxidation products of polyamines may function as growth inhibitors comes from the observation that Ehrlich ascites tumors transplanted in mice grow much slower after an intraperitoneal injection of immobilized diamine oxidase[18] (See fig. 4).

The hyperthermic treatment, as used in cancer therapy, is generally performed in combination with radiotherapy and/or chemotherapy. These techniques interfere with the cell metabolism, and those cells particularly affected appear to be those which are characterized by an active proliferative metabolism like the cancer cells. The cellular target of radiation and antiblastic drugs are many, but more important appears to be the damage that these physical and chemical agents produce to the nucleic acids, and to DNA in particular. All living organisms possess specific enzymes which repair the structural damage of DNA; the heat treatment largely affects the DNA repair synthesis[19], and this characteristic is of primary importance in the clinical applications.

Fig. 5 shows DNA repair synthesis inhibition after UV irradiation, that causes the dimerization of adjacent pyrimidine residues with consequent inability of DNA to be replicated and transcripted into RNA, unless the above mentioned enzymes (which probably represent the target of the heat treatment in this particular case) repair the damage[19].

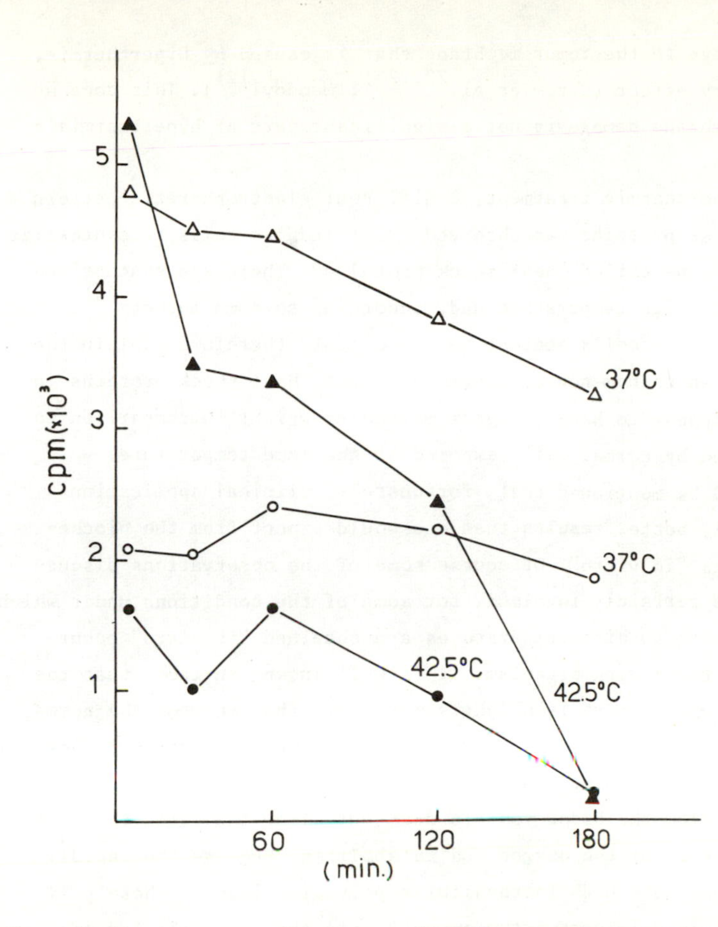

Fig. 5. Chinese hamster fibroblasts irradiated with UV light (10 $J.m^{-2}$) and exposed for 1,2 and 3 hours at 42.5°C and 37°C. (△ and ▲): irradiated cells; (O and ●): control cells (From ref. 19).

The role of plasma membranes as possible targets of hyperthermia is not established. The membranes at first seemed to be a primary target (Strom et al.[21]); however, subsequent results indicate that the irreversible damage to the tumor membrane that is caused by hyperthermia, is a secondary effect (Strom et al.[22, 23]; Mondovì[12]). This does not mean that membrane damage is not a significant part of hyperthermia's effectiveness.

After hyperthermic treatment, a different electrophoretic pattern of the cellular proteins was observed. Heat induces cells to synthesize abnormal proteins called "heat shock proteins". These are synthesized in response to high temperature and, according to some authors[24], should protect the cells against heat and could therefore explain the above mentioned phenomenon of thermotolerance. Heat shock proteins in tumor cells appear to have a higher molecular weight in comparison to those produced by normal cells exposed at the same temperature.

It should be mentioned that, fortunately, clinical applications give generally better results than one would expect from the biochemical findings "in vitro". Of course some of the observations discussed above are certainly involved, for some of the conditions under which better responses to high temperatures are obtained "in vitro" occur naturally in the living organism. It is well known, in fact, that the tumor tissue environment is slightly more acid than that of the normal tissue and therefore becomes more thermosensitive, as mentioned above.

It is also ascertained that the tumor tissue becomes necrotic when its distance from the blood vessels does not allow the cells to be efficiently reached by the oxygen and metabolites. Because the rapidly growing tissues have high intracellular polyamine levels, these polycations are released in noteworthy amounts into the extracellular environment and could therefore function as heat sensitizers according to the mechanism discussed above.

Moreover, it sometimes happens that, because of decreased thermodispersion, in part attributable to vascular alterations of the tumor due to the hyperthermic treatment, the neoplastic tissue is heated more than the normal surrounding one (Lee Veen[25]).

It should be stated that the hyperthermic treatment "in vivo" appears to enhance the immunogenicity of the tumors, as demonstrated by Mondovì et al.[26] on Ehrlich ascites cells.

Finally hyperthermia, when performed "in vivo", enhances the natural killer (NK) cells' activity[27]. It is still unclear why human NK activity should be depressed "in vitro" as the temperature is raised over

37°C while their activity is enhanced by hyperthermia "in vivo".

It is extremely hard to summarize the main findings on hyperthermia among the thousands of experiments that have been made in this field. We can say that the hyperthermic treatment, associated to radiotherapy and/or chemotherapy, is so far an effective method to be used in cancer therapy.

The future efforts have to be directed towards the choosing of the best antiblastic drugs to be used, and the best conditions to be adopted when ionizing radiations are employed, in combination with the heat treatment. The rapid development of sophisticated techniques for heating deep tissues, such electromagnetic irradiation, may be the goal for the next years.

REFERENCES

1) Bush, W. (1866): Verhandl naturh. Preuss. Rhein. Westphal. 23: 28-30.

2) Bruns, P. (1887): Beitr. Klin. Chir. 3: 443-466.

3) Coley, W.B. (1893): Am. J. Med. Sci. 105: 487-511.

4) Nauts, H.C., Fowler, G.A., and Bogatko, F.H. (1958): Acta Med. Scand. Suppl. 338: 5-47.

5) Cavaliere, R., Ciocatto, E.C., Giovanella, B.C., Heidelberger, C., Johnson, R.O., Margottini, M., Mondovì, B., Moricca, G. and Rossi Fanelli, A. (1967): Cancer 20: 1351-1381.

6) Giovanella, B.C., Stehlin, J.S. and Morgan, A.C. (1976): Cancer Res. 36: 3944-3950.

7) Mondovì, B., Strom, R., Rotilio, G., Finazzi-Agrò, A., Cavaliere, R. and Rossi Fanelli, A. (1969): Europ. J. Cancer 5: 129-136.

8) Cavaliere, R., Mondovì, B., Moricca, G., Monticelli, G., Natali, P.G., Santori, F.S., Di Filippo, F., Varanese, A., Aloe, L., and Rossi Fanelli, A. (1983): Hyperthermia in Cancer Therapy 369-399, G.K. Hall Medical Publishers, Boston, Mass.

9) Mondovì, B., Finazzi Agrò, A., Rotilio, G., Strom, R., Moricca, G. and Rossi Fanelli, A. (1969) Eur. J. Cancer 5: 137-146.

10) Rofstad, E.K. and Brustad, T. (1982): Cancer 50: 1304-1308.

11) Kamura, T., Nielsen, O.S., Overgaard, J. and Andersen, A.H. (1982): Cancer Res. 42: 1744-1748.

12) Mondovì, B. (1980): Ann. N.Y. Acad. Sci. 335: 202-203.

13) Fisher, G.A., Li, G.C. and Hahn, G.M. (1982): Rad. Res. 92: 530-540.

14) Ben Hur, E., Prager, A. and Riklis, E. (1978): Int. J. Cancer 22: 602-606.

15) Fuller, D.J.M. and Gerner, E. (1982): Rad. Res. 92: 439-444.

16) Bachrach, U. and Persky, S. (1969): Biochim. Biophys. Acta 179: 484-493.

17) Mondovì, B., Guerrieri, P., Costa, M.T., and Sabatini, S. (1981). Adv. Polyamine Res. 3: 75-84, Raven Press, N.Y.

18) Mondovì, B., Gerosa, P. and Cavaliere, R. (1982) Agents and Action 12: 450-451.

19) Bozzi, A., Mariutti, G., Mondovì, B., and Strom, R., to be published.

20) Mavelli, I., Mondovì, B., and Rotilio, G. (1982) Strahlentherapie 158: 386.

21) Strom, R., Caiafa, P., Mondovì, B. and Rossi Fanelli, A. (1969): FEBS Lett. 3: 343-347.

22) Strom, R., Crifò, C., Rossi Fanelli, A. and Mondovì, B. (1977): Recent Results Cancer Res. 59: 7-35.

23) Strom, R. Scioscia Santoro, A., Crifò, C., Bozzi, A., Mondovì, B. and Rossi Fanelli, A. (1973): Eur. J. Cancer 9: 103-112.

24) Alahiotis, S.N. (1983): Comp. Biochem. Physiol. 75B: 379-387.

25) Le Veen, H.H. (1979): Conference on thermal characteristics of tumors: Applications in detection and treatment, March 14-16, 1979, N.Y.

26) Mondovì, B., Scioscia Santoro, A., Strom, R., Faiola, R. and Rossi Fanelli, A. (1972): Cancer 30: 885-888.

27) Zanker, K. and Lange, J. (1982) The Lancet 1 (1982): 1079.

PROCEEDINGS OF THE II INTERNATIONAL CONFERENCE ON APPLICATIONS OF PHYSICS TO MEDICINE AND BIOLOGY
edited by Ž. Bajzer, P. Baxa & C. Franconi

MICROWAVES AND ULTRASOUND IN CLINICAL HYPERTHERMIA: SOME PHYSICAL ASPECTS OF HEATING AND THERMOMETRY

Jeffrey W. Hand
Medical Research Council Cyclotron Unit
Hammersmith Hospital
Du Cane Road
London W12 OHS

ABSTRACT

Early applications of hyperthermia are underway in many countries but many questions remain to be answered before its potential in cancer therapy can be fully assessed. Amongst these are the questions of how tissues are to be heated and their temperatures monitored.
A review of microwave and ultrasonic techniques covering external, interstitial and intracavitary applicators is given, followed by a discussion of invasive methods of thermometry currently used in hyperthermal therapy.

1. Introduction

The development of suitable methods to heat tissues and to measure the resulting temperature distributions within them is crucial to the future of hyperthermia in the treatment of human cancer. The use of microwaves for local hyperthermia has attracted much attention and descriptions of a number of applicators have appeared in the literature. Likewise, ultrasound has been used as a heating agent. In contrast to the modest success in developing heating techniques, adequate solutions to the problems of temperature measurement remain elusive. This paper gives a brief review of some physical aspects of electromagnetic and ultrasound techniques used in local hyperthermia together with a discussion of currently available methods for thermometry.

2. Heating by electromagnetic fields

When a plane electromagnetic wave propagates through uniform tissue, the magnitude of the E-field is reduced by a factor 1/e in a distance D, where,

$$D = c\sqrt{2} / \omega \, [(\epsilon'^{2} + (\sigma / \omega\epsilon_{0})^{2})^{1/2} - \epsilon')^{1/2}] \qquad (1)$$

and c = velocity of light, ω = angular frequency, ϵ' = real part of the relative permittivity of the tissue (dielectric constant), σ = conductivity of the tissue and ϵ_0 = permittivity of free space. D is known as the (plane wave) penetration depth and the absorbed power density is reduced to approximately 13.5% within this distance. An alternative parameter is the half-value distance, $d_{1/2}$; this is the distance in which the absorbed power density is reduced by a factor of one-half. The relationship between these factors is

$$d_{1/2} = 0.346\ D \qquad (2)$$

Figure 1 shows that, at a given frequency, the penetration in tissues with low electrolyte content (eg fat and bone) is greater than into those with high electrolyte content (eg skin, muscle) and that for both groups, the penetration increases as the frequency of the electomagnetic wave decreases.

In practice, the electromagnetic fields associated with real applicators differ considerably from plane waves. For example, microwaves are often transmitted from a waveguide structure with an aperture close to or in contact with the tissues to be heated or by means of coaxial cables terminated in small antennas which are inserted in the tissues either interstitially or through body cavities. At best, the plane wave penetration depth gives only an indication of the penetrating ability of practical applicators used to heat inhomogeneous tissues. In these cases, the curvature of the tissues, thicknesses of tissue layers and size of applicator must all be considered in relation to the wavelength of the electromagnetic field. In particular, the structure of the fields close to the applicator is important. In this region of the near field of the applicator, there are strongly reactive field components which vary rapidly in space. Unlike plane waves, the electric- and magnetic-fields, E, and H, are not orthogonal and the ratio E/H is different from that of free-space and may vary from point to point. In a medium with zero conductivity, the region is merely one of energy storage (Power flux = 0), energy being transmitted from the applicator by radiation in the far field. When, like tissue, the medium has finite conductivity, energy is also extracted from the near field. Since the fields in this region are relatively intense, excessive heating of the tissue results. As Johnson et al [1] have shown, near field heating effects depend upon the electrical size of the applicator and the conductivity of the tissue. Close to the applicator, the rate of decrease of absorbed power with distance from the device is much greater than the exponential decrease expected for plane waves and so observed penetration depths are smaller than those calculated for plane waves. Figure 2 shows some results reported by Turner and Kumar [2] on the variation with frequency for different sizes of aperture sources. It can be seen that penetration depths do not always improve at lower frequencies.

3. External microwave applicators

Microwave techniques (300 MHz to 3000 MHz) are restricted to the treatment of superficial sites. Techniques involving electromagnetic fields with frequencies in the range 10 MHz to 100 MHz are more appropriate for deep heating. At frequencies above 3000 MHz, electromagnetic waves are absorbed in the most superficial layers of tissue; they are therefore of little use in hyperthermia. Ideally, the dimensions of the aperture of a microwave applicator would be determined by the size of the area to be treated. However, restrictions are imposed for a given operating frequency. For example, if a waveguide applicator is used its dimensions must be such that wave propagation can take place. Thus for a rectangular waveguide, the largest dimension must be greater than half a wavelength. For a circular waveguide the radius must exceed 0.3λ. An upper limit on the dimensions must also be imposed to ensure that the applicator operates in the desired mode. Dimensions may be reduced from those of a simple (rectangular or circular cross-section) air-filled device by loading the applicator with a suitable dielectric material and/or using cross-sections of greater complexity

Dielectric loading can achieve an order of magnitude reduction in linear dimensions whilst ridged waveguides permit a reduction by a factor of approximately 5. However, it should be remembered that due to near field effects electrically small apertures have relatively poor penetration.

Another factor which must be considered in the choice of a direct contact applicator is the electric field distribution in the plane of the aperture. The dominant TE_{10} mode is frequently chosen for rectangular applicators although the electric field distribution across the aperture (figure 3A) is not ideal. Other modes and polarization have been used to increase uniformity [3] [4]. Another approach has been the use of rectangular applicators loaded with dielectric slabs. The inhomogeneously filled waveguide shown in figure 3B supports a simulated plane wave in the central region at a frequency f_{TEM} given by [5]:

$$f_{TEM} = \frac{c}{4\,t\sqrt{\epsilon_1/\epsilon_2 - 1}} \qquad (3)$$

When this condition is satisfied, the electric field, which has a component in the y-direction only, is uniform over the central region. To avoid other modes

$$b < \lambda/2 \quad \text{and} \quad d < (\lambda/\pi)\tan^{-1}\left\{-\sqrt{\frac{\epsilon_1}{\epsilon_2}}\cdot\tan\left(\frac{\pi}{2}\sqrt{\frac{\epsilon_1}{\epsilon_1-\epsilon_2}}\right)\right\} \qquad (4)$$

where λ is the wavelength of a plane wave in an unbounded medium of relative permittivity ϵ_2 [6]. In practice the electric field probe used to excite the applicator must be placed sufficiently distant from the aperture to prevent the higher-order evanescent modes near the probe disturbing the fields at the aperture.

The ridged waveguides shown in figures 3C and 3D exhibit a lower cut-off frequency and a wider band free from higher order mode interference than simple rectangular waveguides with the same internal dimensions. According to Chen [7], the cut-off frequency f_c for the dominant TE_{10} mode is

$$f_c = \frac{c}{\pi}\left(\left\{\frac{a_2}{b_2} + \frac{2}{\pi}\left[\frac{x^2+1}{x}\cosh^{-1}\left(\frac{1+x^2}{1-x^2}\right) - 2\ln\left(\frac{4x}{1-x^2}\right)\right]\right\}(a_1-a_2)b_1\right)^{-\frac{1}{2}} \qquad (5)$$

where $x = b_2/b_1$.

This can be compared with the cut-off frequency of $c/2a_1$ for a rectangular waveguide of width a_1. The lowest values of f_c occur when a_2/a_1 ~0.4 - 0.5 [8]. The aspect ratio of the waveguide b_1/a_1 should be ~0.4. This technique has been adopted for applicators operating in the frequency range 200 MHz to 400 MHz. The greatest heating produced by such an applicator occurs at the ridge and the uniformity across the aperture is indicated in figure 3D.

Another class of microwave applicator is based on microstrip techniques. The simplest form of microstrip applicator consists of a dielectric substrate (usually with $\epsilon' < 10$) which has a metallic patch on one side and a ground plane on the other side. Microstrip has a number of favourable characteristics which include small size and weight, low cost and the possibility of constructing devices with low profiles. A particular advantage of the microstrip structure is that the wave is mainly Transverse Electromagnetic (TEM) and no cut-off frequencies exist, unlike waveguide applicators. A detailed account of the design and performance of a ring-type microstrip antenna has been given by Bahl et al [9]. The modes excited in the applicator depend not only on its dimensions and substrate but also on the nature of the tissues in contact with it. An interesting application of a ring applicator which takes advantage of small size is the hyperthermal treatment of retinoblastoma [10]. Rectangular microstrip patch applicators have been investigated by Sandhu and Kolozsvary [11] who report that the bandwidth of such applicators is small and that the resonant frequency is sensitive not only to the patch dimensions and the permittivity of the substrate but also to the dielectric properties of the tissues being heated. A different type of low profile applicator whose performance is relatively insensitive to loading conditions has been decribed by Johnson et al [1]. Figure 4 illustrates the basic design of the applicator. A metallic patch is sandwiched between two slabs of dielectric material, one of which has a ground plane formed on one face. The dimensions of the patch are chosen to be resonant, eg for a circular patch

$$L = \frac{1.841\,\lambda_0}{\pi\sqrt{\mu_r \epsilon'}} \quad . \qquad (6)$$

The substrate dimensions should be approximately twice those of the patch. The permittivity of the substrate is chosen to be higher than that of microstrip (eg- $\epsilon' = 30$ to 90) and the presence of a cover of the same material over the patch improves impedance matching and decreases sensitivity to different loading conditions. Equation (6) shows that the dimensions of such applicators also depend upon the relative permeability of the substrate, μ_r. By introducing magnetic material into the applicator, $\mu_r > 1$, a further reduction in size compared with simple dielectric loading is possible. In this way, devices with dimensions of 15cm can be operated at frequencies as low as 100 MHz and penetration depths of 4-5cm can be achieved, a considerable improvement over other devices of comparable size.

4. Invasive microwave applicators

A limitation of the external applicators discussed so far is that intervening tissues tend to be heated more than deep ones. In some cases, the use of interstitial or intracavitary applicators may be appropriate. Interstitial applicators are usually needle like sections of miniature semi-rigid coaxial cable (available with diameters as small as 200μm) terminated in a radiating element [12]. This may be a monopole formed by removing a length of the outer conductor of the coaxial cable. de Sieyes et al [13] have carried out a study aimed at optimising this type of antenna and report that a completely insulated antenna has a number of advantages over partially insulated monopoles. The heating characteristics along the antenna are primarily determined by the length of the monopole provided the depth of insertion is greater than about one wavelength. The radial extent of the heating pattern is determined by the near field and a theoretical model of a bare probe [14] predicts that most of the power from this type of antenna is deposited within a distance of the order of the radius of the outer conductor. Increasing the diameter of the device by adding an insulating sleeve displaces the tissue to regions where the fields fall off less rapidly and so better radial penetration is achieved. This radial penetration is relatively insensitive to frequency over a wide range (300 MHz to 3000 MHz) and is 6 - 10 mm for an insulated antenna approximately 2mm in diameter.

An alternative approach in using coaxial applicators is to insert them through natural orifices into body cavities. Sites such as rectum, colon, cervix may be heated in this way. Mendecki et al [15] have described coaxial applicators which radiate into air-filled cavities. The terminations of these applicators usually include a quarter wavelength choke in addition to the quarter wave monopole. The radial extent of the heating pattern produced by these applicators is somewhat greater than that for interstitial antennas, since a greater thickness of dielectric sleeve can be tolerated by the patient. As an example, a half-wavelength dipole encapsulated in a polytetrafluoroethylene sleeve with outer diameter 8mm can produce a heating pattern in vivo in the cervix of a pig in which the temperature rise observed 20mm from the applicator is half that seen adjacent to the device (J.W. Hand and J.W. Hopewell, unpublished data).

5. Multiple applicator systems

Since the applicators described in this paper are limited to treatment of sites which are superficial and easily accessible, the question arises as to whether interference patterns produced by an array of applicators can improve the situation. Preliminary experiments to investigate the effects of varying phase relationships between two direct contact applicators have been reported by Guerquin-Kern et al [16]. A significant improvement in heating at depth was achieved in a phantom when the applicators, placed orthogonally to each other, were driven in phase rather than with a phase difference of 180^0. Absorbed power density is proportional to the square of the electric field, so superposition of the electric fields from N applicators driven coherently and in phase may produce up to N times the absorbed power density achieved when the same applicators are driven incoherently. Preliminary studies suggest that this principle may be used to extend the application of microwave hyperthermia [17] but a number of practical problems have yet to be solved. For example, at microwave frequencies it is difficult to place a sufficient number of applicators around the body so that the gain achieved at depth using the array overcomes the loss due to attenuation in the intervening tissue. Furthermore, inhomogeneities in the body will make optimisation and control of the phase relationships between applicators very difficult. Calculations have shown that although some field enhancement can be produced in a target volume, greater enhancement may be produced simultaneously in other locations [18].

The use of arrays of invasive antennas appears more promising. The limitation in using a single interstitial antenna is that only a small volume around the radiating element is heated. Trembly et al [19] have used 4 such antennas arranged in a square array to heat phantoms, cadaver tissue and highly vascularised tissue and have achieved therapeutic temperatures in a region approximately 5cm in diameter. A discussion of how blood flow and the number of antennas in the array determine the volume heated is given elsewhere [20].

6. Interstitial heating by radiofrequency techniques

Doss and McCabe [21] describe a technique for heating tissue in which RF currents of relatively low frequency (0.1 MHz - 1 MHz) are passed through the treatment volume by means of electrodes in direct contact with the tissues. The electrodes may be external, consisting of metal plates or wire mesh coupled to the skin with a conducting gel, or interstitial in which case stainless-steel needles such as those used in brachytherapy are employed. Unlike the techniques previously described, no radiation of electromagnetic energy is involved (dimensions of cables and electrodes are small compared to the wavelengths at these frequencies); indeed, the tissue forms part of a circuit in which current flows and a 'lumped parameter' description is valid. At these frequencies, $\omega\epsilon'$ is much smaller than σ, the conductivity. Thus currents are predominantly resistive and the displacement current is very small. The absorbed power density is given by

$$P = J^2 \varrho \qquad (7)$$

where J is the current density at a given point and ϱ is the resistivity of the tissue. The risk of burns caused by the high current density around the edges of external metallic electrodes can be minimised if the part of the electrode in contact with the skin has a resistivity equal to that of the tissue. In the interstitial technique, temperature distributions are strongly dependent upon the spacing of the needles. Often, needles are spaced at distances of 1 to 2cm. Absorbed power density is greatest adjacent to the electrodes although the intensity of hot spots can be reduced by cooling the electrodes or by increasing the area through using more needles [22]. Astrahan et al [23] report that improvement in temperature uniformity can be achieved by switching the RF currents between sets of electrodes within and/or surrounding the target volume.

Advantages of RF interstitial hyperthermia include the relatively easy control of temperature within the heated volume, sparing of surrounding normal tissue and the ability to combine the method with conventional interstitial radiotherapy techniques.

Another invasive method of achieving local hyperthermia is by selective heating of ferromagnetic material [24]. If a coil carrying a radiofrequency current is placed near to or around tissues, an electric field is induced within them which gives rise to heating. The frequency dependence of the absorbed power in the tissue is quadratic and at frequencies of 10 MHz to 30 MHz significant heating of tissue occurs. In contrast, the power deposition in a ferromagnetic rod varies as $\sqrt{\text{frequency}}$. If an array of ferromagnetic tubes is implanted in a volume of tissue and subjected to a magnetic field alternating at a frequency of about 1 MHz, direct induction heating of the tissue is negligible whilst the power absorbed by the ferromagnetic material will be redistributed throughout the surrounding tissue by the effects of thermal conduction and blood flow.

7. Heating by ultrasound

Ultrasound is a form of mechanical energy that propagates through tissue as a pressure wave; its ability to heat tissue has been known and utilised for many years [25] [26]. Frequencies in the approximate range 0.3 - 3 MHz have been used to induce hyperthermia. This section of the paper gives a brief review of the characteristics of applicators employing unfocused and focused ultrasound which may be used in clinical hyperthermia.

Table 1 lists some physical characteristics of ultrasound for several tissues.

Table 1. Ultrasonic properties at 1 MHz (after Frizzell and Dunn [27])

Tissue	Attenuation Coefficient α db cm^{-1}	Penetration Depth cm	Velocity m sec^{-1}	Impedance 10^5 rayls
Blood	0.12	70	1566	1.63
Fat	0.61	14	1478	1.36
Nerve	0.88	10		
Muscle	1.2	7.1	1522	1.62
Blood vessel	1.7	5	1530	1.65
Skin	2.7	3.2	1519	
Tendon	4.9	1.8	1750	
Cartilage	5.0	1.7	1665	
Bone	13.9	0.6	3445	6.27

Two general statements can be made from these data. Firstly, at the frequencies of interest in hyperthermia, the penetration of ultrasound is greater than that associated with microwave techniques. Secondly, since the wavelength of ultrasound is of the order of 1mm, well defined beams can be produced by conveniently sized transducers (with diameters of several cm). The combination of short wavelength and good penetration are features unobtainable with most electromagnetic techniques. As an ultrasonic beam travels through tissue it is attenuated and, in an homogeneous tissue far from the transducer (plane wave assumption), the intensity I at a position x in the tissue is:

$$I = I_0 e^{-2\alpha x}$$

where I_0 is the intensity at x=0 and α is the ultrasonic attenuation coefficient. In an inhomogeneous medium, not only absorption (which gives rise to tissue heating) but other processes such as scattering contribute to the attentuation of the beam. Data available for absorption coefficients are sparse but estimates of the contribution of absorption to total attentuation vary from 30% [28] to 70% [29]. In general, α varies approximately linearly with frequency for most tissues over the range 0.5 MHz to 10 MHz. Notable exceptions are bone and lung. For bone, the frequency dependence is approximately quadratic up to 2 MHz and somewhat more linear thereafter [36] whilst the attenuation in lung tissue is high and dependent upon the degree of inflation. The frequency dependence is exponential between 1 MHz and 5 MHz [31].

Sound travels through soft tissues as longitudinal waves but in bone both transverse (shear) and longitudinal waves are possible. Conversion from longitudinal to shear wave propagation may occur when a wave passes from soft tissue to bone. This is strongly dependent upon the angle of incidence at the interface and for a range of angles (approximately 45° to 60°) [32] the greatest contribution to heating in the bone is from shear waves. The reflection of ultrasound at an interface between two media is determined by the characteristic acoustic impedances (density x speed of sound) of the media. For detailed discussions of the acoustic properties of tissue the reader is referred to the works of Dunn and O'Brien [33] and Wells [30].

Ultrasound transducers suitable for hyperthermal therapy are usually cut from a piezo-electric material such as lead zirconate titanate 4 (PZT4). An alternative material is quartz which, having lower losses than the ceramic material, can withstand higher power levels but is more expensive and requires a greater driving voltage. When a RF voltage is applied to the faces of a disc of such a material the thickness varies. This leads to the propagation of an ultrasound wave if suitable coupling medium is provided between the transducer and tissue. For resonance, the thickness of the disc must be an odd number of half-wavelengths and maximum conversion of electrical to acoustic power occurs at the fundamental resonance. An aqueous gel or degassed water is necessary to couple the ultrasound from the transducer to the tissue. The efficiency of a transducer depends not only on the material from which the disc is made but also on the acoustic impedances of the material behind the disc (often air) and of the coupling medium. In practice, efficiencies of ceramic transducers are often >~50%.

Several groups have used transducers in the form of circular discs with plane faces driven at frequencies between 1.0 MHz and 3.0 MHz to treat tumours in superficial sites [34) 35)]. Figure 5a illustrates an applicator of this type used at Hammersmith Hospital, London, whilst figure 5b shows calculated distributions of absorbed power density and temperature for this transducer for conditions similar to those used clinically. The use of plane transducers does not allow selective heating at depth but reasonable penetration may be achieved without undue surface heating. Where deeper heating is required focused ultrasound may be used.

The methods commonly used to focus ultrasound for hyperthermia involve transducers with a spherical concave face (bowl transducers) or a plane (unfocused) transducer in combination with an acoustic lens. Penttinen and Luukkala [36)] have described a calculation of the field distribution from a bowl transducer. From figure 6 it can be seen that the focus of such a transducer lies on the central axis, near the centre of curvature of the bowl. As the depth of the shell, h, increases, the focal point approaches the centre of curvature of the bowl. The intensity at the focus is given approximately by the intensity at C, which ignoring attenuation in the intervening tissue, is:

$$\frac{I_c}{I_o} = \left(\frac{2\pi h}{\lambda}\right)^2 \qquad (8)$$

where I_o is the average intensity at the transducer surface. The ratio I_o/I_c is known as the gain factor, G, and values of several hundreds can be achieved even when attenuation in the intervening tissue is accounted for. The lateral width of the focal volume, w, is approximately

$$W = 1.22 \frac{R\lambda}{a} . \qquad (9)$$

The focal volumes are characteristically ellipsoidal and are typically a few millimetres in diameter with a length/diameter ratio of approximately 10. They are therefore considerably smaller than the tumour volumes to be treated. A similar problem occurs when focusing is achieved by using an ultrasonic lens. If such focused systems are used to heat realistic tumour volumes to acceptable minimum temperatures, undesirable temperature distributions with very large thermal gradients are produced as a consequence of the extremely high energy deposition within the small focal volume. One method of overcoming this problem is to move the focal volume within the tissue by moving the transducer mechanically over the tissue [37]. It should be noted that the gain factor G of a focused system is a result of the ultrasound energy (accounting for attenuation where appropriate) passing through the aperture of the transducer compared with the considerably smaller area at the focus. If the transducer is simply scanned uniformly over distances comparable with the diameter of the transducer, the areas at the surface and at depth through which the ultrasound passes become comparable. The consequence is a reduction in G. Thus, if selective heating at depth is to be achieved, the transducer should be scanned over a large area ('window') at the surface but should remain directed at the target volume [38].

A different approach to the selective heating of volumes of clinical relevance is to arrange for weakly focused beams from several transducers to overlap at depth. Fessenden et al [39] have placed 6 transducers, each 7cm in diameter and driven at its own resonant frequency close to 350 kHz, on a spherical surface. The system has produced selective heating of perfused volumes of approximately 200cm^3 at depths of about 10cm.

It is reported that some patients have experienced pain when ultrasound has been used to heat tumours which overlay bone or nerve tissue. The problem appears to be particularly limiting in the case of the 350 kHz multi-transducer system referred to above where there can be considerable 'overshoot'. Scanned focused transducers can be expected to minimise 'overshoot' and therefore may prove useful in

cases, measurements may be obtained intermittently during short periods in which the heating fields are turned off. If this method of measuring temperature is adopted, some knowledge of the rate of cooling is required to enable an estimate of the true temperature at the moment the power is switched off to be made. The common practice of using T-type thermocouples (copper/constantan) is not ideal since the considerable difference in the resistivities of these materials can lead to heating of their junctions in the presence of RF fields [43]. The use of chromel/constantan or manganin/constantan pairs should reduce this source of error. In addition, the lower thermal conductivities of these materials may be advantageous for their use in multijunction probes and this is currently under investigation in our laboratory. An additional problem with thermocouples is that one of the materials is often a ferromagnetic alloy (eg constantan) which tends to be heated in strong RF magnetic fields [44].

Over recent years a wide range of temperature dependent physical properties have been investigated in the development of new types of thermometer for use in the presence of electromagnetic fields. Some of these are listed in Table 2.

Table 2. Thermometers for use in electromagnetic fields.

Thermometric principle	Author
Reflected light from liquid crystal	Rozzell et al [45]
Birefringence of $LiTaO_3$	Cetas [46]
Absorption of light by GaAs	Christensen [47]
Decay time of the luminescent response of zinc cadmium sulphide phosphors	Samulski and Shrivastava [48]
Ratio of intensities of lines in emission spectrum of europium-activated gadolinium oxysulfide phosphor	Wickersheim and Alves [49]
Magnetic resonance of ferrimagnetic material (indium substituted yttrium-iron garnet)	Weiss et al [50]
Miniature thermistor with high resistance leads	Larsen et al [51]
Miniature thermistor with high resistance leads	Bowman [52]
Change in vapor pressure of trichloro-fluoromethane.	Szwarnowski [53]

Thermometers based on the work of Christensen, Wickersheim, Bowman and Larsen et al are commercially available. Early versions use probes with single sensors but multi-sensor probes for the optical thermometers will become available. A minimally perturbing multi-sensor electrical thermometer consisting of an array of miniature diodes has been described by Barth and Angell [54]. Tests on some of these new thermometers have been reported by Hochuli [55].

Since significant temperature gradients are produced in tumours and normal tissues during hyperthermal therapy, the interpretation of the relatively few measurements provided by invasive thermometry is difficult. As indicated earlier, the ideal thermometry system would record spatial and temperature variations in temperature throughout the treated volume and some systems based on microwaves, ultrasound and nuclear magnetic resonance which may provide a step in this direction are under development.

Microwave radiometry has been used as a diagnostic tool in medicine for a number of years. Several groups of workers are now investigating its potential for monitoring and controlling local hyperthermal therapy [56][57][58]. Major limitations to this method of thermometry are that an average temperature is obtained from a volume of tissue and, in order that tolerable spatial resolution is achieved, the depths over which the averaging process takes place are relatively shallow. One proposed ultrasonic technique is based on computerised reconstruction of the temperature dependent speed of sound in tissue [59]. Although many tissues exhibit similar slopes for the increase in acoustic velocity with increasing temperature, a major limitation appears to be that tissues with fat content show a decrease in velocity with increasing temperature within certain ranges. In addition most techniques of velocity reconstruction require transmission through the tissue volume, thus limiting their application for a small number of sites.

Temperature measurements in phantoms using X-ray tomography [60] and NMR imaging [61] have been reported, but as yet are not capable of mapping temperature changes in living tissue. A major problem appears to be that the parameters imaged are very dependent on physiology as well as temperature.

In the absence of suitable non-invasive thermometry, mathematical modelling of heat transfer within tissues using data from the limited number of invasive probes may prove useful as the basis of practical dosimetry in clinical hyperthermia.

9. Conclusion

This paper has discussed physical aspects of some techniques currently used in local hyperthermia therapy. A tabular summary of the discussion is given below.

Table 3. Summary of techniques.

	RF		Microwaves	Ultrasound
Typical frequency range (MHz)	0.1 - 1		300 - 1000	0.3 - 3
Technique	Localised current fields	RF Heating of implants	Direct contact applicators Interstitial & intracavitary applicators	Plane transducers Focused fields (bowl transducer or lens)
Localisation	Very good	Very good		Excellent
Penetration	-	-	Superficial - 2 - 3cm	Plane transducer 3-4cm Focused fields 10 - 12 cm
Thermometry (Invasive probes)	Small thermocouples or thermistors		Non perturbing probes for continuous monitoring	Small thermocouples
Disadvantages	Often invasive	Invasive		Coupling medium required. Large reflections at gas/tissue and bone/soft tissue interfaces. Rapid attenuation in bone.

10. Acknowledgements

I am grateful to Dr R.J. Dickinson for discussions on several of the topics covered in this paper and in particular for agreeing to the inclusion of unpublished data. Thanks are also due to Mrs. D. Wishart for preparation of the manuscript.

11. References

1) Johnson, R.H., James, J.R., Hand, J.W., Hopewell, J.W., Dunlop, P.R.C. and Dickinson, R.J. 1984. New low profile applicators for local heating of tissues. IEEE Trans. Biomed. Eng. BME-31 in press.

2) Turner, P.F. and Kumar, L. 1982. Computer solution for applicator heating patterns, National Cancer Institute Monograph 61, 521-523.

3) Stuchly, S.S. amd Stuchly, M.A. 1978. Multimode square waveguide applicators for medical applications of microwave power. In Proc. 5th European Microwave Conference, Paris, 4-8 Sept, 1978. (Microwave Exhibitions and Publishers Ltd., Sevenoaks, U.K.) 553-557.

4) Kantor, G., Witters, D.M. and Greiser, J.W. 1978. The performance of a new direct contact applicator for microwave diathermy. IEEE Trans. Microwave Theory & Tech., MTT-26, 563-568.

5) Hudson, A.C. 1957. Matching the sides of a parallel plate region. IRE Trans. Microwave Theory & Tech., MTT-5, 161-162.

6) Van Koughnett, A.L. and Wyslouzil, W. 1972. A waveguide TEM mode exposure chamber. Journal of Microwave Power, 7, 381-384.

7) Chen, T.S. 1957. Calculation of the parameters of ridge waveguides. IRE Trans. Microwave Theory & Tech., MTT-5, 12-17.

8) Hopfer, S. 1955. The design of ridge waveguides. IRE Trans. Microwave Theory & Tech., MTT-3, 20-29.

9) Bahl, I.J., Stuchly, S.S. and Stuchly, M.A. 1980. A new microstrip applicator for medical applications. IEEE Trans. Microwave Theory & Tech., MTT-28, 1464-1468.

10) Lagendijk, J.J.W. 1982. A microwave heating technique for the hyperthermic treatment of tumours in the eye, especially retinoblastoma., Phys. Med. Biol., 27, 1313-1324.

11) Sandhu, T.S. and Kolozsvary, A. 1983. Effect of bolus/tissue inhomogeneities on the resonance frequency of MW Microstrip applicators. Radiation Research, 94, 594.

12) Taylor, L.S. 1980. Implantable radiators for cancer therapy by microwave hyperthermia. Proc. IEEE, 68, 142-149.

13) de Sieyes, D.C., Douple, E.B., Strohbehn, J.W. and Trembly, B.S., 1981. Some aspects of optimization of an invasive microwave antenna for local hyperthermia treatment of cancer. Med. Phys., 8, 174-183.

14) Swicord, M.L. and Davis, C.C. 1981. Energy absorption from small radiating coaxial probes in lossy media. IEEE Trans Microwave Theory & Tech., MTT-29, 1202-1209.

15) Mendecki, J., Friedenthal, E., Botstein, C., Sterzer, F., Paglione, R., Nowogrodzki, M. and Beck, E. 1978. Microwave induced hyperthermia in cancer treatment: apparatus and preliminary results. International Journal of Radiation Oncology Biology Physics, 4, 1095-1103.

16) Guerquin-Kern, J.L., Palas, L., Gautherie, M., Fourney-Fayas, C., Gimonet, E., Priou, A. and Samsel, M. 1980. Étude comparative d'applicateurs hyperfréquences (2450MHz, 434MHz) sur phantômes et sur pièces opératoires en vue d'une utilisation thérapeutique de l'hyperthermie microonde en cancerologie. Proc. URSI Symposium 'Ondes Electromagnetiques et Biologie, July, 1980., Jouy-en-Josas, p 241-247.

17) Melec, M., Anderson, A.P., Brown, B.H. and Conway, J. 1982. Measurements substantiating localised microwave hyperthermia within a thorax phantom. Electronics Letts., 18, 427-428.

18) Arcangeli, G., Lombardini, P.P., Lovisolo, G., Marsiglia, G. and Piatelli, M. 1982. Focussing of 915 MHz electromagnetic power in deep human tissues: a mathematical model study. Strahlentherapie, 158, 378.

19) Trembly, B.S., Strohbehn, J.W., de Sieyes, D.C. and Douple, E.B. 1982. Hyperthermia induction by an array of invasive microwave antennas. National Cancer Institute Monograph, 61, 497-499.

20) Strohbehn, J.W., Trembly, B.S. and Douple, E.B. 1982. Blood flow effects on the temperature distributions from an invasive microwave antenna array used in cancer therapy. IEEE Trans. Biomed. Eng. BME-29, 649-661.

21) Doss, J.D. and McCabe, C.W. 1976. A technique for localized heating in tissue: an adjunct to tumor therapy. Medical Instrumentation, 10, 16-21.

22) Cetas, T.C. and Connor, W.G. 1978. Thermometry considerations in localized hyperthermia. Medical Physics, 5, 79-91.

23) Astrahan, M.A. and George, F.W. 1980. A temperature regulating circuit for experimental localized current field hyperthermia systems. Medical Physics, 7, 362-364.

24) Stauffer, P., Cetas, T.C. and Jones, T.C. 1982. System for producing localized hyperthermia in tumors through magnetic induction heating of ferromagnetic implants. National Cancer Institute Monograph 61, 483-487.

25) Freundlich, H., Sollner, K. and Rogowski, F. 1932. Einige biologische Wirkungen von Ultraschallwellen. Klinische Wochenschrift, 11, 1512-1513.

26) Pohlman, R., Richter, R. and Parow E. 1939. Uber die Ausbreitung und Absorption des Ultraschalls in meschlichen Gewebe und seine therapeutische Wirkung in Ischias und Plexus-neuralgia. Deutsche Medizinische Wochenschrift, 65, 251-254.

27) Frizzell, F.A. and Dunn, F. 1982. Biophysics of Ultrasound. In Therapeutic Heat and Cold (3rd Edition), Ed: J.F. Lehmann (Williams and Wilkins, Baltimore, 1982). pp. 353-385.

28) Goss, S.A., Frizzell, L.A. amd Dunn, F. 1979. Ultrasonic absorption and attenuation in mammalian tisues. Ultrasound in Medicine and Biology, 5, 181-186.

29) Hill, C.R., Chivers, R.C., Huggins, R.W. and Nicholas, D. 1978. Scattering of ultrasound by human tissue. In Ultrasound: Its applications in Medicine and Biology, 1. Ed: F.J. Fry (Elsevier, Amsterdam, 1978), pp. 441-493.

30) Wells, P.N.T. 1977. Biomedical Ultrasonics (Academic Press, London).

31) Dunn, F. 1974. Attenuation and speed of sound in lung. Journal of the Acoustical Society of America, 56, 1638-1639.

32) Chan, A.K., Sigelman, R.A. and Guy, A.W. 1974. Calculations of therapeutic heat generated by ultrasound in fat-muscle-bone layers. IEEE Trans. Biomed. Eng., BME-21, 280-284.

33) Dunn, F. amd O'Brien, W.D. 1978. Ultrasonic absorption and dispersion. In: Ultrasound: Its applications to Medicine and Biology, 1. Ed: F.J. Fry, (Elsevier, Amsterdam), pp. 393-439.

34) Marmor, J.B., Pounds, D., Postic, T.B. and Hahn, G.M. 1979. Treatment of superficial human neoplasms by local hyperthermia induced by ultrasound. Cancer, 32, 188-197.

35) Marchal, C., Bey, P., Metz, R., Gaulard, M.L. and Robert, J. 1982. Treatment of superficial human caancerous nodules by local ultrasound hyperthermia. British Journal of Cancer, 45 (Suppl. V), 243-245.

36) Penttinen, A. and Luukkala, M. 1976. Sound pressure near the focal area of an ultrasonic lens. Journal of Physics D: Applied Physics, 9, 1927-1936.

37) Lele, P.P. and Parker, K.J. 1982. Temperatutre distributions in tissues during local hyperthermia by stationary or steered beams of unfocused or focused ultrasound. British Journal of Cancer, 45 (Suppl. V), pp. 108-121.

38) Dickinson, R.J. 1984. An ultrasound system for local hyperthermia using scanned focused transducers. IEEE Trans. Biomed. Eng., BME-31, in press.

39) Fessenden, P., Anderson, T.L., Marmor, J.B., Pounds, D., Sagerman, R. and Strohbehn, J.W. 1982. Experience with a deep heating ultrasound system. Radiation Research, 91, 415.

40) Magin, R.L., Mangum, B.W., Statler, J.A. and Thornton, D.D. 1981. Transition temperatures of the hydrates of Na_2SO_4, Na_2HPO_4 and KF as fixed points in biomedical thermometry. Journal of Research of the National Bureau of Standards, 86, 181-192.

41) Division of Quantum Metrology, National Physical Laboratory, Teddington, Middlesex TW11 OLW, England.

42) Gibbs, F.A. 1983. 'Thermal mapping' in experimental cancer treatment with hyperthermia: description and use of a semi-automatic system. International Journal of Radiation Oncology Biology Physics, 9, 1057-1063.

43) Chakraborty, D.P. and Brezovich, I.A. 1982. Error sources affecting thermocouple thermometry in RF electromagnetic fields. Journal of Microwave Power, 17, 17-28.

44) Cetas, T.C. 1982. Invasive thermometry, in Physical Aspects of Hyperthermia, Ed: G.H. Nussbaum, (American Institute of Physics, Inc., New York), pp. 231-265.

45) Rozzell, T.C., Johnson, C.C., Durney, C.H., Lords, J.L. and Olsen, R.G. 1974. A nonperturbing temperature sensor for measurements in electromagnetic fields. Journal of Microwave Power, 9, 241-249.

46) Cetas, T.C. 1976. A birefringent crystal optical thermometer for measurements in electromagnetically induced heating. in Proc. 1975 USNC/URSI Symposium Vol. II, Boulder, Co., Oct. 1975. Eds: C.C. Johnson and J.L. Shore (HEW Publications (FDA) 77-8011, Rockville, Md). pp. 338-347.

47) Christensen, D.A. 1977. A new nonperturbing temperature probe using semiconductor band edge shift. Journal of Bioengineering, 1, 541-545.

48) Samulski, T. and Shrivastava, P.N. 1980. Photoluminescent thermometer probes: temperature measurements in microwave fields. Science, 208, 193-194.

49) Wickersheim, K.A. and Alves, R.B. 1979. Recent advances in optical temperature measurement. Industrial Research/Development, 21, 82-89.

50) Weiss, J.A., Hawks, D.A. and Dionne, G.F. 1981. A ferrimagnetic resonance thermometer for microwave power environment. In Proc. 1981 IEEE MTT-S International Microwave Symposium, Ed: J.E. Rane (IEEE Inc., New York) pp. 290-292.

51) Larsen, L.E., Moore, R.A. and Acevedo, J. 1974. A microwave decoupled brain temperature transducer. IEEE Transactions Microwave Theory & Tech., MTT-22, 438-444.

52) Bowman, R.R. 1976. A probe for measuring temperature in radiofrequency heated material. IEEE Transactions Microwave Theory & Tech., MTT-24, 43-45.

53) Szwarnowski, S. 1983. A thermometer for measuring temperature in the presence of electromagnetic fields. Clin. Phys. Physiol. Meas., 4, 79-84.

54) Barth, P.W. and Angell, J.B. 1982. Thin linear thermometer arrays for use in localized cancer therapy. IEEE Transactions on Electron Devices, ED-29, 144-150.

55) Hochuli, C.U. 1981. Procedures for evaluating nonperturbing temperature probes in microwave fields. HHS Publication (FDA) 81-8143 (May 1981), Rockville, Md.

56) Chive, M., Nguyen, D.D. and Leroy, Y. 1982. Une nouvelle application des microondes en génie biologique et médical: l'hyperthermie locale contrôlée par thermographie microonde à 2.45 GHz. L'onde Électrique, 62, 66-70.

57) Sterzer, F., Paglione, R., Woznaiak, F., Mendecki, J., Friedenthal, E. and Botstein, C. 1982. A self-balancing microwave radiometer for non-invasively measuring the temperature of subcutaneous tissues during localized hyperthermia treatments of cancer. In Proc. 1982 IEEE MTT-S International Microwave Symposium (IEEE Publication 82CH1705-3, IEEE Inc., New York) pp. 438-440.

58) Mamoumi, A., Van de Velde, J.C. and Leroy, Y. 1981. New correlation radiometer for microwave thermography. Electronics Letters, 17, 554-55.

59) Sachs, T.D. and Tanney, C.S. 1977. A two-beam acoustic system for tissue analysis. Phys. Med. Biol. 22, 327-340.

60) Fallone, B.G., Moran, P.R. and Podgorsak, E.B. 1982. Noninvasive thermometry with a clinical X-ray CT scanner. Medical Physics, 9, 715-722.

61) Parker, D.L., Smith, V., Sheldon, P., Crooks, L.E., and Fussell, L. 1983. Temperature distribution measurements in two dimensional NMR imaging. Medical Physics, 10, 321-325.

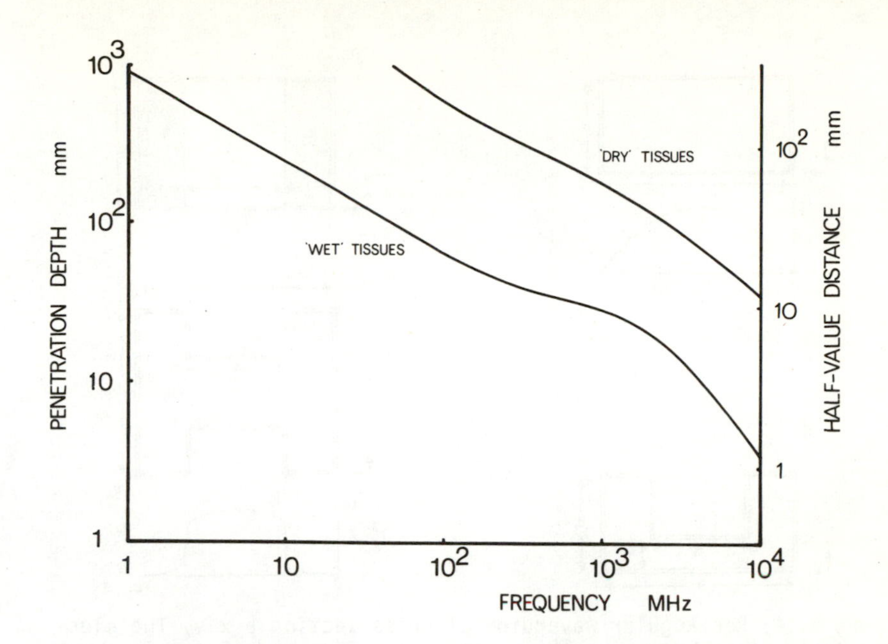

Figure 1. Plane wave penetration depth in tissues with high and low water content as a function of frequency.

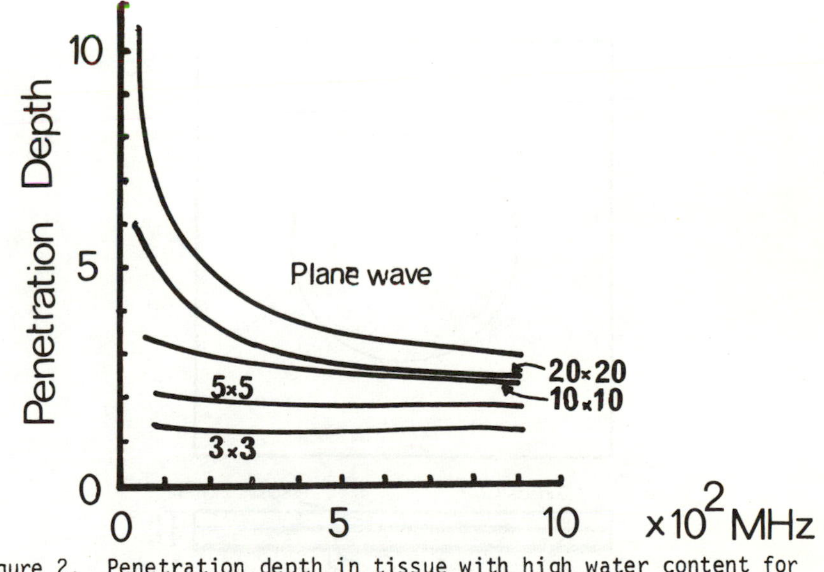

Figure 2. Penetration depth in tissue with high water content for finite apertures (TE_{10} mode) as a function of frequency (After Turner and Kumar [2]).

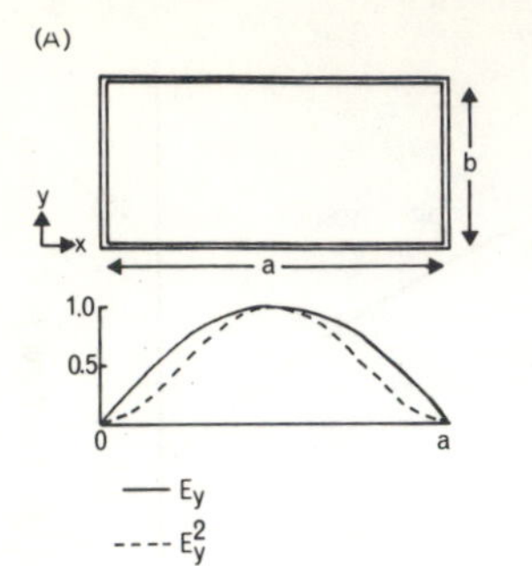

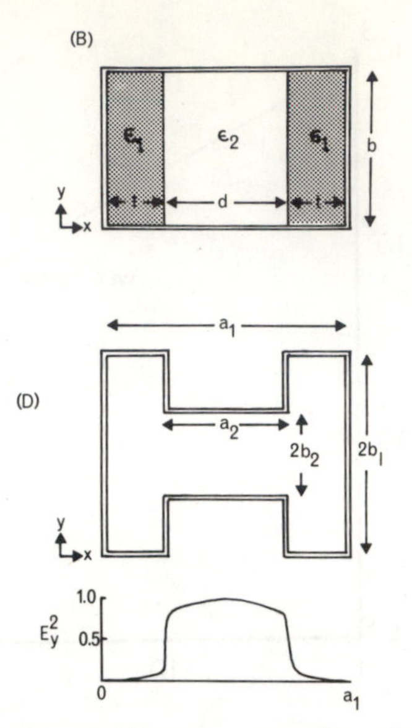

(C)

Figure 3.(A) Rectangular waveguide of cross section a x v. The electric field distribution on the aperture for TE_{10} mode Ey(x) is shown. (B) Rectangular waveguide loaded with dielectric slabs. (C) Single ridge waveguide. (D) Double ridged waveguide. An approximate distribution of $Ey^2(x)$ across the aperture for TE_{10} mode is shown.

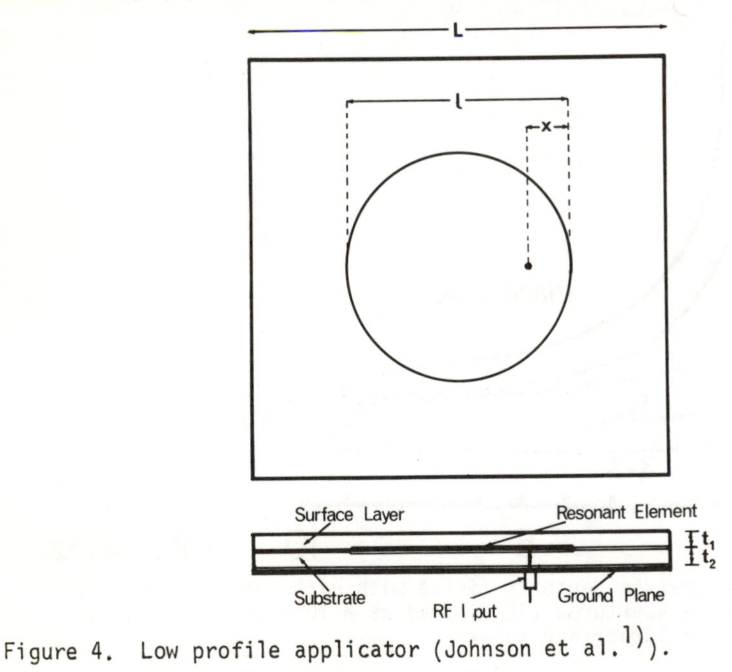

Figure 4. Low profile applicator (Johnson et al.[1]).

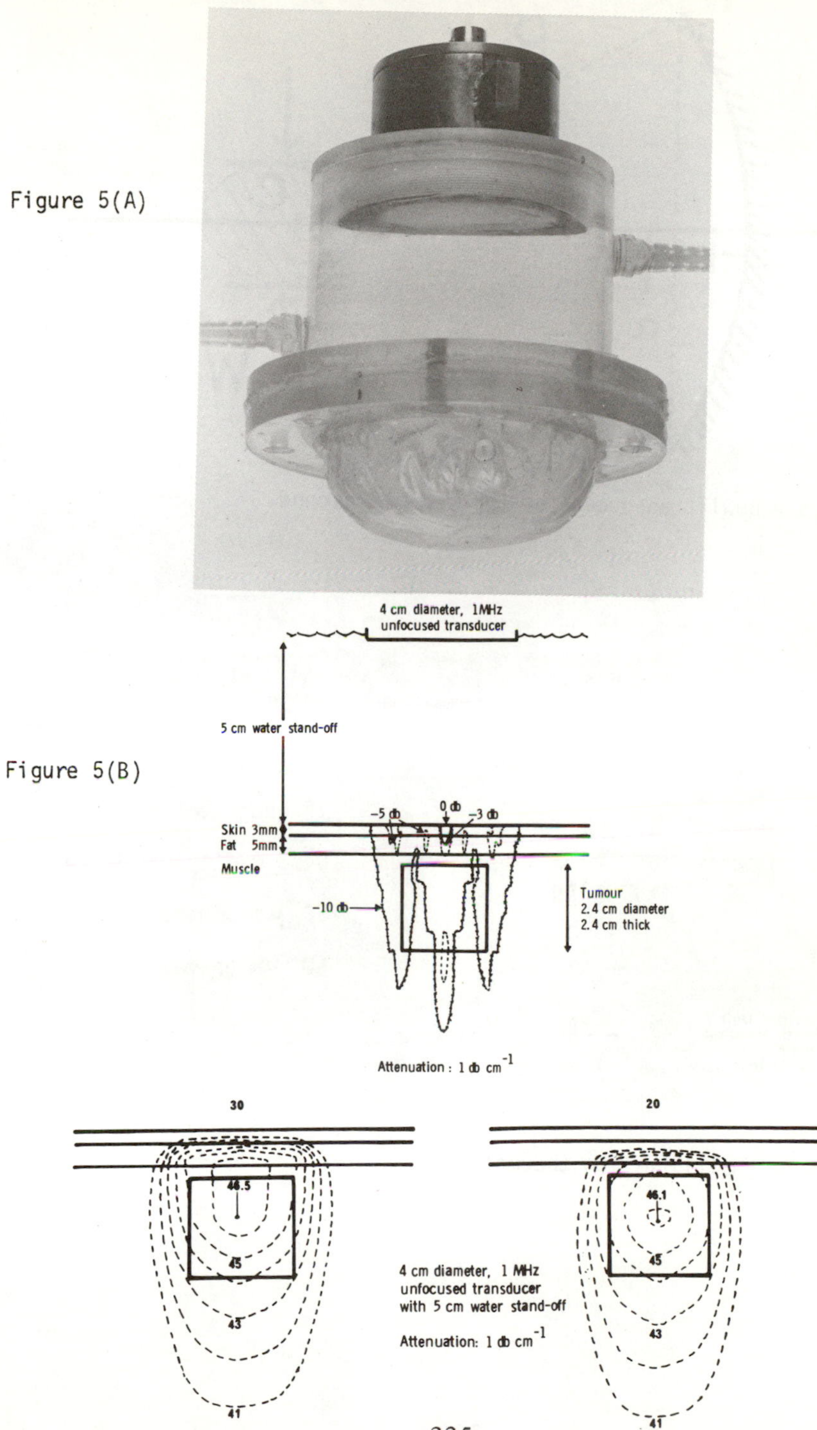
Figure 5(A)
Figure 5(B)
4 cm diameter, 1MHz
unfocused transducer
5 cm water stand-off
Skin 3mm
Fat 5mm
Muscle
0 db
-5 db
-3 db
-10 db
Tumour
2.4 cm diameter
2.4 cm thick
Attenuation : 1 db cm^{-1}
30
46.5
45
43
41
20
46.1
45
43
41
4 cm diameter, 1 MHz
unfocused transducer
with 5 cm water stand-off
Attenuation: 1 db cm^{-1}

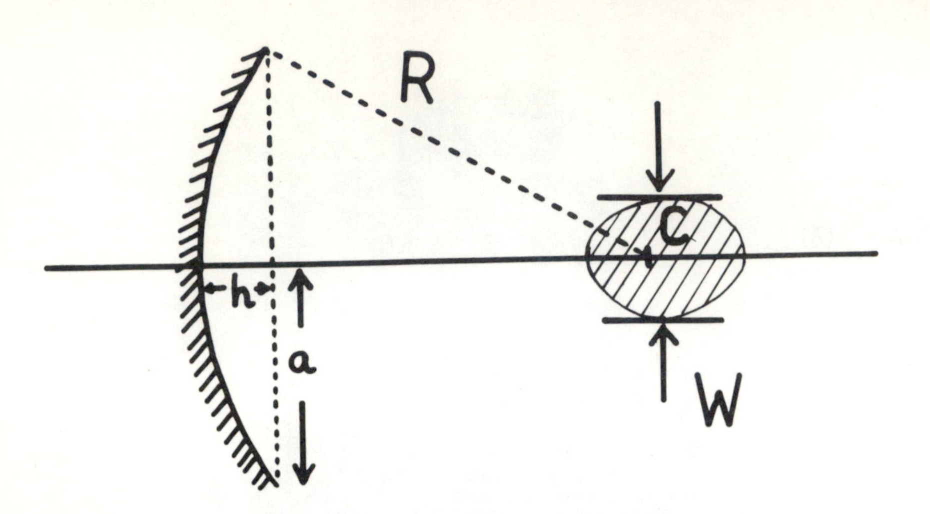

Figure 6. Bowl transducer for focused ultrasound.

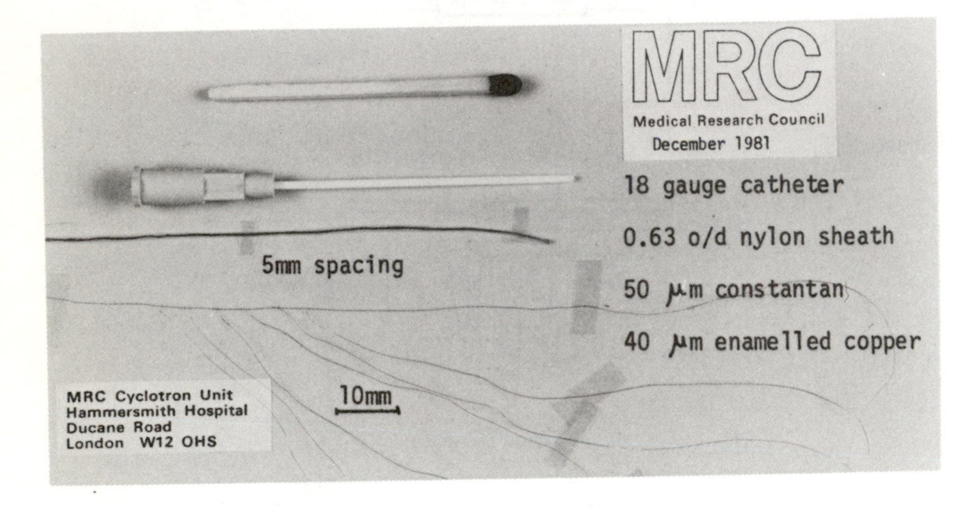

Figure 7. Multijunction thermocouple

PROCEEDINGS OF THE II INTERNATIONAL CONFERENCE ON APPLICATIONS OF PHYSICS TO MEDICINE AND BIOLOGY
edited by Ž. Bajzer, P. Baxa & C. Franconi

ELECTROMAGNETICS OF HYPERTHERMIA

J. Bach Andersen
Aalborg University
9000 Aalborg, Denmark

ABSTRACT

Basic electromagnetic theory of various heating modalities will be reviewed and the fundamental limitations for focussing in a lossy medium will be discussed. A cylindrical model of the body serves as a test case for creating an electronically steerable hot spot in muscle tissue.

1. Introduction

The purpose of the paper is to present a tutorial overview of basic concepts related to electromagnetic heating of tissue. The application in mind is hyperthermia treatment of tumors imbedded in healthy tissue, where the goal is to reach a certain temperature distribution, tumors above a certain temperature, say 42.5°C, and healthy tissue below this temperature. Considering the inhomogeneity and nonlinearity of the thermal problem this is a major task, which we shall not enter, but limit ourselves to considering the electromagnetic power deposition, which is the source for the thermal distribution. In general it is fair to state that we should attempt to confine the electromagnetic power density as much as possible to the tumor tissue, so we do not have to rely on thermal diffusion to create the necessary distribution. In the following the thermal problem will not be discussed.

The physical quantity of interest is the local power density

$$P(\bar{r}) = \sigma(\bar{r})|\bar{E}(\bar{r})|^2 \quad \text{W/m}^3 \qquad (1)$$

where σ is the local tissue conductivity and E is the local electric field. $\bar{E}$ is dependent on the source configuration and the distribution of ϵ and σ in the body, where ϵ is the relative permittivity, so in general we have a complicated synthesis problem: How do we choose frequency and source configurations in such a way that P(r) is maximum in the tumor and small everywhere else? We shall not discuss the mathematical synthesis problem but by way of examples achieve an overview over the relevant possibilities.

Considering non-invasive techniques the medium properties set some basic limitations. The losses in muscle tissue or other "wet"-tissue are so high that a penetration vs. focussing compromise must be made, i.e. a low frequency is chosen in order to have deep plane-wave penetration, but a long wavelength forbids a narrow focus or spot. In order to obtain a clear understanding of the phenomena the discussion is divided into two parts, a section on low-frequency applicators where the principles of operation are based on quasistatic theory, i.e. all dimensions are small compared with the wavelength in the medium, and a section on high-frequency applicators, where wave phenomena are dominant. The division is not clear-cut, most applicators will have elements of both, but it seems to be a natural dividing line as far as the physics is concerned.

2. Low-frequency applicators

2.1 Capacitive or resistive applicators

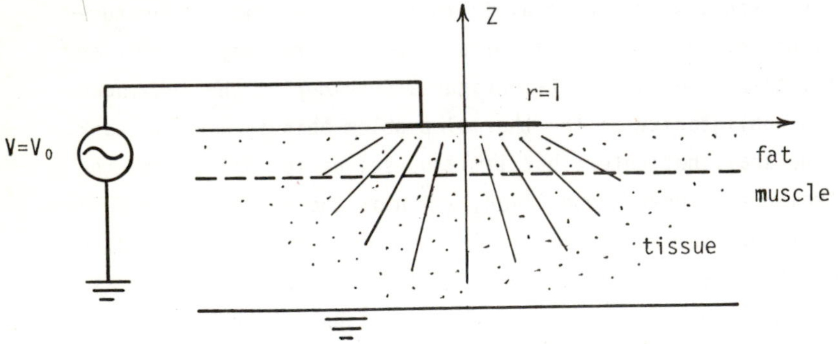

Fig. 1. Resistive electrode on tissue with pick-up electrode (ground) far away.

For capacitive or resistive applicators the tissue or body is part of a quasi-static electric circuit. As illustrated in Fig. 1 a circular electrode is placed on a conducting medium and raised to a voltage, V_0, with respect to ground, which is supposed to be far away. In practice ohmic contact is avoided, and some spacing to the surface is introduced, in which case we call it a capacitive applicator. The advantage of the situation in Fig. 1 is that it may be simply evaluated, assuming that the medium contrast to air is large, so that we can neglect capacitive currents in the air flowing to the top side of the plate. The solution to this boundary value problem is given in ref. 1,

$$V = \frac{2V_0}{\pi} \sin^{-1}\left\{\frac{2}{((r-1)^2 + z^2)^{\frac{1}{2}} + ((r+1)^2 + z^2)^{\frac{1}{2}}}\right\} \qquad (2)$$

where V is the potential. From (2) $\bar{E}$ is easily found as

$$\bar{E} = -\nabla V \qquad (3)$$

and the power density may be obtained from eq. 1. The power density is shown in Fig. 2. with equal steps between the contour lines of constant power density. Two important conclusions may be drawn, the rapid decay away from the surface and the singular peaks near the rim of the metal.

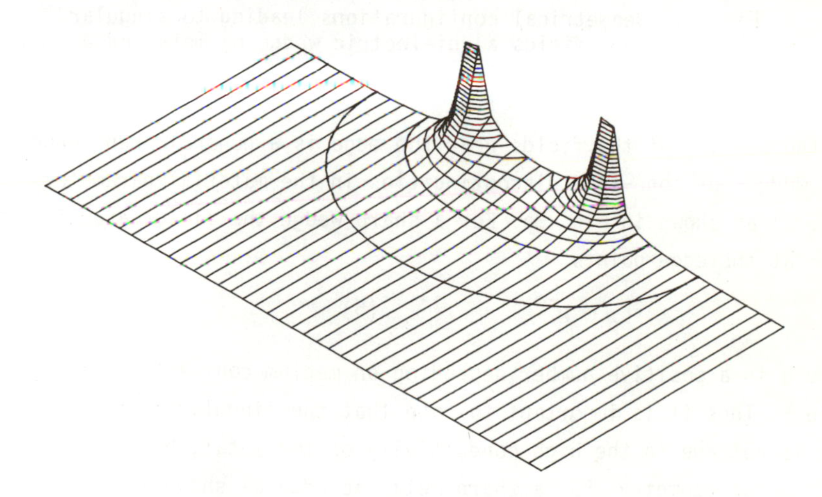

Fig. 2. Power distribution below a circular electrode resting on a conducting material corresponding to Fig. 1.

The asymptotic decay rate may easily be obtained from eqs. 2 and 3, i.e. for a distance, R, large compared with 1 (the radius of the applicator)

$$P(R) \sim \frac{1}{R^4} \qquad R >> 1. \qquad (4)$$

similar to the decay rate for a charged monopole. Note that this is independent of the direction, so the decay rate at depth is the same as along the surface. In case the pick-up electrode is closer to the primary electrode some confinement will take place, similar to the field distribution between two condenser plates. Note that the distribution is independent of frequency and tissue parameters as long as the medium is homogeneous.

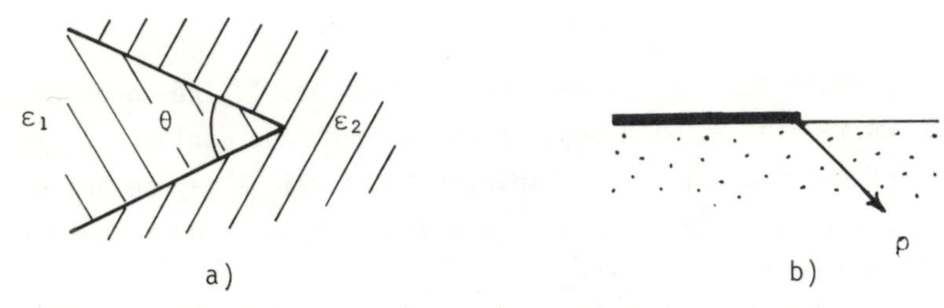

Fig. 3. Geometrical configurations leading to singular edge fields a) dielectric wedge b) metal edge on dielectric.

The peaking of the fields near the edge is a general phenomenon near wedge- or cone-like discontinuities in the material parameters (ref. 2) as shown in Fig. 3a. For a sharp wedge the fields are singular at the edge points

$$E_\rho \sim \rho^{-\alpha} \qquad (5)$$

where α is a positive number depending on medium contrast and the angle θ. Thus it is important to note that the singularity at the edge is not due to the high conductivity of the metal, but rather a function of geometry. For a sharp metallic edge as shown in Fig. 3b, $\alpha = 0.5$, explaining the peaks in Fig. 2(for numerical reasons the plot starts at $z = 0.01$ to avoid very large numbers). In practice the corners will be rounded leading to a high, but finite field strength. The singularities are also less pronounced for the capacitive version as shown f.ex. by Brezovich et al (ref.3), where the intervening layer is water,contained within a rubber surface which may be molded to follow the body contour. The water may also serve the purpose of stabilizing the surface temperature.

One problem is common for the resistive and capacitive applicator, the excessive heating of a subcutaneous fatty layer. Such a layer is indicated in fig. 1 along with a scetch of the current lines. The fat has a higher resistivity than muscle tissue, so it may happen that most of the voltage drop falls over the fat and most of the power will be depositied in the fat.

Concluding this section we may state that the quasistatic electric applicators have problems with small penetration, edge fields and fat-heating.

2.2 Inductive applicators

2.2.1 Magnetic loop

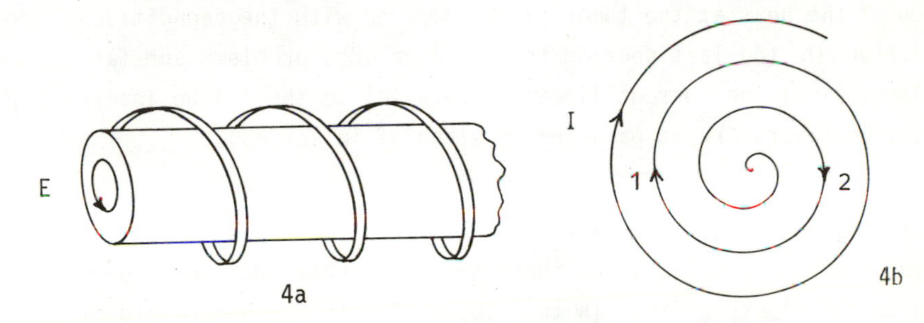

Fig. 4. Different inductive applicators a) Solenoid surrounding a cylindrical lossy medium. b) Flat spiral (pancake coil).

A long solenoid will have an axially directed magnetic field and an azimuthally directed electric field connected via Maxwell's equation. At radius r we have

$$\oint \bar{E} \cdot d\bar{l} = -j\omega \mu_0 \int \bar{H} \cdot d\bar{S} \qquad (6)$$

$$E_\varphi(r) \cdot 2\pi r = -j\omega \mu_0 H \pi r^2 \qquad (7)$$

or

$$E_\varphi(r) = -\tfrac{1}{2} j\omega \mu_0 H r \qquad (8)$$

where we have assumed H to be independent of r. This will be approximately true in the quasistatic approximation where the radius of the cylinder is less than the penetration depth. Note that the tissue parameters do not appear in eq. 8, so the equation will still be valid when a cylindrical body is inserted, i.e.

$$P(r) = \frac{1}{4} \sigma \omega^2 \mu_0{}^2 H^2 r^2 \qquad (9)$$

More information, also on experimental results, may be found in refs. 4 and 5. The distribution in eq. 9 is somewhat unfortunate for deep seated tumors since the surface is heated the most, and there is a null on the axis. Furthermore there is no way of giving preference to one side relative to the other.

In practice (ref. 4) the thermal differences between tumor and normal tissue may lead to a specific tumor heating even though the power density is non-specific. However, the applicator does not satisfy the criterion mentioned in the Introduction, of having the source of the heat at the tumor site. Compared with the capacitive applicator in the last section there are no edge problems and fat problems, since the current lines are parallel to the tissue interfaces (the layers are is parallel in stead of in series).

2.2.2 Planar loop

The planar loop, also called a pancake coil, is difficult to analyze (refs. 6 and 7), since there are in general two contributions to the electric field in the tissue. One is the near field of the voltage between the turns, which may lead to superficial heating. The other is the induced component from the current distribution. We may immediately see from fig. 4b that in the low-frequency range, the contribution from (1) will cancel the contribution from (2) on the axis, so the power density has a minimum below the center of the coil and is in general of a toroidal shape. Since the near-fields decay rather rapidly away from the sources the penetration in practice is limited to 1-2 cms, at a frequency of 27 MHz.

3. High-frequency applicators

3.1 Medium properties and plane wave propagation

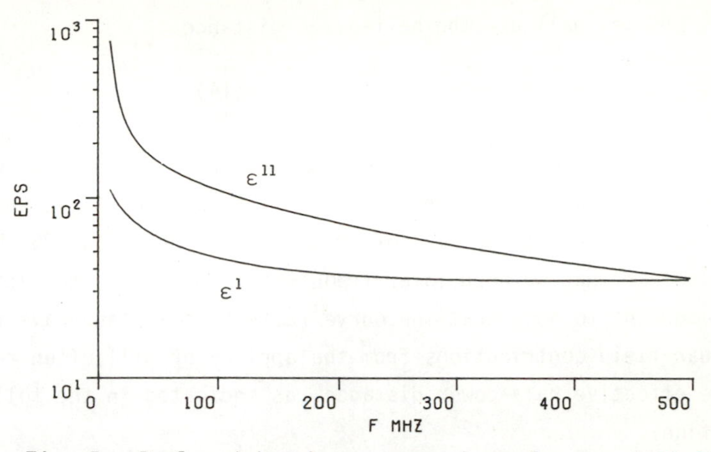

Fig. 5. Real and imaginary part of complex permittivity of averaged muscle tissue.

The permittivity $\varepsilon = \varepsilon^1 - j\varepsilon^{11}$ of tissue is a complicated function of frequency due to the various mechanisms involved. In this section we are concentrating on the frequency range above 10 MHz, where the real and imaginary part of ε for muscle tissue are approximately as shown in Fig. 5. The data (ref. 8) are smaller by a factor 2/3 than for tissue alone, taking into account averaging over a larger volume. Note that the medium is more a conductor than a dielectric below 500 MHz. The relative permittivity appears in wave propagation connections as an important factor for the phase shift β and attenuation α. A plane wave propagates as

$$e^{-j\gamma z} = e^{-j\beta z} e^{-\alpha z} \qquad (10)$$

where

$$\gamma = k_0 \sqrt{\varepsilon^1 - j\varepsilon^{11}} = \beta - j\alpha \qquad (11)$$

and k_0 is the free space wavenumber $2\pi/\lambda$.ε may also be expressed as

$$\varepsilon = \varepsilon^1(1 - j \tan \delta) \qquad (12)$$

where $\tan \delta$ is the loss factor.

Since $\tan \delta$ is of the order unity it is not possible to make approximations for β and α, but we must use the full expressions, i.e.

$$\alpha = k_0 \text{ Im } (\sqrt{\varepsilon^{1}-j\varepsilon^{11}}) \qquad (13)$$

Usually the attenuation is expressed as a penetration depth, $\delta = 1/\alpha$, but we shall use the half-power distance

$$\delta_{\frac{1}{2}} = \frac{\ln 2}{2\alpha} \qquad (14)$$

i.e. at a distance of $\delta_{\frac{1}{2}}$ into the medium, the rate of increase of the temperature is half of the value at the surface. $\delta_{\frac{1}{2}}$ is shown in Fig. 6, it is about 1-2 cms for frequencies higher than 250 MHz, so it is clear that we need lower frequencies for deep penetration. It is important to note that the curve represents a plane-wave result, near-field contributions from the applicator will often reduce the effective half-power distance, as indicated in the following section.

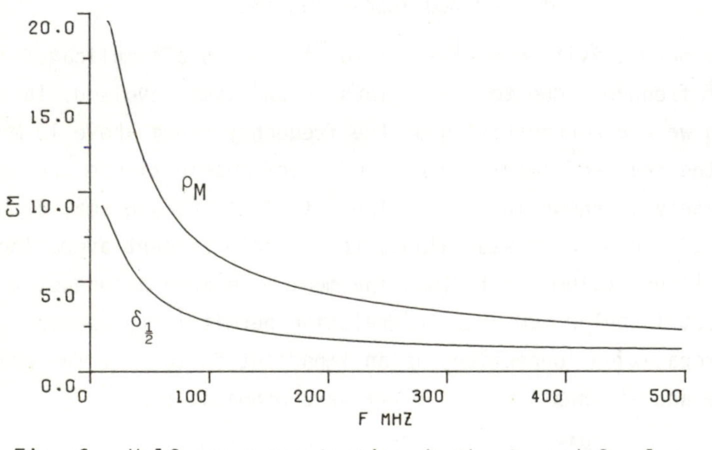

Fig. 6. Half-power penetration depth, $\delta_{\frac{1}{2}}$ and focal spot size ρ_M.

3.2 Line sources and aperture sources

In order to model the influence of nearfields several configurations involving line sources near a lossy cylinder have been analyzed (ref. 9). Consider the situation in Fig. 7 with two closely spaced line currents (a magnetic dipole line source) at a distance s from the surface. The full curves show the power density in the medium for various values of s. The free-space decay of the axial electric field is s^{-1}, and it is clearly seen how rapidly the power decays for small values of s, much faster than indicated by $\delta_{\frac{1}{2}}$. If we consider the contribution from the charges on the lines (or the potential difference between them) the situation is even worse, since the electric field in this case decays as s^{-2}. The conclusion is that, as expected, near-field applicators are sensitive to the spacing from the surface.

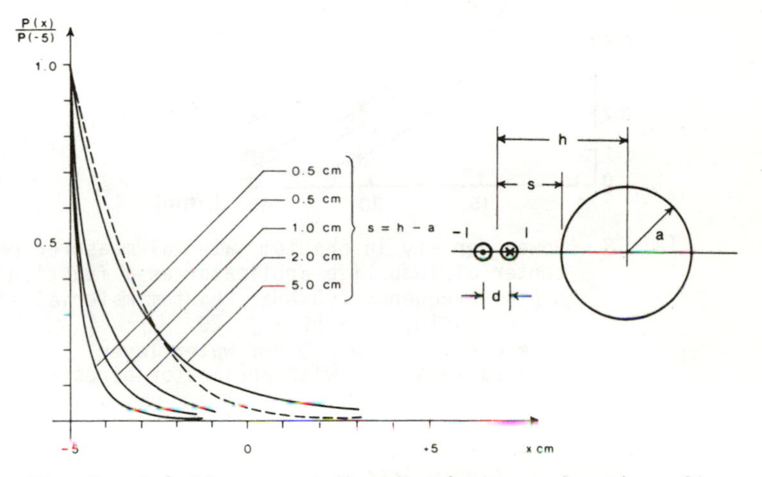

Fig. 7. Relative power distribution as a function of source position for y-directed magnetic dipole line source: f = 100 MHz, a = 5 cm;--------distributed source $2\alpha = 80^0$; ———— δ function source.

The line sources described are δ-function sources, but by creating a continuous distribution of such sources we get an aperture source, similar to a waveguide opening, a horn antenna or other continuous aperture radiators. This has a marked effect on the effective penetration depth since the aperture creates a more beam-like illumination in the medium where the decay rate will approach plane wave results. This is illustrated in Fig. 7 (broken curve) where the aperture source has the same distance from the surface (s = 0.5 cm) as the line source with the most rapid decay. It is difficult to estimate how large an applicator should be to avoid the rapid near-field decay since it depends on the precise design, but experience from ordinary antenna design indicates that the linear dimensions should be larger than a quarter of a wavelength in the lossy medium for a contact applicator.

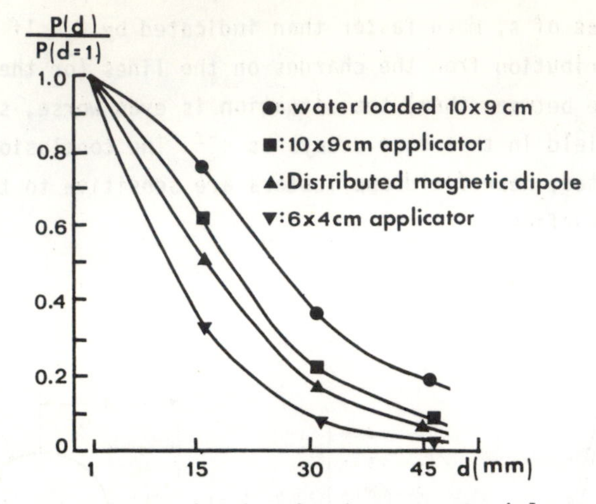

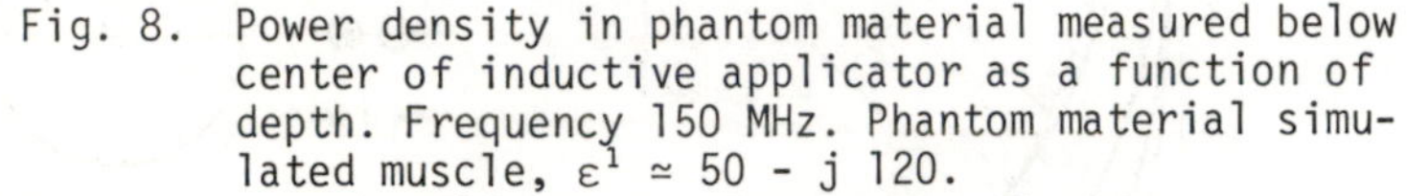

Fig. 8. Power density in phantom material measured below center of inductive applicator as a function of depth. Frequency 150 MHz. Phantom material simulated muscle, $\varepsilon^1 \simeq 50 - j\,120$.
- ● L = 10 cm, W = 9 cm, water layer
- ■ as in a, but with applicator directly on surface
- ▼ L = 6 cm, W = 4 cm, directly on surface
- ▲ theoretical curve for infinite magnetic dipole (ref. 9).

An applicator based on the principles stated above has been designed and is described in detail in ref. 10. It consists of parallel current-carrying strips above a groundplane and radiates a linearly polarized beam into the medium. Experimental results for phantom material are shown in Fig. 8 verifying the theories, i.e. the smallest applicator has the most rapid decay and the largest applicator removed slightly from the surface has the deepest penetration. Note that the frequency is 150 MHz so the applicator in itself is far from resonance, but this is considered as an external circuit problem, which may be solved by other means. Compared with the planar coil the distribution is favourable since the maximum occurs below the center of the applicator and there are no voltage differences between the individual wires or strips.

3.3 Focussing in muscle tissue

An obvious solution to the penetration problem is to use multiple applicators such that the fields from the individual applicators may interfere constructively, i.e. what in optical terms is called a focus. In order to derive at optimal solutions we must study the exact solution for a focus in a very lossy medium, since the attenuation per wavelength is not small. The model we choose is a circular cylinder with no variation of the fields along the axis and around the cylinder, and the cylinder is illuminated by in-phase axial electric fields at the surface. The inside electric fields will then be given exactly by

$$E_Z = A\ J_0\ (k_1\rho) \qquad (15)$$

where J_0 (z) is the zero-order Besselfunction of complex argument, ρ is the distance from the axis, and k_1 is the complex wavenumber satisfying

$$k_1{}^2 = k_0{}^2\ (\varepsilon^{I} - j\varepsilon^{II})\ . \qquad (16)$$

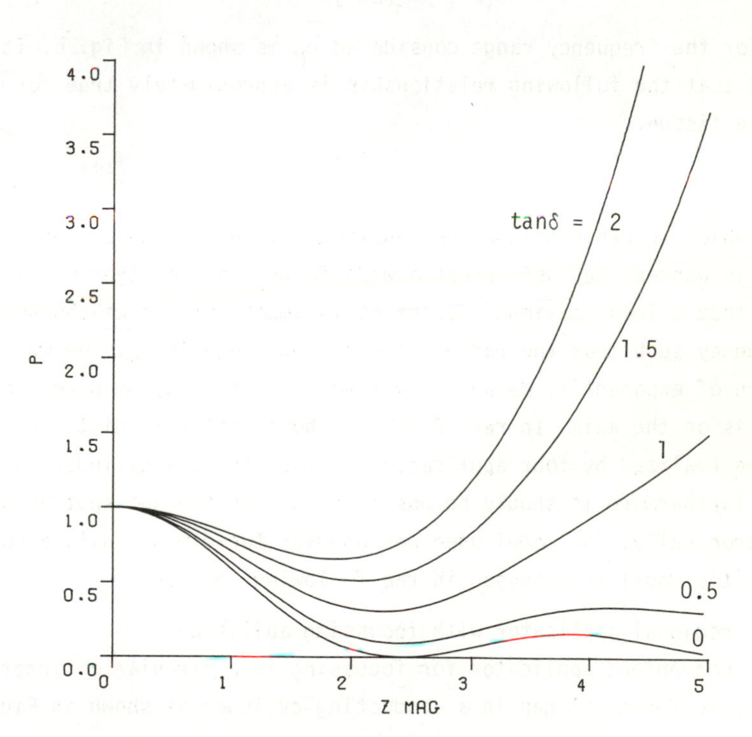

Fig. 9. Power distribution near two-dimensional focus.

A plot of the corresponding power density is shown as a function of $|k_1\rho|$ in Fig. 9 for various values of the loss factor. For $\tan\delta=0$, a lossless medium, the well-known focal distribution for a two-dimensional case is seen with a focal radius to the half-power point of 0.17λ. When $\tan\delta$ increases the focal distribution changes character completely, so we essentially have two regions, a focal region with a local maximum and an exponential decay region dominated by exponential decay as given in eq. 10. It may be argued what one should choose as a spot size for such a distribution, the author has chosen the radius ρ_M where the density is a minimum. An approximate expression for ρ_M may be obtained by expanding the Bessel-function for small arguments,

$$(k_0\rho_M)^2 = \frac{8\varepsilon^1}{3(\varepsilon^1)^2 + (\varepsilon^{11})^2} \tag{17}$$

For the frequency range considered ρ_M is shown in Fig. 6. It is noted that the following relationship is approximately true for muscle tissue,

$$\rho_M \sim 2\,\delta_{\frac{1}{2}} \tag{18}$$

from which we can conclude, not unexpected, that if deep penetration is wanted, we must accept a wide focus. On the other hand, the fact that a local maximum does exist is important. If we choose the frequency such that the radius of a cylinder equals ρ_M, the whole region of exponential decay is avoided, and the maximum power density is on the axis. In ref. 9 it was shown that this distribution may be realized by four applicators surrounding the cylinder, and that furthermore it should be possible to move the hot spot around electronically. The model used was however two-dimensional, a more realistic model is dicussed in the following section.

3.5 A regional applicator with focussing abilities.

A convenient applicator for focussing in a circular cylinder is a cirumferential gap in a conducting cylinder as shown in Fig. 10.

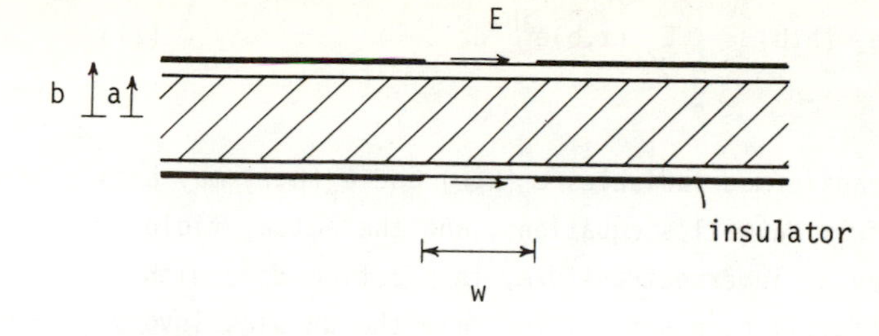

Fig. 10. An electric field is impressed in a gap in a conducting cylinder to create a focus in the lossy medium.

The effective source for the illumination is the electric field in the gap and the metal will tend to shield the volume not illuminated. The basic principles mentioned previously must be obeyed, i.e. the aperture width should be sufficiently large to avoid near-field dominance, and the metal edges should be removed from the lossy medium. The latter is satisfied by having a thin dielectric layer between the metal and the tissue.

The exact form of the axial electric field in the gap depends on the way the gap is excited externally, so the problem is not defined until the external sources are specified. However, a good first-order approximation may be obtained by assuming that the gap field has the same shape as the quasi-static field in a planar gap, i.e.

$$E_z\,(z,b) = \begin{cases} \dfrac{E_0}{\sqrt{1 - \left(\dfrac{2z}{w}\right)^2}} & |z| < \dfrac{w}{2} \\ 0 & |z| > \dfrac{w}{2} \end{cases} \qquad (19)$$

which has the right edge singularity.

The problem is analyzed by a Fourier transformation in the z-domain

$$e_z\,(h,r) = \int_{-\infty}^{+\infty} E_z\,(z,r)e^{jhz}\,dz \qquad (20)$$

where the source term is determined by

$$e_z\ (h,b) = \int_{-\frac{w}{2}}^{+\frac{w}{2}} E_z\ (z,b) e^{jhz} dz\ . \qquad (21)$$

The transformed variables $e_z(h,r)$ and $e_r(h,r)$ may easily be determined from Maxwell's equations, and the actual fields may then be found by an inverse transform, in practice done with an FFT-algorithm. Fig. 11 a, b and c illustrate the physics involved when the frequency is varied. In Fig. 11a we have a solution for low frequencies close to the quasistatic case with a typical saddle point at the center. For 150 MHz (Fig. 11b) we are close to the case where ρ_M equals the radius, there is a focus at the center. At 450 MHz (Fig. 11c) the spot size is smaller, but the exponental decay starts to dominate. There may be situations where this is acceptable since

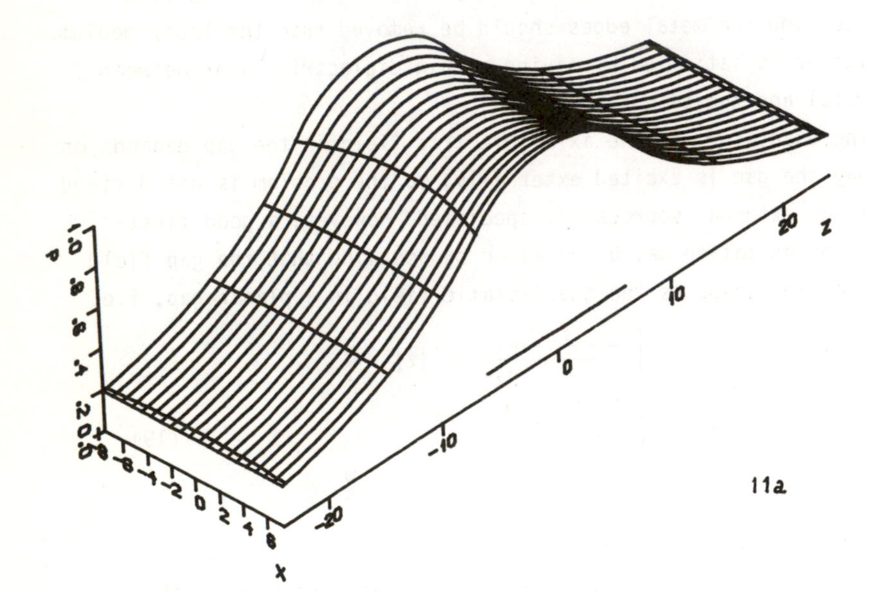

11a

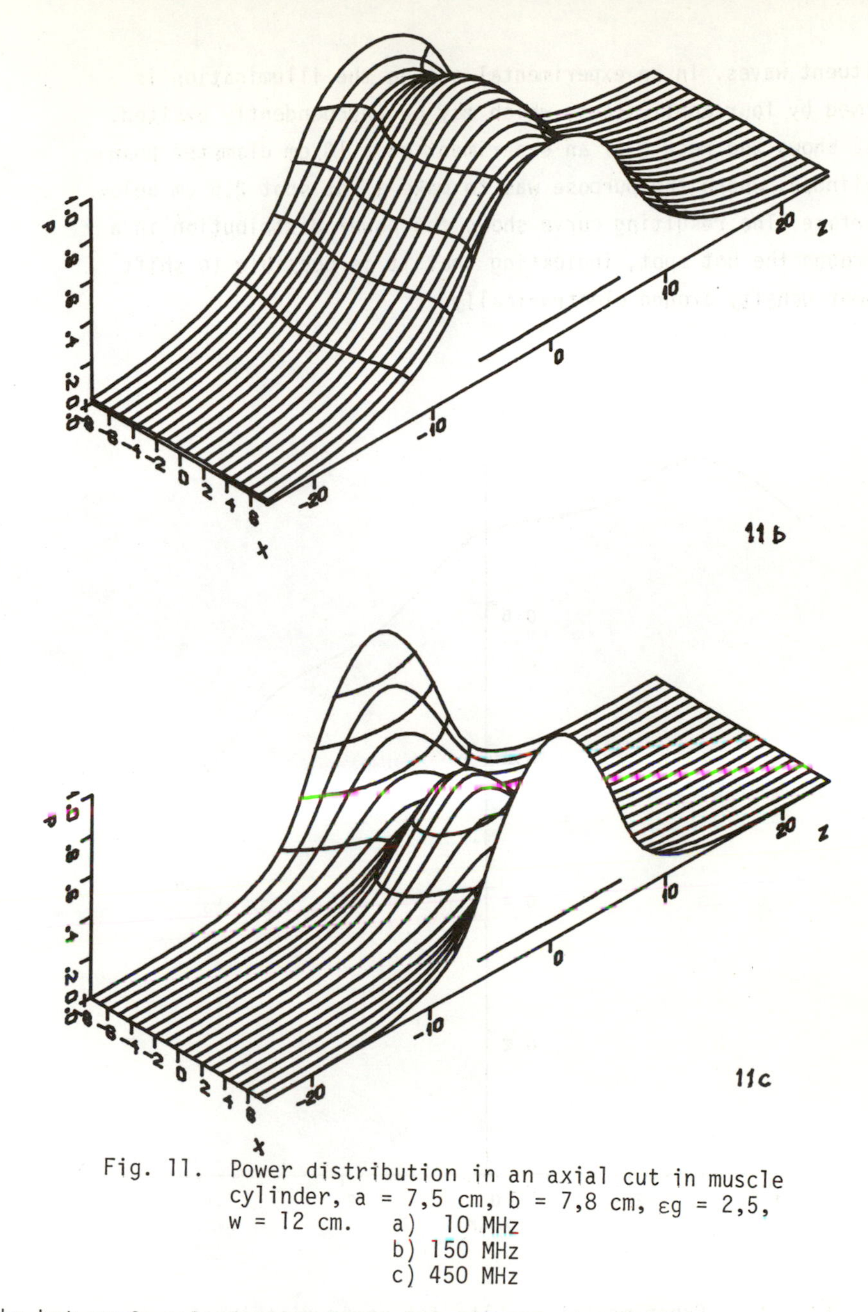

Fig. 11. Power distribution in an axial cut in muscle cylinder, a = 7,5 cm, b = 7,8 cm, εg = 2,5, w = 12 cm. a) 10 MHz
b) 150 MHz
c) 450 MHz

the hot surface layers may be cooled externally.

Since the focus is a result of interfering waves it is possible to shift the focus around by proper amplitude and phase shift of the constituent waves. In an experimental set-up the illumination is performed by four applicators, which may be independently excited. Fig. 12 shows the result of an experiment on a 10 cm diameter phantom cylinder, where the purpose was to make a hot spot 2.5 cm below the surface. The resulting curve shows the power distribution in a cut through the hot spot, indicating that it is possible to shift the power density around electronically.

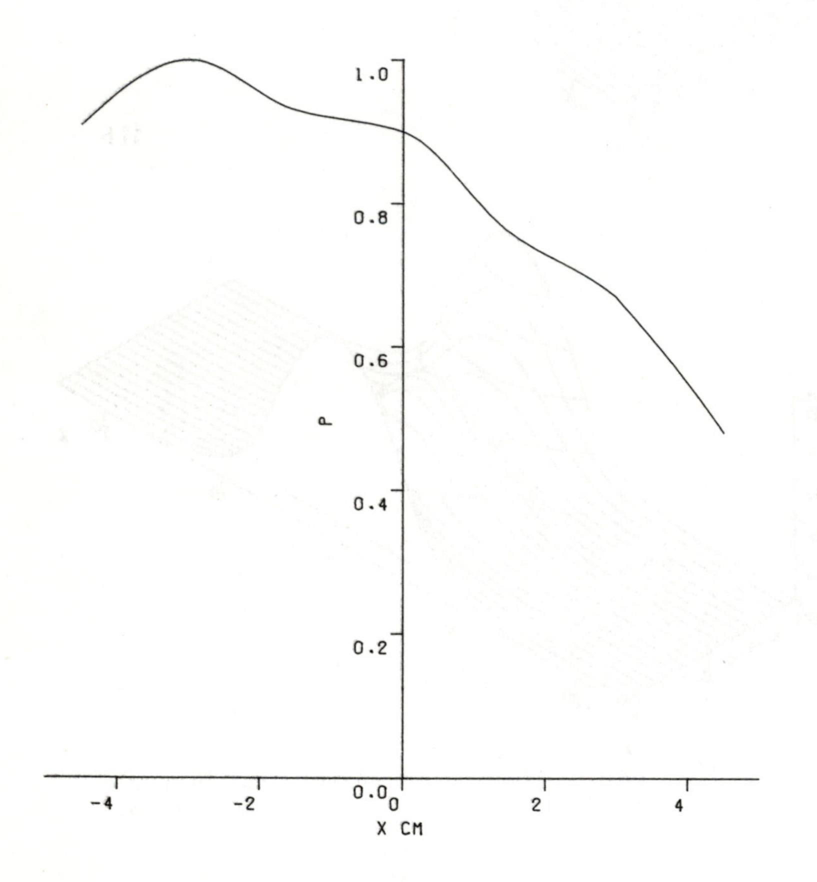

Fig. 12. Experimental results for power distribution in phantom cylinder, a = 5 cm, f = 145 MHz. The intended maximum was at x = -2.5 cm.

4. Acknowledgements

The author would like to acknowledge the indispensable help from his colleagues Povl Raskmark, Leif Heinzl, Kristian Harmark and Aage Baun for the experimental verifications.

5. References

1. Wiley, J.D. and Webster, J.G., "Analysis and Control of the Current Distribution under Circular Dispersive Electrodes", IEEE Trans., Biomedical Engineering, vol. BME-29, May, 1982, pp. 381-385.

2. Bach Andersen, J. and Solodoukhov, "Field Behavior near a Dielectric Wedge", IEEE Trans. Antennas and Propagation, vol. AP-26, No. 4, July 1978, pp. 598-602.

3. Brezovich, I.A. et al, "A Practical System for Clinical Radiofrequency Hyperthermia", Int. J. Radiation Oncology Biol. Phys., vol. 7, 1981, pp. 423-430.

4. Elliott, R.S., Harrison, W.H. and Storm, F.K.,"Hyperthermia: Electromagnetic Heating of Deep-Seated Tumors", IEEE Trans. Biomedical Engineering, vl. BME-29, No. 1, January 1982, pp. 61-64.

5. Paliwal, B.R. et al, "Heating Patterns Induced by a 13.56 MHz Radiofrequency Generator in Large Phantoms and Pig Abdomen and Thorax", Int. J. Radiation Oncology Biol.Phys., vol. 8, 1982, pp. 857-864.

6. Guy, A.W. et al, "Therapeutic Applications of Electromagnetic Power", Proc. IEEE, vol. 62, 1974, pp. 55-75.

7. Hand, J.W. et al, "Considerations of Radio-Frequency Induction Heating for Localized Hyperthermia", Phys. Med. Biol., 1982, vol 27, No. 1, 1-16.

8. Durney, E.H. et al, "Radiofrequency Radiation Dosimetry Handbook", May 1978, USAF, School of Aerospace Medicine.

9. Morita, N. and Bach Andersen, J., "Near-Field Absorption in a Circular Cylinder from Electric and Magnetic Line Sources", Bioelectromagnetics, vol. 3, 1982, pp. 253-274.

10. Bach Andersen, J. et al, "A Hyperthermia System Using a New Type of Inductive Applicator", IEEE Trans. Biomedical Engineering, BME-30, January 1984 (Special issue on hyperthermia).

PROCEEDINGS OF THE II INTERNATIONAL CONFERENCE ON APPLICATIONS OF PHYSICS TO MEDICINE AND BIOLOGY
edited by Ž. Bajzer, P. Baxa & C. Franconi

PLANNING AND DOSIMETRY IN THERMAL THERAPY

T. C. Cetas*, R. B. Roemer*#, and J. R. Oleson*

*Division of Radiation Oncology, Arizona Health Sciences Center,
Tucson, Arizona, USA 87524

and

#Department of Aerospace and Mechanical Engineering, University of
Arizona, Tucson, Arizona, USA 85721

ABSTRACT

Thermal therapy must be placed upon a quantitative basis if it is to become a routine means of treating cancer. The proper prognostic parameters of therapy must be elucidated and then combined to form a unit of thermal dose and a set of rules regarding maximum normal tissue tolerances and minimal requirements for tumor therapy. Thermal dosimetry is the methodology for establishing that adequate therapy has been achieved technically. It encompasses several tasks which include heating equipment characterization, treatment planning, real time treatment monitoring, treatment evaluation, and determination of the thermal dose received everywhere within the treated volume.

1. Introduction

Thermal therapy has been used sporadically to treat cancer for decades but has come under development with increasing intensity over the last several years as is evidenced by a series of symposia proceedings, compendia, and other publications (1-8). Three related but distinct problems in the field are (i) the need for quantitative indications of the prognostic parameters of the treatment which will predict efficacy against the tumor and insure the safety of the normal tissue environment, (ii) the absence of a suitable definition of a unit of dose which properly accounts for the equivalency of treatments of various temperature-time regimens and which can be used as a quantitative guide for prescribing treatments, and (iii) the lack of adequate methods of determining that dose everywhere in the treated volume. In other words, we must determine dose response and toxicity relationships for use in the clinic, establish the unit of thermal dose, and develop sophisticated thermal dosimetry. While not complete, progress is occuring in each of these areas. The broad outlines are becoming more refined and the questions more specific. Speculation gradually is giving way to data and clarified concepts. Our primary purpose here is to discuss the methodology of thermal dosimetry. Nevertheless, we will mention briefly the efforts towards defining a unit of thermal dose and then indicate some of the clinical results from which the prognostic variables and hence the technical requirements for therapeutic heating can be inferred.

2. Implications on Thermal Dosimetry from Clinical Studies

Analysis of thermometric results in the series of 163 patients treated at the University of Arizona (1977-1982) (9) has yielded information relating to thermal dose and thermal doismetry considerations. Groups of patients were heated with interstitial radiofrequency (RF) currents, external microwave fields, external RF plates, or magnetic induction techniques. Thermometry practice varied somewhat by technique. During interstitial and magnetic induction heating, temperature mapping usually was performed giving confidence the intratumoral temperatures near the sites of actual minimum and maximum temperatures were sampled. In the case of tumors treated with other techniques, thermometers usually were fixed at the center of the tumor, and the lateral and superficial edges. Since many of the microwave treatments were performed early in the series, before adequate instrumentation was available, the deepest aspect of the tumor frequently was not sampled. Temperatures at the sites of maximum and minimum intratumoral temperature were averaged over the 30 minute interval. Chi-squared tests as well as the logistic regression model then were used to correlate these time-temperature measures, outcomes of therapy (defined as tumor volume response), and initial tumor volumes. Other variables were examined as well for prognostic importance.

Pertinent results of this analysis are that within the group of patients treated with interstitial hyperthermia and radiation, minimum intratumoral temperature (T(min)) ($p=0.005$), radiation dose ($p=0.007$), and tumor volume ($p=0.001$) were correlated significantly with CR, which sugggests that a thermal dosimetric measure having prognostic importance should be based upon T(min).

Logistic regression analysis was important in accounting for the many obvious differences in patient and treatment variables among patients treated with different techniques. Statistically significant differences in T(max) and T(min) were found versus technique even after accounting for tumor volume. T(min) was the single most important variable predicting CR ($p<0.0005$). Tumor volume, technique and radiation dose composed an alternative set of three variables whcih correlated significantly with CR. Interestingly, T(min) correlated negatively with tumor volume (p=0.014) and T(max) correlated positively with tumor volume (p=0.026). This finding probably relates to the presence of larger necrotic cores in larger tumors, as well as to extension of well-perfused peripheral parts of large tumors outside the volume receiving significant SAR with the particular device used.

In summary, the variables, T(min) and tumor volume, are of prognostic importance in human studies and thus are appropriately the focus for thermal treatment planning and dosimetry. These results are entirely consistent with those found in randomized studies using spontaneously occuring tumors in pet dogs and cats (10,11).

3. Definition of Thermal Dose

At present, no unit of thermal dose has been accepted by a wide range of investigators in the field and this has contributed to the difficulty of comparing results between laboratories and in reducing a plethora of data into manageable form. The most straightforward approach is to pick some threshold temperature for thermal effect and then account for the time that the system under treatment is at that temperature. Accordingly, some have used the temperature increase multiplied by the time, or extended the concept to include Simpson rule integrations for fluctuating temperatures, to define a unit of dose in degree-minutes (e.g. 11). Dewey et al (12) and Gerner et al (13) have noted that when the results of cell survival curves from hyperthermia experiments are analyzed in terms of an Arrhenius plot the activation enthalpy is constant for a wide range of mammalian cells for temperatures between 42 and about 50°C. Consequently, Sapareto and Dewey (14) used this effect by defining a unit of dose in terms of equivalent minutes at 43 °C (14,15). The dose then is expressed as:

$$Eq\ 43 = \Sigma(\Delta t)\ R^{(43-T)} \tag{1}$$

where: Δt is the time interval, and

$$R = 0.25 \text{ for } T \leq 43$$
$$R = 0.50 \text{ for } T \geq 43\ .$$

Recently, Gerner (16) has proposed a thermal dose based upon general thermodynamic arguments especially as they relate to chemical reaction kinetics. His analysis leads to a function of the form:

$$\text{Dose} = \int T(t) \exp \left[- \frac{H}{R} \left(\frac{1}{T(t)} - \frac{1}{Tc} \right)\right] dt \tag{2}$$

in units of kelvin seconds (K s). For the present, he sets the enthalpy H=600 kJ/mole, the compensation temperature, Tc=310 K (37°C) and the gas constant R=8.32 J/K/mole. The degree of thermal damage, which relates to dose, will be a function of the net thermal energy which in turn is proportional to temperature T. The effects are weighted exponentially according to the activation enthalpy of the reaction, the absolute temperature and some compensation temperature to account for normal processes occuring at normal body temperatures. The activation enthalpy is constant for a specific chemical reaction or set of reactions. It may be possible to use an average value for mammalian cell systems as Gerner presently is considering. On the other hand, specific values for specific effects may be required in the future. Clearly, this unit is not in final form and the thermodynamic reasoning needs to be more rigorously developed. Nevertheless, Gerner has been able to show that this form does permit the reduction of data from many different treatment regimens on a given cell line to one single dose related curve. Furthermore, biological effects such as cell line dependence or thermotolerance induction are clearly displayed. This unit, as it stands, is not to be taken as a proposal. It is only mentioned as an indication of the nature of the discussions and difficulties in defining a unit. Because mammals are homeothermic and can respond to thermal challenges, the nature of the dose definition cannot follow the same approach as was used for ionizing radiation where the energy deposited per unit mass of tissue was the basis of the unit. Similarly, the specific absorption rate, SAR, (17) which describes the rate and distribution of energy deposition, is an effective parameter for the characterization of heating equipment, methodology, and configurations. It is not an appropriate unit for thermal dose since it deals only with power deposition and does not address thermal dissipation.

4. Methodology of Thermal Dosimetry

Thermal dosimetry, the methodology of determining the dose, can be distinquished as a physical or engineering discipline somewhat distinct from the definition of thermal dose or the determination of a biological or clinical dose response relationship. From the available data, as well as intuition, the time that a given volume is at some absolute temperature or increment of temperature above some baseline will relate to the thermal dose delivered at that point. We presume from the information available at present, that non-thermal effects are not important. If they are, then the therapy by definition becomes something other than hyperthermia and the relevant dosimetry also would not be thermal. The parameters to be measured, then, are temperature, time, and position. At its most basic level, thermal dosimetry is presenting the first as a function of the latter two. We will see from consideration of the bioheat transfer equation below that measurment of the derivatives of temperature with respect to time and space are useful as well in determining power deposition and heat dissipation parameters.

We find it convenient to consider four different aspects of thermal dosimetry (18,19) as they might be confronted chronologically throughout the course of a patient's therapy. Comparative thermal dosimetry deals with establishing the power deposition characteristics of various equipment so that an appropriate method of heating can be selected. Prospective thermal dosimetry represents the treatment planning aspect for the particular tumor and patient in question. Concurrent thermal dosimetry is the monitoring and control of the treatment session. Retrospective thermal dosimetry combines the information acquired during the planning phase with the data taken during the treatment session in order to establish the dose given throughout the heated region.

The expression for energy balance in an incremental volume of tissue subjected to external power deposition can be approximated by the bioheat transfer equation (20):

$$\rho C \frac{\partial T}{\partial t} - k\nabla^2 T - WC_b\,(T-T_a) = Q_p + Q_m \qquad (3)$$

where is the tissue density (kg/m^3), C is tissue specific heat (J/kg/°C), k is tissue thermal conductivity (W/m/°C), T is temperature (°C), t is time (s), W is blood flow ($kg/m^3/s$), C_b is blood specific heat, T_a is arterial temperature, Q_p is absorbed power per unit volume (W/m^3) and Q_m is the metabolic heat (W/m^3). The first term on the left gives the rate of thermal energy accummulation, the second is the power into the volume from thermal conduction, and the third is the Pennes (20) expression for the effects of convective heat exchange from blood flow. Q_p is the power deposited per unit volume and while Q_m is the metabolic heat generation rate. The bio-heat equation which under the Pennes approximation assumes that the thermal effects of blood flow can be adequately characterized by a scalar heat sink term, has been challenged on the logical basis that no provision is made for directionality, warming of the blood prior to entering the incremental volume, and lack of sufficient detail in the vicinity of major vessels. Studies are continuing towards developing alternative formulations. Unfortunately, greater sophistication there will lead undoubtedly to the need to specify more parameters experimentally and clinically. For the short term, we find it convenient to use the simple formulation but allow the local value of blood flow to serve as a mathematical parameter that can be adjusted force calculated temperatures to agree with measured ones at those specific locations. The extent to which this method is legitimate is under study. Thus, thermal dosimetry can be reduced to solving this expression, or some alternative, for temperature as a function of time and position. From the solution, temperature distributions can be plotted and integral thermal dose distributions can be computed in a straightforward manner.

A solution to the bioheat transfer equation depends upon the boundary conditions; geometry and anatomy; the thermal properties of the tissues in the field; the blood flow patterns, magnitudes, and regulatory responses; and the power deposition pattern which in turn is dependent again upon geometry, and the electromagnetic or ultrasonic field absorption and propagation properties. The manner with which each of these issues are dealt is somewhat different for each of the four aspects of thermal dosimetry. In general, however, normal tissue properties and geometries can be determined adequately from the literature and from diagnostic information such as CT scans. Tumor properties are more difficult to obtain, although estimates can be based upon the constiuent matter. Uncertainties in the blood flow by far cause the greatest difficulty in computing the temperature patterns and hence the thermal dose. Nevertheless, if it is remembered that obtaining the resultant temperature field is the primary objective of thermal dosimetry rather than describing other parameters, then techniques can be developed which permit some, if not perfect, knowledge of the thermal dose. The following sections elaborate on the techniques as they apply to each of the four chronological aspects of thermal dosimetry.

5. Comparative Thermal Dosimetry

The purpose of comparative thermal dosimetry is to determine which modality or configuration would be most appropriate for heating certain classes of tumors. Both experimental and theoretical/numerical techniques are necessary. The general approach is to determine the power deposition pattern of various heating systems with respect to the region of interest, and then use this information in a numerical model of the bioheat equation and calculate the resulting temperature distributions. Since we are interested in comparing various types of equipment or specific configurations in general classes, typical but standardized values of tissue properties can be used. Only the nature of the thermal model of the tumor, which is highly dependent upon blood flow conditions remains a significant unknown.

Nevertheless, this can be treated systematically as well.

The first task is to determine the power deposition distribution. Guy (21) introduced the use of tissue equivalent electromagnetic phantoms along with thermography to study these distibutions experimentally. Since then, his system has been used by many other workers. Simple homogeneous or layered phantoms can be used to verify the principles of power deposition in a given media and to compare one instrument against another, while more complicated phantoms (22,23) using realistic dimensions and anatomical features can be used to study the effect of heterogeneities on the power deposition field. With proper registration, the specific absorption rate (SAR) distribution as determined experimentally can be superimposed on a thermal model of the tissues and the temperature patterns computed (24).

Similarly, the SAR distributions can be computed, analytically in some symmetrical cases, and numerically in general. Examples of these calculations have been performed for magnetic induction (25-29), capacitive induction (28), phased array electromagnetic induction (30-32), interstitial current fields (33), interstitial microwaves (34), ferromagnetic implants (35-38), and ultrasound (39,40).

Criteria for adequate heating must be established in order to properly evaluate the capabilities of a given method. Based upon the clinical results in both animals and humans discussed in the introduction a minimum temperature of 42°C is required for the entire tumor. In order to avoid the possibilites of anomalies in the computer runs, we actually set the criterion that 97.5% of the tumor be above 42°C. In any tumor, theoretical or real, temperatures are not likely to be uniform. To characterize the nonuniformity, we define a maximum acceptable temperature within the tumor as well. Clinically, this is often expressed by a concern over complications resulting from rapid tissue necrosis that can result in infection prone voids. Depending upon the modality, we have used 48, 53 and 60 °C as upper limits. To avoid normal tissue damage, maximum temperatures in muscle and skin are 44°C and in viscera, 42 °C. Finally, a maximum total power into the patient is limited to avoid systemic heating effects. We have used 1000 W and 2000 W in various models.

Figure 1 shows how temperatures might appear in some typical tumor as a function of applied power. In easily heated tumors Fig. 1a, a range of applied powers Q(acc) (41) exists such that the entire tumor is between the acceptable temperature limits. The greater this range, the lower the restrictions on precision and stability of the power generator and the easier the tumor is to heat acceptably. On the other hand, in some cases the nonuniformities will be so severe that maximum temperatures are exceeded in the tumor or normal tissues before the minimum temperature is achieved Fig. 1b. In such cases, the method cannot give an acceptable heating and so would not be considered appropriate for the tumor in question.

The use of a computer to calculate the temperature profiles and thus to determine the acceptability of the method depends upon the existence of a thermal model of a tumor. Since tumors can have a very wide range of properties and blood flows, the designation of a single general model is impossible. However, it is possible to conceive of a thermal model that is easy to heat by a given method. Similarly, it is possible to conceive of a tumor model that has thermal characteristics that make it difficult to heat by that method. Thus, we can

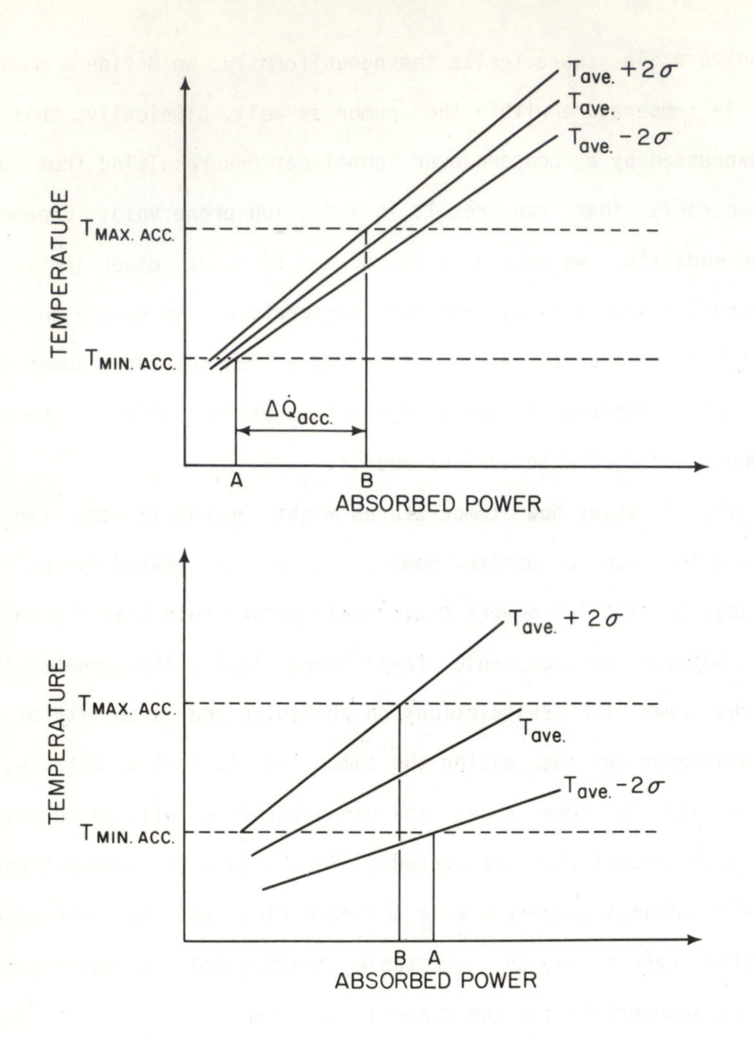

1. Sketch indicating concept of Acceptable Power Range parameter Q(acc) for evaluating equipment. At some level of applied power all of the tumor is above some predetermined minimum temperature while at some other level of applied power, portions of the heated region are above some predetermined maximum temperature. In the upper figure, Q(acc) is positive, indicating that the method gives acceptable heating. In the lower figure, Q(acc) is negative, thus the method will not result in acceptable heating.

construct limiting case models (42) that bracket the characteristics of a real tumor. For many situations the uniform blood flow (UBF) model shown in Fig 2 represents an easy to heat method and the annular blood flow (ABF) model is a hard to heat model. The reason for this difference is that we have placed a limit on the degree of uniformity of temperatures. If the power is deposited uniformly throughout the tumor, then a uniform heat dissipation will lead to uniform temperatures while nonuniform cooling, especially enhanced at the periphery while inhibited in the center leads to large thermal gradients. To get an indication of the validity of the approach, we (42) generated many tumor models in which the blood flow pattern was set randomly by computer. Except in a few cases in which extreme blood flow patterns resulted, the results of the acceptability of heating as characterized by the Q(acc) parameter were intermediate between the limiting case models of UBF and ABF. We have performed studies for several modalities including magnetic induction heating (25), uniform power deposition (43), ultrasonic heating (41) and inductively heated ferromagnetic implants (37) placed interstitially in the tumor. Strohbehn has considered interstitial microwave techniques (34) and interstitial radiofrequency currents (44) although he did not use the Q(acc) parameter.

The validity of this approach is becoming apparent clinically as well. For example, large necrotic tumors often have very nonuniform temperatures. If temperatures are measured at only a few points, the therapist may feel that successful heating was achieved and may even observe rapid regression of the tumor. Nevertheless, the likelihood of a long duration of response is small as was discussed in the introduction.

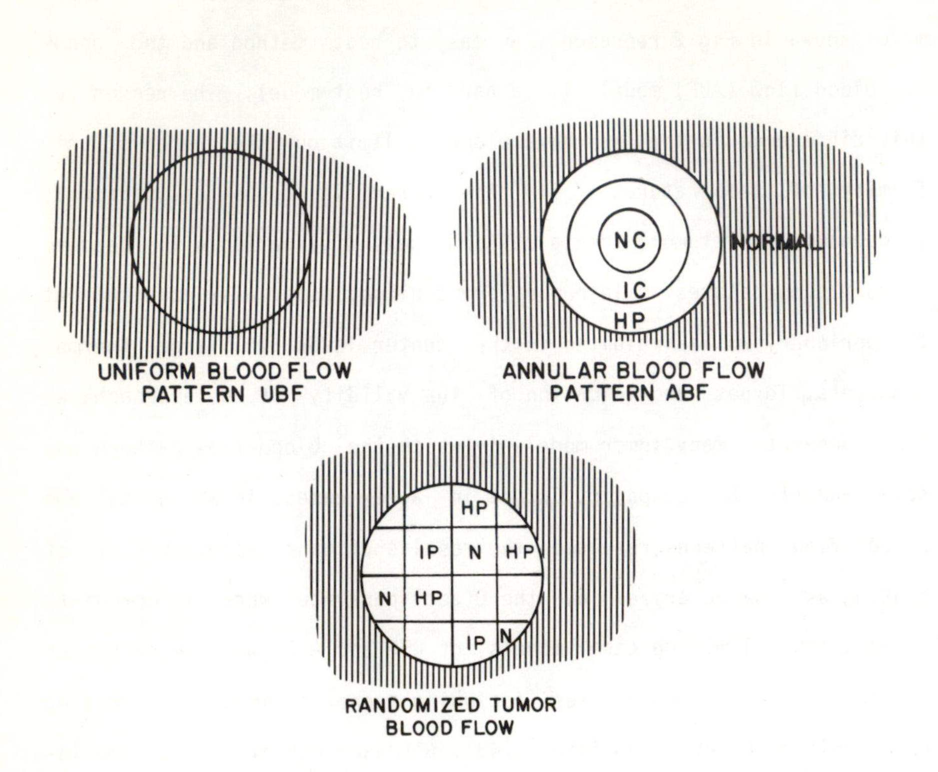

2. Illustration of the three tumor models used in the discussion. See the text for details.

As an example of the method, Figure 3 shows the temperature pattern to be expected from uniform power deposition in the pelvis with an annular blood flow tumor. Similarly, Figure 4 shows the temperature pattern to be expected from magnetic induction heating at 13.56 MHz for a similar case. The modeling which is based upon experimentally as well as theoretically determined power deposition patterns suggests that only tumors within 8 cm of the surface would be acceptably heated. Gibbs (45) and Oleson (46) have compared directly temperatures measured along radial paths from the deepest aspects of tumors to the surface for cases which have been heated both with magnetic induction (Magnetrode) and with a method of relatively uniform power deposition (BSD Annular Phased Array). Figure 5 gives one example and shows that the temperature patterns are consistent with the general characteristics of the models.

6. Prospective Thermal Dosimetry

At this point we become interested in individual treatment planning. The concepts remain the same as in the previous category except that now we require the characteristics to be specifically those of the patient rather than for generalized typical models. Parametric analysis is used for tailoring the treatment to the patient rather than for elucidating the sensitivity of the model to errors or for comparing the general features of various modalities.

The first task is to locate the tumor and the general anatomical features of the surrounding normal tissues through use of CT scans or other imaging techniques. Typical values from the literature can be assigned to the normal tissues. Properties of tumor tissues can be estimated based upon any information that might be available on its

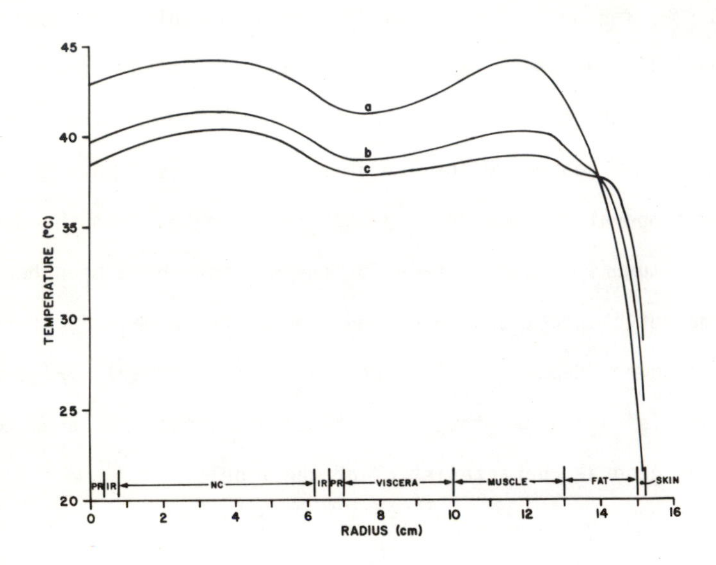

3. Predicted temperature distribution for an annular blood flow (ABF) tumor model located just off the body axis and heated by a concentrically aligned magnetic induction coil.

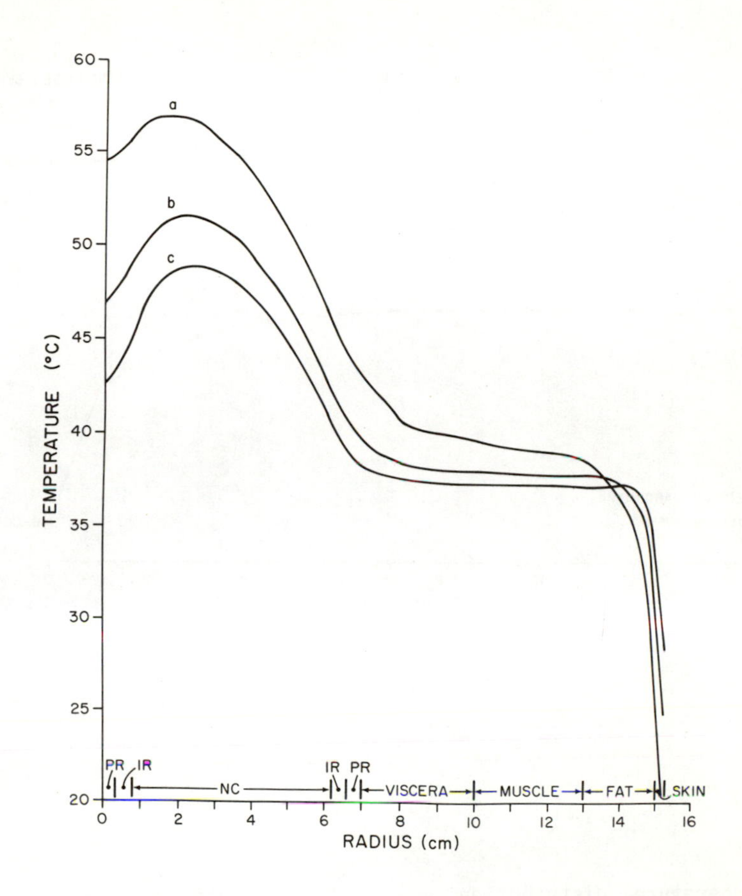

4. Predicted temperature distribution for and annular blood flow (ABF) tumor model located just of the body axis and heated by a method that gives uniform power deposition.

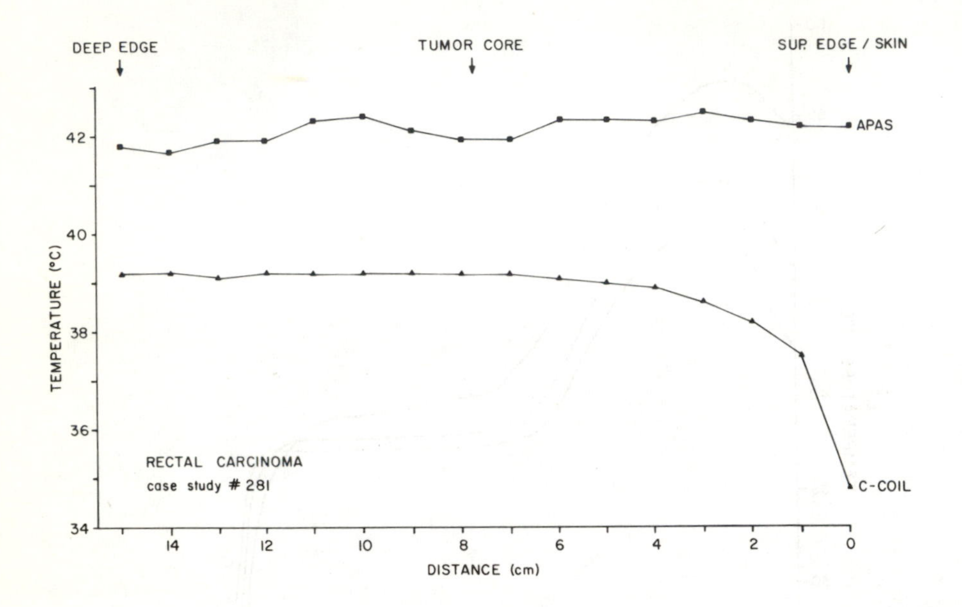

5. Temperature distribution measured in a patient with a large bronchogenic carcinoma heated by three different methods: C-coil, a concentrically aligned magnetic induction coil (Magnetrode); APAS, the BSD annular phased array; and H-coil, a coaxially aligned pair of coils arranged with one anterior to the patient and one posterior. This latter method is not discussed here.

constituents such as the concentration of water or lipids. The power deposition pattern then must be determined for the region to be treated. Detailed phantoms based upon the CT scans can be constructed and the specific absorption rate (SAR) can be determined by using the thermographic technique introduced by Guy (21) and discussed above. Examples are the work of Emery et al (24) and of Oleson (23), although undoubtedly others have done similar work. Alternatively, and preferably once the programs are written, the SAR distribution can be computed numerically. This has been done for high frequency heating using both capacitive and magnetic induction by Armitage, et al (28), by Hill, et al (24) for magnetic induction, and by Iskander, et al (30) for the annular phased array by BSD. Arcangeli, et al (31) have used optimization techniques to refine the heating patterns from an array of applicators at 915 MHz.

In the treatment plan, the actual blood flow in the normal tissues as well as in the tumor are still unknown and so it is still impossible to compute the isotherms that will result. Nevertheless, it is possible to use the bracketing case tumor model concept and hence establish if the heating technique in question has a reasonable chance of producing an acceptable heating of the tumor. Initially, both the uniform blood flow model and the annular blood flow model could be used to represent the tumor. If additional information is available on the location of major vessels, it could be incorporated into the models as well. If the computer run indicates that the most difficult to heat thermal model can be heated acceptably, then it is probable that the method will work on a real tumor located similarly in the patient. Conversely, if the easiest to heat thermal model cannot be heated by the method, then it is unlikely that the method would work in a real case. Of course, if the results of the model computations are mixed, then either further development of the models are necessary or the physician and physicist together must exercise their judgement as to the level of risk. While definite answers cannot be given as in radiation dosimetry, at least some basis for judgement is provided.

From calculations such as these, it will be possible to optimize the treatment technique prior to therapy so that the patient is likely to receive maximum benefit at minimal risk. One additional feature of this approach is that the models will be able to indicate the places where thermometer probes should be placed in order to properly monitor the treatment procedure. The risk of not measuring at points of high temperature in normal tissues or noting cool locations within the tumor, such as near the periphery or a major vessel will be reduced.

7. Concurrent Thermal Dosimetry

Concurrent thermal dosimetry represents the monitoring and control of the actual treatment. We presume that for some time to come, noninvasive tomographic thermometry will not be feasible. A real difficulty to be faced by any such technique is the necessity of adequate space around the patient to have sophisticated multi-aperature heating equipment as well as temperature monitoring equipment. In addition, the noise rejection requirements on any tomographic thermometry system will be extremely high if it is to perform in the presence of intense heating fields. Furthermore, temperature distributions cannot be predicted a priori with accuracy due to the lack of knowledge of the detailed blood flow distributions and the manner in which this will change with tissue temperature and time throughout the treatment. Thus, temperatures must be monitored with invasive probes.

The general principles of thermometry as they relate to hyperthermia have been reviewed in a number of places (e.g. 47-52). An excellent compilation of recent advances in thermometry is given in Temperature (53). Measurements in intense electromagnetic or ultrasonic heating fields require special considerations to avoid serious errors, although the former is in general more difficult that the latter. Again, these are discussed in most of the reviews mentioned as well as in a number of other places (e.g. 54). The problem which we address here is related to the need to sample many places using invasive methods in order to characterize the complete thermal field. Since patients cannot be expected to tolerate more than a few interstitial probes, it becomes necessary to sample several points along a track. Two methods have been used in recent years. The first is to use pull-back techniques within a catheter. This is convenient and once steady state is reached, it can be performed manually with simple probes and a readout instrument. Significant errors can occur with this technique especially in the presence of steep gradients. Presumably, with proper thermal contact media introduced into the catheter, this can be minimized. Alternatively, multiple sensor probes are being used in a number of places. These are extremely helpful in that their location can be determined precisely and they remain fixed throughout the treatment. An automated system is required to monitor the large number of sensors that can be used in such a system.

Proper thermal monitoring and control in the clinic thus should be configured as in Figure 6. Up to three multiple sensor probes should be placed within the region to be heated along with single sensor probes to monitor other specified points such as core temperatures or critical points within the field. Placement of the probes under CT or other diagnostic scanning devices will help both in ensuring that the probe is in the desired location as well as assist in the later analysis of the data.

In addition to the temperature information, the computer should be programmed to monitor forward and reflected power from all applicators and to keep track of scattered or leakage fields at appropriate places within the room. All primary information such as temperatures from all probes should be displayed during the treatment, preferably, some of them graphically. All data must be recorded in nonvolatile storage with frequent memory updates. If a unit of dose is decided upon, it can be calculated and displayed for each of the temperature sensor locations.

Control of hyperthermia treatments in most centers to date consists of simple gain control of power by reference to a single temperature sensor. Lee, et al (55) recently described a multiple array configuration of interstitial microwave antenna with four independent amplifiers controlled by four sensors. A more general approach would be to use many sensors to control the amplitude and phase of several applicators in order to achieve dynamic control of the heating field. Arcangeli, et al (31) have described the use of phase and amplitude optimization methods to arrive at a configuration of 915 MHz applicators for heating a lesion on the thoracic wall. While their paper is more properly part of the previous two aspects of thermal dosimetry, it suggests the power and versatility possible if an analogous

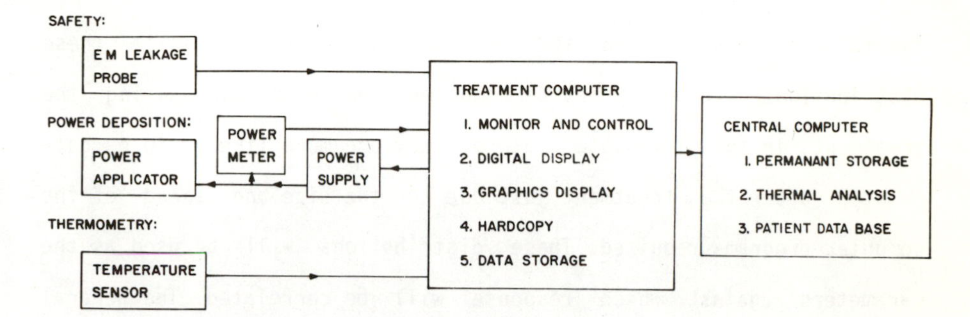

6. System diagram indicating the characteristics required for a computerized data acquisition and control system for hyperthermia therapy which is coupled to a larger mainframe computer for data analysis and data base management.

optimization were carried out under real time computer control of the heating apparatus.

8. Retrospective Thermal Dosimetry

Retrospective thermal dosimetry is the computation of the isothermal distributions and the accumulated given dose. While these distributions eventually may be computed and displayed during the treatment, in the immediate future, they are more likely to be calculated after the treatment just due to the size and speed of the computer programs required. These distributions will be used as the parameters against which response will be correlated. The general approach is to use experimental data taken during the treatment to normalize the predictions of the model calculations in the treatment planning phase. This approach is outlined in number of papers.

The first step is to refine the SAR distributions by using the initial heating transient to determine the SAR at the location of the thermometer sensors using:

$$\rho C\ dT/dt = \rho(SAR). \qquad (4)$$

It can be shown (56) that the same information can be obtained from cooling transients. This information then can be used in the numerical model of the bioheat transfer equation to compute temperature distributions. The actual blood flow values are still unknown, but the blood flow parameter can be adjusted arbitrarily in the model to in order to force the computed temperatures to agree with the real values at the measurement locations. As an illustration (57) of the technique, we consider the case of a dog thigh heated with microwaves at 2.45 GHz with a waveguide aperature (Fig. 7). The skin surface was cooled with ambient air blowing through the waveguide. Thermometer

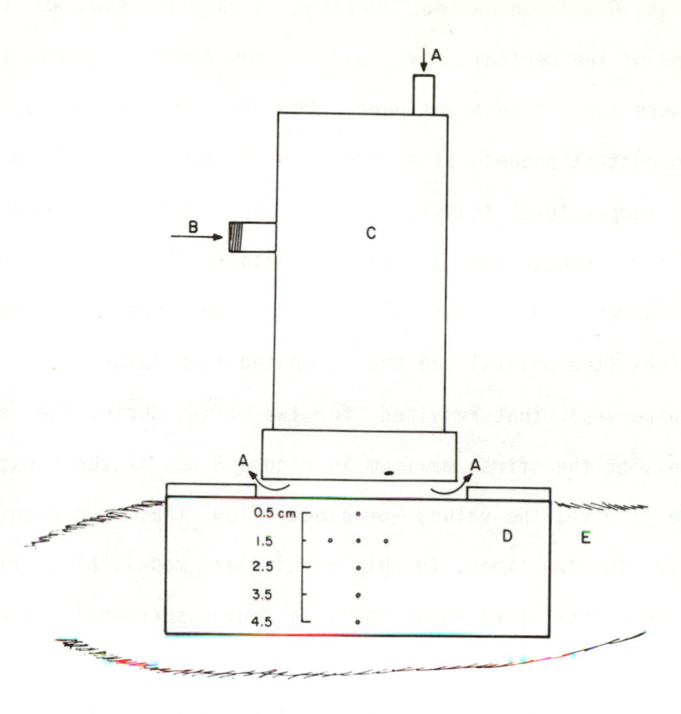

7. Sketch showing the basic experimental configuration for a series of experiments of heating of normal dog thighs. A. Forced air input for surface cooling. B. RF power input. C. Slab dielectric loaded WR430 microwave applicator operating at 2.45 GHz. D. Template used to position catheters for thermistors and insure proper geometry with waveguide applicator. E. Dog thigh. The dots and scale show the location of the thermistors.

probes were inserted along the axis of symmetry at 1.0 cm intervals beginning at 0.5 cm below the surface, in parallel rows at 1 cm on either side of the central row, and in the femoral artery. The experiments were run at constant power. Measured temperatures are given for the two hottest probes as a function of time in Figure 8. From the initial temperature transients, the local SAR was computed and was fitted to a smooth function versus depth in the tissue, usually an exponential (Fig. 9). Figure 10 shows the measured temperatures at five locations (the points) and the computed temperature distribution (the smooth curves) that resulted for two times during the experiment, namely at the first maximum in Figure 8 and at the first minimum. Figure 11 gives the values of blood flow that were required by the model for the two times. In this particular model, blood flow was taken as a constant throughout the 1 cm region surrounding each temperature probe. This is one of the simplest physical situations to treat and so in itself is not exciting. Nevertheless, it does suggest that reasonable numerical models can be used for interpolation between measured points. Emery, et al (24) represented the blood flow by a series of linear functions of temperature, with different proportionality constants required at different times of the treatment in order to make the model fit the temperature data. The approach was different in details from ours, but again supports the optimism expressed.

To evaluate the validity of this method, Divrik et al (57) have developed a state (temperature) and parameter (blood flow) estimation algorithm which minimizes the differences between specified (measured or simulated) and computed temperatures from the model. The first test of the method is based upon well defined experiments such as

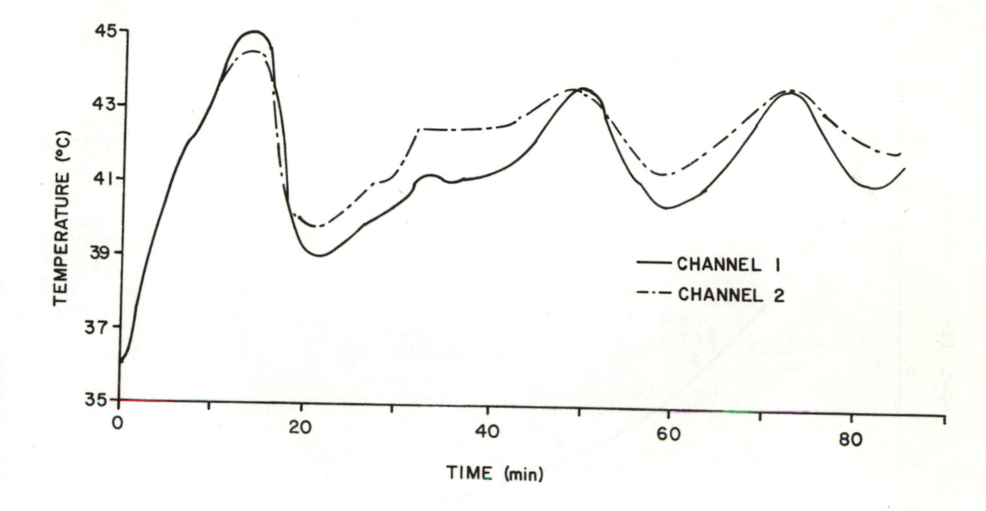

8. Temperature versus time for two thermometers in a typical heating experiment. The power level was maintained constant for the entire period. The discussion in the text centers on two times: the first at the first maximum at about 13 min. and the second at the first minimum at about 21 min.

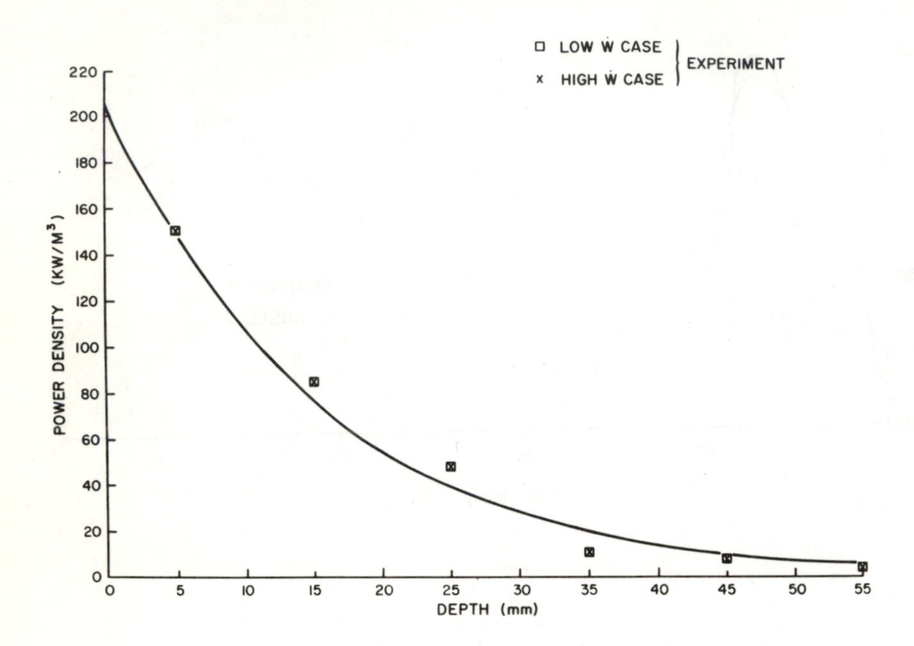

9. Power density versus depth for the experiment shown in Fig. 8. The symbols indicate values calculated from the initial warming transient, dT/dt, for each of the probes. The curve is an exponential fit to the measured points.

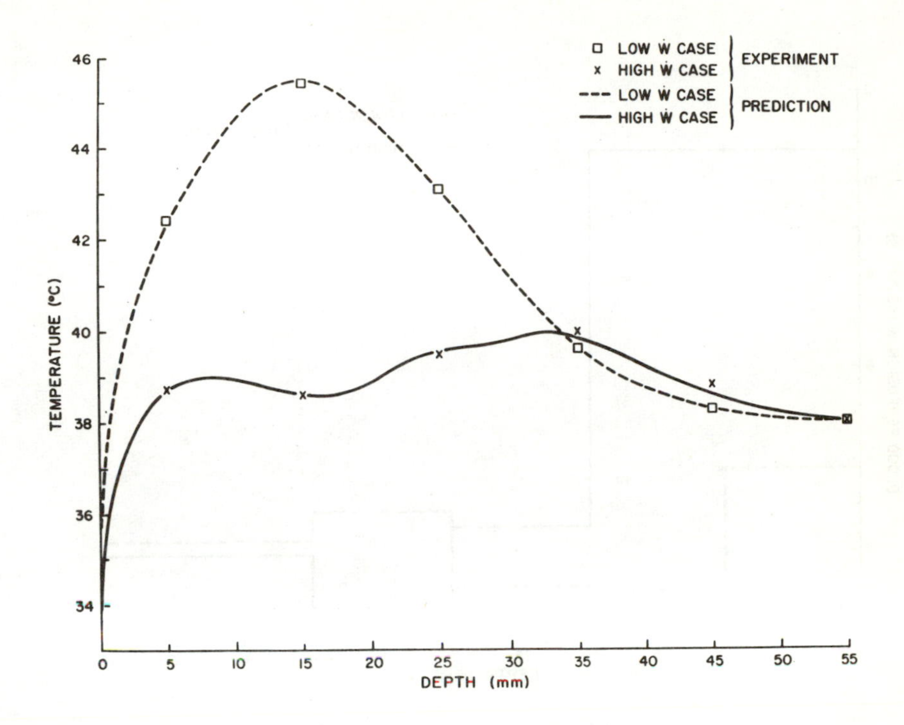

10. Temperature versus depth for the experiment shown in Fig. 8. The squares give the measured temperatures at the time of the first maximum in Fig. 8, while the X's give measured temperatures at the time of the first minimum. A state and parameter iteration algorithm was used to compute a smooth curve that satisfied the boundary conditions, applied power deposition pattern, and yielded temperature values in agreement with measured ones at the measurement points. The curves thus serve as a means of interpolation.

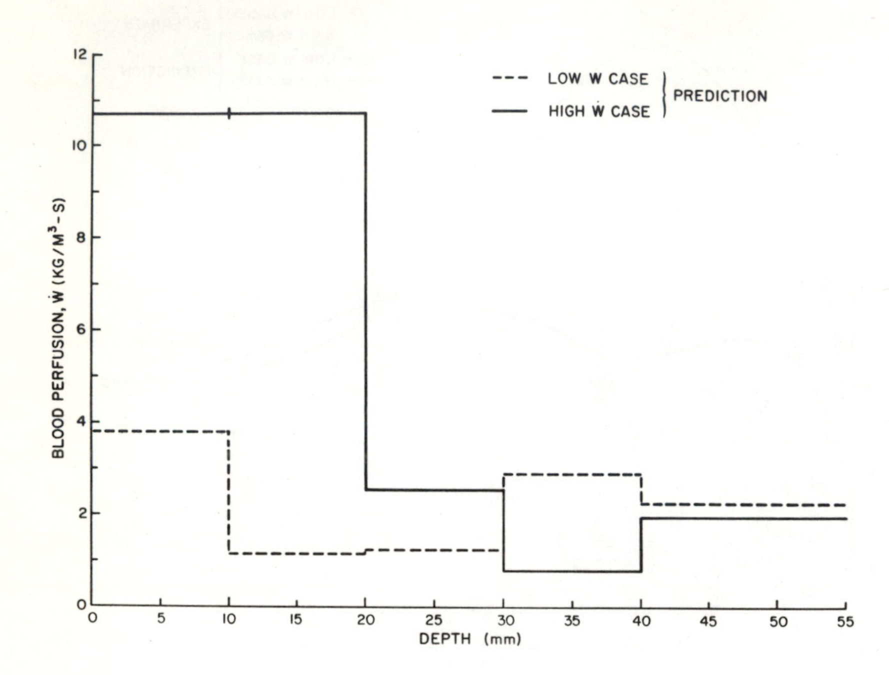

11. Values of blood flow, w, required for the state and parameter iteration algorithm to give proper temperatures at the measured points. The dashed line is for the low blood flow case (first temperature maximum in Fig. 8). The solid line is for the high blood flow case (first temperature minimum in Fig. 8).

that just described. The second test makes use of a specified temperature profile which is based upon the bioheat equation and defined, but typical, tissue and blood flow properties. The estimation algorithm is given a set of trial values for blood flow, and a set of tissue parameters and then is left to compute the temperature profiles. The extent to which it can reproduce the simulated temperature distribution gives an indication of the consistency of the method of estimation. Since all tissue parameters have uncertainty, erroneous values of these various parameters can be used in the estimation algorithm and the routine run to see how well the simulated temperature profile can be reproduced. Under these latter circumstances, the blood flow values will be in error such that the effects of erroneous tissue properties are just compensated. Since the temperature distribution is the desired endpoint for thermal dosimetry rather than knowledge of properties or blood flows, this compensation effect is very encouraging. Of course, to the extent that the correct parameters are used, the blood flow patterns will be calculated accurately.

Using a different state and parameter estimation algorithm, some studies in progress here suggest that temperature maxima and minima can be determined at other than the measured points. Thus, the methodology appears to be capable of extrapolation as well.These ideas need to be developed and tested further, both numerically and experimentally, but we are optimistic that complete temperature fields can be characterized reasonably well from knowledge of temperatures measured at several points, knowledge of the anatomical and geometrical configurations and modestly sophisticated numerical models.

Finally, once isotherms are computed and once a definition of thermal dose is selected, it is possible to compute isodose lines for the treatment. For example, Diller and Hayes (58) have computed cummulated thermal injury lines based upon the Arrhenius relation.

9. Acknowledgements

This work was supported in part by the National Cancer Institute under grants CA29653, CA17343, and contract NO1-CM-17480-22. Discussions with our colleagues, Drs. Mark W. Dewhirst and Eugene W. Gerner have been most helpful.

10. References

1. Dethlefson, L.A., Ed. Third International Symposium: Cancer Therapy by Hyperthermia, Drugs, and Radiation. Natl. Cancer Inst. Monogr. v61, 1982

2. Jain, R.K. and Gullino, P. M., Eds. Thermal Characteristics of Tumors: Applications in Detection and Treatments. Ann NY Acad. Sci. v335, 1980

3. Streffer, C. Ed. Cancer Therapy by Hyperthermia and Radiation, Urban and Schwarzenberg, Baltimore-Munich, 1978

4. Wizenberg, M.J. and Robinson, J. E., Eds. Proceedings of the International Symposium on Cancer Therapy by Hyperthermia and Radiation. Amer. Coll. of Radiology, Bethesda, 1976

5. Gautherie, M. and Albert, A., Eds, Biomedical Thermology, Progress in Clinical and Biological Research v107, Alan R. Liss, Inc. New York 1982

6. Storm, F.K., Ed., Hyperthermia in Cancer Therapy, G. K. Hall Medical Publishers, Boston, 1983

7. Nussbaum, G.K., Ed., Physical Aspects of Hyperthermia, Am.

Assoc. Phys. Med. Monogr. 8., Am. Inst. of Physics, New York, 1982

8. Oleson, J.R. and Dewhirst, M.W., "Hyperthermia: An Overview of Current Progress and Problems", Current Problems in Cancer, (Accepted) 1983

9. Oleson, J.R., Sim, D.A., and Manning, M.R., "Analysis of Prognostic Variables in Hyperthermia Treatment of 163 Patients" Inter. J. Rad. Onc. Biol. Phys. 1984 (submitted).

10. Dewhirst, M.W., Connor, W.G., Sim, D.A. "Preliminary Results of a Phase III Trial of Spontaneous Animal Tumors to Heatand/or Radiation: Early Normal Tissue Response and Tumor Volume Influence on Initial Response" Int. J. Rad. Onc. Bilo. Phys. v8 1951-1961, 1982

11. Dewhirst, M.W., Moon, T., Carlin, D., "Analysis of Tumor Volume and Thermal Dosimetric Effects on Tumor Response to Heat, Radiation and Heat Plus Radiation: Results of a Phase III Randomized Clinical Trial in Pet Animals" in Physical Aspects of Hyperthermia, Medical Monog. 8, Ed. G.H. Nussbaum, Am. Inst of Physics, New York, 1982 pp.495-510

12. Dewey, W.C., Hopwood, L.E., Sapareto, S.A., Gerwick, L.E., Cellular Responses to Combinations of Hyperthermia and Radiation" Radiology, v123, 463, 1977

13. Leith, J.T., Miller, R.C., Gerner, E.W., Boone, M.L.M., "Hyperthermic Potentiation: Biological Aspects and Applications to Radiation Therapy" Cancer v139, 766, 1977

14. Sapareto, S.A., Dewey, W.C., "Thermal Dose Determination in Cancer Therapy" Int. J. Rad. Onc. Biol. Phys. (accepted) 1984

15. Dewhirst, M.W., Sim, D.A., Sapareto, S., Connor, W.G., "The Importance of of Minimum Tumor Temperature in Determining Early and Long Term Responses of Spontaneous Pet animal Tumors to Heat and Ra-

diation" Cancer Res. Jan. 1984 (accepted).

16. Gerner, E.W. "A General Concept of Thermal Dose" (submitted) 1984 (personal communication)

17. Guy, A.W., Correspondence, J. Microwave Power, v10, 358, 1975

18. Roemer, R.B. "Thermal Dosimetry: Calculation of Tissue Temperature Fields", Presented at Third Annual Meeting of the North American Hyperthermia Group, San Antonio March, 1983

19. Cetas, T.C., Roemer, R.B. "Thermal Dosimetry: Four Aspects" Front. Radiat. Ther. Onc. v18, 1983

20. Pennes, H.H., "Analysis of Tissue and Arterial Blood Temperatures in the Resting Human Forearm" J. App. Physiol. v1, 93-122, 1948

21. Guy, A.W., "Analysis of Electromagnetic Fields Induced in Biological Tissues by Thermodynamic Studies on Equivalent Phantom Models" IEEE Trans. Microwave Theo. and Tech. MTT-19, 205, 1971

22. Lehmann, J.F., Guy, A.W., Stonebridge, J.B., deLateur, B.J., "Evaluation of a Therapeutic Direct-Contact 915-MHz Microwave Applicator for Effective Deep-Tissue Heating in Humans" IEEE Trans. Microwave Theo. and Tech. MTT-26, 556-563, 1978

23. Oleson, J.R., "Hyperthermia by Magnetic Induction: I. Physical Characteristics of the Technique. Int. J. Rad. Onc. Biol. Phys. v8, 1747-1756, 1982

24. Emery, A.F., Sekins, K.M., "The Use of Heat Transfer Principles in Designing Optimal Diathermy and Cancer Treatment Modalities" Int. J. Heat Mass Transfer v25, 823-834, 1982

25. Halac, S., Roemer, R.B., Oleson, J.R., Cetas, T.C. "Magnetic Induction Heating of Tissues: Numerical Evaluation of Tumor Temperature Distribution" Int. J. Rad. Onc. Biol. Phys. v9, 881-891, 1983

26. Strohbehn, J.W. "Theoretical Temperature Distributions for So-

lenoidal-type Hyperthermia Systems" Med. Phys. v9, 673-682, 1982

27. Brezovich, I.A., Young, J.H., Wang, M.-T., "Temperature Distributions in Hyperthermia by Electromagnetic Induction: A Theoretical Model for the Thorax" Med. Phys. v10, 57-65, 1983

28. Armitage, D.W., LeVeen, H.H., Pethig, R., "Radiofrequency Induced Hyperthermia: Computer Simulation of Specific Absorption Rate Distributions Using Realistic Anatomical Models" Phys. Med. Biol. v28, 31-42, 1983

29. Hill, S.C., Christensen, D.A., Durney, C.H., "Power Deposition Patterns in Magnetically Induced Hyperthermia: A Two-dimensional Quasistatic Numerical Analysis" Int. J. Rad. Onc. Biol. Phys., v9, 893-904, 1983

30. Iskander, M.F., Turner, P.F., DuBow, J.B., Kao, J. "Two-dimensional Technique to Calculate the EM Power Deposition Pattern in the Human Body" J. Microwave Power, v17, 175-185, 1982

31. Arcangli, G., Lombardini, P.P., Lovisolo, G., Marsiglial, G., Piattelli, M. "Focussing of 915 MHz Electromagnetic Power on Deep Human Tissues: A Mathematical Model Study" IEEE Trans. Biomed. Eng. BME-31, 1984. (In press)

32. Morita, N., Bach Andersen, J. "Near-field Absorption in a Circular Cylinder from Electric and Magnetic Line Sources" Bioelectromagnetics v3, 253-274, 1982

33. Doss, J.D., "Calculation of Electric Fields in Conductive Media" Med. Phys., v9, 566-573, 1982

34. Strohbehn, J.W., Trembly, B., Douple, E.B., "Blood Flow Effects on the Temperature Distributions from an Invasive Microwave Antenna Array Used in Cancer Therapy" IEEE Trans. Biomed. Eng. BME-29, 649--661, 1982

35. Stauffer, P.R., Cetas, T.C., Jones, R.C. "Magnetic Induction Heating of Ferromagnetic Implants for Inducing Localized Hyperthermia in Deep Seated Tumors" IEEE Trans. Biomed. Eng. BME-31, Feb. 1984 (In press)

36. Stauffer, P.R., Cetas, T.C., Fletcher, A.M., DeYoung, D.W., Dewhirst, M.W., Oleson, J.R., Roemer, R.B., "Observations on the Use of Ferromagnetic Implants for Inducing Hyperthermia" IEEE Trans. Biomed. Eng. BME-31, Jan. 1984 (In press)

37. Matloubieh, A.Y., Roemer, R.B., Cetas, T.C., "Numerical Simulation of Magnetic Induction Heating of Tumors with Ferromagnetic Seed Implants" IEEE Trans. Biomed. Eng. BME-31, Feb. 1984 (In press)

38. Atkinson, W.J., Brezovich, I.A., Chakraborty, D.P., "Useable Frequencies in Hyperthermia with Thermal Seeds" IEEE Trans. Biomed. Eng. BME-31, Jan. 1984 (In press)

39. Swindell, W., Roemer, R.B., Clegg, S.T., "Dynamic Tumor Scanning with Focused Ultrasound: A Numerical Sensitivity Analysis" Presented at the North American Hyperthermia Group Annual Meeting, San Antonio, March, 1983

40. Swindell, W., Roemer, R.B., Clegg, S.T., "Temperature Distributions Caused by Dynamic Scanning of Focused Ultrasound Transducers", Proc. IEEE Symp. on Sonics and Ultrasonics, 82CH1823-4 pp.750-753,1982

41. Roemer, R.B., Cetas, T.C., Oleson, J.R., Halac, S., Matloubieh, A.Y., "Comparative Evaluation of Hyperthermia Heating Modalities: II Applications of the Acceptable Power Range Technique" Submitted. 1983

42. Roemer, R.B., Cetas, T.C., Oleson, J.R., Halac, S., Matloubieh, A.Y., "Comparative Evaluation of Hyperthermia Heating Modalities: I Numerical Analysis of Thermal Dosimetry Limiting Cases" Radiation

Res. 1984 (Accepted)

43. Halac, S., Roemer, R.B., Oleson, J.R., Cetas, T.C. "Uniform Regional Heating of the Lower Trunk: Numerical Evaluation of Tumor Temperature Distributions" Intl. J. Rad. Onc. Biol. Phys. 1984 (Accepted)

44. Strohbehn, J.W., "Temperature Distributions from Interstitial RF Electrode Hyperthermia Systems: Theoretical Predictions" Int. J. Rad. Onc. Biol. Phys. v9, 1655-1667, 1983

45. Gibbs, F.A., "Thermal Mapping in Experimental Cancer Treatment with Hyperthermia: Description and Use of a Semi-automatic System" Int. J. Rad. Onc. Biol. Phys. v9, 1057-1063, 1983

46. Oleson, J.R., Private communication to be published.

47. Cetas, T.C., Connor, W.G., "Thermometry Considerations in Localized Hyperthermia" Med. Phys., v5, 79-91, 1978

48. Cetas, T.C., Connor, W.G., Manning, M.R., "Monitoring of Tissue Temperature During Hyperthermia Therapy" Annals NY Acad. Sci., v335, 281-297, 1980

49. Cetas, T.C., "Thermometry" Chapt. 2 in Therapeutic Heat and Cold, 3rd Edition, Ed. J. F. Lehmann, Williams and Wilkins, Baltimore, 1982, pp.35-69

50. Cetas, T.C., "Invasive Thermometry", Chapt 13 in Physical Aspects of Hyperthermia, Ed. G. K. Nussbaum, Am. Assoc. Phys. Med. Monogr. 8., Am. Inst. of Physics, New York, 1982, pp. 231-265

51. Christensen, D.A., "Thermal Dosimetry and Temperature Measurements", Cancer Res. v39, 2325-2327, 1979

52. Christensen, D.A., "Thermometry and Thermography", Chapt. 10 in Hyperthermia in Cancer Therapy, Ed. F. K. Storm, G. K. Hall Medical Publishers, Boston, 1983, pp. 223-232

53. Schooley, J.F., Ed. "-Temperature: Its Measurement and Control in Science and Industry" vol. 5, Am. Inst. Physics., New York 1982

54. Johnson, C.C., and Guy, A.W., ""Nonionizing Electromagnetic Wave Effects in Biological Materials and Systems", Proc. IEEE, v60, 692, 1972

55. Lee, D.-J., Cheung, A., Mayo, G., O'Neill, M., Lam, W.-C., "An Interstitial Microwave Hyperthermia System with Multiple Temperature Sensors for Automatic Control" Presented at the North American Hyperthermia Group, Annual Meeting, San Antonio, March 1983

56. Roemer, R.B., Fletcher, A.M., Cetas, T.C., "Obtaining Local SAR and Blood Perfusion Data from Temperature Sensors: Steady State and Transient Techniques Compared", (Submitted) 1983

57. Divrik, A.M., Roemer, R.B., Cetas, T.C., "Inference of Complete Tissue Temperature Fields from a Few Measured Temperatures: An Unconstrained Optimization Method" (Submitted) 1983

58. Diller, K.R., and Hayes, L.J., "Finite Element Analysis of Hyperthermic Heat Transfer in Biologic Media", Presented at North American Hyperthermia Group Annual Meeting, San Antonio, March 1983

PROCEEDINGS OF THE II INTERNATIONAL CONFERENCE ON APPLICATIONS OF PHYSICS TO MEDICINE AND BIOLOGY
edited by Ž. Bajzer, P. Baxa & C. Franconi

BIOLOGICAL BASES FOR CLINICAL APPLICATION OF COMBINED HYPERTHERMIA AND RADIATION

Giorgio Arcangeli[1], Francesco Mauro[2] and Jens Overgaard[3]

[1]Istituto Medico e di Ricerca Scientifica, Via S. Stefano Rotondo 6, 00184 Rome, Italy.
[2]ENEA, Casaccia, 00060, Rome, Italy.
[3]The Institute of Cancer Research, Radiumstationen, Norrebrogade 44, DK - 8000, Aarhus C, Denmark.
Supported by the Danish Cancer Society (grant nº 24/79) and the Danish Technical Scientific Research Council.

INTRODUCTION

Among the different aspects of hyperthermic cancer therapy, the potential use of hyperthermia as an adjuvant to radiation therapy has been the most widely studied. However, the interaction between heat and radiation is complex and not yet completely understood. If we consider that some treatment parameters such as the time-temperature relationship, the sequence of the two modalities, the fractionation interval, the size of radiation fractions, and some physiological factors such as blood flow, pH, and thermotolerance, may have a large relevance on the response of tumors and normal tissues, then the final result seems that a certain grade of confusion and uncertainty can be produced in the design of clinical scheduling.

The aim of this paper is to analyse the relevance of such parameters and factors for the most appropriate combination of heat and radiation.

EXPERIMENTAL STUDIES

The interaction between heat and radiation involves two principally, not mutually exclusive, different mechanisms:

1) Hyperthermic citotoxicity and 2) Hyperthermic radiosensitization.[6,7,20-23,25,26,32]

Firstly, heat has a direct cytotoxic effect. Although the heat sensitivity varies among the different cells and tissues, the intrinsic heat sensitivity does not seem to be higher in malignant cells than in their normal counterparts. However, the cytotoxicity is strongly enhanced under certain environmental conditions such as insufficient nutrition, chronic hypoxia, and especially, increased acidity which are typically found in large areas of poorly vascularized tumors and that hyperthermia itself, may, furthermore, enhance by causing collapse of the tumor microvascularization.[29] This implies that a moderate heat treatment is able to almost selectively destroy a large proportion of those clonogenic cells that, for their situation in a chronic hypoxic environment, may be the most radioresistant and indirectly influence the response to the combined treatment.[23]

Secondly, hyperthermia has a radiosensitizing effect which manifests itself in several ways: 1) Direct radiosensitization (a decreased Do) 2) Decreased repair of sub-lethal and potentially lethal damage and 3) Sensitization of cells in relatively radioresistant phases of the cell cycle. Furthermore, heat has been reported to sensitize hypoxic cells more than oxygenated cells, thus causing a decreased oxygen enhancement ratio; however the data supporting this hypothesis are ambiguous.

The Influence of Sequence

Recently, some measure of agreement among investigators seems to emerge on optimum treatment sequence.[4-7,21,22,25,26] Experimental animal data by several authors on normal and tumor tissues are summarized and plotted in figure 1. With a single treatment, at 42-43°C for 60 minutes, the maximal thermal enhancement in several normal tissues and tumors is obtained when heat and radiation are given simultaneously or in very close sequence.[6,7,8,22,25] This is probably the consequence of the hyperthermic radiosensitization, which is found to be most prominent with simultaneous application. The introduction of an interval between the two modalities decreases the thermal enhancement to a different extent in tumors and in normal tissues. When heat is applied after irradiation (on the right), there is a sharp separation between tumor and normal tissue thermal enhancements with increasing time, the thermal enhancement in normal tissues disappearing with intervals greater than 4 hours.[7,25]

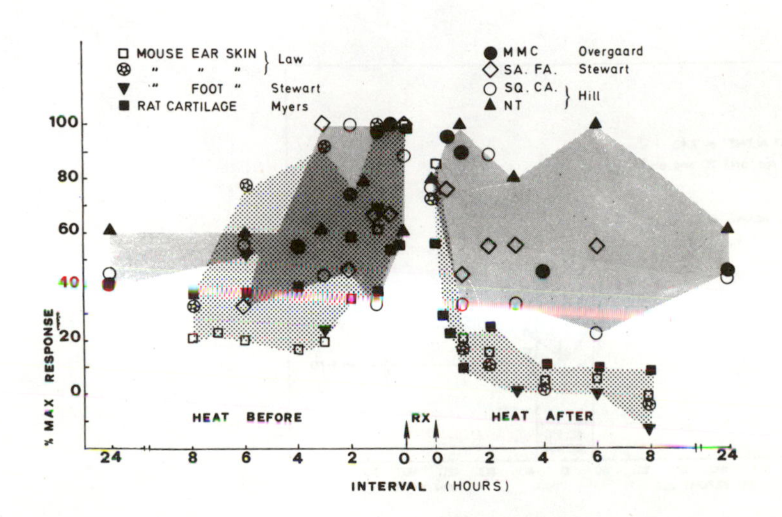

Fig. 1: The time course of the decay of heat radiosensitization in different normal and tumor tissues for hyperthermia given either before or after irradiation. Values have been normalized to the percentage of the maximum response.

Normal tissues:

Mouse ear skin, 42°C/60 min[12]
Mouse ear skin, 43°C/60 min[13,14]
Mouse foot skin 42.5°C/60 min[30]
Rat cartilage 43°C/60 min[16,17]

Tumors:

MMC[22]
SA FA[31]
Sq. Ca[10]
NT[10]

Overlapping

The opposite sequence, i.e. heat given before radiation, causes a higher and longer thermal enhancement in the normal tissue resulting in a large area of overlapping of the data. In most solid tumors a moderate thermal enhancement persists even with long intervals between the modalities and independent of the sequence, a phenomenon which is probably a consequence of the selectively hyperthermic destruction of acidic and chronically hypoxic tumor cells. Furthermore, in vitro data (Fig. 2) show that changes in survival as a function of varying the sequence between heat and radiation are also dependent on the pH.[6)]

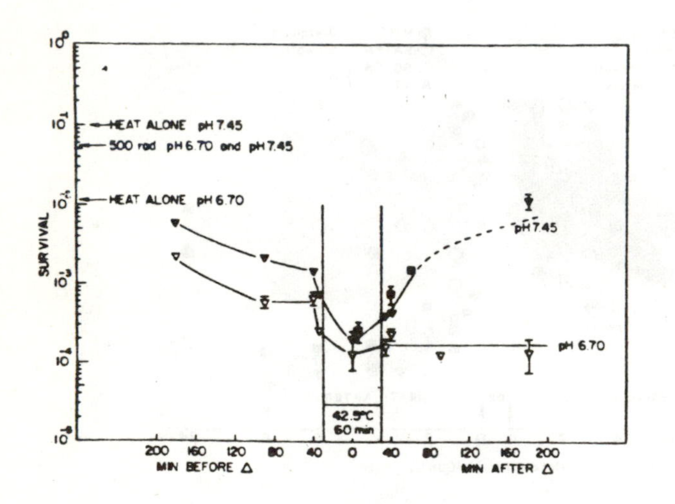

Fig. 2: The effect on CHO survival of varying the sequence between heat and radiation. Radiation (500 rad) was administered either during hyperthermia (points within the 60-minute period at 42.5°C), before heating (points to the left), or after heating (points to the right). The pH was maintained (with CO_2) either at 7.45 or 6.70 before and during the treatments.(From Dewey et al., ref. 6).

Assuming that the response of in vitro cells at low or normal pH represents the response of tumors or normal tissues respectively, the in vitro and in vivo data are in agreement when irradiation precedes thermal treatment, i.e. thermal enhancement is higher for tumors or cells at low pH than for normal tissues or cells at normal pH. With the opposite sequence, i.e. heat given before irradiation, or when the two modalities are given simultaneously, the enhancement observed in vitro at low pH is not observed in vivo for tumors. This could be related to heat-induced physiological changes such as alterations in blood flow, oxygen concentration and pH. However such studies indicate that when heat is applied after radiation damage is repaired in both tumors and normal tissues, a therapeutic advantage may derive from the increased thermal killing on acidic and chronically hypoxic tumor cells.

The Problem of "Heat Dose"

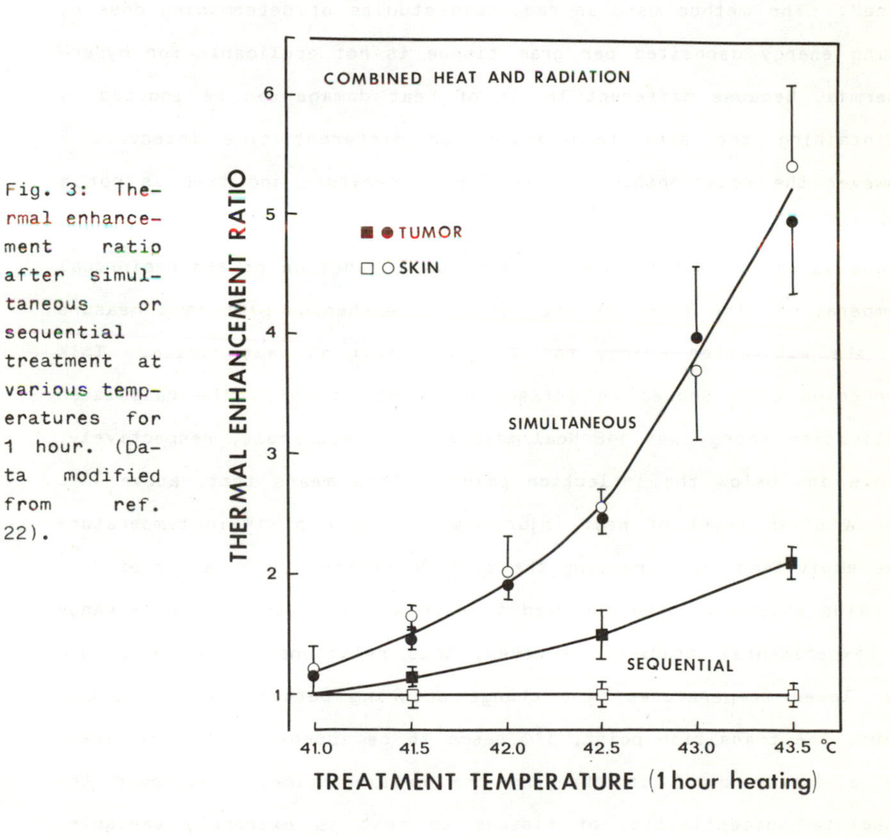

Fig. 3: Thermal enhancement ratio after simultaneous or sequential treatment at various temperatures for 1 hour. (Data modified from ref. 22).

Both the thermal cell killing and radiosensitization further depends on the temperature and the treatment time in that an increase in either produces an increase of both cell killing and radiosensitization. This is clearly shown in animal experiments reported in figure 3. After simultaneous treatment, there is a steep increase in the thermal enhancement ratio (TER) with increasing temperature at a given heating time, but to the same extent in tumor and normal tissue. The enhancement after sequential treatment, where heat is given 4 hours after radiation is smaller, but occurs only in the tumor.

At present, there are no definite methods of determining "heat dose". The method used in radiation studies of determining dose by using energy deposited per gram tissue is not applicable for hyperthermia, because different levels of heat damage can be induced by maintaining the same temperature for different time intervals. However the relationship between the temperature and time is not a single linear product of the two, as shown in figure 4, where the slope values in minutes were plotted as a function of the reciprocal temperature. The slope of the resulting Arrhenius plot is a measure of the activation energy for a given level of heat damage. This Arrhenius plot showed an inflection point at 43°. The calculated activation energy was 148 Kcal/mole and 365 Kcal/mole, respectively, above and below the inflection point. This means that, above 43°, for a given level of heat injury, an increase of 1° in temperature was equivalent to increasing the time of heating by a factor of 2,[5)] a value which has been observed in vivo and in vitro in a wide range of experimental studies. However, this relationship does not hold for lower temperatures, the change occuring between 42 and 43°C. Below the transition point, 1°C seems to be approximately equivalent to a factor of 6 in changing the heating time.[9)] Although the absolute susceptibility of tissues to heat is extremely variable, covering a factor of 200 in heating times,[9] the relationship between eating time and temperature appears reasonably uniform for different tissues and it may be used to relate treatments at different temperatures or to derive an effective treatment time at a given temperature during a hyperthermic treatment.

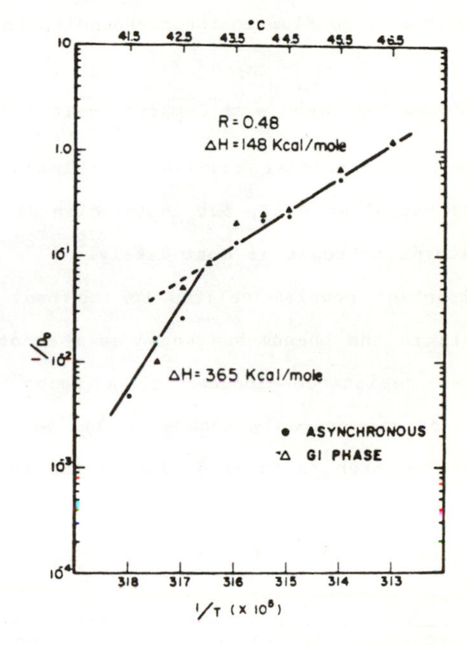

Fig. 4: An Arrhenius plot for heat inactivation of CHO cells. On the ordinate, the reciprocal of the D_o values (inactivation rates) are plotted versus the reciprocal of the absolute temperatures. Separate data points are shown for asynchronous cells and for cells heated in the G_1 phase. (From Dewey et al., ref. 5).

This relationship has been recently successfully applied in clinic to relate radiation dose and heat dose, expressed as equivalent time at 43° for several levels of tumor response.[33] It was found that for each level of tumor response, an increase of the equivalent time at 43° was necessary to achieve the same response with the use of a lower radiation dose.

From the data shown above, it is evident that if tumor and critical normal tissues are heated at the same temperature, the simultaneous or the close sequence of heat and radiation are unlikely to result in an improvement of the therapeutic effect. This implies that a clinical utilization of the simultaneous treatment is only suitable if a tumor can be heated at a temperature higher than that of the surrounding tissues. Although the current heating techniques are far from sufficient in this regards, there are indications that heat treatment itself may cause collapse of the microvascularization especially in central areas of the tumor, whereas it will result in an increased blood flow in the surrounding normal tissue.

Fractionated Treatment with Combined Heat and Radiation

Most experimental studies are related to single application of heat and radiation only, but in the clinical situation, a fractionated treatment schedule is most likely.[2,3,24-26] In this situation, a very important problem related to the heat treatment becomes apparent. This is the phenomenon known as thermotolerance (i.e. a temporary heat resistance induced by a prior heat treatment).[8,27,28] Thermotolerance probably occurs in all normal tissues and tumors,[8] and has also been found to influence the effect of the combined heat and radiation when given either simultaneously or sequentially.[19,28] It has been shown that thermotolerance is most prominent for the interaction between heat and radiation, if these are given according to the sequential principle,[19] since hyperthermia in such a schedule acts due to its direct cytotoxic effect, and therefore will express the same amount of thermotolerance as observed with heat alone. It means that if the time intervals between fractions are such that significant thermotolerance is present, very long heating times are required to increase the thermal destruction after heat alone, compared to the effect of a single fraction.[11,18] Figure 5 illustrates this problem, and shows that in this specific tumor, heated at 42.5°C for one hour per fraction, a daily fractionation schedule did not improve the hyperthermic response, since the tumor growth time after 5 fractions was not statistically significant different from that observed after a single fraction. However, if no interval or a

5-day-interval was allowed between fractions, then no thermotolerance was present, and the hyperthermic damage accumulated with increasing number of fractions.

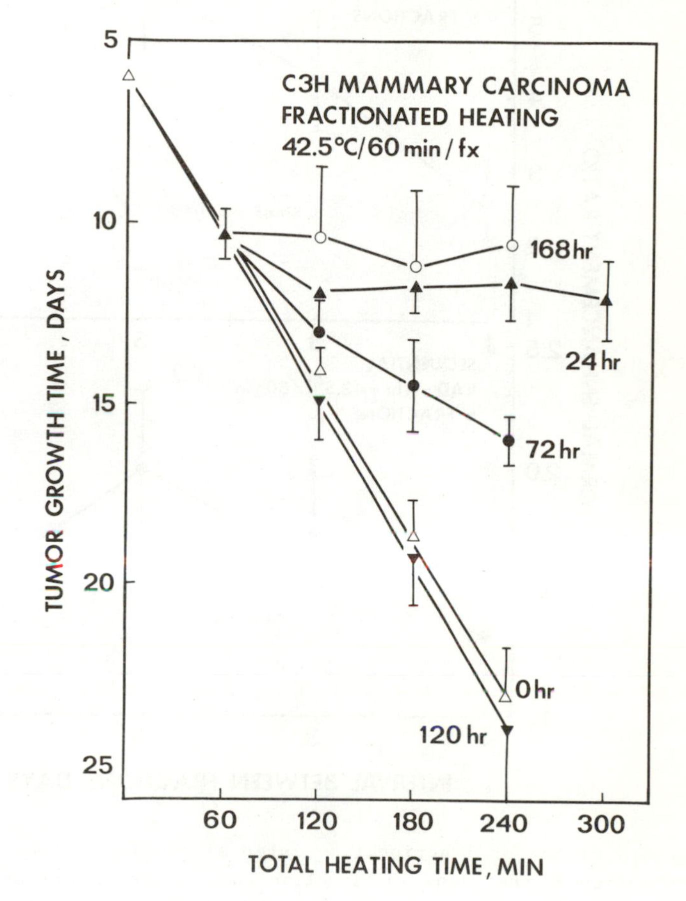

Fig. 5: Effect of fractionation intervals on the tumor growth time (time to increase the tumor volume 5 times). For details of the study see ref. 11 and 28.

Figure 6 shows that a similar daily fractionation schedule with sequential combined heat and radiation did not result in an improvement of the thermal enhancement ratio in the tumor compared to single treatment. However, if the intervals between the fractions were extended to allow the thermotolerance to disappear (in the present example about 5 days) the thermal enhancement improved significantly in comparison with the TER value achieved with single dose. A

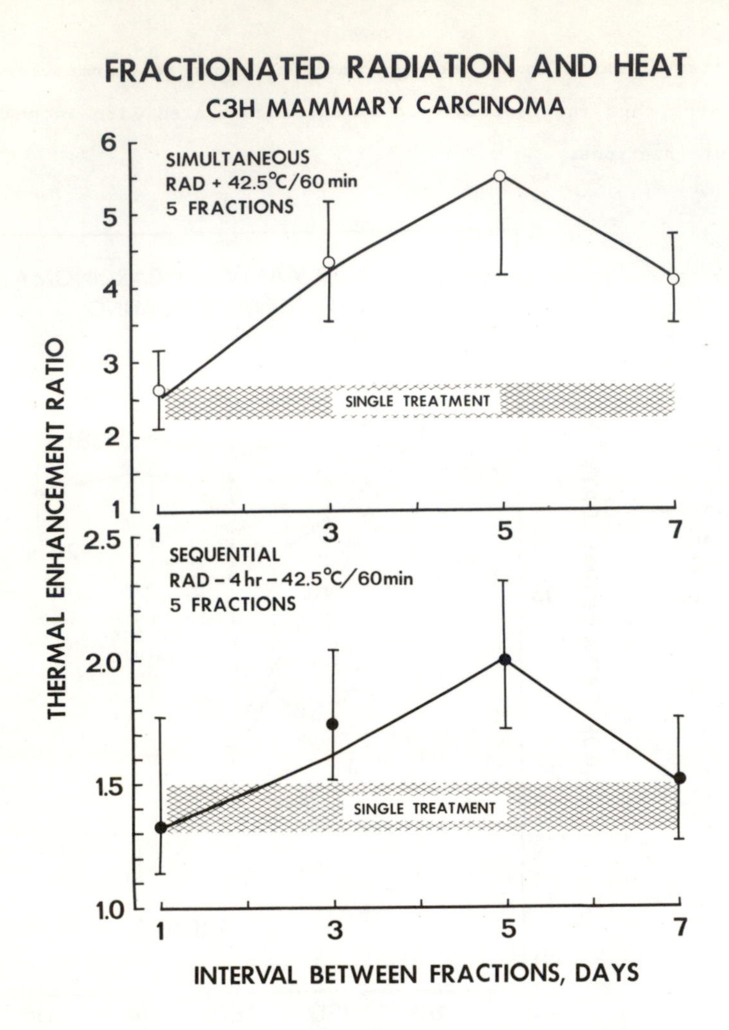

Fig. 6: Effect of fractionation interval on the thermal enhancement ratio after 5 fractions of simultaneous (upper panel) or sequential (lower panel) radiation and hyperthermia (42.5°C for 1 hour per fraction). The hatched area represents the 95% confidence limits of TER values obtained with a similar single treatment. For details of the study see ref. 11 and 28.

3-day-interval between fractions resulted in a somewhat intermediate result, due to only partly presence of thermotolerance, whereas a 7-day fractionation interval again reduced thermal enhancement ratio, not due to the presence of thermotolerance but because the fractionation intervals now were so long that the tumors were able to grow between the fractions, which naturally is likely to reduce the outcome of the treatment (see ref. 28 for details). Thus, the optimal spacing between fractions seems to be extremely critical for the outcome of therapy.

Similar results were found with simultaneous radiation and hyperthermic treatment (Fig. 6). As seen, the kinetic of the thermal enhancement ratio was the same as after sequential treatment, and it therefore seems that the kinetics for development and decay of thermotolerance and for the influence of thermotolerance on the combined heat and radiation treatment follow the same pattern in the described tumor model.[19,28] This may be different in other tissues.[14]

It is very important to emphasize that the intervals given here only apply to the specific tumor treated at 42.5°C for 60 minutes or with similar heat treatment resulting in the same amount of hyperthermic damage. In any other tumor or normal tissue or with a heat treatment resulting in a different heat damage, the quantitative magnitude of thermotolerance will probably be different. Thus, the data shown bear no quantitative relationship to the clinical situation, and is only presented in order to illustrate the enormous problems which may be related to the phenomenon of thermotolerance.[28]

It is evident that if optimal thermal enhancement ratio should be achieved, thermotolerance must be avoided in the tumor. On the other hand, the presence of thermotolerance would be a preferable situation in normal tissues, since it may reduce the amount of damage. As mentioned, there is a considerable variation in the kinetics and magnitude in different tissues,[8,19,28] and it is there-

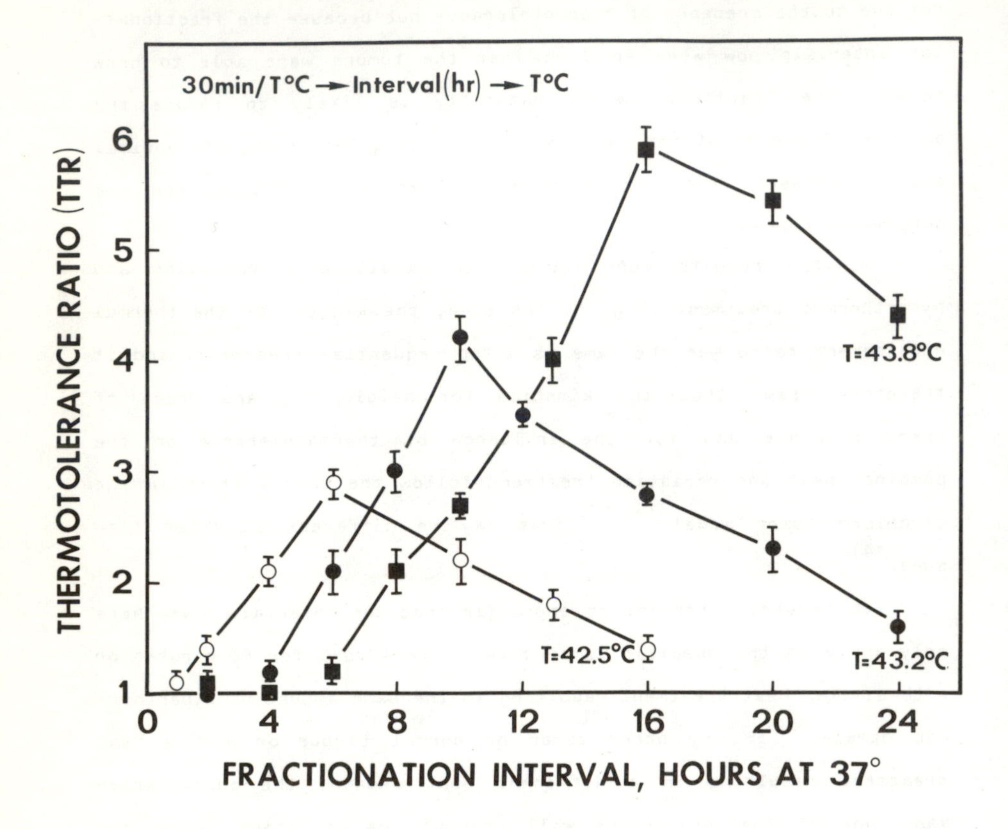

Fig. 7: Effect of different pre-heating temperatures given for 30 min. on the kinetics and magnitude of thermotolerance in L1A2 cells. For details of the study see ref. 11 and 28.

fore at present impossible to predict how thermotolerance will develop in a given tumor or in normal tissue.[28)] The only known relationship is that both the magnitude and the kinetics of thermotolerance appear to depend on the heat damage induced by the primary heat treatment.[15,19,28)] Thus, in a given tumor or normal tissue, thermotolerance will develop later and reach a higher maximum if the primary heat treatment is large (figure 7). In the clinical settings, this means that if a not completely homogenous heat treatment can be applied to a tumor, the different parts of the tumors which have received different heat treatments will develop thermotolerance with different kinetic patterns.[28)] Thus, at the time of the subsequent fraction, thermotolerance is likely to be expressed to a different extent in different areas of the tumor. This problem which clinically seems enormous, is probably the main reason why the response to clinical fractionated hyperthermia treatments seems to be almost the same despite different heat treatment and fractionation intervals.[26)]

CLINICAL STUDIES

Considerations of all difficulties and problems, including the technical problems, which may considerably affect the response of normal tissues and tumors makes it natural to ask the question: is hyperthermia useful as adjuvant to radiotherapy to treat cancer? In spite of the large number of clinical studies published in the last decade, only few studies have been done with comparable treatments on comparable lesions (Table 1). They demonstrate that percent tumor response increases from an average of about 21% after radiotherapy alone to an average of about 65% after combined heat and radiation.[26)]

However, nothing has been clearly reported on the effect of the combined treatment on normal tissues, although in most clinical studies heat has been delivered simultaneously or in close sequence with radiation but frequently associated with active skin coolings.

This makes it difficult to know whether the addition of heat to radiation can give an increase of the therapeutic gain. So the present knowledge about the potential usefulness of hyperthermia as adjuvant to radiotherapy is sparse and it deserves to be more carefully focussed.

In an effort of applying an optimal treatment, a series of studies have been carried out at the Istituto Medico e di Ricerca Scientifica testing several combination schedules on patients with 2 or more lesions. This allowed us to compare, in the same patients, the response of both radiotherapy alone and combined treatment. Many of these tumors were failures to conventional treatments, including radiotherapy, and most of the patients had very advanced lesions and/or disseminated disease.[3)]

TABLE 1- EFFECT OF ADJUVANT HYPERTHERMIA ON THE RADIATION RESPONSE IN HUMAN TISSUE

STUDY	NUMBER OF PATIENTS OR TUMORS	FREQUENCY OF COMPLETE RESPONSE		
		RADIATION ALONE[a]		RADIATION[a] + HEAT
Arcangeli et al.	123	16%	→	67%
U et al.	7	14%	→	85%
Overgaard	23	44%	→	71%
Johnson et al.	14	36%	→	86%
Kim et al.	159	33%	→	80%
Bede et al.	19	0%	→	9%
Kochegarov et al.	161	16%	→	63%
Lindholm et al.	44	11%	→	40%
Corry et al.	33	0%	→	62%
ALL	583	21%		65%

[a]Identical radiation dose in both arms.

Modified from Overgaard (ref. 26)

Radiation was given with electrons of various energies or with a 6 MeV photon beam by linear accelerators. Heat was delivered by means of various microwave or radiofrequency external applicators described in previous papers.[1-2] The treatment protocols, employed in this study are summarized in Table 2.

TABLE 2 - TREATMENT PROTOCOLS

TREATMENT PROTOCOLS	TEMPERAT. °C/MIN	HEAT SES. TOT.(WK)	RADIAT. FRACTN.	FRACTION SIZE(GY)	TOTAL DOSE
1. Simultaneous/ Sequential Heat	42.5/45	7(3/w)	3/d	1.5-2	60
vs					
Radiation Alone	-	-	3/d	1.5-2	60
2. Simultaneous Heat	43.5/45	5(1/w)	1/d	2	50
vs					
Radiation Alone	-	-	1/d	2	50
3. Sequential Heat	42.5/45	8(2/w)	2/w	5	40
vs					
Simultaneous Heat	42.5/45	8(2/w)	2/w	5	40
vs					
Radiation Alone	-	-	2/w	5	40
4. Simultaneous Heat	45/30	5(2/w)	2/w	6	30
vs					
Radiation Alone	-	-	2/w	6	30

In the first protocol, radiation was delivered as 3 fractions per day of 1.5 to 2 Gy each, at 4 hour intervals up to a total of 60 Gy; heat at 42.5°C for 45 min. was applied every other day, immediatley after the second daily fraction of radiation, for a total of 7 sessions. In the second protocol, radiation was given as conventional fractionation (i.e. 1 daily fraction of 2 Gy) up to a total of 50 Gy; heat at 43.5°C was applied once a week immediately after the last weekly fraction of radiation for a total of 5 sessions. In the third protocol, radiation was given as 2 weekly fractions of 5 Gy up to a total of 40 Gy; heat at 42.5°C for 45 min. was applied with all radiation fractions either immediatley after (simultaneous treatment) or 4 hours later (sequential treatment) for a total of 8 sessions. Finally, in the fourth protocol, radiation was delivered as 2 weekly fractions of 6 Gy up to a total of 30 Gy; heat at 45°C for 30 min. was applied immediatley after each radiation fraction for a total of 5 sessions; in this case, the skin around the lesion was cooled by means of circulating cold water.

Results

The results are shown in Table 3. Tumor response to combined modality was constantly better in respect to radiation alone, whichever treatment schedule was employed, the highest improvement being obtained with the use of large radiation fractions and intense heating employed in protocol n° 4.

TABLE 3 – RESPONSE TO TREATMENT

TREATMENT	COMPLETE TUMOR REGRESSION		SKIN MOIST DESQUAMATION	
PROTOCOLS	RT + HT	RT Alone	RT + HT	RT Alone
1. Simultan/ Sequential	19/26(.73)[a]	11/26(.42)	11/26(.42)	10/26(.38)
2. Simultan.	7/10(.70)	4/10(.40)	4/10(.40)	4/10(.40)
3. Simultan. Sequential	10/13(.77) 8/12(.67)	6/16(.37)	9/14(.64) 6/13(.46)	5/14(.36)
4. Simultan.	13/15(.87)[a]	5/15(.33)	5/15(.33)[b]	4/15(.27)

[a]Statistically significant; b Skin cooling

However, by employing large radiation fractions, the addition of heat resulted also in an increase of skin reaction, especially after simultaneous treatment. Because of an active skin cooling system employed in protocol n° 4, no increase of skin reaction was observed even with the use of intense heating simultaneously combined with large dose fractions.

The failure pattern of the lesions treated with the protocol number 1 (i.e. radiotherapy given as 3 small fractions per day combined with moderate heat) is shown in figure 8.

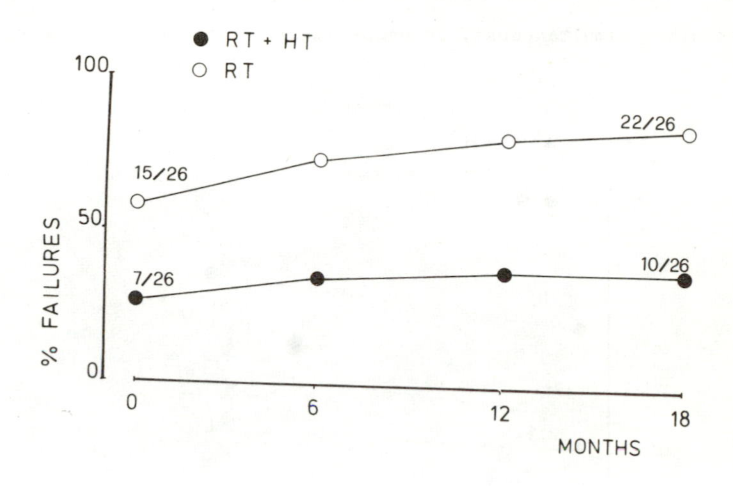

Fig. 8: Failure rates of the lesions treated with the protocol n° 1 (i.e. radiotherapy given as 3 small fractions per day combined with moderate heat): ○ RT alone; ● Combined modality. (From Arcangeli, ref. 3).

During the follow-up period, percent total failures after radiation alone increased from 58% at the end of treatment to 85% at 18 months, in contrast with the lesions treated by the combined modality in which, after a small initial increase, percent total failures remained almost at a plateau level of about 35%. Therefore, the addition of heat seems to be not only more effective in achieving a higher response rate but also in maintaining tumor control during follow-up. This is more clearly shown by the combined to the single modality failure ratio, which represents the actual thermal enhancement of tumor response, as different lesions were treated, in the same patients, with different modalities. In this trial, the ratio had a constant value of 0.46 through the whole follow-up period, indicating that, during this period, in more than a half of patients, recurrences occured only in the group of lesions treated with radiotherapy alone (Fig. 11).

Figure 9 shows the failure pattern of the lesions treated with the protocol number 3 (i.e. radiotherapy given as 2 large weekly fractions combined simultaneously or sequentially with moderate heat).

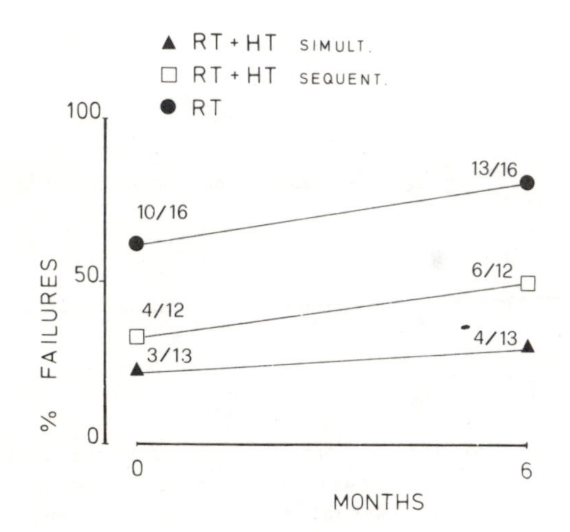

Fig. 9: Failure rates of the lesions treated with the protocol n°3 (i.e. radiotherapy given as 2 large weekly fractions simultaneously or sequentially combined with moderate heat): ● RT alone; □ Sequential treatment; ▲ Simultaneous treatment. (From Arcangeli, ref. 3).

Percent total failures at the end of treatment and at 6 months, ranged from 23 to 31% with simultaneous treatment, and from 33 to 50% with the sequential schedules. This range, in contrast, was 62 to 81% after radiotherapy alone. The combined to the single modality failure ratios ranged from 0.53 to 0.62 for the sequential and from 0.37 to 0.38 for the simultaneous treatment, indicating the higher effectiveness of the last schedule in controlling tumors (Fig. 11).

Finally, in the trial employing the protocol number 4 (i.e. radiotherapy given as 2 large weekly fractions combined simultaneously with intense heating) the failure rate ranged from 67% at the end of treatment to 87% at 6 months after radiotherapy alone. After combined modality, only 13% lesions failed and no one recurred during the follow-up period (Fig. 10). In this case, the combined to the single modality failure ratio was the lowest at the end of treatment (0.19) and decreased to 0.15 at 6 months, indicating again the higher effectiveness of the combined versus the single treatment modality, and suggesting that this protocol could be the most effective in controlling the tumors (Fig. 11).

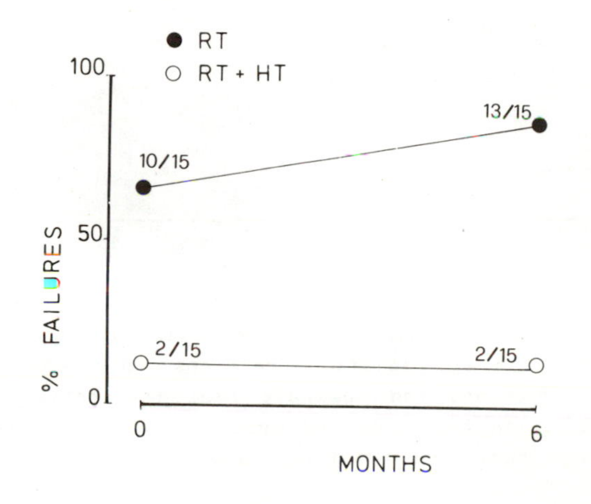

Fig. 10: Failure rates of the lesions treated with the protocol n° 4 (i.e. radiotherapy given as 2 large weekly fractions simultaneously combined with intense heat): ● RT alone; ○ combined modality. (From Arcangeli, ref. 3).

The increase of the fraction size results in an increase of percent tumor control after simultaneous treatment. In contrast, because of the lower total dose administered with the use of large radiation fractions, percent tumor control tends to decrease after radiation alone and after sequential treatment (Fig. 12a). However, by increasing the fraction size there is also an increase of percent skin moist desquamation, especially after simultaneous treatment;

sequential treatment induces only a small increase of percent skin reaction. Again, because of the active skin cooling, no increase of percent moist desquamation is observed with the use of large radiation fractions and intense heating (Fig. 12 b).

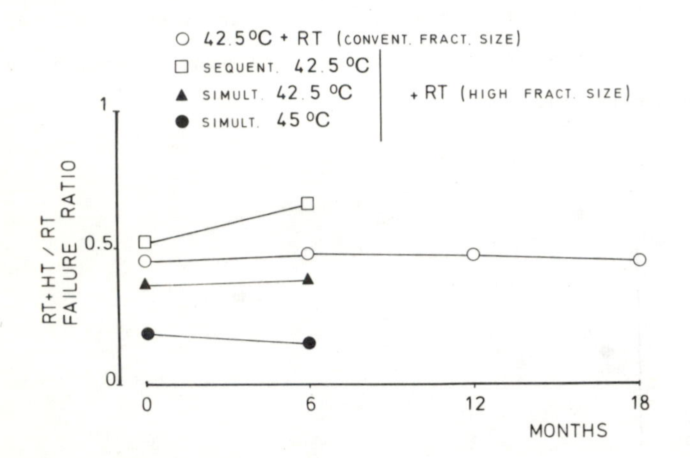

Fig. 11: Combined to single modality failure ratios after different treatment protocols: ○ Small radiation fractions and moderate heat; □ Large radiation fractions and moderate heat (sequential treatment); ▲ Large radiation fractions and moderate heat (simultaneous treatment); ● Large radiation fractions and intense heat. (From Arcangeli, ref. 3).

The same trend is likely to be observed with the late skin reaction as suggested by the example of a patient with multiple melanoma nodules treated with the simultaneous treatment on the leg and with the sequential treatment or radiation alone on the thigh: the acute reaction in the leg was more pronounced than that in the thigh (Figs. 13a and 13b). Some months later the highest degree of late skin damage is observed in the area treated with simultaneous modality (Figs. 14a and 14b).

These results confirm the biolcgical observation that, heating the tumor and the surrounding normal tissue at the same temperature, a differential effect can be obtained by increasing the interval between the 2 modalities or by decreasing the radiation fraction size.

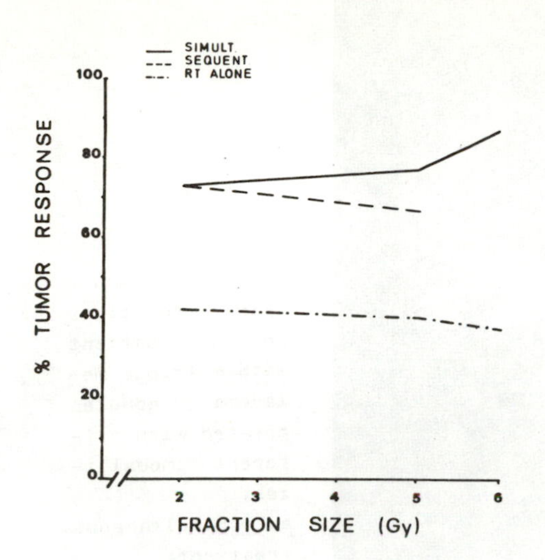

Fig. 12: Percent response as a function of the radiation fraction size.
a. Tumor: —— Simultaneous treatment; - - - Sequential treatment; -.-. RT alone.
b. Skin: —— Simultaneous treatment; - - - Sequential treatment; -.-. RT alone. (Modified from Arcangeli et al., ref. 3).

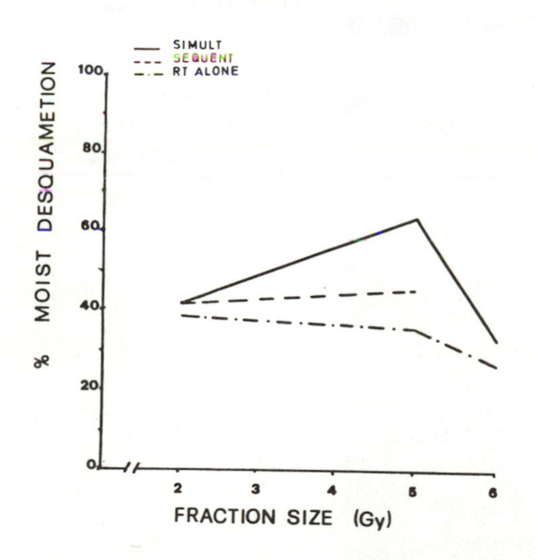

If heat is delivered 3 to 6 hours after radiation, at that time the thermal enhancement has disappeared in the normal tissue but still persists moderately in the tumor. On the other hand, at small doses, radiation cell killing mainly results from single-hit lethal events

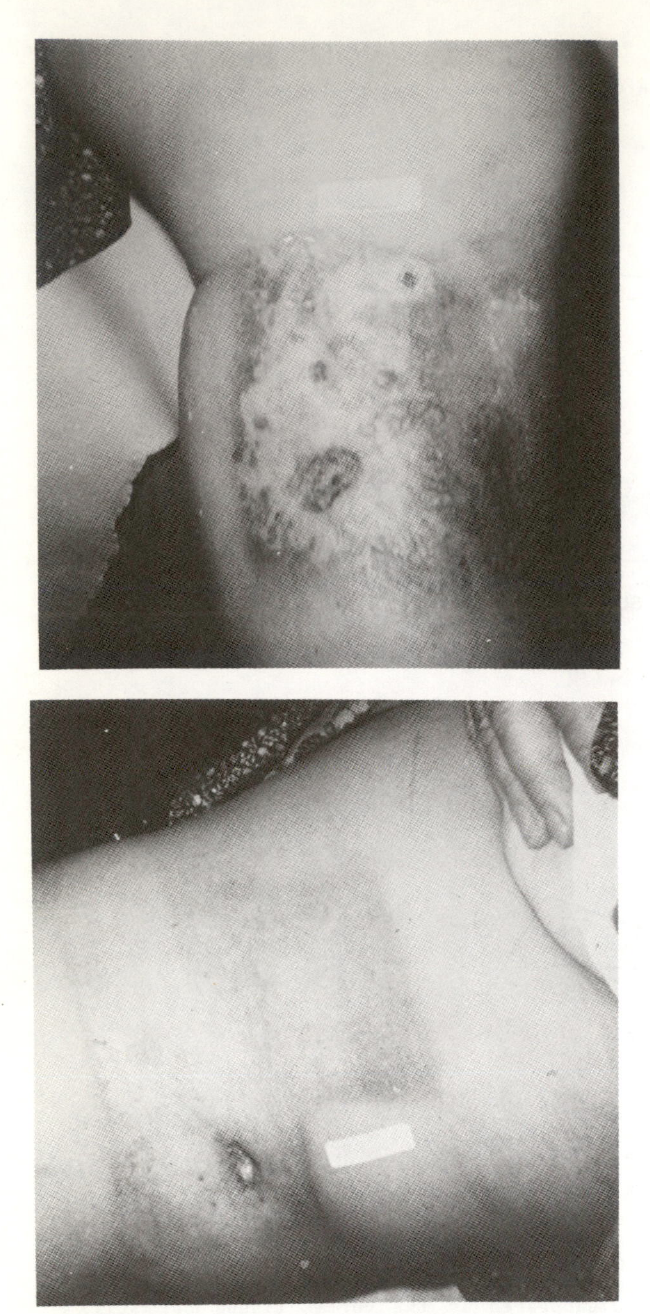

Fig. 13: Acute skin reactions in a patient with multiple melanoma nodules treated with different modalities.
a. Simultaneous treatment
b. Sequential treatment (upper part of the thigh), RT alone (lower part of the thigh). (From Arcangeli et al., ref 3).

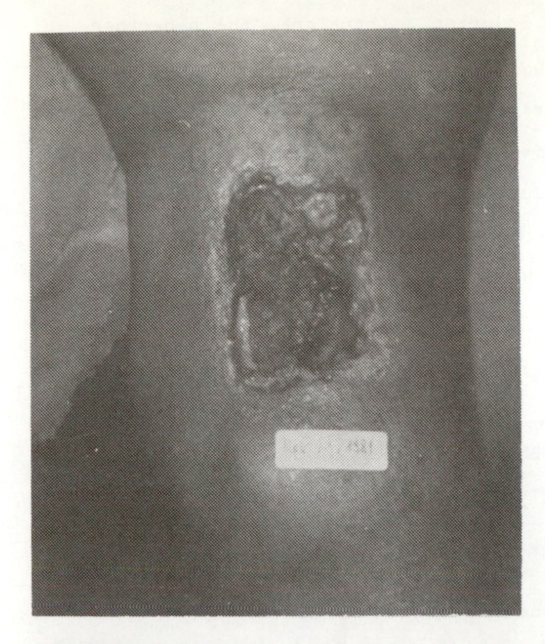

Fig. 14: Late skin reactions in the same patient as in figure 13.
a. Simultaneous treatment
b. Sequential treatment (upper part of the thigh). RT alone (lower part of the thigh). (From Arcangeli et al., ref. 3).

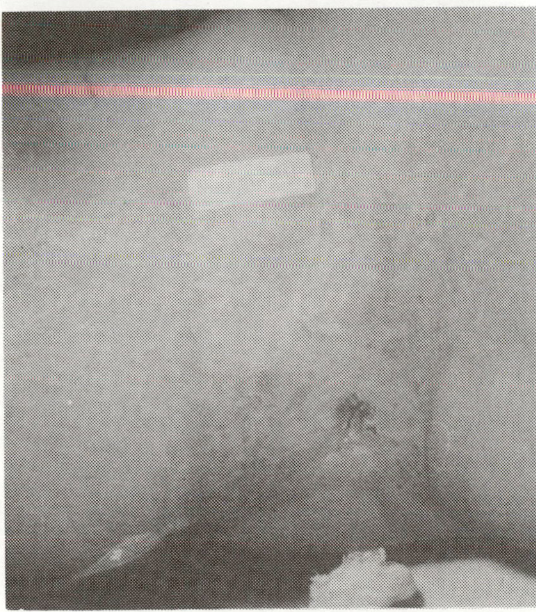

than from accumulated sublethal injuries. Therefore, the hyperthermic radiosensitization induced by the impairment of sublethal damage repair by heat is progressively decreased. In this case, a differential effect may be due to the fact that cell killing in tumors may still result from accumulated sublethal injury and/or from the direct hyperthermic cytotoxic destruction of the nutritionally deprived, acidic and chronically hypoxic radioresistant tumor cells.

TABLE 4 - EFFECT OF HEAT ON THE RADIATION RESPONSE OF TUMOR AND SKIN

PROT.	TREATMENT	TREATMENT	TUMOR RESP.		SKIN RESP.		T.G.F.
Nº	PROTOCOLS	MODALITY	%C.R.	T.E.R.	%M.D.	T.E.R.	
1 & 2	Small rad. fract/ Moderate heat	RT + HT	72	1.71	41.6	1.07	1.60
		RT Alone	42		38.8		
3	Large rad. fract/ Moderate heat	RT + HT (sequent.)	67	1.79	46	1.28	1.40
		RT + HT (simult.)	77	2.05	64	1.78	1.15
		RT Alone	37.5		36		
4	Large rad. fract/ Intense heat	RT + HT	87	2.63	33.3	1.25	2.10
		RT Alone	33		26.6		

From these results, an estimation of the therapeutic gain factor (TGF: ratio of tumor to skin TER) was attempted by defining TER as the ratio of percent response after combined modality to percent response after radiation alone (Table 4). Heating tumor and normal tissue at the same temperature, the best therapeutic gain factor was obtained by using small radiation fractions or by introducing an interval between the two modalities. When the tumor can be preferentially heated in respect to the surrounding normal tissue, the therapeutic gain factor was remarkably good even with the use of large radiation fractions combined simultaneously with intense heat.

However, it must be noted that the number of radiation fractions associated with hyperthermia is different in the 4 schedules, and so are the interval and magnitude of the heat treatments. Thus, there may be a different (but unknown) infulence of thermotolerance in the different treatment response which could influence the response. The problems can only be clarified when more clinical data are available.

CONCLUSIONS

Some useful suggestions can be derived from the previous results and used for designing clinical treatment schedules. In cases where radiotherapy has never been given before, an optimal treatment should result by simply adding once or twice a week a simultaneous session of moderate heat to a full conventional fractionation radiotherapy course.

In heavily irradiated tissues or when fast treatments of multiple lesions are required for palliation, an optimal treatment should result by delivering 5-7 bi-weekly fractions of 5-6 Gy in combination with hyperthermia. If the tumor can be selectively or preferentially heated in respect to normal tissue, a simultaneous application of intense heating can be safely administered with a significant probability of tumor control. On the contrary, the sequential administration of moderate heat should provide an effective treatment with a lower incidence of normal tissue damage.

A further development of better heating techniques is now necessary to employ heat in all clinical situations as adjuvant to radiotherapy.

REFERENCES

1. Arcangeli G, Barni E et al. Heating patterns after 27 MHz local hyperthermia. Comparative results in piglet normal tissue and in phantom. In: Arcangeli G, Mauro F, "Hyperthermia in Radiation Oncology", Milano, Masson Italia Editori. 69-78, 1980.

2. Arcangeli G, Cividalli A et al. Tumor control and normal tissue damage for several schedules of combined radiotherapy and local hyperthermia: an updated study. Strahlentherapie 158: 378, 1982.

3. Arcangeli G, Cividalli A, et al. Tumor control and therapeutic gain with different schedules of combined radiotherapy and local external hyperthermia in human cancer. Int J Radiat Oncol Biol Phys 9: 1125, 1983.

4. Dethlefsen LA, Dewey WC (eds). Proceedings of the Third International Symposium on Cancer Therapy by Hyperthermia, Drugs and Radiation. Natl Cancer Inst Monogr 61: 1982.

5. Dewey WC, Hopwood LE, et al, Cellular responses to combinations of hyperthermia and radiation. Radiology 123: 463, 1977.

6. Dewey WC, Freeman ML et al. Cell biology of hyperthermia and radiation. In: Meyn RE, Withers HR, "Radiation Biology in Cancer Research," New York, Raven Press, 589-621, 1980.

7. Field SB, Bleehen NM. Hyperthermia in the treatment of cancer. Cancer Tret Rev 6: 63, 1979.

8. Field SB, Anderson RL. Thermotolerance: a review of observations and possible mechanisms. Natl Cancer Inst Monogr 61: 193, 1982.

9. Field SB. "The Biological bases for hyperthermia" In: Mirard EA, Hutchinson WB, Mikich E, "13th International Cancer Congress, Part D, Research and Treatment," New York, Alan R Liss, Inc, 195-204, 1983.

10. Hill SA, Denekamp J. The response of six mouse tumors to combined heat and x-rays: implications for therapy. Brit J Radiol 52: 209, 1979.

11. Kamura T, Nielsen OS et al. Development of thermotolerance during fractionated hyperthermia in a solid tumor in vivo. Cancer Res 42: 1744, 1982.

12. Law MP. Some effects of fractionation on the response of the mouse ear to combined hyperthermia and X-rays. Radiat Res 80: 360, 1979.

13. Law MP, Ahier RG, et al. The response of mouse skin to combined hyperthermia and X-rays. Int J Radiat Biol 32: 153, 1977.

14. Law MP, Ahier RG, et al. The effect of prior heat treatment on the thermal enhancement of radiation damage in mouse ear. Brit J Radiol 52: 315, 1979.

15. Law MP, Coultas PG et al. Induced thermal resistance in the mouse ear. Br J Radiol 52: 308, 1979.

16. Myers R, Field SB. The response of the rat tail to combined heat

and X-rays. Brit J Radiol 50: 581, 1977.

17. Myers R, Field SB. Hyperthermia and oxygen enhancement ratio for damage to baby rat cartilage. Brit J Radiol 52: 415, 1979.

18. Nielsen OS, Overgaard J. Importance of preheating temperature and time for the induction of thermotolerance in a solid tumour in vivo. Brit J Cancer 46: 894, 1982.

19. Nielsen OS, Overgaard J. Influence of thermotlerance on the interaction between hyperthermia and radiation in a solid tumour in vivo. Brit J Radiol 56: 267, 1983.

20. Overgaard J. Effect of hyperthermia on malignant cells in vivo. A review and hypothesis. Cancer 37: 2637, 1977.

21. Overgaard J. The effect of local hyperthermia alone and in combination with radiation, on solid tumors. In Streffer C et al. "Cancer Therapy by Hyperthermia and Radiation," Urban and Schwarzenberg, Baltimore, Munich, 49-61, 1978.

22. Overgaard J. Simultaneous and sequential hyperthermia and radiation treatment of an experimental tumor and its surrounding normal tissue in vivo. Int J Radiat Oncol Biol Phys 6: 1507, 1980.

23. Overgaard J. Effect of hyperthermia on the hypoxic fraction in an experimental mammary carcinoma in vivo. Brit J Radiol 54: 245, 1981.

24. Overgaard J. Fractionated radiation and hyperthermia. Experimental and clinical studies. Cancer 48: 1116, 1981.

25. Overgaard J. Influence of sequence and interval on the biological response to combined hyperthermia and radiation. Natl Cancer Inst Monogr 61: 325, 1982.

26. Overgaard J. Hyperthermic modification of the radiation response in solid tumors, in Fletcher GH, Nervi C et al. "Biological Bases and Clinical Implications of Tumor Radioresistance." Masson Publishing USA, New York 337-352 1983.

27. Overgaard J, Nielsen OS. Pre-clinical studies in hyperthermia. In: Williams CJ, Whitehouse JMA (eds) "Recent Advances in Clinical Oncology," Churchill Livingstone, Edinburg, London, Melbourne and New York, 19-34, 1982.

28. Overgaard J, Nielsen OS. The importance of thermotolerance for the clinical treatment with hyperthermia. Radiotherapy and Oncology (in press) 1983.

29. Song WS, Rhee JG et al. Effect of hyperthermia on hypoxic cell fraction in tumor. Int J Radiat Oncol Biol Phys 8: 851-856, 1982.

30. Stewart FA, Denekamp J. Sensitization of mouse skin to X-ir-

radiation by moderate heating. Radiology 123: 195, 1977.

31. Stewart FA, Denekamp J. The therapeutic advantage of combined heat and X-rays on a mouse fibrosarcoma. Brit J Radiol 51: 307, 1978.

32. Suit H, Gerweck LE. Potential for hyperthermia and radiation therapy. Cancer Res 39: 2290, 1979.

33. Van der Zee J, van Rhoon GL, et al. Thermal enhancement of radiotherapy observed in a clinical study. Strahlentherapie 159: 385, 1983 (abstr.).

PROCEEDINGS OF THE II INTERNATIONAL CONFERENCE ON APPLICATIONS OF PHYSICS TO MEDICINE AND BIOLOGY
edited by Ž. Bajzer, P. Baxa & C. Franconi

SOME BIOLOGICAL EFFECTS OF LOW FREQUENCY MAGNETIC AND ELECTRIC FIELDS

Anthony T. Barker
Department of Medical Physics and Clinical Engineering
Royal Hallamshire Hospital
Sheffield S10 2JF ENGLAND

1. Introduction

The electromagnetic spectrum is traditionally classified into ionising and non-ionising radiation based on the frequency or energy of that radiation. The biological interactions of ionising radiation have been extensively studied and are widely used in traditional Medical Physics. It is only recently, however, that the effects of non-ionising radiation have been seriously investigated. Considerable interest has arisen, particularly in the frequency range of d.c. to 100 kHz, and equipment with outputs in this range is now becoming available for therapeutic purposes. This paper makes no attempt to be comprehensive but describes selected examples of claimed low frequency bioelectric phenomena, ranging from the well established and understood to the much less well understood.

2. Neuromuscular stimulation by time-varying magnetic fields

The mechanisms of neuromuscular tissue stimulation by the alteration of transmembrane potential are well known. The normal potential across a neural membrane is approximately 70 mV and if this is altered by more than a few millivolts for at least 20 μsec the nerve fibre will be stimulated. This usually results in a self-propagating nerve action potential and direct electrical stimulation is often used in physiological measurement, for example to determine nerve conduction velocity.

Time varying magnetic fields induce electric fields and currents in conducting media (for example biological tissue) and for a given conducting loop the induced voltage is given by:

$$V = A \frac{dB}{dt}$$

where $\frac{dB}{dt}$ is the rate of change of magnetic field and A is the loop area.

Because the transmembrane potential needs to be reduced for at least 20 μsec if neuromuscular stimulation is to occur, magnetic fields with frequencies above 25 kHz will not normally cause direct stimulation. However, at frequencies lower than this it is possible for the induced voltage during one half cycle to stimulate the tissue before $\frac{dB}{dt}$ changes sign.

2. (i) Magnetophosphenes

The most sensitive example of direct perception of neuromuscular stimulation due to time-varying magnetic fields are Magnetophosphenes. This is the phenomenon whereby magnetic fields applied to the head cause faint flickering lights to appear, usually in the periphery of the visual field. Lovsund et al [1)] found that the threshold flux density was as low as 10 mT for frequencies of 20-30 Hz. It is thought that the induced currents polarise retinal synaptic membranes and cause changes in synaptic transmission similar to those caused by light. No hazards have been shown to be associated with Magnetophosphenes and the fields needed to produce them can sometimes be found in the domestic environment e.g. near hair driers or tape recorder heads, or in industry near high current cables.

2. (ii) General neuromuscular stimulation

More general neuromuscular tissue, such as the major peripheral nerve trunks, require much higher magnetic fields and rates of change of magnetic field to stimulate them than do the cells of the retina. However, it is still possible to stimulate such tissue by the use of single magnetic field pulses. Polson et al [2)] have described a capacitor discharge system designed to non-invasively stimulate human peripheral nerve trunks. By discharging an energy of 480 Joules (6000 μF x 400V) into a 35 mm diameter 3 μH coil, giving a peak field of 2.2 Tesla after 180 μsec they were able to stimulate nerves such as the ulnar or median at the wrist. Stimulation can be verified, either by direct observation of the neuromuscular twitch, or by recording the resultant e.m.g. at the hand or nerve action potential at the elbow. Peak coil current was 6800 Amps and was handled by 12 thyristors, each switching the energy from a 500 μF capacitor. The present system is only able to produce supermaximal stimulation in superficial nerves due to limited field strength but an improved version, using the same energy storage but at 3 kV is being developed which, because of higher efficiency, is hoped to produce fields of 4 Tesla. Magnetic stimulation is subjectively less painful than electrical stimulation through surface electrodes due, it is thought, to the surface pain fibres not being subject to the very high electric fields present in the vicinity of electrodes. The ability of magnetic fields of this frequency to penetrate biological tissue with only minimal attenuation suggests that the technique may be able to stimulate deep nerves non-invasively. This is not possible with conventional electrical stimulation due to the high current densities that occur immediately beneath the surface electrodes.

3. Acute hazards from exposure to magnetic fields

The recent advent of NMR imaging systems in the clinical environment has lead to renewed interest in the possibility of hazards associated with exposure to magnetic fields. Present NMR scanners expose patients to high level static magnetic fields (up to 2 Tesla) in conjunction with pulsed fields of up to 10^{-3}T having rates of change of 20 Ts^{-1} and maximum spatial gradients of 10^{-2} Tm^{-1}. In addition NMR scanners generate pulsed r.f. fields which are beyond the scope of this paper to consider in detail although they are unlikely to cause acute hazards based on the experience of physiotherapy departments, who routinely use much higher power densities for therapeutic purposes in both pulsed and continuous shortwave diathermy treatment. The possibility of non-acute hazards being associated with long term exposure to magnetic fields cannot be discounted and this topic will be mentioned in a later section.

The most likely acute hazard to be caused by the interaction of magnetic fields and biological tissue is ventricular fibrillation due to direct stimulation of neuromuscular tissue as described in the previous section. No data exists concerning the magnetic fields required to actually cause ventricular fibrillation although the problem has studied by Polson et al[3]. In experiments specifically designed investigate possible acute hazards from NMR systems they exposed thorax of anaesthetised rats to the output from their magnetic neuromuscular stimulator whilst monitoring the e.c.g. With fields rising to a peak of 2.6 Tesla after 180 μsec, measured at the ches surface, and maximum rates of change of field of 2.5×10^4 Ts^{-1} they were unable to demonstrate any disruption of sinus rhythm or in e.c.g. shape even when the magnetic field pulse was applied dur the most sensitive phase of the heart cycle. The e.c.g's of rats pre-sensitised with digitalis to lower their cardiac thresholds we also unaffected, although major stimulation of the chest muscles occurred in both groups.

Whilst care must be taken in extrapolating these results to hu they suggest that the fields present in an NMR system under normal circumstances are several orders of magnitude below those needed to cause ventricular fibrillation.

A catastrophic failure of the main magnet in a high field NMR s resulting in a rapid collapse of the static field may cause neuro-muscular stimulation. However, for this to occur, the fields would to decay in a period comparable with, or less than, the membrane ti constants involved, which are typically less than 1 msec. It is th unlikely that the energy stored in a whole body magnet would be rel in this time scale, even under fault conditions, because of the resultant high back e.m.f's.

4. D.C. fields within the body

Whilst the transmembrane potentials of nerve and muscle fibres has been studied in detail, little work has been carried out to investigate the significance of extracellular electric fields, eithe naturally occurring or deliberately applied. One example of natural fields are the so called 'skin batteries'.

4. (i) D.C. fields and wound healing

The ability of several species to regenerate lost body parts is a well documented but little understood phenomenon. Recent work [4,5] has shown the presence of natural d.c. currents in the vicinity of regenerating amphibian limbs, driven by skin batteries which maintain a potential difference across the epidermis. It has been shown that these currents are essential for normal regeneration and that they may help to initiate regeneration if artifically introduced into species that are not natural regenerators [6].

Illingworth [7] has reported that children fingertips will regenerate with excellent cosmetic and functional results if treated conservatively. Currents similar to those observed in the regenerating amphibians with peak densities of 35 $\mu A/cm^2$ have also been shown to exist during this remarkable human regeneration [8]. It has been confirmed that these currents also come from skin batteries, the electrical properties of which have been studied in the guinea pig [9] and to a lesser extent in humans [10].

Mammalian skin appears to maintain a voltage of 10 mV - 60 mV across itself, with the stratum corneum negative with respect to the dermis. This has been shown to drive currents of 1 μA/mm of wound perimeter into damaged tissue and to generate voltage gradients of up to 200 mV/mm in the intact skin at the wound edge [9]. One of the main mechanisms of wound healing and tissue repair is 're-epitheliasation', the process by which cells migrate from the intact skin at the wound edges to form a new surface layer over the wound. Studies using pigs have shown that the migrating cells come from a strip of intact skin approximately 0.5 mm wide around the wound and it is in this area that the substantial voltage gradients exist. The mechanism by which the cells migrate is not understood but the movements of some cells are known to be affected by voltage gradients as low as 7 mV/mm [11]. Hence it is possible that the signal for the cells to migrate into the wound is the presence of a voltage gradient at the wound edge. If wound healing is mediated by electrical signals, at least in part, the possibility exists of electrically stimulating wounds to enhance this process.

4. (ii) D.C. field and bone healing

Ever since the discovery of the piezoelectric properties of bone in 1953 attempts have been made to explain the natural remodelling of bones under stress, described by Wolff's law, as a response to the generated electrical signals. This in turn has lead to attempts, using a wide range of electromagnetic signals, to stimulate fractured bones to unite. In particular the type of fracture which fails to unite of its own accord, known as a non-union, has received the greatest attention because of the serious disability it can cause. The initial experiments used d.c. currents delivered through implanted electrodes and this technique is still used clinically with a considerable degree of success [12]. Recently the use of pulsed magnetic fields has become the more popular form of treatment and this will be discussed in the next section.

Clinical stimulation of fracture healing with d.c. is carried out by implanting up to four stainless steel wire cathodes at the non-union site. They are driven via separate electronic constant current generators with currents of up to 20 μA d.c. continuously for typically 12 weeks. Success rates of 70 - 90% are usually claimed to be achieved. The presently accepted clinical procedure is summarised by Brighton [12]. The stimulation of new bone growth (osteogenesis) with d.c. currents has been demonstrated by many workers in animal studies. Whilst the mechanisms by which it occurs are not yet proven the evidence for its existence does appear to be soundly based (see Barker and Lunt [13] for a brief review and references). Perhaps the most likely theory is that of Brighton et al [14] who suggest the effect is due to electro-chemical changes at the electrodes resulting in a decrease in oxygen tension and an increase in the number of hydroxyl radicals present, which in turn favours bone growth and calcification.

5. Magnetic stimulation of bone healing - a low level effect ?

The literature contains many reports of biological effects of low level electromagnetic fields, both detrimental and beneficial, that are claimed to be non-thermal and cannot be explained by direct stimulation of neuromuscular tissue resulting in nerve or muscle action potentials. These report vary from biochemical responses in cell culture systems to animal and human behavioural studies (see for example Perry et al[15] a study showing correlation between incidence of suicide and 50 Hz magnetic fields from domestic wiring and also includes a useful bibliography of behavioural effects). Few, if any, of these studies have been independently duplicated, the author himself having been involved in studies using systems as diverse as E.Coli generation rate, newt limb regeneration and noradrenaline release from nerve cells in culture, without as yet obtaining the results published for similar systems. The National Radiological Protection Board has also failed, so far, to confirm observations of behavioural and brain wave frequency effects with exposure frequencies in the 15 - 20 Hz range [16]. The general literature on tissue interactions with non-ionising radiation has been reviewed by Adey [17].

Whilst, in most cases, the reports of low level or 'non-thermal' electromagnetic field effects have remained scientific curiosities there is one area in which these claimed effects are significantly influencing clinical practice - that of electromagnetic stimulation of bone healing.

Because of the invasive nature of the d.c. stimulation of bone non-unions described in the previous section, investigators quickly turned their attention to the possibility of applying stimuli by using pulsed magnetic fields to non-invasively induce time varying electric fields in the tissue. The first reported case of a human non-union treated with time varying magnetic fields was in 1975 and it is estimated that over 12000 patients world-wide have now been treated with success rates of greater than 80% claimed by proponents of the technique. Many papers have been published showing biological effects of the type of fields used clinically, in cell cultures and animal studies as well as uncontrolled clinical trials. The literature has recently been reviewed by Barker and Lunt[13], who include detailed information on the waveforms and magnetic field amplitudes used by the most widespread systems.

All the clinical systems expose the fracture site to a more or less spatially uniform magnetic field generated at right angles to the limb, either by a pair of air cored coils on opposite sites of the limb or using a C-shaped electromagnet with the limb in the airgap. Peak fields vary from 1.1 mT to 53 mT and pulse repetition rates from 0.7 Hz to 4 kHz depending on the system used, with the higher frequency signals usually having correspondingly lower peak field strengths.

The system in by far the most widespread clinical use is that originally developed by Bassett and co-workers (see for example Bassett et al [18] for its clinical application) and has been used to treat over 11000 patients worldwide to date. The sytem has been patented and the equipment, which consists of a small, mains powered pulse generator and set of coils is commercially available. (Electro-Biology International (UK) Ltd). Patients use the equipment in their own home for 10-16 hours per day as part of a well defined regime of conservative orthopaedic management which places great emphasis on lack of movement at the fracture site and complete absence of weight-bearing. Median treatment time is approximately six months. The waveform used in the Bassett/EBI system is complex and claimed to be absolutely critical to achieve good results. It consists of pulses in bursts of 5 msec duration repeated at 15 Hz. Within each burst the pulses are asymmetric with the field rising, approximately linearly, to a peak of 1.5 mT after 200 μsec in the centre of the pair of coils and then decaying over 28 μsec. Full technical details may be found elsewhere [13,19]. Some animal studies appear to demonstrate waveform dependant effects (see for example Christel et al [20]) although a remarkable study by Smith and Pilla [21] in which newt limb regeneration was drastically influenced, ranging from a 200% increase over the natural rate to total inhibition of regeneration with only slight changes in stimulus waveshape, has yet to be duplicated. Independent verification of this study is important because of its implications for developmental biology and because of the possible danger of inhibition of natural healing processes.

The overall success rate quoted for the 11,000 patients treated with the Basset system is 75% [22] and for a study of 127 ununited tibial fractures 87% [18]. Despite these impressively high success rates and the large numbers of patients involved, controversy still surrounds the technique. A critical editorial in the Lancet [23] regarded the technique as clinically unproven, the laboratory studies to be 'floundering' and much of the animal work to be 'irrelevant'. It described as essential the need for a controlled double blind clinical study to separate any possible effect of electromagnetism from those of immobilistion or placebo. Bassett et al have replied that, to achieve a success rate of over 75% in a patient group with disability times longer than 24 months, many of whom have had several failed operations and in whom 'spontaneous healing is unlikely', leaves 'little room for rational doubt that the method is effective' [24].

A clinical double-blind trial

In an attempt to shed some light on this controversy a double blind clinical trial of tibial non-unions was instigated in January 1981 at the Royal Hallamshire Hospital, Sheffield, jointly between the departments of Medical Physics and Clinical Engineering, Orthopaedics, and Community Medicine. Interim results are now available and represent the first data to be produced from a controlled clinical trial of electromagnetic bone stimulation. Considerable delay is likely before suitable patient material becomes available to increase the group numbers significantly.

Sixteen patients with tibial non-unions, whose original injury had been present for at least 12 months, were randomly allocated either a working or dummy stimulator. Entry criteria included absence of any operative procedure in the previous six months or radiological changes at the fracture site in the previous three months. Neither the patient nor the clinical staff were aware of the machine type, both dummy and active machines being externally identical. Active machines generated the 5 msec burst 15 Hz waveform developed by Bassett and are fully described elsewhere [19]. The patients treated themselves for 10 - 16 hours per day in their own homes. The orthopaedic management described by Bassett et al [18] was followed and basically consists of total immobilisation of the fracture site in a long leg plaster and complete absence of weight-bearing. Patients were treated for 24 weeks, the median treatment time found necessary by Bassett et al. Union was defined as absence of movement at the fracture site under mechanical stressing, confirmed by independent observers, using image intensification techniques and stress radiographs in two planes. Static radiographs were not used as the orthopaedic surgeons felt that lack of mobility was the primary requirement of union. Clinical examinations were carried out at 12 and 24 weeks and included subjective pain and

and tenderness measurements as well as the mobility measurements. All patients remained in plaster for the first 24 weeks using their machines even if they were thought to have united at week 12. Patients who had not united at the key assessment point of 24 weeks were guaranteed working machines for the next 24 weeks. When union was achieved, at week 24 or beyond, the patients started on axial compression exercises followed by a controlled weightbearing regime as described elsewhere [18].

Results

Nine patients were allocated active machines and at week 24 five of them had united. (Group median age 38 years, median duration of injury 23 months.) Seven patients were allocated dummy machines and at week 24 five of these had united. (Group median age 30 years, median disability time 17 months.) All the patients in each group who were united at week 24 were bearing weight unprotected by week 36 and were still united at week 48. The average pain and tenderness scores decreased in each group over the first 24 weeks with no significant differences in the scores between the groups. Achieving union at 24 weeks does not appear to be linked to duration of injury with successes in both active and dummy groups distributed throughout the range of duration values and the subject with by far the longest duration injury (133 months) uniting on a dummy machine at 24 weeks.

The limited patient numbers to date in this trial make it impossible to detect, with statistical confidence, small improvements in success rate due to the presence of pulsed magnetic fields in the overall conservative management regime. However, the large increases in success rates due to magnetic fields claimed by proponents of the technique were not present. Statistical analysis enables probabilities to be placed on the true success rate of patients treated with active machines exceeding that of those treated with dummy machines by a specified figure. From the results of five success out of nine patients in the active group and five successes out of seven patients in the dummy group it can be shown that any increase in true success rate, due to the magnetic field is unlikely to be greater than 27% ($p = 0.05$) or at most 41% ($p = 0.01$) and it is quite possible that there is no difference, or even a difference in favour of the dummy treatment, between the two groups. All previous studies of the efficacy of pulsed magnetic field therapy have made the assumption that the success rate for conservative management of long-standing non-union would be very low.

The orthopaedic literature gives little information on the success rates of long term conservative management, largely because surgical intervention is the preferred form of treatment. However Nicoll [25)], in an often misquoted study of 705 tibial fractures, shows a minimum success rate of 12½% with conservative management of those patients who had not united at 52 weeks and the true figure may well be higher because the design of study allowed an unspecified number of subjects to withdraw before week 52 and become part of the failure group. Watson-Jones [26)] in a study that include 417 tibial fractures states that all fractures will unite if immobilised for extended periods.

Possible explanations for the high success rate in the dummy group include the efficacy of good long term immobilisation of the limb and non-weightbearing status, the placebo effect of patients 'treating' themselves for extended periods each day, the enforced limits on patient activity caused by being connected to the machine with a cable approximately 2 metres long, the additional attention they receive on their frequent visits to the clinic, or any combination of the above.

Whilst pulsed magnetic field therapy may be effective in the treatment of bone healing the case is not yet proven. It is arguable that more scientific evidence is provided by a double blind trial with just 16 patients than from the results of 11,000 patients treated in uncontrolled studies. A possible MRC multi-centre double blind trial has been under discussion for some time. Such studies are essential if the true therapeutic value of such techniques is to be assessed and should be encouraged.

Conclusions

High level low frequency electromagnetic fields produce readily demonstrable biological effects by direct neuromuscular stimulation. Natural d.c. fields have been shown to be essential for some regeneration processes and may be an important mechanism in wound healing. Osteogenesis can be stimulated in the vicinity of an implanted cathode passing a few microamps d.c., probably by an electrochemical mechanism, and is successfully used for the treatment of human bone non-unions.

No easily reproducible low level effects of electromagnetic fields have yet been demonstrated. The use of pulsed magnetic fields to stimulate human bone healing is becoming increasingly widespread although no controlled clinical studies have yet been carried out to demonstrate its efficacy. Preliminary results from a small double-blind clinical trial have not shown a higher success rate in those patients treated with magnetic fields compared to the control group.

The physicist and engineer have an important role to play in the development of this subject. Many machines have become available in the last decade which claim therapeutic effects that have never been scientifically demonstrated. It is essential that the physicist becomes involved in this area, both to develop new methods and, just as importantly, to assess objectively the merits of commercial systems. To fail to do so will allow us to return to the era, not long past, in which electricity was regarded as the panacea for all ailments.

References

1. Lovsund P., Oberg P.A. and Nilsson S.E.G. Magnetophosphenes: a quantitative analysis. Med. Biol. Eng. and Comput., 1980, 18: 326-334

2. Polson M.J.R., Barker A.T. and Freeston I.L. Stimulation of nerve trunks with time-varying magnetic fields. Med. Biol. Eng. & Comput., 1982, 20: 243-244

3. Polson M.J.R., Barker A.T. and Gardiner S. The effect of rapid rise-time magnetic fields on the ECG of the rat. Clin. Phys. Physiol. Meas., 1982, 3: 231-234

4. Borgens R.B., Vanable J.W. and Jaffe L.F. Bioelectricity and regeneration: large currents leave the stumps of regenerating newt limbs. Proc. Natl. Acad. Sci. (USA) 1977, 74: 4528-4532

5. Jaffe L.F. and Nuccitelli R. Electrical controls of development. Ann. Rev. Biophys. Bioeng., 1977, 6: 445-476

6. Borgens R.B., Vanable J.W. and Jaffe L.F. Bioelectricity and regeneration: Initiation of frog leg regeneration by minute currents. J. Exp. Zool., 1976, 200: 403-416

7. Illingworth C.M. Trapped fingers and amputated fingers in children. J. Pediatr. Surg., 1974, 9: 853-858

8. Illingworth C.M. and Barker A.T. Measurement of electrical currents emerging during the regeneration of amputated fingertips in children. Clin. Phys. Physiol. Meas., 1980, 1: 87-89

9. Barker A.T., Jaffe L.F. and Vanable J.W. The glabrous epidermis of cavies contains a powerful battery. Am. J. Physiol., 1982, 242: 358-366

10. Foulds I.S. and Barker A.T. Human skin battery potentials and their possible role in wound healing. Brit. J. Dermat., 1983 (in press).

11. Hinkle L., McCaig C.D. and Robinson K.R. The direction of growth of differentiating neurones and myoblasts from frog embryos in an applied electric field. J. Physiol., 1981, 314: 121-135

12. Brighton C.T. The treatment of non-unions with electricity. J. Bone. Jt. Surg., 1981, 63A: 847-851

13. Barker A.T. and Lunt M.J. The effects of pulsed magnetic fields of the type used in the stimulation of bone fracture healing. Clin. Phys. Physiol. Meas., 1983, 4: 1-27

14. Brighton C.T., Friedenberg Z.B., Mitchell E.I. and Booth R.E. Treatment of non-union with constant direct current. Clin. Orthop. Rel. Res. 1977, 124: 106-123

15. Perry F.S., Reichmanis M. and Marino A.A. Environmental power-frequency magnetic fields and suicide. Health Physics, 1981, 41: 267-277

16. Proposals for the health protection of workers and members of the public against the dangers of extra low frequency, radiofrequency and microwave radiations: a consultative document. National Radiological Protection Board 1982, Didcot, England.

17. Adey W.R. Tissue interactions with non-ionising electromagnetic fields. Physiological Reviews. 1981, 61: 435-514

18. Bassett C.A.L., Mitchell S.N. and Gaston S.R. Treatment of ununited tibial diaphyseal fractures with pulsing electromagnetic fields. J. Bone Joint Surg., 1981, 63A: 511-523

19. Barker A.T. The design of a clinical elecromagnetic bone stimulator. Clin. Phys. Physiol. Meas., 1981, 2: 9-16

20. Christel P., Cerf G. and Pilla A.A. Modulation of rat radial osteotomy repair using electromagnetic current induction. In: Mechanisms of Growth Control, 1981, Ed. Becker R.O. Pub. C.C. Thomas, Springfield, Illinois.

21. Smith S.D. and Pilla A.A. Modulation of newt limb regeneration by electromagnetically induced low level pulsing fields. In: Mechanisms of Growth Control, 1981, Ed. Becker R.O. Pub. C.C. Thomas, Springfield, Illinois.

22. Goldberg A.A.J., Gaston S.R. and Ryaby J.P. Computer analysis of data on more than 11,000 cases of ununited fracture submitted for treatment with pulsing electromagnetic fields. Abstract 61, 2nd Annual Bioelectrical Repair and Growth Society Meeting, Oxford 1982

23. Electromagnetism and Bone. Lancet 1981, 1: 815-816

24. Bassett C.A.L., Mitchell S.N. and Gaston S.R. Pulsed electromagnetic field treatment in ununited fractures and failed arthodeses. J.A.M.A. 1982, 247: 623-628

25. Nicoll E.A. Fractures of the tibial shaft. J. Bone Joint Surg., 1974, 46B: 373-387

26. Watson-Jones R. Slow union of fractures with a study of 804 fractures of the shafts of the tibia and femur. Brit. J. Surg., 1943, 260-276

PROCEEDINGS OF THE II INTERNATIONAL CONFERENCE ON APPLICATIONS OF PHYSICS TO MEDICINE AND BIOLOGY
edited by Ž. Bajzer, P. Baxa & C. Franconi

PROGRESS IN TELEMEDICINE

Irving A. Lerch

Division of Radiation Oncology

New York University Medical Center

566 First Avenue, New York, NY 10016

United States of America

ABSTRACT

The communication of images, data, information, and knowledge permits the sharing of medical resources on a scale unprecedented in the history of medicine. The new media exploit computers; local area and long distance networks based on cable technology and telephone lines; data compression technology; computer mediated store-and-forward conferencing; networked audio conferencing supported with slow- and fast-scan video, computer graphics, electronic blackboards and similiar imaging devices; Videotex databases; expert systems derived from artificial intelligence studies; automated and computerized branch exchanges; and satellite transmission. The activities which are now under study include teaching, routine screening of medical images, professional and scientific deliberations, sharing of computer peripherals and other resources, and the publishing of data and information.

1. Introduction

Modern telecommunications promises to solve one of society's most pressing problems: the sharing of a limited number of resources among a large number of users. By resources, I mean not only machines and services, but expertise--the wisdom and knowledge of our civilization. What I wish to describe, then, is the emergence of a vast collective intelligence, given coherence by the innovative use of existing technologies and the development of new technologies.

The heart of such telecommunications is the network, the electronic pathways which interconnect users and resources. The network may be local, for the transmission of voice, picture, and computer data within a medical center or community, or it may be extended, encompassing sites separated by seas and continents. The local network is usually proprietary with committed cables and electronics. Long-distance communication is accomplished through telephone lines, satellite, microwave transmission, and special leased high-capacity lines.

In general, the long-distance networks (LDN's) have limited transmission capabilities and are best suited for exchanging textual, graphic, some video, and voice information. The modest performance of long-line facilities is due to the cost of providing the central retransmission facility for doing the necessary frequency division multiplexing and the inadequate bandwidth of some circuit elements. The local area networks (LAN's),on the other hand, increasingly use broadband transmission technology capable of mediating a great range of transfers: analog and digital real-time video, high-density computer exchanges, digitally processed images (radiographs, CT, NMR, ultrasound, PET, scintigraphy), extended databases (with high resolution graphics), and for sharing of computer peripherals and software.

These distinctions are blurring, however. The deployment of optical fibre communications lines--with their almost unlimited information-carrying capacity--will give LDN's great broadband transmission potential, and data compression technology has already increased the information-carrying capacity of long-lines. However, for the immediate future, there will be a difference between LDN's and LAN's which will dictate their respective uses in medicine, science, business, and government.

To illustrate the scope of the new telecommunications technology, the following is a sample of programs well underway in the United States:

1.1 A group of experts--scientists and physicians--had been brought together by the National Library of Medicine for the purpose of defining a body of authoritative literature on hepatitis. The Hepatitis Data Base of the National Library of Medicine was made available to the experts as a Videotex service which they accessed through computer terminals in their offices. They recorded their views and engaged in discussions on a computer conferencing service. The NLM participants moderated the expert transactions in the computer conference, abstracted the final version of the database, and in discussion with the experts, developed a computer program designed to handle inquiries from physicians who were not hepatitis specialists. The Hapatitis Knowledge Base (HKB) was then tested by thc experts who again recorded their views in the computer conference. The final system was scrutinized by non-experts who used a computer conference to critique the HKB. 80% claimed adequate responses from the system. From inception to conclusion, the project required less than two years [1-3].

1.2 Radiologists in the American middle west are dispersed over a large geographical area, at varying distances from any major medical center. Each physician is equipped with video terminals capable of receiving slow-scan images over the telephone lines. They perform routine screening and have those x-rays which may indicate special problems set aside for direct inspection by a physician. Studies demonstrate diagnostic error rates equivalent to those found in the medical center, despite the fact that the images are not produced by state-of-the-art high-resolution technology [4,5].

1.3 The Universities of Texas and Wisconsin initiate large audio conferencing systems for the purpose of providing continuing medical education programs for physicians and other health-care professionals. In addition, special-topic seminars are conducted. The audio conferences are supported with slow-scan video, computer graphics, and other aids. Numerous programs are offered throughout the year with the participation of faculty and students distributed over thousands of square miles [6,7].

1.4 Increasing numbers of medical centers, laboratories, and university administrations are tying their computer facilities together in local area networks. The initial goal of these LAN's is data communications, but more and more benefit is being derived from resource sharing--access to computer peripherals, databases, and software.

In these and other examples, the story of the evolution of biomedical telecommunications is unfolding. It is the purpose of this paper to outline the important features of that technology, its uses, and its potential in spreading the benefits of modern medicine.

2. The Long-Distance Network

The three most important parameters which describe a network are topology, speed of data transmission, and reliability. The topology characterizes the pattern of interconnections via which information may be transferred from one point to another. The two extremes of such patterns are the fully distributed network--such as found in any national telephone system--and the "star" or centralized network in which all access is passed to a central facility. Another important pattern is the circle in which each user is connected to no more than two other users--all of whom form a closed loop.

Obviously, the "safest" topology is the distributed network since any single failure can be bypassed. Just as clearly, however, the management of such a network--getting the information from the point of origin to its destination--can be tortuous, demanding sophisticated equipment and software. Information management for the "star" configuration can be quite simple but the entire network is vitally dependent on the health of the central facility for its integrity. The circle configuration--while free of the need for a central master facility--is dependent on the well-being of all its member nodes. (The term "node" is used to characterize each entity in a network that is located on or at the terminus of a communications line or at a communications junction or branch-point. Thus a node can be a switching computer, a user's terminal, or a computer resource such as a program or database.) LDN's employ each of these network designs. In reality, no single LDN operates configured with a "pure" topology. Star and ring networks are usually accessed from distributed services [figure 1].

2.1 The star network is most often found in computer telecommunications. Videotex databases are supported on single computer facilities or computer arrays which are accessed by the user directly, via a dedicated communications link, or through a phone line. Similiarly, computer-mediated store-and-forward conferencing and messaging services are usually managed on a central facility (although there are distributed conferencing and messaging systems). Audio conferencing is another example of the star networking configuration. All participants dial into a central facility or "bridge" which interconnects the conferees. As a general rule, teleconferencing--computer mediated, audio, and video--are all examples of the star network. The network may be serviced by lines varying in transmission rates from 60 bps (bits per second) to over 50 Kbps (provided as part of the national telephone network). The 60-100 bps rates are used for slow-speed teletypewriter terminals, 2000-9600 bps for voice and computer data, and special-leased 50 Kbps lines are used to support computer networks and limited video transmission. Most Videotex services are accessed at 2400-9600 bps over the normal telephone lines. In such a case, it is usual for the subscriber to gain access to the host computer by dialing into a commercial distributed network (the packet-switching networks described below) which permits bypassing the more expensive long-distance telephone access. The error-rate for such communications is on the order of one bit per million (although this way of describing the expected error rate is not valid for transmission of computer-generated data blocks) [8]. Single-frame or slow-scan supported video conferencing can be provided by communications links operating at from 1200 bps to 500 Kbps--affecting the amount of time needed to reconstruct the transmitted image. Real-time video conferencing using data compression techniques requires transmission rates varying from 56 Kbps to 1.4 Mbps and higher.

2.2 Distributed networks are the principal telecommunications highways. In the United States, most exchanges of computer data are performed in network facilities which access the telephone long-lines and superintend the flow of information between millions of subscribers and thousands of resources. The technology used is packet switching--the breaking up of a large and complex array of data into standard packets of information, each packet carrying the control, address, and origin details necessary to reassemble the data at its destination [9,10]. This strategy permits the network to handle large numbers of packets simultaneously, each multiplexed onto a given channel as they are received, transmitted to the address coded in the packet, then put into order for delivery [figure 2]. The data transmission rates are essentially from 2400-9600 bps if commercial services which access the telephone long-lines are used. Special lines capable of 50 Kbps transmission are used for proprietary networks such as the Department of Defense Advanced Research Projects Agency Network (ARPANET) [figure 3].

2.3 Examples of ring networks are the Distributed Computer system at the University of California, Irvine, and the Triangle University Computing Center which is a homogeneous network of IBM 360-370's at three universities in North Carolina [9]. Actually, it is not technically correct to class these configurations as LDN's since the loops are constrained to a limited geographical area. It is not yet feasible or desirable to build such loops linking very distant geographical sites. Extremely high transmission rates on the order of Mbps are required for the network to operate efficiently (owing to the large volumes of information which must be shared among and between facilities).

3. The Local Area Network

The LAN is the invention of necessity, born from the need for numerous users to share the peripheral machines which service a community. There are two kinds of LAN: proprietary and general. A proprietary network is one for which the manufacturer provides the software and equipment needed to tie together the terminals, mass storage devices, printers, and computer programs into an integrated unit. In principle, no other component or program is compatible with a proprietary network. A general network is designed to be universal, to enable users of many different kinds of computers and peripherals to share their resources. In truth, not all proprietary networks are exclusive and few general networks are universal. And many manufacturers have recognized the need to give users the option of linking formerly incompatible components into the community network.

The first LAN's were not designed for multiple-channel, high-rate data transmission. The original purpose of the LAN was simply the transferring of limited volumes of data between relatively small devices. For medical applications, however, especially those involving video images, the data transmission rate must be high, broad, and accurate. Since much LAN architecture is structured around the coaxial cable, which has an intrinsic broadband capability of roughly 400 MHz and transmission rates approaching 200 Mbps, almost any range and volume of data transmission is accessible.

The LAN made its appearance with the bus topology, the stringing together of computer elements along a cable which was an extension of the internal organization of the central processor. The bus could be made to branch to accomodate a structure approximating a distributed network but this imposed an inflexible requirement on the development of the system: all components and tasks had to be compatible with the heart of the network--the original machine around which the network coalesced.

3.1 The operational characteristics of networks can be classified as either broadband or baseband. With baseband operation, the encoded data is transmitted directly into the carrier--either a twisted-pair, coaxial, or fiberoptic cable. Only one signal is handled at any given time. The rate of such data transmission is limited to the speed with which data can be assimilated by each network terminal or peripheral, and the bandwidth of the cable. Many current baseband networks, such as Xerox's Ethernet, operate at bandwidths of 20 MHz in order to provide a 10 Mbps communications channel [11]. Such proprietary LAN's as Ethernet, Ungermann-Bass, and Zilog--although capable of very high-speed, single-channel operation--really transmit data at rates lower than this. The basic capacity of the network peripherals and processors found in most small offices is less than the network potential. Since the LAN's are limited in extent (compared to the telephone networks), the error rate is very low, on the order of one bit per thousand million.

3.2 Cable television provides bandwidth transmission of up to 400 MHz and LAN's using this broadband technology are capable of high-speed data transmission rates, broken up among numerous communications channels. However, to exploit this capacity, multiplexing techniques are needed to provide simultaneous transmission. The expense of the central retransmission facility for amplifying and modulating signals can cost several thousand dollars, thus making the technology suitable only for very large network applications. In addition to the central or "head-end" facility, each user interface costs much more than the equivalent interface to a baseband system. Representative examples of broadband LAN's are manufactured by Sytek, Amdax, as well as Ungermann-Bass [11,12]. Ultimately, fiber-optic communications links will greatly enlarge the broadband transmission of LAN's. At present, the central equipment and interfaces are very expensive and unsuited to many applications.

3.3 The capacity of a broadband network is best described by the number of links and the required bandwidth per link as a function of the data transmission rate. Thus, if no more than 1200 baud (1200 bytes/second--9600 bps) were needed on a typical 105 MHz bandwidth network, then the required bandwidth per link is 10 KHz and 10,500 links could be accomodated. For 56 Kbps transmission, the 60 KHz required bandwidth will accomodate 1,750 links. For 3.088 Mbps compressed full-motion video transmission, the 3,200 KHz required bandwidth will only permit 32 links. However, if high-resolution, full-motion, color--analog video and sound--is to be transmitted, then only eight links will be accomodated. As a matter of design, the cable is used to transmit one bandwidth into the retransmission facility (the return channel--105 MHz, from 5 to 110 MHz) and another bandwidth out (the forward channel of 140 MHz--from 160-300 MHz). This design permits full duplex operation--the simultaneous transmission and receipt of data [figure 4].

3.4 A serious alternative to baseband computer networking, is the existing telephone system within a building. The private branch exchanges (PBX) in most hospitals can be upgraded with additional-lines to handle computer traffic. More sophisticated equipment may be integrated into the PBX's to automate (PABX) or computerize (CBX) the switching of processors and peripherals to available lines. The choice between selecting a baseband, broadband, or PABX/CBX facility will depend on the transmission data-rate required and the total cost of the facility. Obviously, PABX/CBX supported networks are totally unsuited to the transmission of full-motion video but would be adequate for most computer communications.

4. Computer Teleconferencing

Projecting the voice, image, and information of numerous, widely-dispersed people into an arena where discussion can take place, has usually required face-to-face encounters for limited periods at some facility distant from the place of business of most of the participants. Telecommunications is beginning to change this.

Consider the ways we communicate--talking, writing, gesturing--and how we enhance that communication: telephone, gatherings, print, video, signs, radio. Such exchanges are limited by environment, by the need for synchrony in time and place for dialog to occur. With the computer as communications device, the bonds of time and place have been cut.

The most apparent difference between computer-moderated communications and face-to-face meetings or real-time teleconferences, is in the sequence of exchanges--messages are forwarded in time until a correspondent reads and replies. Correspondents may review, respond to specific messages, enter new information, ask questions of any participant, or acquire data at any time, day or night. Since each participant has a terminal attached to the central computer via a normal telephone or leased communications line, the interchange is independent of place as well as time.

Because the computer maintains an accessible record of all transactions, there is no need for memoranda or synopses. The final documentation can simply be edited from the computer memory. Add a feature for privacy--to protect proprietary exchanges--and the system begins to develop a useful shape.

To a limited degree, we already have a measure of this power. The telephone enables us to contact people anywhere, at almost any time, and to carry on group conversations. We can even use the telephone to transmit printed and graphic information. But, such teleconferencing must be conducted in the heat of the here and now; the spoken word is fragile and notoriously elusive. What is transparent to one mind can be opaque to another. Much information is invariably lost. Many meetings are expensive, of limited duration, and difficult to organize--to find a time suitable and convenient for all participants.

All of this is about to change in a way so fundamental, so profound, that the whole business of exchanging, collating, and synthesizing information will cease to exist as a separate activity, divorced from communication.

Problem solving is the most glamorous and apparent activity in a constellation of communications tasks. The Communication Studies Group of the University College, London, has catalogued seven such tasks: Information exchange; Problem solving (associated with decision making); Idea generation (brainstorming); Presentation (with lectures and training); Familiarization (in interviews and personnel meetings); Persuasion (by lobbying, debating, and negotiation); and Mediation (conflict resolution). This type of analytical breakdown leads us to the fine structure of communications [13]. Problem solving itself has been broken down by the Institute for the Future, a private research and development firm based in Menlo Park California, into conceptualizing, searching, structuring, implementing, evaluating, and documenting [14]. Hence, any new medium must facilitate and not impede these fundamental activities.

In 1970, the White House Office of Emergency Preparadness, set up by the President to administer the Wage-Price Freeze, put into use the first electronic conferencing system, the Emergency Management Information System and Reference Index (EMISARI). The advent of this unique, computer-moderated conferencing system was the first attempt to automate the management of a major information-exchange. Murray Turroff, one of EMISARI's developers, and Starr Roxanne Hiltz, a sociologist investigating the impact of computer conferencing on human communications, summarized the unique features of the new medium as follows [15]:

> The primary innovation in EMISARI was the ability to set up alternative communication forms, such as collections of numeric estimates, tables of numbers, and situation report forms, and have these assigned as a permanent responsibility to some member of the communication group who would supply the information on a regular basis. ... All this was under the control of a human monitor who could tailor the communication structure and the responsibilities as a function of the problem at any time.

As of this writing, there are fewer than a dozen commercial conferencing services available in the United States, some of which are run on central computers accessible via packet switching networks such as TYMNET and TELENET, and some which may be licensed directly to the user. Most of these services are summarized in appendices 1 and 2 [16,17].

It is difficult to intercompare the various conferencing systems because many were implemented with different tasks in mind. This has led to certain structural and operational quirks characteristic of each one. Since some features are shared, there are undeniable family resemblences.

All systems are implemented to facilitate many-to-many communications. PARTICIPATE differs somewhat in that it was designed to accomodate what Harrison Chandler Stevens, president of PSI refers to as, "Inquiry Networking," [18]. Users of PARTICIPATE may branch off a main conference and develop their own subconference, usually based on an inquiry generated in the trunk discussion. This multiple branching capability is analogous to the organization of subcommittees in a plenary session. The subcommittees are brought into being to examine specific topics and then report back to the full committee. This type of structure facilitates the "problem solving" aspect of conferencing.

Of course, any system which provides such opportunities for multiple-branched conferencing must also supply the means for maintaining user access and control. This requirement imposes a more complex set of instructions which can frustrate the unsophisicated subscriber. Initially, the user may have difficulty knowing where he is within the conference hierarchy. While most computer conferencing systems have limited branching capabilities, they have attempted to maintain as simple a procedure as possible in order to achieve "user friendliness." This does not mean that such systems are simple--far from it. EIES furnishes elaborate subdivisions for specialized information transfers, such as an "electronic marketplace" for buying and selling information and a "paper fair"--a repository of research documents--whereas PARTICIPATE relies on its branching capability which the user must learn in order to tailor the system to his or her own needs. Perhaps the most elaborate package is AUGMENT, which gives the subscriber maximum control over all textual operations, from processing through multi-color compositing. Such a system requires a high degree of user knowledge and skill. COM, the Swedish entry, has an extremely flexible structure which enables participants to track related messages via the links which interconnect them as a discussion grows. A single command is all that is needed to trace the complex branches of a mature discussion.

Very important are the resources available to the user. Obviously it is necessary to have an editor to process text entered into conference discussions. Data bases, programs, graphics packages, instructional texts, are all significant enhancements.

This support may be achieved with one of three strategies: either by putting the resources on the conferencing computer, accessing a resource computer over a network link, or connecting a resource directly to a user computer or terminal. In this last case, the conference computer may append a user's resource for the benefit of other users, although this is not the most important feature of such an arrangement. Obviously, line costs may be minimized if the user has the editing and programming resources capable of supporting his conference activities off-line. This will be discussed further.

EIES uses a microprocessor to dial out through TELENET to specific resources such as the New Jersey Educational Computing Network which provides software packages for data analyses. In developing HUB, the designers decided to support four basic activities: graphics, programming, documenting, and balloting. Upon conclusion of the developmental and testing phase of the system, Hubert Lipinski and Richard Adler of IFTF concluded that a more general method of achieving user-mediated access to resources is through a programmable microcomputer so that connection to the conference computer and the resource computer could occur simultenously [14].

PARTICIPATE, as implemented on Source, relies on the resources of the host computers in Mclean Virginia. This includes various compilers, mathematical, statistical, and financial packages, and management modeling programs. However, the data must be processed outside of the conference, the result placed into a file, and the file transferred into the conference.

Differences in operating characteristics are usually not too important to the user. For example, the majority of systems have document workspaces to enhance the drafting of manuscripts. In PARTICIPATE, the user must call up a scratchpad and then send the contents to a conference in packets of 12,000 characters or less.

PARTICIPATE, as it appears on Source, has facilitated conferencing among disparate groups who subsequently organize themselves into communities. An interesting example of the syntheses achieved in such a structure was the design and implementation of a project by a group of approximately 20 participants within a three month period. Starting with a concept developed by David Hughes, an early exploiter of computer telecommunications as a teaching tool, Jim Rutt, then manager of Product Development for Source, suggested that PARTICIPATE provide popular electronic lectures--ELECTURES--for subscribers. A conference was started in which the members examined the scope, organization, and operation of an electure.

Despite the fact that many key members of the conference had to periodically absent themselves for extended periods, first the mechanism and procedures, then the electure itself was defined and put on-line. The electurer was Starr Roxanne Hiltz and the subject was the computer revolution--how the "Electronic Cottage" is changing the home. It appeared in seven segments, each segment followed by a ballot and a separate discussion conference [19].

Within one week, with very little advance warning, over 65 discussants joined the electure, adding more than 100 comments, queries, complaints, and revelations. A second electure on cancer was organized with participation by physicians and laymen. The databases of the National Library of Medicine were accessed to support responses to specific inquiries concerning the location of special treatment programs and the specifics pertaining to research and clinical studies. This demonstrated the flexibility inherent in combining computer assisted multi-user conferencing with external Videotex services.

In 1978, a conference was organized on EIES called POLITECHS which included federal laboratories, state legislatures, local governments, public interest groups, technical professional societies, and the White House. The conference was dedicated to promoting exchanges on the impact of science and technology on public policy. As a result of this study, proposals have been made aimed at developing permanent interchanges for various departments of the Federal government. After the Three Mile Island incident in 1979, the nuclear industry responded by founding an Institute of Nuclear Power Operators which uses NOTEPAD to tie together each utility with consultants, vendors, and numerous organizations.

To reduce on-line charges, an intelligent terminal, capable of supporting text processing and a wide variety of communications tasks, will become an essential tool. A long document can be edited off-line, composited, then transmitted to the conference. Conversely, documents can be downloaded and read at leisure.

Ultimately, multi-tasking multi-user operating systems running on the new generation of microprocessors, will enable conferees to access many discussions, data bases, and computer resources simultaneously. Without question, this will elevate telecommunications into a realm which we can only speculate about.

Limited communications software exists for linking personal computers to Videotex susbscription database services.

Integrating text editing and formatting, data base management, and asynchronous communications packages into a single system is a recent innovation. One such system, Microcomputer Information Support Tools--MIST--was developed by Peter and Trudy Johnson-Lenz and is available from New Era Technologies for licensing on CP/M 2.0 operating systems (it requires 56K RAM and two 250 KB floppy disks). The basic necessities for the communications package are,

4.1 A transparent talk utility which relegates the terminal to the "dumb" role of inputting key strokes and displaying information sent back by the host.

4.2 The ability to record into mass storage all information received in the talk mode.

4.3 The ability to upload files in the talk mode.

4.4 The ability to upload a file with a "stop and wait" protocol. Files will be uploaded until the host machine sends a wait signal--usually an ASCII DC3 (equivalent to the <CONTROL>S on most keyboards)--and will resume uploading when the host sends a continue--usually an ASCII DC1 (the <CONTROL>Q on most keyboards). Thus large files may be uploaded without exceeding the buffer capacity of the recipient machine.

4.5 The ability to divert incoming data to the printer if a hard copy is needed.

Many other utilities are useful, and much depends on factors such as the baud rate. With slow data transmission, the "stop wait" protocol is usually not needed. If accurate transmission is essential, such as when sending binary or numerical data, asynchronous transmission will not be satisfactory.

The first computer communications network, the Advanced Research Projects Agency Network--ARPANET--began operations in 1969 by the Department of Defense and rapidly grew into an important medium which academic and business contractors used to access specialized and general libraries, computation services, and programming resources [20]. Since that time, general purpose networks have been made available by commercial firms. All that is needed is for the user to dial a local number which enters him into the network. The user then pays for the amount of time he spends on the line, sometimes at a cost far below that for most long-distance telephone calls. In this way, the user can connect his terminal to any computer in the network. Although few foreign computers are accessible, satellite connections will soon be available.

Some large corporations, primarily those involved in computer technology, have begun to experiment with automated conferencing as a tool to facilitate their developmental work. MIT's Center for Information Systems Research, has been using PARTICIPATE to run a conferencing project for a number of corporations. The MIT program has been particularly successful in using PARTICIPATE to organize meetings and to explore specific conference topics. NOTEPAD was used to organize the Spring 1982 Office Automation show in San Francisco.

In the past, the organization of a major meeting required careful planning and volumes of communication. The administration for national conferences usually consumes one to two years and depends on the energies of dozens of people. The cost in time, communications, and effort is considerable and limited to a relatively few active participants within the sponsoring group. But by performing the preparatory work in a computer conference, the content and scope of a meeting is easily developed. Computer conferencing cannot substitute completely for all meetings, of course, but it can greatly increase the productivity of meetings when they do take place. There is also a democratizing feature: the organization is open to criticism and discussion among all conferees.

This immediately points to a potentially important new application. Computer conferencing may be ideally suited to filling certain gaps in publishing.

Consider this: With a computer terminal connected to a network, the user has potential access to thousands of information libraries--data bases--from the American Chemical Society's CA SEARCH to the Library of Medicine's numerous databases. There is no substantial publisher of data which does not put at least part of its information into computer storage. Some commercial information services such as DIALOG, a subsidiary of Lockheed, command access to over 45,000,000 records in more than 120 data bases in business, science, medicine, technology, the humanities, and politics.

The computer eases the search for meaningful information by providing highly selective search strategies. It is no longer true that the proliferation of information is threatening to deluge us with meaningless drivel. The problem is the opposite. Information is not being processed fast enough.

Particularly troublesome is the fact that some scientific and medical articles take a year or more after submission to a journal before emerging in print. The processing of economic statistics can be excrutiatingly slow. We need more information, faster, if we are to be able to correlate and organize the diverse pieces necessary to make coherent and prompt decisions.

Certain scientific organizations have attempted to overcome the publishing logjam by putting out frequent, less formal letters and bulletins in which timely announcements concerning vital information is made available. But this is a mere trickle in the total volume. Technical papers must be submitted to major meetings many months in advance--usually six months or more. Print publication is no longer sufficient to meet our information needs.

One exciting solution is for the scientific societies to supplement their publishing operations with computerized conferences in which information exchanges can be conducted and subjected to the critical scrutiny of their members. Specialized research groups can also conduct exchanges, partially in public or wholly in private, in order to accelerate the interchange of ideas and data. Ongoing dialogs can be developed and nurtured along diverse, interdisciplinary lines, merging with specialists in different groups when necessary.

In late 1982, a new family of electronically distributed journals challenged the vast science publishing establishment. A recently organized New York firm, Comtex Scientific, began providing scientific reports to academic and industrial subscribers within six to eight weeks of acceptance by distinguished scientist-editors. Unlike electronic conferencing, the computer is used to speed the one-way distribution of information and not to establish a communications forum. Needless to say, there is great fear among publishers that Comtex will both increase the "noise" level and accelerate the demise of high-quality small-circulation journals. This fear is based on the severe reductions in journal subscriptions by libraries responding to increased subscription costs and declining budgets [21].

5. Audio Teleconferencing

Audio conferencing--the most established and widely-used technology--has been greatly enhanced with the addition of video and graphic capabilities. Relying on the existing voice long-lines, the cost is much less than that based on sophisticated high-density data transmission technologies. In its purest form, it consists of a central bridge into which all conferees are interconnected. Participants then dial into the bridge and are placed into the conference, either from their office telephone handset or from a central facility designed for group use.

Proprietary network services such as those offered by ISA Communications Services, Inc., (ISACOMM) based in Atlanta, Georgia, are resale common carriers which lease satellite capacity to support voice, facsimile, data, and video telecommunications. Other companies, such as Darome, Inc., Kellogg, and Connex International sell bridging services capable of connecting more than 60 locations via the telephone system. This type of conferencing is referred to as a "meet-me" call service.

For large, routine conferencing within a widely distributed network of facilities and individuals, permanent multiuser rooms are needed, capable of accomodating numerous participants. In addition, such a conference must have a system to manage the audio interchange. There are a variety of tactics used to track the discussion [22] and to guide users on the requirements for audio conferencing facilities [23,24]. Visual support of audio conferencing is usually provided by slow-scan video, facsimile transmission, electromechanical drawing devices, computer graphics, and remote operation of slide projectors--all transmittable on voice-grade longlines. Typically, a 32 connector bridge is used to support a large-capacity audio conference with image support channeled into a parallel communications link.

A video image consists of a two-dimensional matrix made up of anything from 256 rows and 512 columns to 1024 rows and 1024 columns (a 256 X 512 to 1024 X 1024 pixel array). The video compressor must then act as a valve to project a linear stream of pixels into either a dedicated 8 KHz private line or the 1 KHz baseband telephone line (the maximum transmission rate is 2000 pixels/second over the the telephone company voice-lines). Thus, if a 512 X 512 high-resolution image is transmitted, it will take about two and one-half minutes for the image to be received (by the video expander which then reconstructs the image on the monitor). The most usual form of transmission is a reduced resolution 256 X 256 pixel image which requires little more than 30 seconds for transmission.

The maturity of this conferencing technology is illustrated by the scope of such networks as those operated by the Center for Interactive Programs (CIP), University of Wisconsin-Extension at Madison, and the Teleconference Network of Texas, University of Texas Health Science Center at San Antonio. The CIP operates two networks: the Educational Teleconference Network (ETN) and the Statewide Extension Education Network (SEEN). The ETN, which began operations in 1965, is a strictly audio network which links more than 200 locations within the state and services over 40,000 residents in programs dedicated to continuing education, administration, and public service. Additional audio conferencing is supported with 20-line bridges for special direct-connect statewide, national, and international meetings. SEEN employs slow-scan video and interconnects 26 sites for a 100-course program on a variety of academic subjects. On December 7 and 8, 1983, SEEN conducted a seminar which included presentations from CIP, the Telemedicine for Ontario Project, and the University of Texas Health Science Center. The conference, "Teletraining in Health Care/Education," focussed on the use of the technology in medical education. The Teleconference Network of Texas links 90 hospitals in the state and is primarily dedicated to biomedical continuing education programs. The Texas Network uses dedicated communications lines so that individuals require a special link with the network if they elect to participate from their office or home. Visual aids are usually distributed with instructional packets or slides at the site.

There are many other networks. Oklahoma interconnects 110 hospitals across three state boundaries: Oklahoma, eastern Kansas, and northern Texas. The University of South Dakota School of Medicine, in cooperation with the South Dakota Medical Information Exchange, links individual callers and 14 sites into their conferencing network.

Single events are usually serviced by companies which provide the bridge and management for the teleconference. The Colorado firm, Kellog Teleconferencing Service, routinely produces medical conferences for a spectrum of professional and health-care industry sources. And at least one medical specialty, The American Society of Clinical Pathologists, headquartered in Chicago, has sponsored as many as 83 conferences covering 981 sites with almost 20,000 participants in 1982--numbers which are increasing rapidly.

Aside from educational and some scientific and professional activities, audio teleconferencing--supported with limited pictorial peripherals--is not yet extensively employed in medicine. Slow-scan video is used as a separate medium by radiologists and other medical specialists and must be discussed in a different context.

6. Slow-Scan Teleradiology

Radiologists have long recognized the potential of video technology as a tool in clinical communications. But unlike video teleconferencing which is primarily used to enhance human communications, slow- and fast-scan video is used to convey medical information and data to and among doctors and clinics.

In the late 1960's, the UCLA Center for the Health Sciences installed a high-resolution, 875-line camera equipped with a zoom-lens, mounted with a view of an x-ray light-box. A 15 MHz bandwidth LAN was installed to distribute the images to sites as far away as 1000 feet and the console operator was able to control scanning, zooming, detail emphasis, edge and contrast enhancement, brightness, and contrast with an electromechanical "joystick." While the system lacked the ability to display rapid serial sequences of radiographs, numerous physicians were able to access the material. In his evaluation of the project, Dr. Richard J. Steckel concluded [25]:

> To the economy-minded hospital administrator, a high-resolution television system may seem nothing more than a costly plaything, but, as our own experience proves, it is nothing of the sort. We know that given its maximum technical potential, the savings in time, enhancement of staff morale and subsequent sharing of information among specialists can only benefit the patient. This, after all, is what a hospital is all about.

Since the time of Dr. Steckel's report, medical imaging has flowered in a technological explosion and the need for both local and long-distance transmission of images and data has become acute. The central question concerning the usefulness of video transmission has been whether or not the quality of the transmitted images was sufficient to conserve diagnostic accuracy.

In analyzing the characteristics of a 512 X 512 pixel system with 256 gray levels transmitted over a 9,600 baud telephone line, researchers at Johns Hopkins, the Bureau of Radiological Health, and the MITRE corporation, concluded that the technology had promise but was inferior to existing film technologies. Nonetheless, the image quality was found more than adequate for a variety of routine screening tasks--a vital point in extending radiology services to rural areas and small hospitals [26,27]. This study essentially confirmed an earlier 1975 report by investigators at the Massachusetts General Hospital who pointed out that discomfort with the new technology was the principal factor in discriminating between direct-view and teleradiology systems. Both reports expressed confidence that the teleradiology systems could be improved to match the performance of the existing technologies [28].

At the same time as the Johns Hopkins study, radiologists in Canada used the ANIK-B satellite to relay images from Northern Quebec to

Montreal. The study was part of the Quebec Telehealth Project, centered at the Biomedical Engineering Institute of the Ecole Polytechnique and University of Montreal. Although limited in scope and duration, the investigators found very high agreement between the physicians doing direct interpretation of radiographs and those viewing the broadcast video system [29].

Presently, there are numerous long-term programs in teleradiology in the United States and Canada, all of which rely on slow-scan video technology. Routine screening of x-rays, microscope slides, ECG and EEG traces, liver scans, and dental radiographs is performed in a network centered on Toronto, at the Sunnybrook Medical Center. The Toronto project is directed by the head of telemedicine, Dr. Earl V. Dunn and is a cooperative venture of the Joint Telecommunications Project of the Royal College of Physicians and Surgeons [30,31].

The St. Francis Medical Center in Peoria, Illinois, transmits CT and nuclear medicine scans to the radiologist-on-call and two statewide networks in Colorado tie together 11 sites. The Regional Hospital of Torino and the Civil Hospital of Susa inaugurated a telemedicine link in 1979 to support emergency care and remote teleconsultation [32,33]. It has been estimated that approximately two-dozen facilities use slow-scan video compression technology world-wide to service a growing, wide-flung scattering of physicians, hospitals, and patients.

7. Data Compression and Video Technology

With broadband technology, LAN's have the intrinsic capacity to distribute full-motion video images at the expense of occupying significant bandwidths. The cost of interrogative broadband cable installations for LDN's is prohibitive for many applications. The technological link between slow-scan and full-motion video transmission is the method for achieving sufficient compression of data to be able to use baseband or narrow-band lines.

The telephone network provides one KHz lines capable of supporting transmission rates of 9600 bps, and special leased lines or microwave links which can carry data at rates up to 56 Kbps. And until recently, this was adequate only for slow-scan video images, voice, and graphic telecommunication. Full-motion video, primarily for video teleconferencing, has required broadband transmission of at least 1.544 Mbps--restricting the technology to those rather elaborate situations which warrant the expenditure [figure 5].

The simplist form of data compression is information reduction--from 512 X 512 pixels to 256 X 256 pixels, for example. This can be self-defeating, however, if, in the case of a radiological image, the physician is forced to cope with the reduced resolution by scanning and zooming into portions of the radiograph, necessitating multiple transmissions.

Many applications require full-motion video. American analog video signals require a six MHz channel for the transmission of 525-line images and 12 MHz for high-resolution 1024-line images (both voice and picture). The use of analog technology is relatively inexpensive, requiring a standard head-end modulator for the transmission of signals. To transmit a digitized image consisting of 30, 525-line frames per second, (each separated into two fields--one containing the even numbered lines, the other containing the odd numbered lines) the image is encoded and transmitted over an 86 Mbps commercial television channel. Compressed video images are usually relegated to those channels designated T1 or T2 with transmission rates of 1.544 and 6.312 Mbps, respectively. There are many ways to reduce these data rates by simply deleting redundant information and selectively reducing resolution where it is not needed. There are limits, of course. Video teleconferencing would obviously benefit from high-fidelity color imaging which would not be necessary for many medical applications.

The first compression technique is therefore field or frame elimination. Processing within a single frame--intraframe coding--can be used to eliminate redundant information (such as noting that a region is occupied by a single color). Interframe coding compares adjacent frames and can eliminate unneeded information when there is a lack of motion or change. A very sophisticated method of motion compensation coding uses data from successive frames to predict object motion. The interframe and intraframe coding may be further processed with spacial filtration and transforms--such as the cosine transform--to reduce the number of bits needed to describe a pixel [34].

The most radical use of compression technology is made by Widergren Communications, Inc., which achieves full-motion color video transmission at 56 Kbps. They obtain this compression with spatial, temporal, and chromatic filtering--inter- and intraframe coding--which reduces the data by a factor of 19. Further reduction is achieved with frame differencing to exploit motion information (a 12-fold reduction) and transform coding (a 7-fold reduction) [figures 6 and 7].

8. Videotex

Videotex refers to all database services accessible to the user either directly or via packet-switching networks. (For the purpose of this discussion, Teletext--the one-way transmission of data--is ignored in favor of the full-interrogative systems.) The databases may be contacted either with a simple terminal connected to a telephone modem, via a sophisticated processor integrated into a proprietary high-speed network, or through interrogative cable television links. All medical and scientific Videotex services are presently offered to subscribers as government or commercial ventures available through direct lines or via packet-switching networks.

In the United States, the National Library of Medicine (NLM) provides several levels of Videotex services. The primary service is the MEDLARS system which consists of a suite of 28 medical databases accessible from 1900 MEDLARS centers located in various medical institutions and medical libraries around the country. The total system contains over 6,000,000 references to journal articles and books. Any one or combination of 14,000 headings are used to conduct a Boolean search through the mass of data for the purpose of finding a given reference. This is primarily a fee-for-service librarian utility which is available to physicians and other health-care professionals through their cooperating institutions. The vast majority of the data bases contain literature abstracts and data concerning research programs sponsored by facilities affiliated with the National Institutes of Health. Many of the MEDLARS databases are made available to commercial Videotex companies for resale to their subscribers. There are 7 regional centers responsible for coordinating the MEDLARS centers in their area.

In early 1982, the NLM initiated a nationwide cancer Videotex service made up of several databases. These are summarized as follows:

8.1 Protocol Data Query (PDQ) contains descriptions of approximately 700 cancer therapy research programs to include the title and objective of treatment, criteria for patient selection, the names of the investigators, and address of the institution. The treatment programs included in PDQ are derived from cooperative study groups, cancer centers, intramural NCI groups, and Institutions under contract to the NCI (including those groups supported by the NCI in other countries). Ultimately, PDQ will be expanded to contain general information of use to lay people such as cancer patients and their families.

8.2 Cancer Literature (CANCERLIT) contains roughly 300,000 citations and abstracts of the published cancer literature. About 4,000 new abstracts are added monthly. More than 3,000 U.S. and foreign journals are abstracted.

8.3 Cancer Projects (CANCERPROJ) contains approximately 20,000 descriptions of current cancer research projects reported internationally and is updated quarterly. The database is restricted to projects considered as active within two years of being listed.

8.4 Clinical Protocols (CLINPROT) summarizes the more than 3,000 clinical trials of new anticancer treatments and is updated quarterly.

In addition, a number of files are available to the user to assist in searches, notify the user of new developments, and provide other relevent information.

Commercial Videotex services offer a broader range of electronic publication but are generally more expensive to use. Dialog, for example, the information retrieval service of Lockheed Corporation, charges roughly twice to three times the cost incurred with MEDLARS.

Unlike most European systems which have been developed with a view to integrating high-resolution color graphics, the original American Videotex operations were devoted to the transmission of simple text with rare rudimentary graphic garnish. However, a variety of standard Videotex protocols have been developed in several countries and an international standard is currently being negotiated in numerous committees of the International Standards Organization (ISO) [35,36]. Of the nine major commercial Videotex services now in operation in the United States, six provide some degree of graphics support.

A potentially revolutionary development in the marketing of database services is the advent of the videodisk. Capable of ultimately storing 20 Gbytes per side (current disks store roughly 500 Mbytes on a single side), it may prove more economical for the subscription service to distribute disks and updates in view of the expense of network connect time [37]. This would have the effect of relegating the packet switching networks to supporting communications services such as computer conferencing. However, such a vast amount of information can only be managed with sophisticated software and large-capacity operating systems. The present generation of personal computers may not be suited to the use of videodisk storage. This will rapidly change with the new generation of 16-bit machines and multitasking operating systems.

9. Expert Systems

In the discussion of computer mediated teleconferencing, the development of the Hepatitis Knowledge Base by the National Library of Medicine was described. The HKB was one product of a burgeoning line of inquiry aimed at the development of medical expert systems. Expert systems research is that branch of artificial intelligence which seeks to mimic the problem-solving behavior of experts. In a very real sense, this object belongs in the realm of telemedicine since our goal is to amplify expertise as well as to provide access to facilities independent of constraints of both time and space.

One of the first successful interactive consultant programs in medicine was MYCIN. It was designed to assist in determining therapy for patients with systemic infections [38]. MYCIN is a rule-based creation which catalogues the roughly 500 rules that deal with the diagnosis and treatment of bacterial blood infections and meningitis. In the April 1983 issue of Science, MYCIN's developer, Edward H. Shortliffe and a colleague, Richard O. Duda, describe the program's operation [39]:

> Thus MYCIN's strategy in rule selection is goal-oriented, and its inference method is to "reason backward" from its initial goal. It attempts to achieve any goal by applying all the directly relevant rules. ... When the program eventually requests some factual information from the user, the rule that prompted the request may become applicable and, if so, is applied. The application of a rule enters a new fact into the database. ... Thus the line of questioning, the rules that are applied, and the conclusions that are reached are determined by the data obtained for a particular patient.

The user of the program is presented with a series of questions concerning the nature and medical status of the patient. As the data is accumulated, the information is used to inflate a diagnostic database which is operated on by rules in a manner designed to converge towards a diagnosis. But the program operates within a framework of intelligence: as a rule is applied, the patient data base is examined to see if adequate information is available to successfully apply the rule. If not, the rule is either discarded or additional information is requested. The following is an example of a rule expressed in lyrical form [39]:

> If (i) the infection is meningitis and (ii) organisms were not seen in the stain of the culture and (iii) the type of infection may be bacterial and (iv) the patient has been seriously burned, then there is suggestive evidence that *Pseudomonas aeruginosa* is one of the organisms that might be causing the infection.

MYCIN was generalized to a version called EMYCIN (Essential MYCIN) and numerous other programs, ranging in scope from internal medicine diagnosis (INTERNIST-1), glaucoma assessment and therapy (CASNET), to pulmonary function test interpretation (PUFF). Attempts have also been made to develop expert medical systems with standard deterministic programming techniques. An example of this latter approach is the cancer data management system (CDMS) developed at the Boston University School of Medicine by R.H. Friedman and A.D. Frank [40].

CDMS is a "disease-oriented" program that was implemented for clinical oncology research and patient care and essentially uses the standard conditional IF-THEN computer statements. Thus, a physician or nurse can use the system to determine if a specific patient qualifies for any given treatment protocol. The entered data is acted upon by the system's condition dictionary to generate a description of the patient's status. An equivalent action dictionary then leads the interrogator towards an appropriate therapy.

Examples of computer-generated questions pertaining to the patient's static or changeable (dynamic) condition are [40],

> [static] IS THE PATIENT BEING TREATED FOR A BROKEN LEG?
>
> [dynamic] WAS THE PATIENT'S WHITE BLOOD COUNT BELOW 4000/ul AT ANY POINT DURING THE LAST FOURTEEN DAYS?

Hence the program can operate iteratively during the course of examination or treatment. Like MYCIN, CDMS will then apply rules to the condition dictionary such as [40],

> *if* THE PLATELET COUNT TODAY IS LESS THAN 100,000/ul, *then* REDUCE THE DOSE OF CYCLOPHOSPHAMIDE BY 25%.

Few expert systems are in routine use, despite the excellent record of some of them when compared with clinicians' performances. It must be emphasized, however, that the purpose of such systems is not to replace medical specialists, but rather to enlarge the access of the medical community to expertise in a carefully controlled clinical environment. Perhaps what is missing, is a communications link to enable users to discuss their findings with others. In the development of the Birth Defects Information System (BDIS), researchers at the Tufts-New England Medical Center and the Massachusetts Institute of Technology took care to incorporate a communications channel so that users could record comments and inquiries. This parallel human-communications link has played a significant role in the overall development of the system [41].

10. Summary and Conclusions

Telecommunications is a vast arena of technology and operations which cannot be swallowed whole. This is a pity. Because no single element can be viewed as a separate entity. There is a fillagree of interconnections between one realm and all others. Computer teleconferencing is an intimate of Videotex and, to some degree, of expert systems. The reason for this should be clear: as the volume and complexity of information increases, the management of that information--its transformation into knowledge--requires the intervention of rules and expertise.

The transfer of images over slow-scan video links is only a rudimentary first step in the processing of medical data. This, too, must be supported with additional expertise, information, and knowledge. The sharing of the new technology--NMR and PET scanners, digital radiographic images, and the unseen armamentarium of the great urban medical centers--with rural facilities and isolated communities, demands a new and innovative approach to the distribution of medical services.

Telemedicine (for want of a more descriptive term) has emerged as an *ad hoc* collection of technologies and disciplines which seeks to integrate the new diagnostic services, information and databases, communications, networks, expertise, health-care professionals and consumers into one transparent unity. If this can be done, it will have an incalculable impact on society. No community will be too isolated, no individual too poor to obtain rudimentary services. The ultimate promise is universally accessible medical care.

11. Acknowledgments

It is impossible to list all of the experts and innovators who have made substantial contributions to this developing field and who have been so generous with their advice and information. What follows is a partial list--in no particular order. And I apologize to all of those whom I may have inadvertantly omitted: David Conrath, Ken Phillips, Harry Stevens, Chris Bullen, Martin Elton, Glen Southworth, Virginia Ostendorf, Lynn Graham, Phyllis Wood, Stewart Meyer, Elliot Gold, Irving Wendorff, Maxine Rockoff, Paul Debaldo, Robert Cowan, Jim Schimpf, Jim Nelson, Judy Roberts, Earl Dunn, Charles Savage, Bob Thompson, Rick Irving, Lorne Parker, and Elliot Siegel.

[FIGURE 1A]: Diagram of a simple star network.

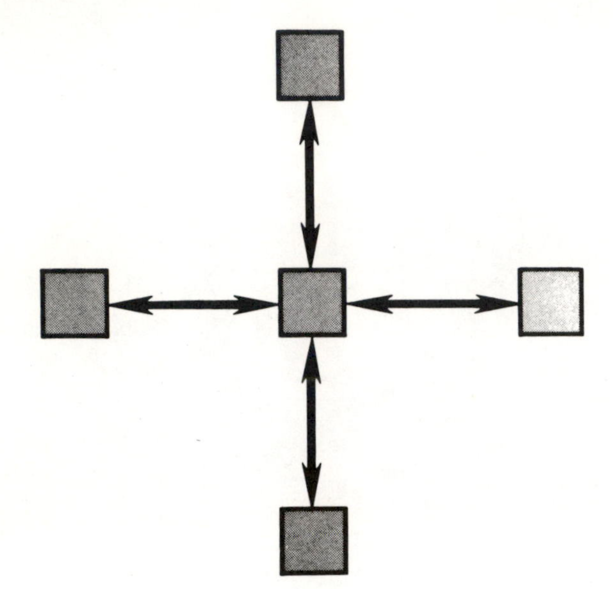

[FIGURE 1B]: Diagram of a simple distributed network.

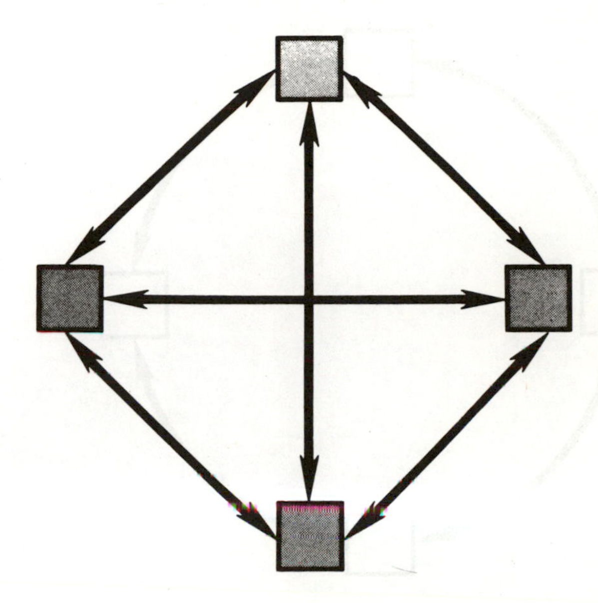

[FIGURE 1C]: Diagram of a simple ring network.

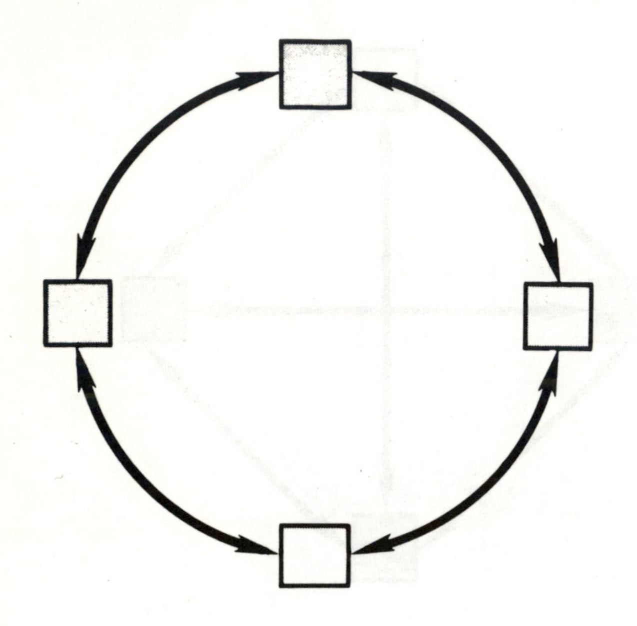

[FIGURE 1D]: Diagram of a simple unbranched bus network.

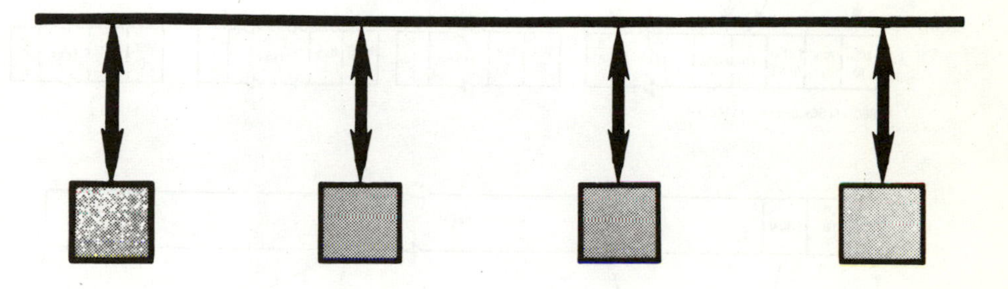

[FIGURE 2]: Diagram of the message-switching format (top) and packet-switching format (bottom). Taken from reference [10].

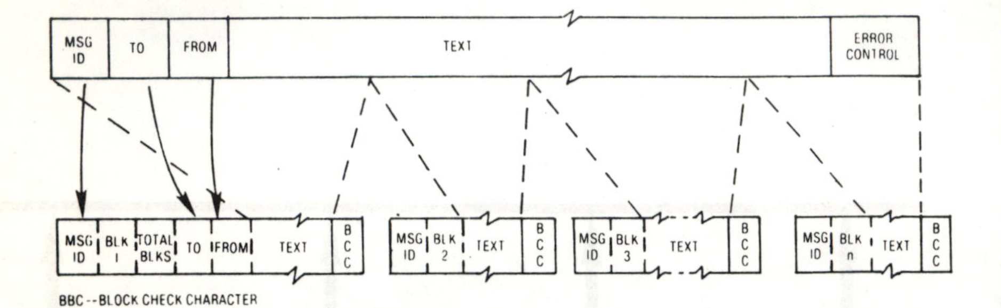

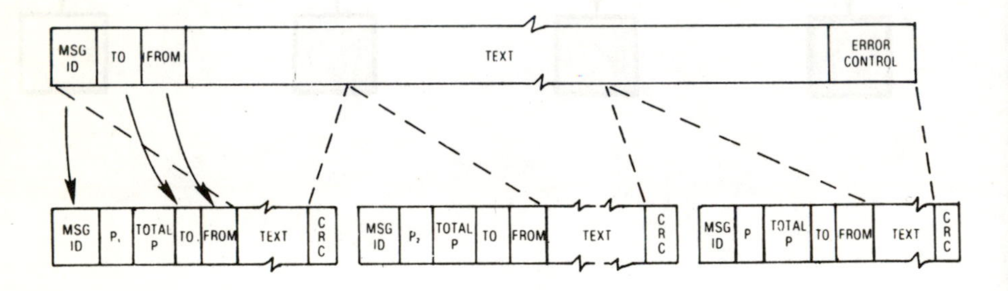

[FIGURE 3]: Geographic map of the ARPA Network. Taken from reference [42].

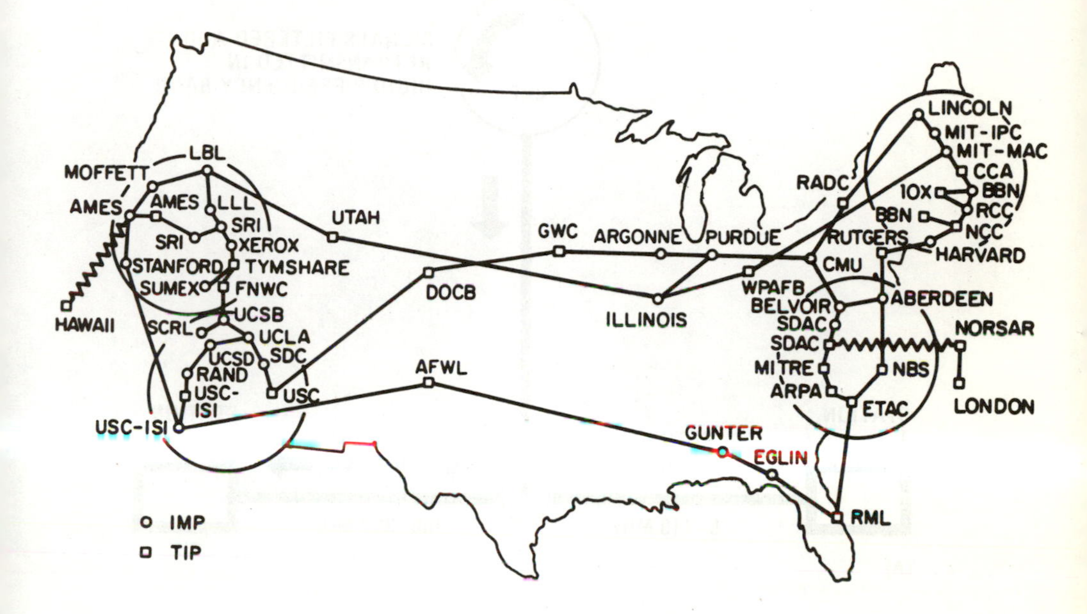

[FIGURE 4A]: Schematic of a broadband head-end facility, illustrating the bandwidth splitting for sending and receiving data. Taken from reference [12].

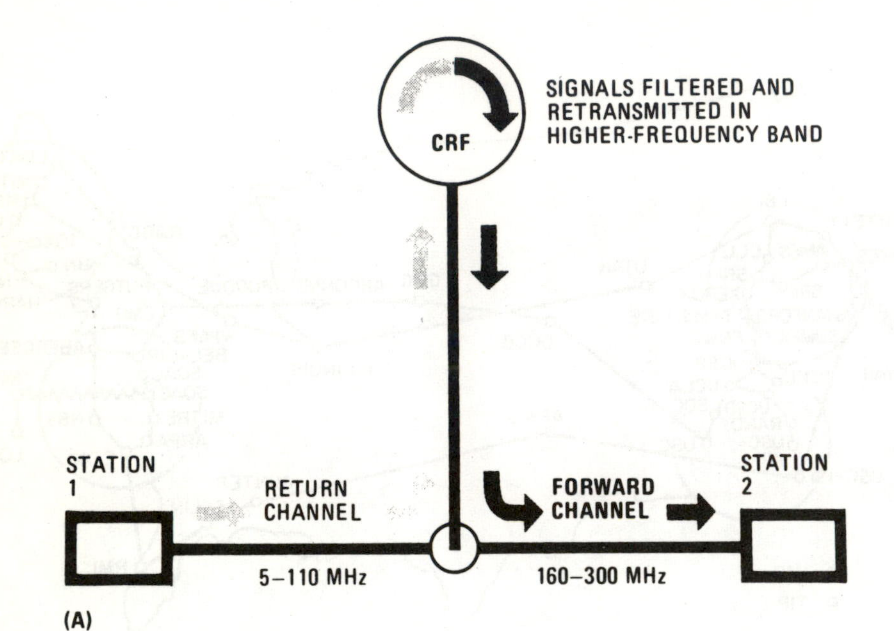

[FIGURE 4B]: Spectral assignments for the broadband forward and return channels. Taken from reference [12].

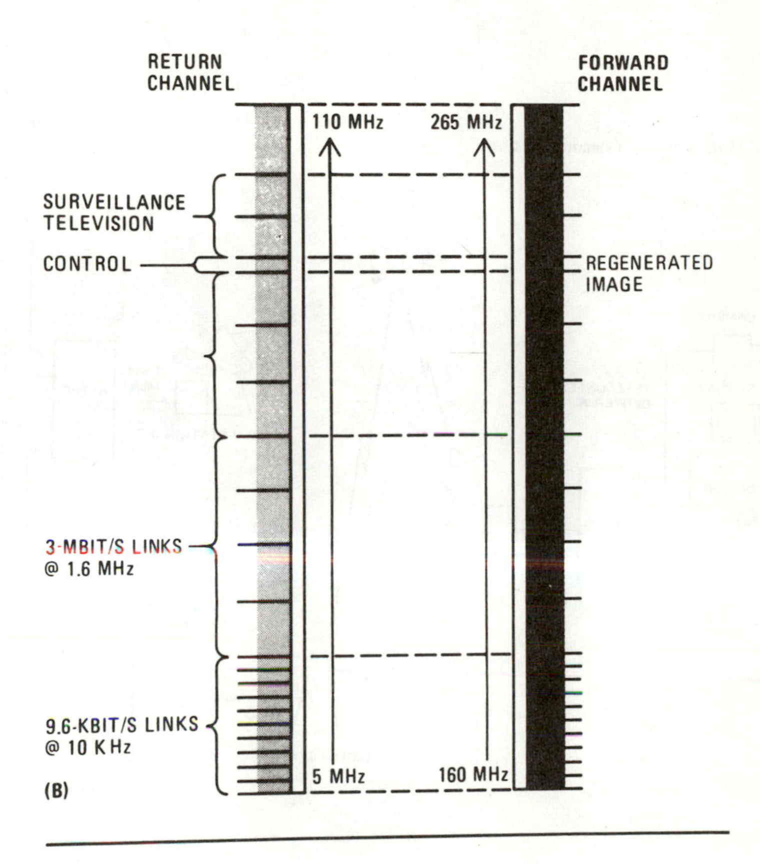

[FIGURE 5]: Schematic of a video teleconferencing network using satellite transmission between ground stations. Illustration courtesy of NEC America, Inc., Radio and Transmission Division.

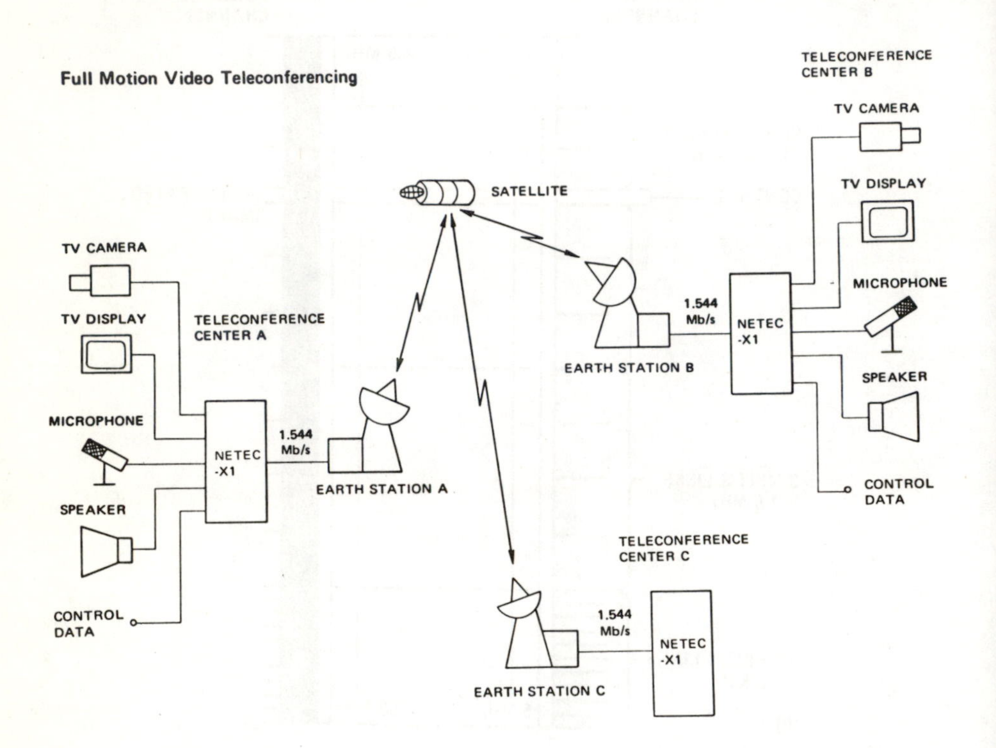

[FIGURE 6]: Schematic of a typical installation for a compressed-video send/receive station. Illustration courtesy of Widergren Communications, Inc.

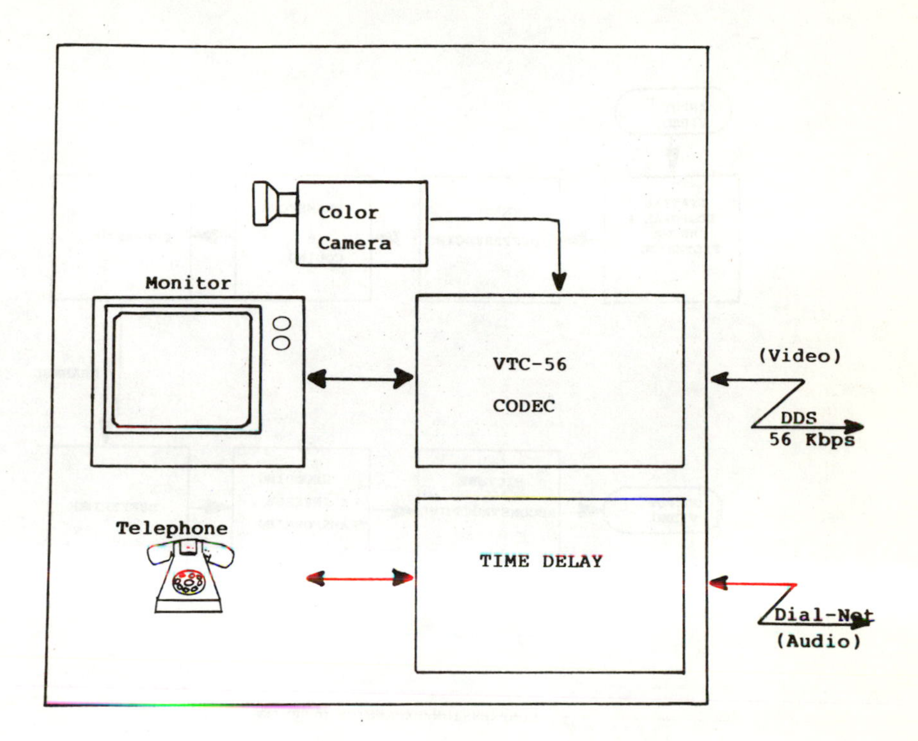

[FIGURE 7]: Schematic of the processing steps in a compressed-video network. Illustration courtesy of Widergren Communications, Inc.

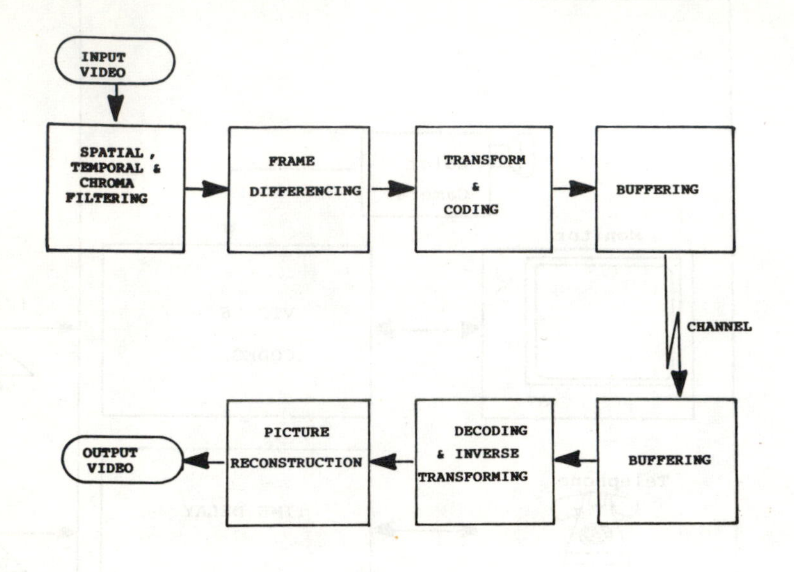

COMPRESSION/EXPANSION OVERVIEW

APPENDIX 1

EXAMPLES OF ELECTRONIC COMMUNICATIONS*

Form	Characteristics	Principal Suppliers
Electronic Mail	Letter and facsimile communication in either real-time (TELEX) or store-and-forward mode. There are two network configurations: centralized and distributed. The centralized system uses a computer to manage and distribute subscriber communications. The decentralized system relies on local or regional facilities to compose, store and forward messages. The communication network then performs distribution.	Centralized: Digital Equipment Corp's EMS; GTE Telenet's HERMES; Tymshare's ON-TYME; ARPANET's MSG; Mohawk Data Sciences' WINC. Decentralized: Xerox's ETHERNET; Public-switched telephone network; TYMNET; TELENET
Computer Conferencing	The focus is on the topic of many-to-many communications rather than on the person or address. Transactions may be in either real time or store-and forward. A central computer coordinates composition, distribution, and retrieval. In addition, a formal structure may be managed by the computer to control privacy, conference membership, branching of discussions, and special utilities such as graphics displays, computer modeling programs, report generation, and Delphi interrogation (questionnaire processing).	See Appendix 2 for a listing of commercial systems and their features.
Stand-Alone Computer Communications	Many mini- and mainframe computers have a limited capability for transferring information between termi-	TALK on DECSYSTEM 20. TOPES System (run on a DEC-10 with

	nals. Terminal linking is primarily in real-time and must be prearranged. Some limited store-and-forward capability exists, but without the conferencing software such communication is cumbersome. There is no protocol for managing communications flow. More than two-way correspondence is not practicable.	appropriate operating system).

*Derived in part from, The HUB Project, Computer-Based Support for Group Problem Solving, by Hubert Lipinski and Richard P. Adler, published by Institute for the Future, January 1982.

APPENDIX 2

COMMERCIAL CONFERENCING SERVICES*

Company	Availability	Processor	Connect Costs
PARTICIPATE (Participation System, Inc. 43 Myrtle Terrace, Winchester, Mass. 01890).	To subscribers of Source, accessible via TELENET or TYMNET. May ultimately be offered by DIALCOM and by the Canadian service, SAGE. May be licensed by users.	Prime, versions for other systems under development.	Subscriber hourly connect fees on Source (300 baud) vary from $20.75 during prime-time to $7.75 after business hours. 1200 baud rates are about 24% higher. Closed user groups of 25 may be established for a one-time fee of $5,000 (plus $100 each additional user) Licensing costs vary from $10,000 plus $5/hour/user to $60,000 plus $1/hour/user.
ELECTRONIC INFORMATION EXCHANGE SYSTEM (EIES, New Jersey Institute of Technology, 323 High St., Newark, N.J. (07102).	To subscribers, accessible via TELENET, Uninet, or direct-dial. Not available for licensing on user machines.	Perkin Elmer, Using a tailored operating system partially written in Interact, a language designed for communications.	Individual accounts, $75/month plus an hourly connect charge which varies with time of day, averaging about $8. Group accounts cost about $200/month plus $10/month per user plus the hourly connect charge.
NOTEPAD (Infomedia, 801 Traeger Ave., Suite 275, San	To subscribers, accessible via TYMNET. May be licensed for use on user machines.	DEC-20 (TOPS-20 operating system)	The "Business Intelligence" demonstration conference costs approximately $10/ hour. Closed user

Bruno, Cal. 94066).			group hourly charges are $60 plus TYMNET charges. There is a user start-up fee of $1,000. The one-time licensing fee is $35,000.
CONFER (Center for Learning and Teaching, University of Michigan, Ann Arbor, Mich. 48109).	To subscribers, accessible via TELENET. Not licensed for use on user machines.	Amdahl 580	There is only an hourly connect charge which varies somewhat with time of day--averaging about $18.
HUB (Institute for the Future, 2740 Sand Hill Road, Menlo Park, Cal. 94025).	Subscription service offered by National Information Systems, accessible via Uninet. May be licensed on user machines.	DEC-20 (TOPS-20 operating system)	The hourly connect charges vary according to time and other parameters but average $30-$40. Purchase price for commercial use is $40,000--a variety of leases available.
COM (Swedish National Defense Research Institute). Stockholm University Computing Center, Box 27322, S-10254 Stockholm, Sweden.	Subscription service available via the international packet-switching networks (TYMNET, TELENET, EURONET) or through a direct line. May be licensed on user machines.	DEC-10 and DEC-20, a new version not yet available, PORTACOM, is being implemented in Pascal on CDC Cyber, Digital VAX-11, Siemens B 2000, Burroughs 7800 and IBM/OS/VS1.	Purchase cost is $7,000 for universities and $14,000 for commercial implementation. Lease charges are $2,000/year for universities and $4,000/year for commercial use.
GENIE (Data Dynamics, Inc. Box 5517, Portland, Ore. 97228).	No subscription service (new users may rent machine time for a three month period). May be licensed for use on user	CDC Cyber 175 and DEC VAX-11/780	Three month rental is $5,000/month. CDC Cyber 175 system costs $65,000 and the DEC VAX-11/780 costs $35,000.

	machines.		
MTX (Matrix Transaction Exchange, Cross Information Co., Suite B, 934 Pearl Mall, Boulder, Colo. 80302).	To subscribers, accessible via TELENET. May be licensed for use on user machines.	DEC systems	50$ membership fee for subscribers, $25 hourly connect charge. Monthly lease fee begins at $1,000. Modular system available to interface with other computers.
AUGMENT (Tymshare, 20705 Valley Green Drive, Cupertino, Cal. 95014).	To subscribers, accessible via TYMNET. May be licensed for use on user machines.	Dec-20 (TOPS-20 operating system)	$14-$18 hourly connect charges. Special terminal leases available at $200 per month. Software costs dependent upon implementation.

*Derived from, "The Movable Conference," by Irving A. Lerch, Byte, Vol. 8, Num. 5, pp. 104-120, May 1983.

REFERENCES

[1] Bernstein, L.M., Siegel, E.R., and Goldstein, C.M., "The Hepatitis Knowledge Base: A Prototype Information Transfer System," Annals of Internal Medicine, Vol. 93 (Part 2), pp. 169-181, July 1980.

[2] Siegel, E.R., "Use of Computer Conferencing to Validate and Update NLM's Hepatitis Data Base," Electronic Communication: Technology and Impacts, edited by Henderson, M.M. and MacNaughton, M. J., AAAS Selected Symposium 52, pp. 87-95, Westview Press, Boulder, Colorado, 1980.

[3] Roderer, N.K., King, D.W., McDonald, D.D., and Bush, C.G., Evaluation of the Hepatitis Knowledge Base System, Report NLM-78-3, Lister Hill National Center for Biomedical Communications, National Library of Medicine, 249 pp., Bethesda, Maryland, October, 1981.

[4] Jelaso, D.V., Southworth, G., and Purcell, L.H., "Telephone Transmission of Radiographic Images," Radiology, Vol. 127, pp. 147-149, April, 1978.

[5] Steeg, R., "Slow-Scan TV Networking Decentralizes Medical Imaging," RNM Images, Vol. 13, Num. 4, pp. 38-39, April, 1983.

[6] For information concerning the Teleconference Network of Texas, write: Lynn E. Graham, Director, Teleconference Network of Texas, The University of Texas Health Science Center at San Antonio, 7703 Floyd Curl Drive, San Antonio, Texas 78284 [Telephone (512) 691-7291].

[7] For information concerning the Center for Interactive Programs, write: Lorne A. Parker, Director, Center for Interactive Programs, University of Wisconsin--Extension, Radio Hall, Madison, Wisconsin 53706 [Telephone (608) 262-4342].

[8] Crowther, W.R., Heart, F.E., McKenzie, A.A., McQuillan, J.M., and Walden, D.C., "Issues in Packet Switching Network Design," Computer Networks and Communication (The Information Technology Series, Volume IV), edited by Korfhage, R.R., pp. 181-195, American Federation of Information Processing Societies, Inc., Press, Montvale, New Jersey, 1978.

[9] Doll, D.R., "Telecommunications Turbulence and the Computer Network Evolution," Computer, Vol. 7, Num. 2, pp. 13-22, February, 1972.

[10] Greene, W., and Pooch, U.W., "A Review of Classification Schemes for Computer Communications Networks," Computer, Vol. 10, Num. 11, pp. 12-21, November, 1977.

[11] Seaman, J., "Local Networks: Making the Right Connection," Computer Decisions, pp. 123-158, June, 1983. NOTE: This article lists 78 suppliers of LAN equipment and systems including PABX vendors. The list was derived from a study of LAN's by Shotwell and Associates, 680 Beach Street, San Francisco, California.

[12] Dineson, M.A., and Picazo, J.J., "Broadband Technology Magnifies Local Networking Capability," Data Communications, Vol. 9, Num. 2, pp. 61-79, February, 1980.

[13] Pye, R., Champness, B., Collins, H., and Connell, S., The Description and Classification of Meetings, Communication Studies Group, University College, Paper P/73160/PY, London, England, 1973.

[14] Lipinski, H. and Adler, R.P., The Hub Project: Computer-Based Support for Group Problem Solving, R-51, 216 pp., Institute for the Future, Menlo Park, California, January, 1982.

[15] Hiltz, S.R. and Turoff, M., The Network Nation: Human Communica-Via Computer, Addison-Wesley Advanced Books, Reading, Massachusetts, 1978.

[16] Lerch, I.A., "The Movable Conference," Byte, Vol. 8, Num. 5, pp. 104-120, May, 1983.

[17] Barney, C. and Cross, T.B., "The Virtual Meeting: A Report on Computer Conferencing," In Depth, pp. 1-16, September 28, 1982.

[18] Stevens, C.H., Many-to-Many Communication, Center for Information Systems Research No. 72, Sloan WP No. 1225-81, 55 pp., CISR, Massachusetts Institute of Technology, Sloan School of Management, Cambridge, Massachusetts, June, 1981.

[19] Gretz, P. and Wolak, R., "Electure: Just a Hobby or a Wave of the Future?" Unpublished Report, available from Christine V. Bullen, Assistant Director, Center for Information Systems Research, Massachusetts Institute of Technology, 77 Massachusetts Avenue, Room E40-197, Cambridge, Massachusetts 02139 [Telephone (617) 253-2930], December 13, 1982.

[20] There are several collections of articles which summarize the ARPANET experience. Two useful references are,

Abrams, M., Blanc, R.P., and Cotton, I.W., editors, Computer Networks: Text and References for a Tutorial Revised and Updated 1980, IEEE Computer Society Catalog No. 297, EH0162-8, IEEE Computer Society Publications Office, Long Beach, California, 1980 Revised and Updated Edition.

See reference [8] for Computer Networks and Communication.

[21] Broad, W.J., "Journals: Fearing the Electronic Future," Science, Vol. 216, Num. 28, pp. 964-966, 968, May, 1982.

[22] Elton, M.C.J., Teleconferencing: New Media for Business Meetings, American Management Associations management briefing, 57 pp., AMACOM, New York, NY, 1982 (adopted from "The Practice of Teleconferencing," published in Telecommunications in the United States: Trends and Policies, edited by Lewin, L., Artech House, Dedham, Massachusetts, 1982).

[23] AT&T, Acoustics of Teleconferencing Rooms, Technical References PUB 42901 and 42903.

[24] Ruedisueli, R.W., "Audio Teleconferencing: Transmission Systems and Considerations," Technical Design for Audioconferencing, Gilbertson, D.A. and Riccomini, B., editors, University of Wisconsin-Extension, Madison, Wisconsin, 1978.

[25] Steckel, R.J., "Daily X-Ray Rounds in a Large Teaching Hospital Using High-Resolution Closed-Circuit Television," Radiology, Vol. 105, pp. 319-321, November, 1972.

[26] Gayler, B.W., Gitlin, J.N., Rappaport, W., Skinner, F.L., and Cerva, J., "Teleradiology: An Evaluation of a Microcomputer-Based System," Radiology, Vol. 140, pp. 355-360, August, 1981.

[27] Rappaport, W.H., Skinner, F.L., Gayler, B.W., A Laboratory Evaluation of Teleradiology, MTR-8028, The MITRE Corporation, McClean Virginia, March, 1979.

[28] Andrus, W.S., Dreyfuss, J.R., Jaffer, F., and Bird, K.T., "Interpretation of Roentgenograms Via Interactive Television," Radiology, Vol. 116, pp. 25-31, July, 1975.

[29] Page, G., Gregoire, A., Galand, C., Sylvestre, J., Chahlaoui, J., Fauteux, P., Dussault, R., Seguin, R., and Roberge, F.A., "Teleradiology in Northern Quebec," Radiology, Vol. 140, pp. 361-366, August, 1981.

[30] Dunn, E., Conrath, D., Acton, H., Higgins, C., and Bain, H., "Telemedicine Links Patients in Sioux Lookout with Doctors in Toronto," CMA Journal, Vol. 122, pp. 484-487, February 23, 1980.

[31] Roberts, J.M., House, A.M., and Canning, E.M., "Comparison of Slow Scan Television and Direct Viewing of Radiographs," Journal de L'Association Canadienne des Radiologistes, Vol. 32, pp. 114-117, June, 1981.

[32] Dogliotti, R., Garibotto, G., and Tamburelli, G., "Telemedicine: State of the Art and Results on Torino-Susa Experimental Link," CSELT Rappaporti Tecnici, Vol. VIII, Num. 1, 4 pp., March, 1980.

[33] Garibotto, G., Garozzo, S., and Micca, G., "Some Test Results on Remote X-Ray Diagnosis by using an Experimental Telemedicine System," Medical Informatics Symposium, Toulouse, France, May, 1980. Available from CSELT, Via G. Reiss Romoli, 274-Torino, Italy, C.A.P. 10148.

[34] Gold, E.M., editor, "Compression Labs Demonstrates T1 Codec for Video Conferencing," The TeleSpan Newsletter, Vol. 1, Num. 5, pp. 1-9, September 15, 1981.

[35] Miller, D., "Videotex: Science Fiction or Reality," Byte, Vol. 8, Num. 7, pp. 42-56, July, 1983.

[36] The February, 1983, edition of Byte (Vol. 8, Num. 2) contains a series of introductory articles on computer-communications standards. Of particular note are,

Card, C., Prigge, R.D., Walkowicz, J.L., and Hill, M.F., "The World of Standards: The Process for Producing Standards is Full of Checks and Balances," pp. 130-142. Abstracted in part from, The World of EDP Standards, 3rd Edition, (same authors), Sperry Univac Corp., Blue Bell, Pennsylvania, 1978.

Witten, I.H., "Welcome to the Standards Jungle," pp. 146-178.

Flemming, J., and Frezza, W., "NAPLPS: A New Standard for Text and Graphics," pp. 203-254.

[37] Sieck, S.K., "Database Publishing Applications of Optical Videodisks," presented at the Fourth National Online Meeting, New York, NY, April 12-14, 1983. Published in the Proceedings by Learned Information, Inc., Medford, New Jersey, pp. 499-502, April, 1983.

[38] Shortliffe, E.H., "Computer-Based Medical Consultations: MYCIN," Artificial Intelligence Series, No. 2, Elsevier, New York, 1976.

[39] Duda, R.O. and Shortliffe, E.H., "Expert Systems Research," Science, Vol. 220, Num. 4594, pp. 261-268, April 15, 1983.

[40] Friedman, R.H. and Frank, A.D., "Use of Conditional Rule Structure to Automate Clinical Decision Support: A Comparison of Artificial Intelligence and Deterministic Programming Techniques," Computers and Biomedical Research, Vol. 16, pp. 378-394, 1983.

[41] Edwards, C.N. and Buyse, M.L., "The Making of BDIS: Marketing an Online Medical Information and Decision Support System," presented at the Fourth National Online Meeting, ibid, pp. 119-125.

[42] Walden, D.C., "Experiences in Building, Operating, and Using the ARPA Network," pp. 75-80, ibid reference [8].

PROCEEDINGS OF THE II INTERNATIONAL CONFERENCE ON APPLICATIONS OF PHYSICS TO MEDICINE AND BIOLOGY
edited by Ž. Bajzer, P. Baxa & C. Franconi

ADVANCES IN CLINICAL EVALUATION OF THE BIOMAGNETIC METHOD

Gian Luca Romani, Roberto Leoni and Ivo Modena+

Istituto di Elettronica dello Stato Solido - C.N.R.

Via Cineto Romano 42, 00156 Roma

ITALY

ABSTRACT

The most significant advances achieved in the clinical evaluation of the biomagnetic method during the last two years are presented. In particular results will be discussed concerning the study of the human heart conduction system, the localization of sources of normal brain rhythms and of epileptic foci.

1. Introduction

During the last few years the biomagnetic method has featured impressive ability to localize souces of physiological and pathological bioelectric activity in the human body (1). This has been due to continuous improvement of the instrumentation used for measuring magnetic fields associated with these biological phenomena and, at the same time, to the parallel development of adequate models and computation procedures for interpreting the measured field distributions. Nowadays complete instruments for biomagnetic investigation are commercially available. Some of them provide high quality performance also in unshielded hospital environments (2). The configuration of these instruments follows the standard setup already described elsewhere (3,4). In this paper, therefore, we will not discuss details of the instrumental state of art. We can anticipate, however, that the new commonly followed approach consists in developing multi-channel apparatuses and using the last generation of microfabricated superconducting magnetometers which is supposed to provide sensitivity adequate to the task and permits the required assembly integration. On the contrary, the practical approach to the modeling problem will be discussed to some extent, with particular emphasis to the cases concerning the clinical investigations on brain and heart presented in the specific sessions.

2. Modeling

The problem of identifying electrical sources in the human body can be faced in two ways, namely the forward and the inverse problem. The forward problem is to calculate the magnetic field and/or the electric potential distribution at the surface of a volume conductor generated by a configuration of sources of known strength and position. The inverse problem consists in identifying the sources in the volume conductor from the experimental distribution of magnetic fields and potentials at the surface. The latter problem has no unique solution unless some specific hypotheses on the number and structure of sources and on the physical properties of the volume conductor are made. In this case an approximate solution can be obtained by means of appropriate iterative numerical minimization procedures.

One of the most used sorce model to account for neural activity is the current dipole, namely a current flow concentrated in a very small volume. The use of such a simple model does not mean, however, to oversimplify the problem: the idea of schematically representing the activity of populations of neurons by means of an "equivalent" dipole provides a tool mathematically accessible and sufficiently realistic. Fig.1 shows the current lines generated by a current dipole immersed in a homogeneously conducting infinite medium. The black arrow represents the "primary" current, i.e. the intracellular ionic source; the outer thin lines are the "volume" currents which, flowing in the surrounding medium close the loop. The heavy-typed transverse circles are the magnetic field lines. Some interesting properties arise when the dipole is placed in a medium with particular geometry: a homogeneously conducting half space, a sphere with homogeneous conductivity or even with conductivity homogeneous in concentric shells. In all these cases the contribution from volume currents to the component of the magnetic field perpendicular to the medium surface is zero (5,6). This is a crucial point as it shows that a magnetic measurement can provide direct information on the primary source, i.e. the intracellurar one. The second point is that only the component of the dipole tangential to the surface produces external magnetic field, while the same is not true for the electrical counterpart (1). Furthermore, the distribution of the component of the magnetic field normal to the surface features particular symmetry properties which simplify the source localization problem, even if sometimes we pay the penalty of magnetically silent configurations.

3. Cardiomagnetism

Biomagnetic investigation of heart physiology and pathology carried out during the last years identified two phenomena for which the clinical contribution of this new approach was significant: the temporal evolution of the His-Purkinje system (HPS) activation (7-10), and the abnormally delayed ventricular depolarization (late potentials) (11). In particular, interesting results have been already obtained in the noninvasive study of the heart conduction system, and will be briefly resumed here. Work on item 2 is in progress in the biomagnetic shielded room at the Physicalisch Technische Bundesanstalt (PTB) Institut in West Berlin and important findings are likely to be gathered in the next future.

Electric surface mapping technique has been used to attempt an inverse solution for the problem of the HPS source localization. Alternatively, magnetic mapping of signals occurring during the PR segment of the cardiac cycle can provide information on the temporal evolution of the HPS activation as well as on the structure of the system itself. An example of an experimental high resolution magnetocardiographic map of a normal subject is shown in fig.2. The map, recorded in the unshielded laboratory at the Istituto di Elettronica dello Stato Solido (IESS) in Rome, contains averaged signals measured at 48 positions of a chest grid (7,12). Apart from information concerning timing - the onset of the "ramp" is commonly related to the activation of the common His bundle - other significant observations can be inferred from this map. Fig.3 shows four isofield contour maps depicting the magnetic field distribution at four successive time instants of the PR segment: a=-38ms, b=-20ms, c=-7ms and d=0ms before ventricular activation (see central insert). The experimental distribution satisfactorily agrees with a dipolar pattern in all cases and will be furtherly used for source localization. These maps were obtained chosing as a reference field distribution that 4ms before of instant a). We want to point out that the choice of the appropriate reference is of crucial importance, in that different contributions to the measured fields can be identified. It has been long discussed whether or not atrial repolarization could significantly modify the signals detected during the PR segment. The presence of an almost steady distribution of high amplitude can be evidenced by chosing as a reference the average distribution occurring during the last 10ms before P wave onset. This portion of the cardiac cycle is commonly believed to be electrically and magnetically silent. Fig.4 shows four isofield maps approximately corresponding to the time instants of fig.3 but obtained on the basis of the pre-P reference: the relatively strong, constant pattern can be interpreted in terms of atrial repolarization dominating on the much weaker HPS activity.

On the basis of the isofield maps shown in fig.3 it is possible to localize the four equivalent dipolar sources which represent the successive propagation of the activation impulse along the HPS. The dipole-in-half-space model identifies the three-dimensional location of the sources (7). We remark that only the component of the equivalent dipole parallel to the boundary surface can be localized. However, information on the 3-d evolution of the system can be inferred from the identification of the equivalent source in each map. It is evident that the described approach is still demanding significant improvements: a better discrimination between various contributions to the measured signals; to account for intrinsic and induced (breathing) heart movements; to consider the presence of inhomogeneities in the actual torso. All these things and probably some others presently limit the quality of the method which, however, already show promise of important applications in the clinical field.

4. Normal spontaneous brain activity

The strongest electric cerebral activity, the alpha rhythm, has been long studied but still doubts exist on the structure and location of its generator(s) (13). The electroencephalographic recording shows strongest alpha potentials over the occipital region of the scalp. Furthermore their amplitude is reduced by visual input, thus supporting the idea that the current sources lie in the occipital cortex. Recently performed neuromagnetic measurements (14) pointed out the actual problem experimentalists are faced with when measuring spontaneous brain activity by means of a single magnetic sensor: the lack of simultaneity, which prevents from getting information on the phase of the measured signals. In order to overcome this problem, a new approach has been developed at the IESS (15). It consisted in simultaneously measuring the MEG and the standard 10-20 system EEG. One EEG channel was used as reference for each of the successively recorded magntic positions, by calculating the covariance between the MEG and the reference. The fundamental idea was that the covariance was positive for in-phase MEG and EEG, and negative for out-of-phase signals. Variation in the covariance should reflect the variation in the alpha MEG at various recording sites, and, to compensate for changes in source intensity the authors divided the calculated covariance by the variance of the electric reference. In this way the Relative Covariance (RC) was obtained, the scalp distribution of which reflected that of the magnetic field.

By applying this procedure to the study of two normal subjects it was possible to get isocovariance contour maps like the one shown in fig.5. The overall dipolar distribution was interpreted as due either to a quite deep, single dipole, or to two simultaneously acting shallower sources. Chapman et al. (15) preferred the latter solution on the basis of four arguments: i) for a single dipole the fastest rate of change of normal magnetic field occurs above the dipole, and this is not the case for the measured data (see fig.5); ii) the brain is divided in two hemispheres; iii) the amounts of alpha MEG on the two sides were selectively suppressed by stimulating left and right visual field separately; iiii) the depth of a single dipole, to account for the observed pattern and field amplitude, would have had to be larger than 6cm beneath the scalp: the corresponding source strength would have been too large to be interpreted in terms of neural currents. All these considerations and a first approximation two-dipole fit lead the author to conclude that two sources, sinchronously acting, and located at a tentative depth of 5cm could be responsible for the major portion of the occipital alpha rhythm. On the basis of an improved data elaboration procedure we could successively perform an accurate best fit (chi-square test) and confirm the conclusions of the authors of ref.15 (see fig.6). This new approach seems particularly promising to study other spontaneous brain rhythms and functions, and has proved to be successful also in pathological situations as shown in the next paragraph.

5. Localization of epileptic foci

It is well known that the investigation of pathological spontaneous activity by means of the neuromagnetic method seem particularly powerful in that generally permits localizing areas of the scalp where abnormal magnetic signals, connected to focal epilepsies, are relatively confined (16). This sort of surface analysis, however, did not allow three-dimensional localization of the epileptic focus. Recently Barth et al. (17) proposed and tested an averaging procedure which proved to be successful in two cases where an high voltage, sustained electrical activity accompanied the magnetic one. These authors could average magnetic signals successively recorded at different sites of the scalp using the simultaneously detected EEG as a trigger, and succeed in localizing the underlying focus. Unfortunately only a limited population of epileptic patients fits the strong requirement on the sustained high voltage electric activity. Quite recently Chapman et al. (18) applied the RC method (see previous paragraph) to the analysis of MEG and EEG detected in a few cases of focal epilepsy. This method permits studying more general situations, where the EEG shows low amplitude signals, not useable as trigger. Chapman et al. first demonstrated the equivalence between the averaging procedure and the RC one, in one case of right temporal lobe epilepsy, which allowed both kind of analyses. Indeed they got quite similar localizations, identifying the source site approximately in the anterior part of the second temporal gyrus at a depth of 2.3cm. On the basis of this "a posteriori" confirmation of the goodness of the approach Chapman et al. presented other cases which could be studied only by the RC method. One example is shown in fig.7. In this case the patient was affected by diffuse occipital epilepsy, with left side dominance; the CAT analysis identified a deformation of the left occipital lobe. The localization obtained by the RC procedure confirmed that the epileptic focus was close to the abnormal brain region, at a depth of about 3cm below the scalp. Results obtained from a wider population of patients support the idea that this method will be soon successfully used in the clinical routine work.

6. Conclusions

The present state of art ensures sufficiently accurate degree of localization, particularly for cerebral sources. The major drawback is the non-simultaneity of magnetic recording at different sites: this causes both lengthy measuring sessions, often unbearable for the patient, and the need for specific procedures to compare successively recorded signals (i.e. the RC method). As a consequence, many interesting cases - for instance of focal epilepsies where an abnormal magnetic activity is not accompanied by an electric counterpart (16) - cannot be presently studied. Only the development of multichannel magnetic probe will overcome this problem and, if coupled to an appropriate data elaboration system, yield a new means for a "functional" imaging of the human body.

REFERENCES

1) Williamson S.J. and Kaufman L.: J. Magn. Magn. Mat., 22, 129 (1981).

2) Fenici R.R. in: Biomagnetism, an interdisciplinary approach (S.J. Williamson, G.L. Romani, L. Kaufman and I. Modena Eds., Plenum, New York, 1983), p.285.

3) Romani G.L., Williamson S.J. and Kaufman L.: Rev. Sci. Instrum., 53, 1815 (1982).

4) Romani G.L., Modena I. and Leoni R. in: Proc. of the Intern. Conf. on Applications of Physics to Medicine and Biology (G. Alberi, Z. Bajzer and P. Baxa Eds., World Scientific, Singapore, 1983), p.187.

5) Grynszpan F. and Geselowitz D.B.: Biophys. J., 13, 911 (1973).

6) Cuffin B.N. and Cohen D.: IEEE Trans. Biomed. Eng., BME-24, 372 (1977).

7) Fenici R.R., Romani G.L., and Leoni R.: Il Nuovo Cimento, 2D, 280 (1983).

8) Patrick J.L., Hess D.W., Tripp J.H. and Farrell D.E.: Il Nuovo Cimento, 2D, 255 (1983).

9) Erné S.N., Fenici R.R., Hahlbohm H.-D., Masselli M., Lehmann H.P and Trontelj Z.: Il Nuovo Cimento, 2D, 248 (1983).

10) Leifer M., Capos N., Griffin J. and Wikswo J.: Il Nuovo Cimento, 2D, 266 (1983).

11) Erné S.N., Fenici R.R., Hahlbohm H.-D, Jaszczuk W., Lehmann H.P. and Masselli M.: Il Nuovo Cimento, 2D, 340 (1983).

12) Saarinen M. and Siltanen P.: I. Morphology Ann. Clin. Res., 10, 21 (1978).

13) Andersen P. and Andersson S.A.: Physiological basis for the alpha rhythm (Appleton, New York, 1968).

14) Carelli P., Foglietti V., Modena I. and Romai G.L.: Il Nuovo Cimento, 2D, 538 (1983).

15) Chapman R.M., Ilmoniemi R.J., Barbanera S. and Romani G.L.: submitted for publication.

16) Modena I., Ricci G.B., Barbanera S., Leoni R., Romani G.L. and Carelli P.: Electroenceph. Clin. Neurophysiol., 54, 622 (1982).

17) Barth D.S., Sutherling W., Engel J. and Beatty J.: Science, 218, 891 (1982).

18) Chapman R.M., Romani G.L., Barbanera S., Leoni R. Modena I., Ricci G.B. and Campitelli F.: Lett. Nuovo Cimento, 38 (1982).

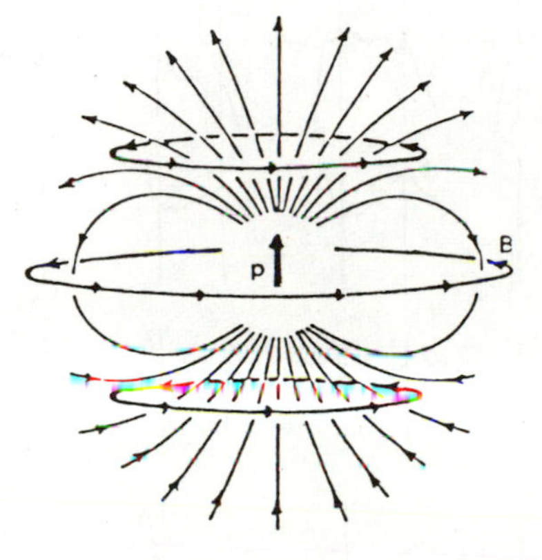

Fig. 1 Current pattern associated with a current dipole P. Magnetic field lines B are circles centered on the dipole direction.

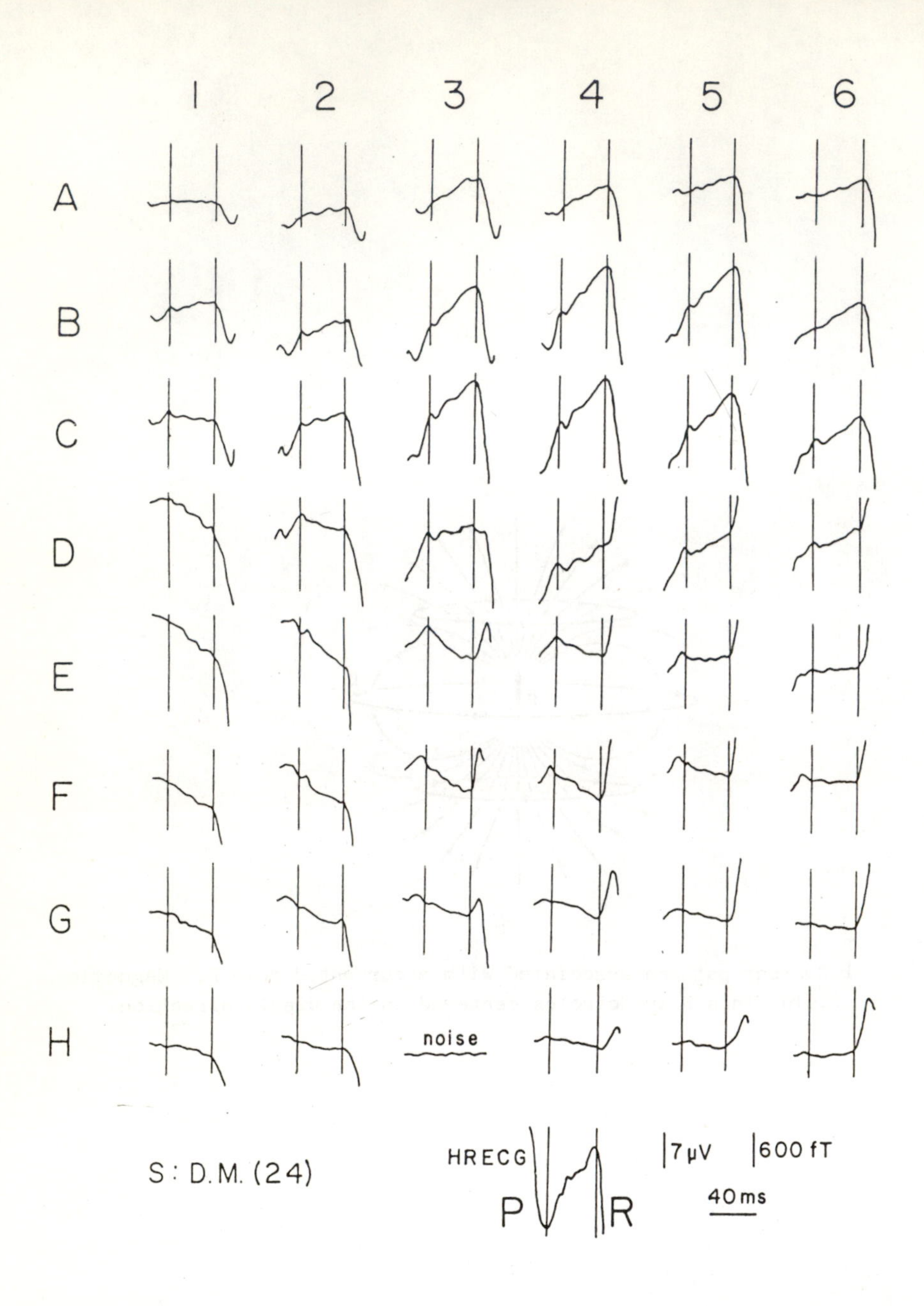

Fig. 2 Experimental high-resolution magnetocardiographic map of a normal subject. The corners of the grid correspond to the two midclavicular points and to intersections of the lowest ribs with the midclavicular lines. The bandwidth was 0.1-150 Hz.

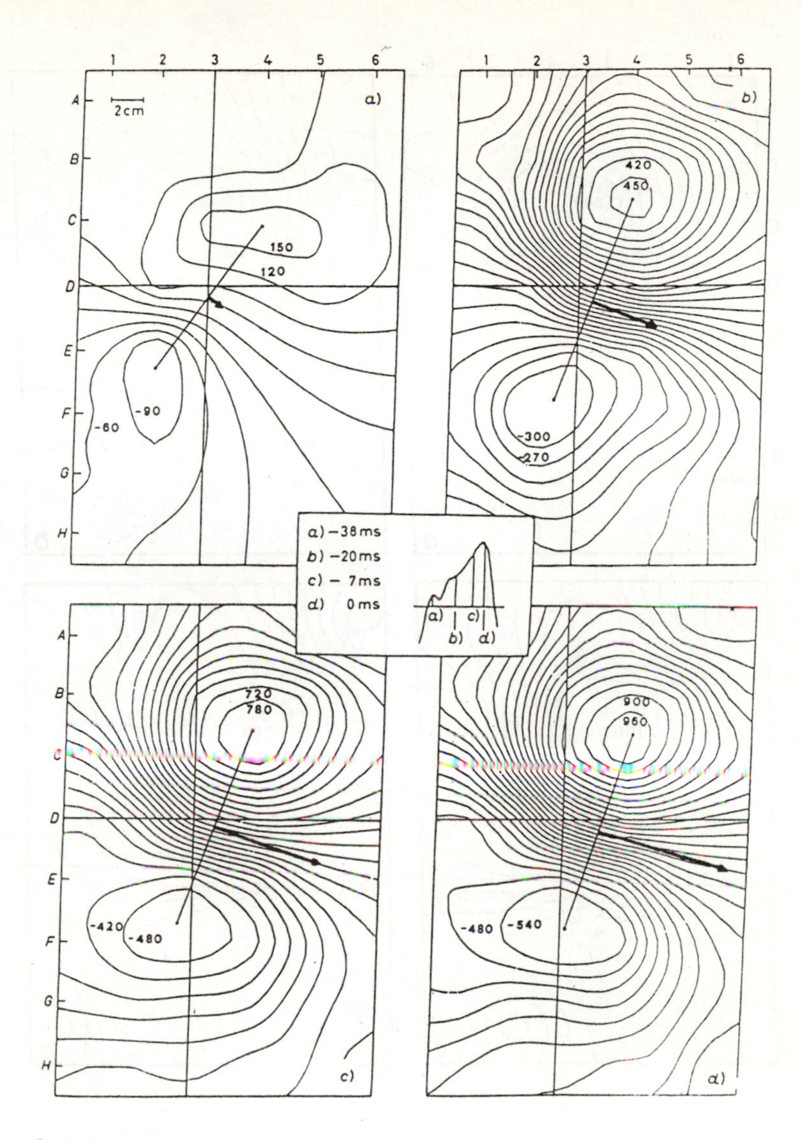

Fig. 3 Magnetic isofield contour maps obtained from the experimental map of Fig.2. The four maps correspond to the four time instants indicated on the centre of the figure. The maps were obtained choosing as reference field distribution the 4 ms before the instant a). The displacement from right to left of the equivalent dipole (black arrow) is evident. The figures on the isofield contour lines represent the values of the magnetic field amplitude in fT.

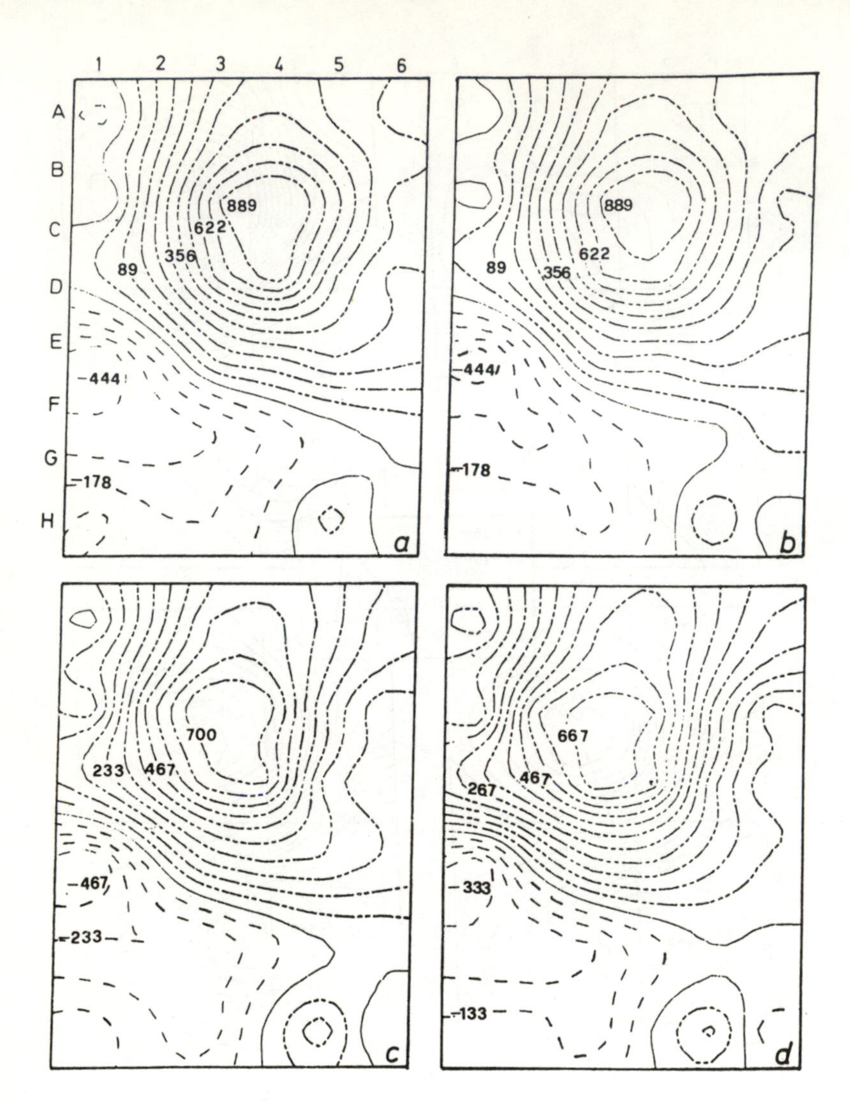

Fig. 4 Magnetic field contour maps corresponding approximately to the time instant of Fig.3, but obtained on the basis of the pre-P reference. It is evident the presence of an almost steady distribution that can be interpreted in terms of atrial repolarization. The figures on the isofield contour lines represent the values of the magnetic field amplitude in fT.

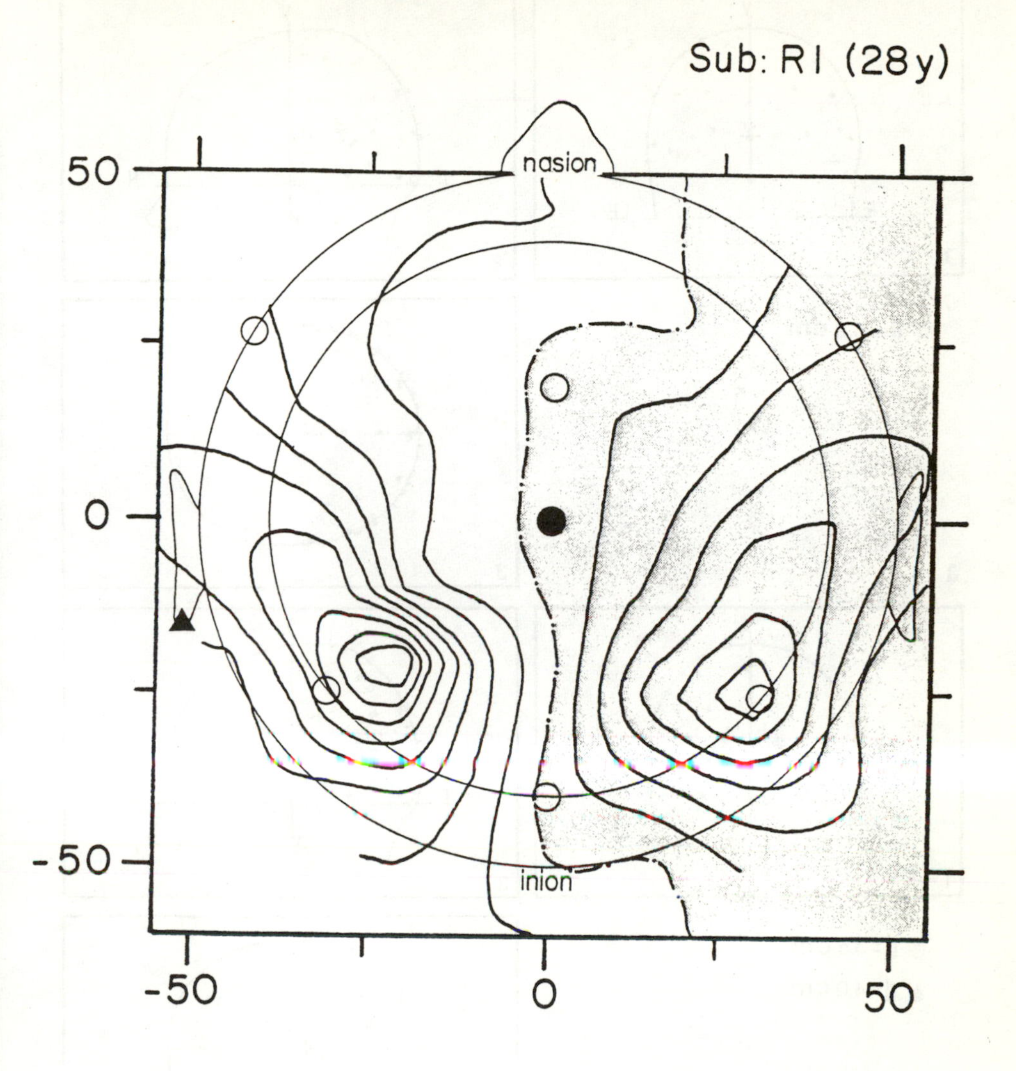

Fig. 5 Magnetic isofield contour map of alpha brain activity based on Relative Covariance method. Filled circle and filled triangle (left ear lobe) indicate the electrodes for these Relative Covariance measures.

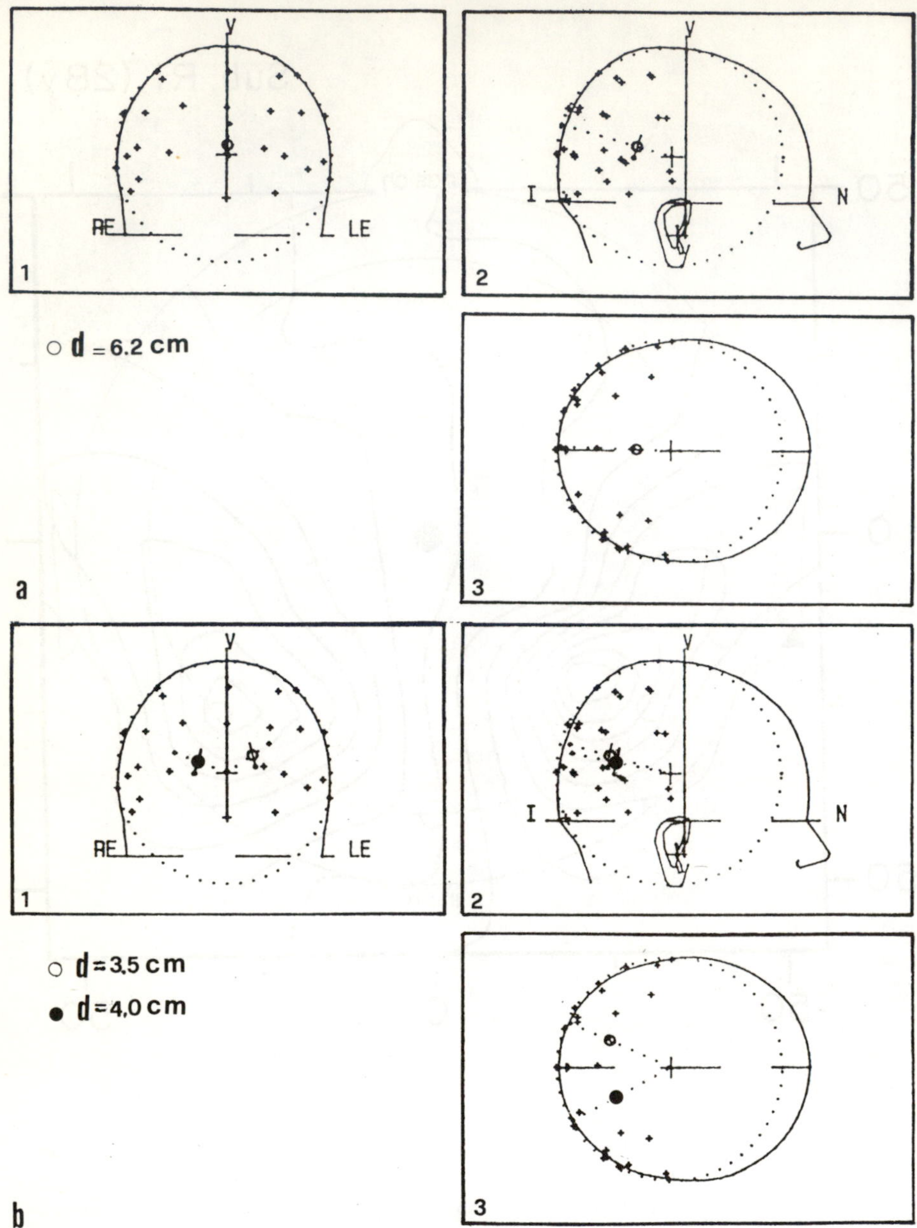

Fig. 6 Best fit localization for data shown in Fig.5 using 1 dipole (a) or 2 dipoles (b) model configuration. The three sections correspond to the sagital,vertical (frontal view) and horizontal maximum section of the actual head of the subject. Crosses are the experimental points used for the fit. The radius of the best sphere is 9.1 cm. d represents the depth of the localized dipolar source.

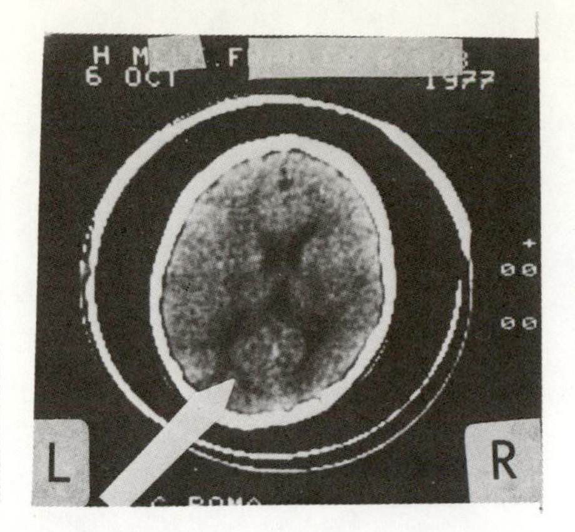

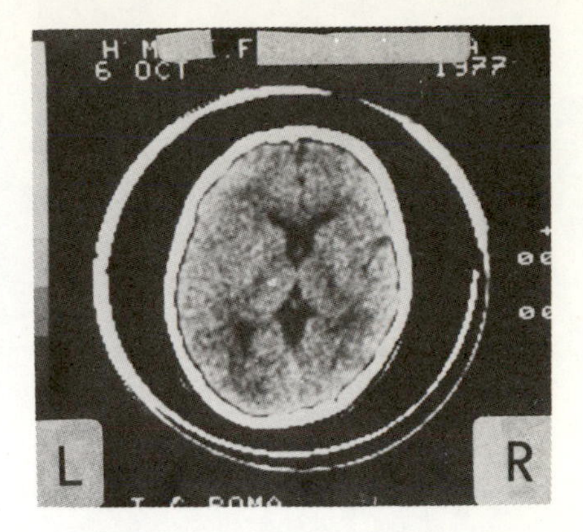

a

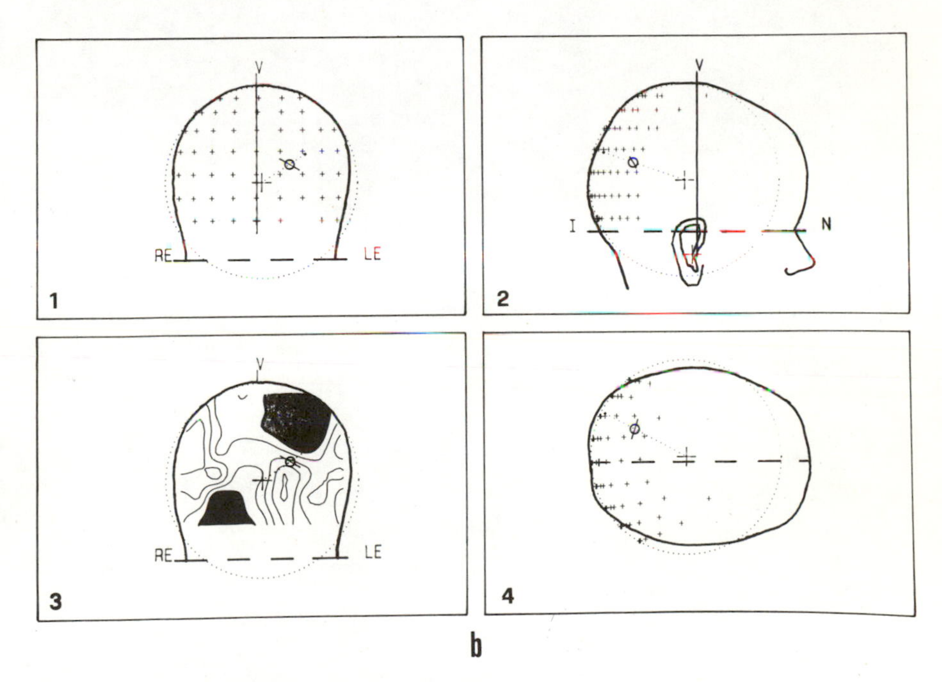

b

Fig. 7 (a) Computerized Tomography (CT) of subject FM. The intracerebral lesion is evident. (b) 3-d localization on the basis of the distribution of the Relative Covariance reported in b3. The localization of the source (depth= 3cm and sphere radius = 8.2 cm) well agrees with the results of CT scan (white arrow). Note the different left-right view of the CT picture with respect to the vertical projection.

CONTRIBUTED PAPERS

Physiological Models and Measurements

Radioisotopes

PROCEEDINGS OF THE II INTERNATIONAL CONFERENCE ON APPLICATIONS OF PHYSICS TO MEDICINE AND BIOLOGY
edited by Ž. Bajzer, P. Baxa & C. Franconi

TRANSMISSION LINE MODEL OF THE ARTERIAL SYSTEM

Fabrizio Locchi
Istituto Med. Nucl. e Fisica Medica - Università degli Studi
V.le Morgagni 85 - 50134 Firenze
ITALY

1. Introduction

We report an improvement of a model (of the systemic circulation) already developed[+]. The correspondence "artery-transmission line" is the basic assumption whereas the continuous variation of the arterial properties is seen as a second order approximation and here neglected. At the present the model consists of three transmission lines.

2. Methodology

Since the model is linear and time-invariant a signal at any point of the lines can be predicted from another "starting" one by means of the convolution product.This is performed as the product between the signal transform (FFT algorithm) and the $H(\omega)$-transfer function theoretically drawn (and numerically evaluated) on the basis of the line characteristic parameters. $H(\omega)$ is then low-pass filtered.

The matching of two predicted signals to the experimental data (performed by numerical non-linear least square fitting techniques) allows the system identification and thus the determination of several hemodynamic quantities (also "internal", such as the input impedance $Z_{in}(\omega)$, although the pressure signals are non-invasively taken).

In the present version of the model, to improve the $Z_{in}(\omega)$ expression we have simulated the lower limb arteries by two lines and introduced the "leakage" parameters (simulated by shunt conductances (G_i)). As a consequence $H(0)\neq 1$, and signal normalization parameters (N_i) have been also introduced because the experimental pressure-signals are taken by piezoelectric transducer and their amplitudes are uncalibrated. These two "lower" lines are supposed to be equal so that the definitive parameter set for the fittings is: $R_1, L_1, C_1, G_1, R_2, L_2, C_2, G_2, R$ (peripheral load) and N_1, N_2.

As before+), the starting signal is the carotid pulse and the matchings are performed by fitting simultaneously the femoral and pedal signals.

3. Results and comments

We performed about 100 fittings (90 sec. each one of run-time on a CDC 7600) on the same subjects of the previous work and found: i) $Z_{in}(0)$ can be computed with continuity and not arbitrary taken equal to peripheral load, as usually done; ii) the final parameter values depend on the initial set, but not the profiles of some derived quantities such as: phase velocities (v_1, v_{23}), input impedance and reflection coefficient; iii) for each set, variations of the cut-off frequency (f_c) or the normalization criterion sensibly modify the signal matching but little the param. values which are normally distributed ($p>20\%$ w/test of Kolmogorov-Smirnov (K.S.)); iv) $v_1(\infty)$, $v_{23}(\infty)$ are normally distributed ($p>20\%$, K.S.) and f_{min}, f_{max} almost constant, whatever are the initial parameters but provided that $f_c \leq 10$ Hz.

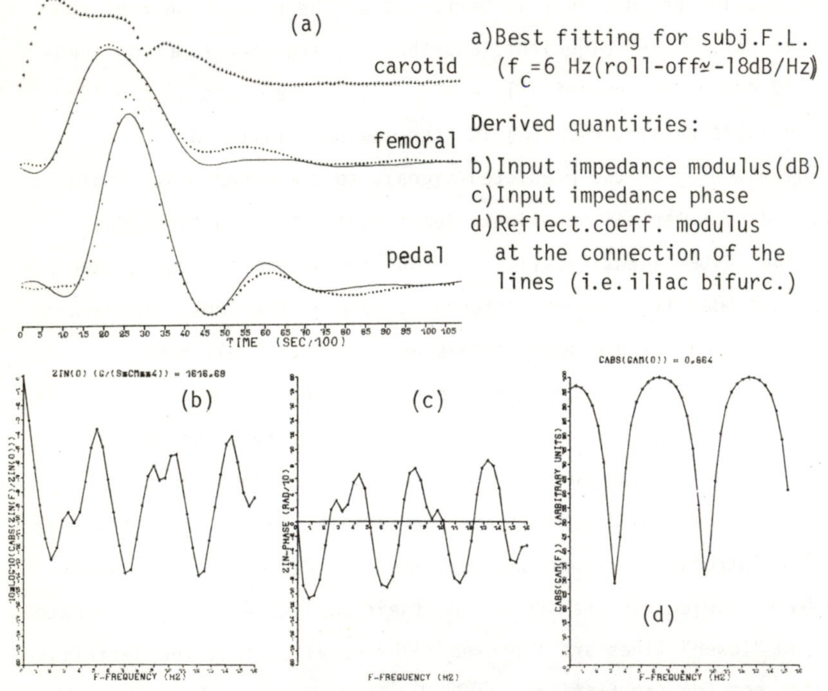

a) Best fitting for subj. F.L. (f_c=6 Hz (roll-off≃-18dB/Hz))

Derived quantities:

b) Input impedance modulus (dB)
c) Input impedance phase
d) Reflect. coeff. modulus at the connection of the lines (i.e. iliac bifurc.)

+) F. Locchi et al. Comm. No. 1.30, World Congress Med. Phys., Hamburg 1982

PROCEEDINGS OF THE II INTERNATIONAL CONFERENCE ON APPLICATIONS OF PHYSICS TO MEDICINE AND BIOLOGY
edited by Ž. Bajzer, P. Baxa & C. Franconi

A MODEL FOR MITRAL STENOSIS

J.Soto, L.S.Quindos, J.L.Ubago
Departamento de Fisica Medica
Facultad de Medicina de Santander
ESPAÑA

Introduction

Arterial and capillary pressures are a very important factor in patients with mitral stenosis, since they are directly related to the production of pulmonary edemas and failures of the right ventricle (1), (2).Therefore, it is of interest to determine what variations occur in these pressures for the different characteristic parameters of the pulmonary circuit. In order to study this problem, we present a model based on the numerical solution of the equations that determine the flow of blood in the pulmonary circuit.

Descripcion of the model

The analytical model basically consists of two blocks: the first represents the pulmonary arterial section and the second, the auricular venous one. Both are connected by a haemodynamic resistance, R, and have a rigidity modules of K_1 and K_z respectively. Using the model, the evolution of pressures P_1 and P_2 is calculated from the systolic and diastolic periods, t_s and t_d.

a) Systolic period

During the systolic period, there are two flows: the first, from the exterior to the pulmoarterial section producing the ejective volume V_i , and the second from the pulmoarterial section to the venoauricular one. From these flows, we can calculate variations in the pressures in these blocks, P_1 and P_2 , through the following expresions:

$$(\Delta P_1)_i = \left[\Delta V_i - \frac{(P_1)_i - (P_2)_i}{R} \right] K_1 \, (P_1)_i$$

$$(\Delta P_2)_i = \left[\frac{(P_1)_i - (P_2)_i}{R} \right] K_2 \, (P_2)_i$$

that allow us determine the evolution of P_1 and P_2 during the whole systolic period from two initial values $(P_1)_o$, $(P_2)_o$.

b) Diastolic period

In the diastolic period the flow of blood from the pulmoarterial section to the venoauricular one, coincides with another from the auricular section to the exterior, controlled by Gorlin´s law.As in the case of the systolic period, we can obtain expresions for variations in the pressures in these two blocks with the following results when these news flows are taken into account:

$$(\Delta P_1)_i = - \frac{(P_1)_i - (P_2)_i}{R} K_1 (P_1)_i$$

$$(\Delta P_2)_i = \left[\frac{(P_1)_i - (P_2)_i}{R} - K' A \sqrt{(P_2)_i - P_v} \right] K_2 (P_2)_i$$

where A:mitral area; P_v: mean ventricular pressure during the diastolic period; K´: Gorlin´s constant.

These expressions make it possible to calculate pressures P_1 and P_2 during the diastolic period from the values obtained at the end of the systolic period. Finally, we hypothesize the same condition for each block, taht the entrance and exit volumes are equal. In other words, we assume that pressures P_1 and P_2 at the end of the diastolic period are the same as the initial values $(P_1)_o$, $(P_2)_o$.

Application of the model

The characteristic parameters have to be introduced into the model were calculated measurements taken during the haemodynamic exploration of a patient with mitral stenosis.The introduction of these values into the model, produced a good simulation of the pulmonary arterial and capillary pressures. For reference, we used the clinical data of the patient in changing haemodynamic situations (rest and overload).

The importance of hte model is that if the characteristic parameters of a patient are known we can simulate the influence of each one on the pulmonary capillary and arteriolar pressures.We specifically studied three situations:

A) Variation of the mean ventricular pressure P_v during the diastolic period.

From the model, we find that an increase in P_v increases the mean values of the pulmonary arteriolar and capillary pressures, in accordance with a growing exponential function. This fact is directly related to the occurrence of failures of the right ventricle.

B) Variations in the arteriolar resistance

A decrease in the arteriolar resistance leads to a decrease in the pulmoauricular gradient and consequently to a greater possibility of pulmonar edema if an increase in the venoauricular pressure, provoked by a mitral stenosis develops.

C) Variations in the venoauricular rigidity

When the value of K_2 increases the model shows an large increase in both the mean capillary pressure and its maximum value. Likewise, there is a decrease in the initial capillary pressure value. In these conditions, there is a high risk of pulmonary edema.

REFERENCES

(1) Marco J.D., Standeven J.D., Barner H.B., Am. Heart, J. 94, (1977).

(2) Hitch D.C., Nolan S.P.. Jour. Surg. Res., 30, 110, (1981).

PROCEEDINGS OF THE II INTERNATIONAL CONFERENCE ON APPLICATIONS OF PHYSICS TO MEDICINE AND BIOLOGY
edited by Ž. Bajzer, P. Baxa & C. Franconi

FINITE ELEMENT ANALYSIS OF VENTRICULAR WALL STRESS

Donatella Badiali*, Leopoldo Conte*, Guido Pedroli*, Edoardo Verna§

Servizio di Fisica Sanitaria* and Divisione di Cardiologia§

Ospedale di Circolo ,Viale L.Borri ,57-I-21100 Varese

ITALY

Quantification of ventricular wall stress(LVWS) is necessary for a full understanding of both normal and pathological ventricular mechanics. Since coronary artery disease(CAD) leads to regional myocardial damage the need for regional rather than global function parameters is generally felt.In such cases the finite-element analysis(FEA)offers an extremely powerfull tool for analyzing regional variation in wall stress.

A simple method of FEA as been developed and is now available for clinical use at the Ospedale di Circolo of Varese. Circumferential wall stresses are determined from single-plane cineangiography and transmural pressure measurements.The mathematical calculations are based on the FEA described by Pao et al.[1] Ventricular borders are digitized into the computer from single-plane cineangiograms.Two shells of revolution are than generated by rotating the left and right hand sides of the angiogram about its longitudinal(apex/base)axis and each shell is divided into 144 coaxial rings of triangular cross sections.(fig.1).

Even if the computer program has been developed in such a way to permit elasticity matrix to vary from element to element,so that the left ventricle may be treated as a filament-wound fiber-reinforced composite material,so far the left ventricle has been considered homogeneous and isotropic.

LVWS analysis has also been carried out only for end-diastolic and end-systolic images. The modulus of elasticity E has been taken equal to

Fig.1: Finite-element left ventricular partitioning

$5x10^5$ dyne/cm^2 in diastole and $5x10^6$ dyne/cm^2 in systole;Poisson's ratio μ has been taken equal to 0.45.

LVWSs have been calculated at the centroid of each triangular element and at each node.However a better approximation is obtained by averaging values of adiacent triangles thus obtaining a quadrilateral mesh, instead of the original triangular one.The computer print-out shows the values of the circumferential wall stress along the ventricular wall from base to apex in three leyers from endocardium to epicardium.

The method has been applied to normal and CAD patients submitted to diagnostic cardiac catheterization.In CAD ventricles it has been observed that akinetic and dyskinetic regions are associated to abnormal stress distribution.Compared to normal ventricles the stress distribution of CAD ventricles is characterized by higher absolute values of stress and by regional variation of stress gradient across the wall from endocardium to epicardium which occurs mainly at the edges of the akinetic and dyskinetic regions(figs.2 and 3).

References

1) Y.C.Pao,E.L.Ritman,E.H.Wood:Finite -element analysis of left ventricular myocardial stress. Biomechanics,7 ,469 (1974)

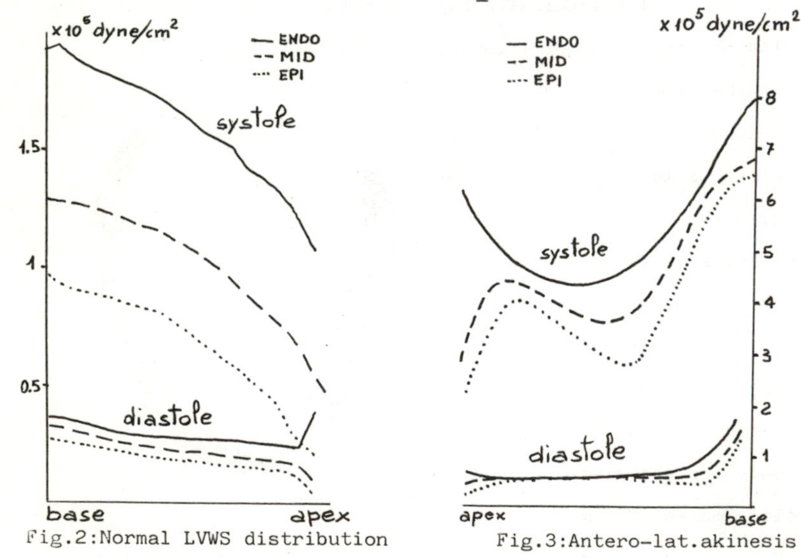

Fig.2:Normal LVWS distribution

Fig.3:Antero-lat.akinesis

PROCEEDINGS OF THE II INTERNATIONAL CONFERENCE ON APPLICATIONS OF PHYSICS TO MEDICINE AND BIOLOGY
edited by Ž. Bajzer, P. Baxa & C. Franconi

MODEL STUDY OF VENTRICULAR DYNAMICS

M. Guerrisi, A. Magrini, and C. Franconi
Istituto di Fisica Medica, II Università, Rome, Italy

Our heart and circulation model[1-3] was modified in order to improve precision in the description of human cardiovascular system. Ventricular end-diastolic pressure-volume curves were obtained from review of published data on human intact heart, thus also including properties of the whole heart, such as effects of pericardial and cardiac fossa constrains, which are not taken into consideration in ventricular models only based on myocardial properties. Various nonlinearities were introduced in the vascular model, thus accounting for several mechanical, neural and humoral factors affecting aortic compliance, systemic venous compliance, and systemic and pulmonary resistances. Effects of varying intrathoracic pressure due to respiration were also included.

Severe physical exercise was simulated and compared to resting conditions. Results, summarized in Fig. 1 and in Table I, are in good agreement with known physiological data, for both heart and vascular variables.

A correct model description of ventricular volume dynamics (Fig.1) was mandatory to obtain such high cardiac output as 25 l/min. Mean intrathoracic pressure had to be decrea

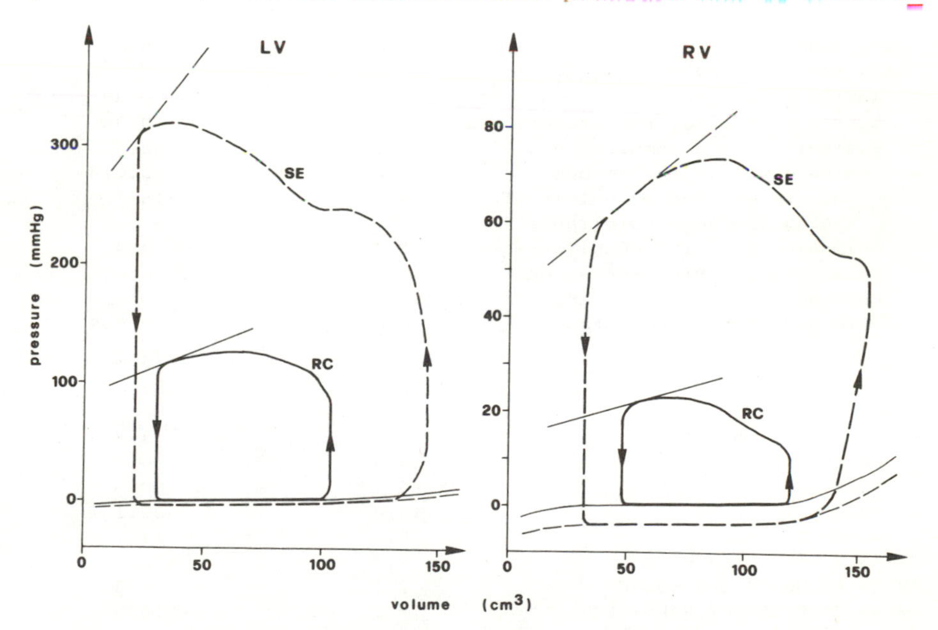

Figure 1. Ventricular P-V loops in simulated resting conditions (RC) and strenuous exercise (SE).

sed 4 mmHg (100%) to obtain slopes of the ventricular functions similar to those quoted in the literature (∼6 l/min per mmHg for RV and ∼2/min/mmHg for LV; see Tab. I). Such a decrease in intrathoracic pressure occurs in physiological conditions due to the increase in depth of breath (increased tidal volume) needed to increase gas exchange in the lungs. The role played by mechanical interaction between circulation and respiration appears to be more important than generally believed.

REFERENCES

1) Arabia M, et al. Trans. ASAIO 26:60-65, 1980
2) Magrini A, et al. Proc. 81st Ann. Meet., 1-9/62,1980
3) Magrini A, et al. Proc. 1st Int. Conf. on Applications of Physics to Medicine &Biology, 277-303, 1982

Table I. Steady state hemodynamics in simulated resting conditions and strenuous exercise (CGS units except pressures in mmHg, HR in min^{-1} and CO in l/min).

	RESTING CONDITIONS	STRENUOUS EXERCISE
HEART PARAMETERS:		
Heart rate	70	200
Left ventricle contractility	0.26	0.78
Right ventricle contractility	0.05	0.16
INTRATHORACIC PRESSURE*	- 4.0	- 8.0
MAIN VASCULAR PARAMETERS		
Aortic compliance	1.88	0.33
Systemic arterial resistance	1.02	0.36
Systemic venous compliance	82.5	82.5
Systemic venous resistance	0.032	0.014
Change in unstreched volume of systemic venous compliance	0	815
Pulmonary arterial compliance	3.4	3.4
Pulmonary arterial resistance	0.051	0.023
Pulmonary venous compliance	10	10
Pulmonary venous resistance	0.042	0.019
CARDIAC OUTPUT	5.0	25.0
STROKE VOLUME	72	125
PRESSURES*:		
Right atrium	2.5	5.8
Left atrium	4.9	16.7
Aorta	95	180
Pulmonary artery	13.4	38.1
Systemic veins	5.2	11.7
MEAN SYSTEMIC PRESSURE	7.2	12.4
MEAN PULMONARY PRESSURE	7.1	22.5
MEAN CIRCULATORY PRESSURE	7.2	13.8

*Mean values in the cardiac cycle.

PROCEEDINGS OF THE II INTERNATIONAL CONFERENCE ON APPLICATIONS OF PHYSICS TO MEDICINE AND BIOLOGY
edited by Ž. Bajzer, P. Baxa & C. Franconi

MEASUREMENT OF LEFT VENTRICULAR BLOOD VOLUME BY EIGHT-ELECTRODE CATHETER: MODELS FOR ELECTRICAL CONDUCTANCE OF THE HEART

HG de Bruin, G Mur, ET van der Velde and J Baan
Clin. Physiol. Lab, Dept. Pediatrics, Univ. Hosp., 2333 AA LEIDEN, NL

INTRODUCTION. An eight-electrode conductance catheter was developed in our lab for continuous determination of left ventricular (LV) volume [1]). LV volume is calculated from five measured four-point conductances and blood specific resistivity (ρ_b) according to a simple model. A second more sophisticated model was implemented to estimate the errors caused by oversimplification in the first model.

MATERIALS AND METHODS.

a. The conductance catheter (cc), is located in the LV as shown in fig. 1. Through electrodes 1 and 8 an alternating current is applied, creating an electric field in the LV cavity. The six other electrodes are used to determine five four-point conductances (G_i, i=2-6).

b. The stacked-cylinder model (SCM). In the SCM (see fig. 1), the heart is divided into five cylinder-shaped segments. Whithin these five segments with height L, current is assumed to be uniformly distributed. At the apex, a conical segment is added to account for blood present beyond electrode 2. Then, the time dependent volume is given by (1):

Fig. 1

$$V = L^2.\rho_b.\{(\sum_i G_i + \tfrac{1}{3}G_2) - G_p\} \quad \ldots\ldots \quad (1)$$

The expression in parentheses will be written as G while G_p represents the parallel conductance of surrounding tissues. To calculate Gp, G is measured at two different LV volumes and measured again at the same volumes after changing ρ_b.

c. The prolate spheroid model (PSM). In this model, the boundaries between the LV cavity, wall and surrounding tissues are represented by two confocal prolate spheroids [2]). A ρ value is assigned to each region. The electrodes may be defined at random along the major axis of the spheroids. The electric fields in the configuration are calculated using eigenfunction expansions of the Laplace equation and the electrode potentials are computed. The values of these potentials are used to compute G as defined above. To simplify the characteristics of the PSM, a slope is defined in the relation between G and V at V = 10 and 50 ml.

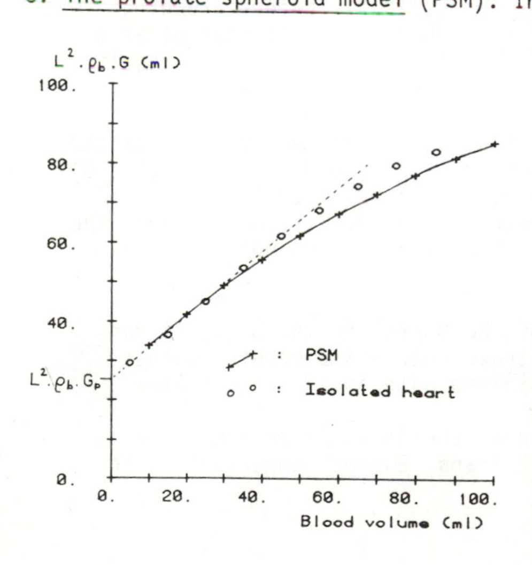

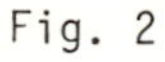
Fig. 2

RESULTS. A plot of $L^2.\rho_b.G$ against actual blood volume

in an isolated *post mortem* canine heart is shown in fig. 2. (L= 1cm, ρ_b= 96 Ω.cm). It shows an Y-axis intercept caused by G_p, a slope < 1, and deviation from linearity above 50 ml.
The latter two findings are not explained by the SCM. The PSM, however, shows all these features when G is calculated for relevant configurations, i.e. ρ_b=100 Ω.cm, myoc.vol.= 63 ml, ρ_m= 316 Ω.cm.
From these and other findings we concluded that:
- Linearity of the G/V relation is good (2%) within the area of interest (10-50 ml), so $L^2.\rho_b.G = a.V + b$
- The equation derived from the SCM (1) might give a good fit to experimental data as well as data derived from the spheroid model, provided it is expanded with a factor to account for the slope:

$$V = 1/a \;.\; L^2 \;.\; \rho_b \;.\; (G-G_p) \quad . \; . \; . \; . \; . \; . \; . \; . \; . \; . \; . \; . \quad (2)$$

According to the PSM, a may be estimated with reasonable accuracy (± .05) from G_p: see fig. 3, in which the relation between a and G_p is given for a large range of values of myocardial resistivities and volumes. This implies that equation (2) may be derived from measurable quantities, i.c. G_p.
Dependence of the G/V relation on the electrode location was also studied. When the catheter is displaced 2 L along the major axis out of the LV the slope in the PSM increases slightly before dropping below 80 % of control at a dislocation of 4 L (results not shown).

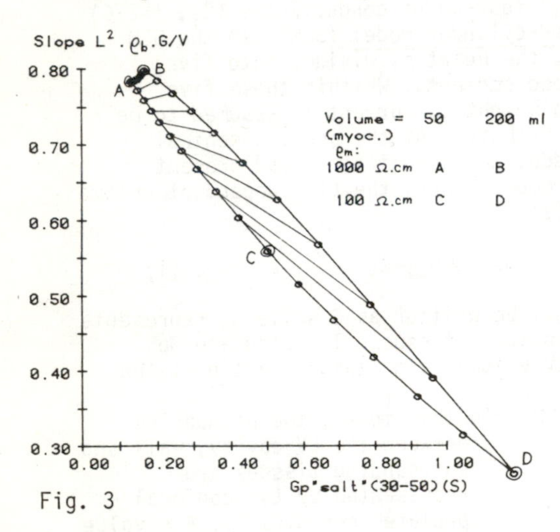

Fig. 3

CONCLUSION. According to the PSM, LV blood volume may be estimated accurately from measurable quantities (ρ_b, G_p, L and time dependent G) although it may be expected that *in vivo*, the relation is characterized by constants and geometry differing from the model. The exact location of the catheter along the LV axis does not seem very critical.

REFERENCES.

[1]) Baan J, TT Aouw Jong, PLM Kerkhof, RJ Moene, AD van Dijk, ET van der Velde and J Koops: Continuous stroke volume and cardiac output from intraventricular dimensions obtained with impedance catheter, Cardiovasc. Res. 15, 328-334 (1981)
[2]) Mur G and J Baan, The computation of the input impedances of a catheter for cardiac volumetry, IEEE Trans. Biomed. Eng, (1983), in press.

PROCEEDINGS OF THE II INTERNATIONAL CONFERENCE ON APPLICATIONS OF PHYSICS TO MEDICINE AND BIOLOGY
edited by Ž. Bajzer, P. Baxa & C. Franconi

SIMULATION OF V.C.G. SIGNALS WITH SIMPLE MATHEMATICAL FUNCTIONS I) ANALYSIS IN THE TIME DOMAIN

L. Bellomonte, R.M. Sperandeo-Mineo and L. Cannizzaro
Istituto di Fisica Università di Palermo, Palermo 90123 Italy

1. Introduction

We set up a mathematical analysis of typical Frank lead vectocardiograms (VCG) in the time domain. The purpose consists in further investigation of the electrical properties of the heart and in the reduction of the amount of data in storage processes. The latter is accomplished by simulating the VCG signal with simple functions that depend on a small number of parameters. Through a best fitting procedure the entire cycle is characterized by 13 undetermined parameters.

2. Acquisition procedure and elaboration

The experimental data were collected from an hp vectocardiograph using a microprocessor controlled system. The acquisition phase lasted 2 sec. The three derivation outputs (X, Y, and Z) were digitized into 3000 points each. In a later operation a full cycle was extracted for each derivation, filtered to reduce noise. Low frequency drifts were removed.

3. Analysis of the P- and T-waves

Both waves were simulated with asymmetric gaussians. For a P-wave we have :

$$f_P(t) = A_P \exp(-\{(t - t_P)/a_{Pj}\}^2) \tag{1}$$

where t_P gives the delay of the wave, a_{Pj} is the halfwidth and has usually different values for $t < t_P$ ($j = 1$) and for $t > t_P$ ($j = 2$). A_P gives the value of the maximum. For a T-wave we replace the subscript P with T.

4. Analysis of the QRS-complex

This complex was simulated by a cosine modulated exponential :

$$f_R(t) = A_R \exp(-|t - t_R|/a_R).\{\cos(\pi(t - t_R)/Dt_R) + b.\sin(\pi(t - t_R)/Dt_R)\} \tag{2}$$

smooth curve, it has low frequency oscillations according to what we said in the introduction. The total power spectrum is not the sum of the power spectra of the isolated waves except for $\nu = 0$. It reproduces the shape of the QRS-complex and its cut-off frequency as a general behaviour. Figure 4 shows a typical full cycle spectrum and its simulated one.

5. Conclusion

The values of the undetermined parameters calculated in the frequency domain are very close to those calculated in the time domain and reported in the Table of I. They are not given here for space reasons. The analysis of the power spectra turns out to be very useful since it has proved that the simulation described in I is realistic. It has also shown in detail the variations that occur in the spectrum when only a part of the temporal signal is considered or when spectra of isolated waves along different directions are compared.

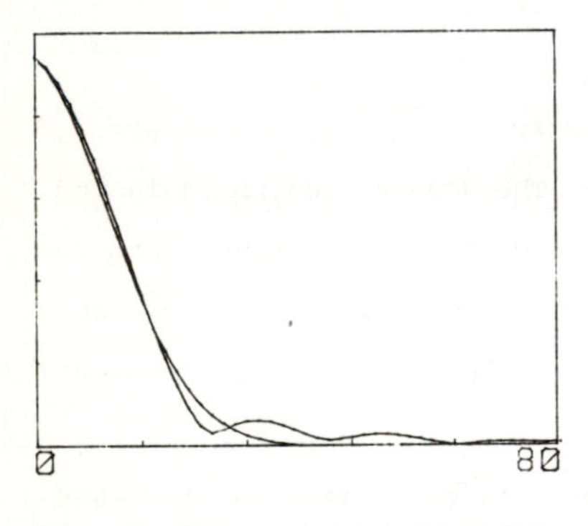

Fig. 1 P-wave power spectrum

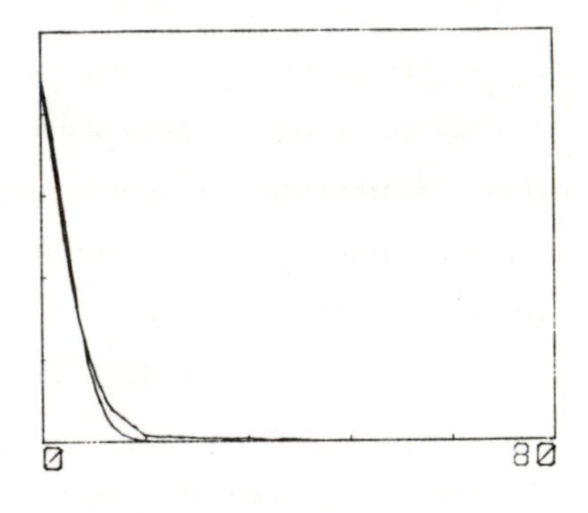

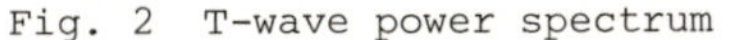
Fig. 2 T-wave power spectrum

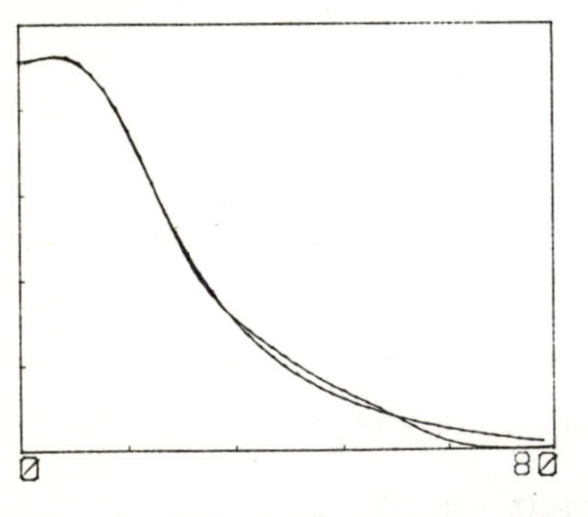

Fig. 3 QRS-complex power spectrum

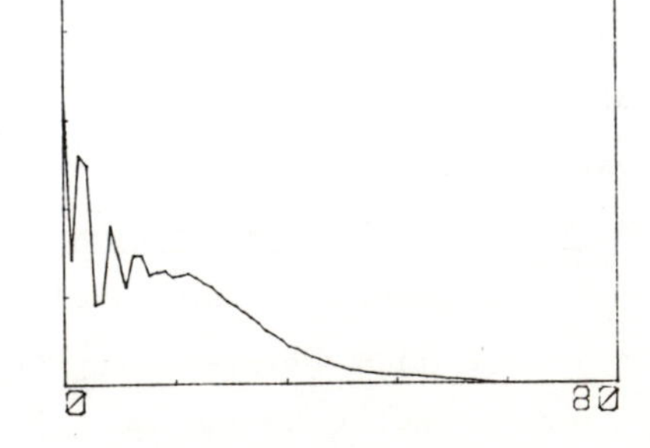

Fig. 4 Full cycle power spectrum

PROCEEDINGS OF THE II INTERNATIONAL CONFERENCE ON APPLICATIONS OF PHYSICS TO MEDICINE AND BIOLOGY
edited by Ž. Bajzer, P. Baxa & C. Franconi

SIMULATION OF VCG SIGNALS WITH SIMPLE MATHEMATICAL FUNCTIONS : II) ANALYSIS IN THE FREQUENCY DOMAIN

L. Bellomonte, R.M. Sperandeo-Mineo and L. Cannizzaro
Istituto di Fisica Università di Palermo, Palermo 90123 Italy

1. Introduction

This work is strictly related to the previous one (to be called I) and uses the same models described therein for the analysis of VCG power spectra. The line shapes in the time domain are entirely different from those in the frequency range. The temporal signal has contributions from various waves that are delayed in time with respect to each other, these contributions superimpose in the frequency spectrum and the time delays give rise to low frequency oscillations due to "interference" factors.

2. Power spectra of P- and T-waves

These waves were simulated with the transform of (1) in I. Figures 1 and 2 show the experimental curves and the simulated ones for a typical P- and T-wave, respectively. These are gaussians whose width is inversely proportional to the corresponding one calculated in I. Consequently the P-wave power spectra are broader than the T-waves by a factor of 3. The transform peak amplitude is proportional to the area of the temporal signal and is greater for the T-waves than for the P-waves by a factor of 10 or more. The approximate cut-off frequencies for P- and T-waves are 30 and 12 cps, respectively, sometimes they depend sligtly on the derivation.

3. Power spectra of QRS-complexes

The QRS-complex was simulated with the transform of (2) in I. Figure 3 shows a typical energy distribution and its simulated one. When the Q- and S-waves are negligible in the temporal signal, the power spectrum has its maximum at $\nu = 0$, otherwise the maximum moves at $\nu > 0$, this value can be calculated and it turns out to depend mainly upon the parameter Dt_R used in (2) of I. The cut-off-frequency is 40 cps, approximately.

4. Power spectra of entire tracings

There is a remarkable difference between the power spectrum of an isolated wave and that of an entire tracing. The latter is no longer a

where A_R gives the value of the maximum, a_R the R-wave half-width, t_R the delay, Dt_R the period of the cyclic term and b the eventual asymmetry.

5. Conclusion

The Table gives the average values of the various parameters used in (1) and (2), calculated in a sample of 35 healthy subjects. The average deviations (Dev) are also given. One of the delays (t_P) is arbitrary. The period T must also be considered. Its average value is 0.78 sec. From the comparative exam of these data, very useful conclusions can be drawn about the variations of the average values when one changes derivation. For each derivation 13 parameters are sufficient to reconstruct the signal as shown in the Figure in which the original signal and the reconstructed one are shown for a typical subject.

T A B L E

A_P (Volt)	a_{P1} (sec)	a_{P2} (sec)	t_P (sec)	Dev (Volt)	
.066	.026	.023	.150	.0045	X-direc.
.105	.029	.023	.150	.0040	Y-direc.
-.020	.022	.015	.150	.0040	Z-direc.
A_T	a_{T1}	a_{T2}	t_T		
.397	.081	.042	.383	.0120	X-direc.
.239	.071	.045	.389	.0144	Y-direc.
-.074	.093	.050	.390	.0133	Z-direc.

A_R	a_R	Dt_R	b	t_R		
1.751	.014	.010	-.437	.279	.1345	X-direc.
1.100	.014	.012	-.270	.286	.0648	Y-direc.
.642	.025	.014	8.950	.305	.0827	Z-direc.

F I G U R E

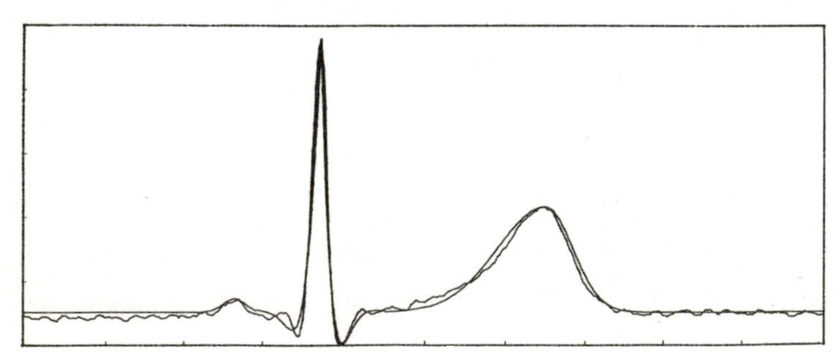

PROCEEDINGS OF THE II INTERNATIONAL CONFERENCE ON APPLICATIONS OF PHYSICS TO MEDICINE AND BIOLOGY
edited by Ž. Bajzer, P. Baxa & C. Franconi

THE TEMPERATURE BEHAVIOR OF BLOOD AS A CARRIER OF RESPIRATORY GASES. A MATHEMATICAL MODEL

M. Brumen and S. Svetina
Institute of Biophysics, Medical Faculty and "J. Stefan" Institute,
E. Kardelj University of Ljubljana, 61105 Ljubljana
Yugoslavia

It is of physiological as well as clinical interest to know the temperature behavior of blood with respect to oxygen binding to hemoblobin. Two types of external conditions can be of particular importance, firstly, when pH value of blood plasma is kept constant, and secondly, when the total CO_2 content in blood remains unchanged. The first example corresponds to the usual "in vitro" experimental conditions, whereas the second may account for conditions in blood vessels. The present communication aims by the use of a simple mathematical model to estimate the temperature effect on the capacity of the blood to carry oxygen in the two cases.

The half saturation oxygen partial pressure (p_{50}) which determines the position of the oxygen binding curve with respect to oxygen pressure, may be taken as a reasonable measure of the capacity of the blood to accept oxygen. A mathematical model is constructed invoking the separation of blood into plasma and red blood cells. The equilibrium conditions between plasma and the cell interior are described by the osmotic equilibrium of water and equilibria of chloride, carbonate and hydrogen ions, whereby other constituents included in the model of the system, i. e. cations, organic phosphates and macromolecules, are enclosed in plasma and/or cells. The system as a whole is assumed to be closed with respect to water and chloride. The reaction between carbon dioxide and water is described by the Henderson-Hasselbach equation. Electric neutrality conditions are assumed separately for plasma and inner cell solution, in which the titration curves of plasma protein, 2,3-diphosphoglycerate and hemoglobin are taken into account. The effect of temperature is exhibited by its action on the blood acid-base status and oxygen binding to hemoglobin. The former effect includes the temperature dependences of the solubility of carbon dioxide, the pK_1 of the Henderson-Hasselbach equation and isoelectric pH value of

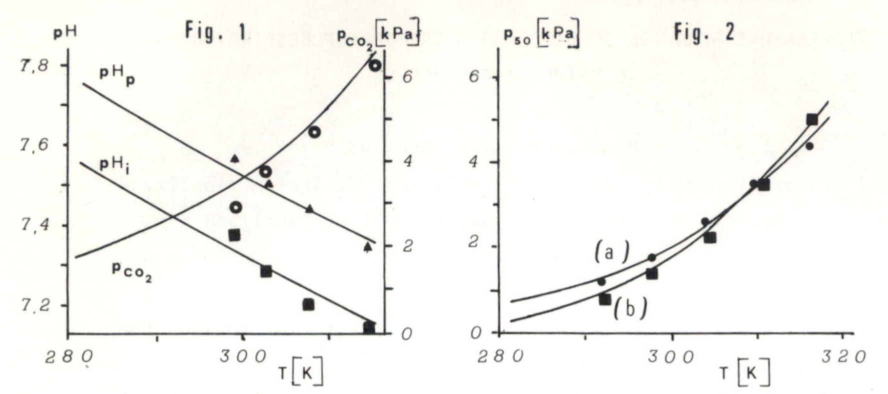

Fig. 1 Predicted temperature dependences of plasma pH, cell pH and CO_2 partial pressure at constant CO_2 content in blood (full lines). Corresponding experimental data (dots)[1].

Fig. 2 Predicted temperature dependences of p_{50} for the two cases: (a) plasma pH is constant, (b) CO_2 content in blood is constant (full lines). Corresponding experimental data (dots)[2].

hemoglobin solution. The latter effect is described by the temperature effect of p_{50} determined for hemoglobin solution. Furthermore, the dependence of p_{50} on pH is included. The interdependence between oxygen binding properties and the blood acid-base status is performed by the connection between hemoglobin charge and p_{50}. The effect of carbon dioxide on oxygen binding is not taken into consideration.

Some predictions of the model are shown in Figs. 1 and 2. Both results are in good agreement with existing experimental data[1, 2]. Predictions in Fig. 2 refer to the two cases mentioned in the introduction. When pH of plasma is kept constant, pH of the cell inner solution remains unchanged also. Hence, the predicted temperature dependence of p_{50} can be attributed to the temperature effect on the oxygen binding to hemoglobin. In the case of the closed system, i. e. at constant blood content of carbon dioxide, an additional slight decrease of p_{50} is observed. This indicates the small contribution of the temperature dependence of the acid-base status to that of the capacity of the blood to carry oxygen.

References

1) M. Castaing and J. J. Pocidalo, Respiration Physiology 38, 243 - 256 (1979)
2) R. B. Reeves, Respiration Physiology 42, 317 - 328 (1980)

PROCEEDINGS OF THE II INTERNATIONAL CONFERENCE ON APPLICATIONS OF PHYSICS TO MEDICINE AND BIOLOGY
edited by Ž. Bajzer, P. Baxa & C. Franconi

TRANSPORT COEFFICIENTS IN N MEMBRANES SYSTEMS

F.C. Celentano and G. Monticelli

Dip. di Biologia and Dip. di Fisiologia e Biochimica Generali
University of Milan, Via Celoria 26, 20133 Milano, Italy

In various situations of biological interest, diffusion barriers may be modelled as nonlinear series arrays of several membranes, including unstirred layers, which behave as non selective membranes. The steady state solute (J_s) and volume (J_v) flows across a nonlinear membrane may be obtained integrating along the membrane thickness the Kedem-Katchalsky practical flow equations (1) written in local form (2), a situation where their intrinsic limitations vanish (3). From such equations, the solute concentration (C) and the hidrostatic pressure (p) in the n-1 compartments separated by n membranes may be computed (2,4). Subsequently, by means of recursive substitutions of C_{i-1} and p_{i-1} into the C_i and p_i expressions, one can obtain two nonlinear relationships between J_v and J_s and the driving forces across the array, hydrostatic Δp and osmotic $\Delta\Pi$ pressure differences:

$$J_s = \frac{J_v}{A^n}\left[\Delta C + \left(1 - \sum_1^n {}_i\, h^i\right) C_1\right]$$

$$J_v = -\Lambda \left\{ \Delta p - \frac{RT\phi\Delta C}{A^n} \sum_1^n {}_i (s^{i+1} - s^i)\, A^i - \frac{RT\phi C_1}{A^n} \sum_1^n {}_i \left[A^n \prod_1^i {}_j\, h^j + A^i \left(1 - \prod_1^n {}_i\, h^i\right)\right] (s^{i+1} - s^i) \right\}$$

where

$$A^i = \prod_1^i {}_k\, h^k \sum_1^i {}_j\, (1 - h^j)/s^j \prod_1^j {}_k\, h^k \quad ; \quad 1/\Lambda = \sum_1^n {}_i\, 1/L_p^i$$

and $RT\phi\Delta C = \Delta\Pi$, $h^i = \exp(s^i J_v/P^i)$, $s^i = 1-\sigma^i$, being P^i the solute permeability and σ^i the reflection coefficient of the i-th membrane. C_1 is the solute concentration above which ΔC is built up. These equations contain as a particular case the ones already obtained for a two membranes array (3,4).

Being the volume flow equation implicit in respect to J_v, the inverse of the hydraulic permeability and of the osmotic flow coefficients may be defined as $1/L_p = (\partial\Delta p/\partial J_v)_{\Delta\Pi=0}$ and $1/L_{pd} = (\partial\Delta\Pi/\partial J_v)_{\Delta p=0}$ Their limiting values for vanishing J_v are

$$\frac{1}{L_p} = -\frac{1}{\Lambda} + RT\phi C_1 \sum_1^n {}_i [\sum_1^i {}_j a^j - \frac{\sum_1^i {}_j 1/P^j \sum_1^n {}_i a^i}{\sum_1^n {}_i 1/P^i}](s^{i+1} - s^i)$$

$$\frac{1}{L_{pd}} = -\frac{1}{L_p} \sum_1^n {}_i 1/P^i \Big/ \sum_1^n {}_i \sigma^i/P^i .$$

These coefficients, as well as J_v and J_s, depend on C_1, indicating that solutions do not behave as pure solvent. In particular, our J_v equation reduces to the linear law by Darcy, with $L_p = \Lambda$, for $C_1 = 0$.

The $-L_{pd}/L_p$ ratio coincides with the limit of $\Delta p/\Delta\Pi$ for vanishing J_v and thus the reflection coefficient for the array is

$$\sigma = -L_{pd}/L_p = \sum_1^n {}_i \sigma^i/P^i \Big/ \sum_1^n {}_i 1/P^i .$$

Also the solute permeability for the whole array, defined as the limit of the $J_s/\Delta\Pi$ ratio for vanishing J_v, assumes the same form as σ, being a weighed mean on the inverse of the individual membranes thicknesses. Thus these two parameters appear to be the only ones characterizing the array independently of the actual experimental conditions, represented by C_1.

References

1 - O. Kedem and A. Katchalsky: Biochim.Biophys.Acta 27, 229 (1958)

2 - C.S. Patlak, D.A. Goldstein and J.F. Hoffman: J.Theor.Biol. 5, 426 (1963)

3 - G. Monticelli and F. Celentano: Considerations on different Thermodynamic models for mass transport across membranes, J. Membr. Sci., in press (1983)

4 - G. Monticelli and F. Celentano: Further properties of the two membranes model, Bull.Math.Biol., in press (1983)

PROCEEDINGS OF THE II INTERNATIONAL CONFERENCE ON APPLICATIONS OF PHYSICS TO MEDICINE AND BIOLOGY
edited by Ž. Bajzer, P. Baxa & C. Franconi

FAST WALKING AND SLOW RUNNING

V. Delgado

Física Médica. Fac. de Medicina

Universidad Complutense. Madrid

SPAIN

1. Introduction

It is a well known fact that, for a given speed, it begins to be easier to run slowly than walk quickly.

It is possible to explain qualitatively this fact as follows:

Wen walking, the legs are (more or less) straight, and we waste energy in accelerating (and decelerating) the extended leg to its maximum angular velocity, ω , in the forward and backward motion.

Wen running, there is, in the forward motion, a sharp bending at the knee, which reduces the moment of inertia.

We waste energy in moving the extended leg in the backward motion, in moving the bent leg in the forward motion and in lifting the calf and foot.

If the energy wasted in lifting the lower half of the leg is lesser than the energy saved in the forward motion of the bent leg, it will be less tiring running than walking at the same speed.

2. The model

It is possible to estimate the velocity for which the energy consumption in walking and running is equal.

We made the following approximations:

- The leg is cylindrical.
- Mass of thigh = Mass of calf and foot = m
- length of thigh = Length of lower half = 0.5 meters

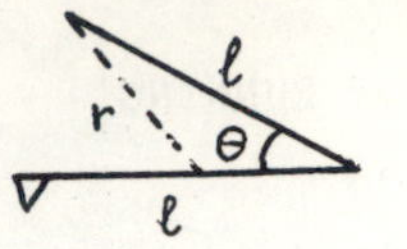

Sketch of a bent leg

$r^2 = l^2(1 - \cos\theta + 1/4)$

The moment of inertia of a cylindrical bent leg is:

$$I_{bent} = ml^2(1/3 + 1 - \cos\theta + 1/4 + 1/12) \quad (1)$$

and the heigth lifted by the center of masses of the lower half of the leg is:

$$h = \tfrac{1}{2}l(1 + \cos\theta). \quad (2)$$

The energy consumed in walking and running can be expressed as follows:

$$E_{walking} = \tfrac{1}{2}(I_{extended} + I_{extended})\,\omega^2 = 8ml^2\omega^2/3 \quad (3)$$

$$E_{running} = \tfrac{1}{2}(I_{extended} + I_{bent})\,\omega^2 + \text{gravitational work} =$$

$$= \tfrac{1}{2}ml^2(9/3 + 1 - \cos\theta + 1/4 + 1/12)\,\omega^2 + \tfrac{1}{2}mgl(1 + \cos\theta). \quad (4)$$

By equating E_w and E_r we can obtain the ω and the corresponding velocity for which the energy wasted is equal.

$$\omega^2 = g/l \quad \text{and} \quad v = 2\sqrt{gl} = 16 \text{ Km/h}. \quad (5)$$

We have a crude estimate, but the functional dependence $v \propto \sqrt{l}$ is exact, and it explains the advantage of taller athletes over shorter ones in long-distance walking.

The same functional dependence can be obtained if the leg is analysed as a double pendulum.

3. References

Topics in classical Biophysics. H.J. Metcalf (Prentice-Hall)

Mecánica. L.D. Landau (Reverté)

PROCEEDINGS OF THE II INTERNATIONAL CONFERENCE ON APPLICATIONS OF PHYSICS TO MEDICINE AND BIOLOGY
edited by Ž. Bajzer, P. Baxa & C. Franconi

NEW METHOD FOR ARTERIAL DYNAMIC ELASTICITY EVALUATION

A. Castellano, A. Neve, G. Palamà & E. M. Staderini*

Physics Department, University of Leece, 73100 LEECE

*Medical Physics Institute, II University of Rome, 00173 ROME

1. Introduction

The external carotid pulse (CP) is the indirect low frequency recording of pressure pulse obtained by applying a proper pick-up device at the neck on the common carotid artery or on the external carotid artery. External carotid pulse tracings show an initial peak or percussion wave (PW), due to the abrupt ejection of blood from left ventricle into the aorta, followed by a second wave or tidal wave (TW), due to systolic contraction. With the stiffening of arterial wall the carotid pressure pulse increases its amplitude but PW in CP decreases and TW is more evident, as the arterial wall is no more able to rapidly transmit internal pressure to external tissues.

2. The system

In order to more objectively study CP to obtain information on arterial wall dynamic elasticity a system has been developed in our laboratory. To perform a preliminary study, a method has been devised to alter non-invasively arterial diameter in order to vary the hydraulic load. An occluding system, using a cuff whose internal pressure may be precisely controlled, is applied on the external carotid artery at the neck and carotid pulse recordings are taken immediately upstream of the occlusion. When occluding pressure is raised from zero to higher values, below max arterial pressure, the pulse of pressure inside the artery goes up due to both diminishing diameter and reflection wave. So, if arterial wall is normal, a clear increment in PW is recorded and on the converse, if arterial wall is stiff, PW change is less evident or not visible at all. Thus the method is based on the revelation of PW changes owing to variations in occluding pressure; the value of this, at which a definite change in PW is detected, may be used to represent an evaluation of arterial dynamic elasticity.

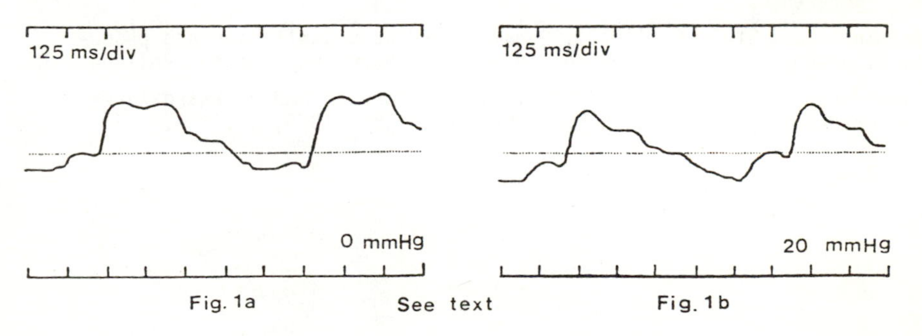

Fig. 1a See text Fig. 1b

3. Results

Figure 1a shows a typical CP tracing from a young male volunteer with no occluding pressure. With only 2670 Pa (20 mmHg) occluding pressure, CP contour quickly changes showing a sharp increase in PW as in Fig. 1b. This is a typical tracing of a healthy young individual. Fiqure 2 shows a couple of recordings taken from a male volunteer suffering from arteriosclerosis. 8000 Pa (60 mmHg) occluding pressure was necessary to see a clear increase in PW peak.

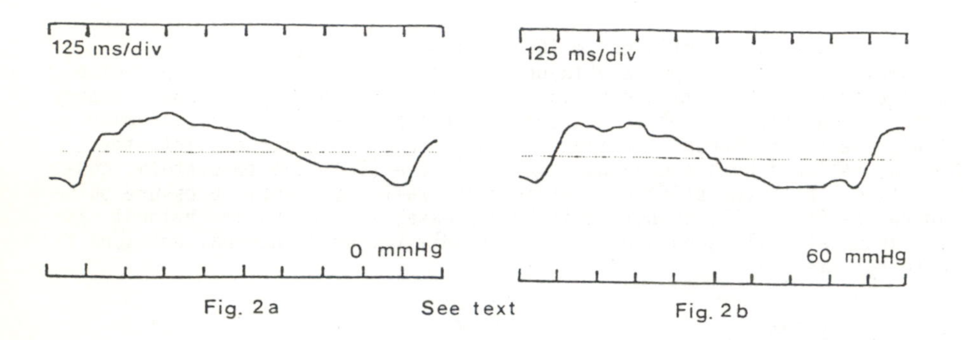

Fig. 2a See text Fig. 2b

4. Conclusion

A new method for arterial elasticity evaluation was devised and a μP-controlled system was developed to study the feasibility of such a technique. The measuring procedure is very simple and not so disconfortable for patient as it seems, on condition that not very high occluding pressures be used and/or applied for more than a few heart beats. Repeatibility and reliability are to be completely verified but in a preliminary clinical application the system proved very effective in evaluating artery damage due to arteriosclerosis so to encourage a complete testing and an improvement of theoretical knowledge underlying it.

5. References

Delman A. J., Stein E. (1979). Dynamic cardiac auscultation and phonocardiography - A graphic guide, 1979, W. B. Saunders Company, Philadelphia.

Laxminarayan S. (1979). The calculation of forward and backward waves in the arterial system, Med. & Biol. Eng. & Comput. 1979, 17, 130.

PROCEEDINGS OF THE II INTERNATIONAL CONFERENCE ON APPLICATIONS OF PHYSICS TO MEDICINE AND BIOLOGY
edited by Ž. Bajzer, P. Baxa & C. Franconi

NON INVASIVE BRAIN MONITORING IN NEAR I.R.: A FIRST ANALYSIS OF HUMAN PHYSIOLOGICAL DATA

Ivo Giannini, Fabrizio Carta, *Marco Ferrari

ASSORENI, Monterotondo, Roma; *Laboratorio di Fisiopatologia,

ISTITUTO SUPERIORE DI SANITA', Roma

ITALY

1. Introduction

A large number of non-invasive measurement techniques have been developed and some have reached high levels of sophistication in the medical field. In the context of optical methods we have developed and evaluated near infrared spectroscopy for non invasive monitoring of brain function [1,2]. The purpose of this comunication is to present a computer-based near I.R. technique with examples of signal changes on volunteers brain, indicating the possibility of obtainig informations on local blood content, hemoglobin oxygenation level and redox state of cytochrome-c-oxidase (cyt a,a_3).

2. Method

Skin and bone tissues are mostly transparent to near I.R. photons and only a few biological chromophores of the brain absorb light in the 700-900 nm w.l. range, namely the heme of hemoglobin (Hb) and the Cu ions of some metal protein, cyt a,a_3 being the most abundant Cu protein in brain tissue. Cyt a,a_3, the terminal member of the respiratory chain, catalyzes about 95% of cellular O_2 consumption and its redox shifts are signal of oxidative metabolism for understanting the electrophysiological activity of brain tissues. The near I.R. spectra of Hb-HbO_2 are well known [3,4] and recently we have provided direct evidence of the possibility to measure <u>in vivo</u> the band of oxidized cyt a,a_3 [5]. On these bases a low cost instrument suitable for near I.R. measurements at 4 different w.l. has been realized, using a small acquisition system based on 6502 microprocessor to collect and elaborate near I.R. data together with other physiological measurements.

3. Results

The preliminary experiences have been carried out on 25 volunteers' brains. The typical responses during the respiration of different gas mixtures and variation of respiratory activity (hyperventilation-apnoea) as shown in Fig. 1 can be easily explained on the basis of physiological and spectral properties of hemoglobin and cyt a,a_3. These data are attractive for a medical use of a continuous, non invasive monitoring of Hb content, Hb oxygenation and of the redox state of cytochrome-c-oxidase in brain tissue.

4. Aknowlegments. This work is supported in part by C.N.R. P.F. Tecnologie Biomediche.

5. References

1) Giannini I, Ferrari M, Carpi A, Fasella P. Physiol Chem Phys 1982; 14: 295-305.
2) Ferrari M, Giannini I, Carpi A, Fasella P J. Cerebral Blood Flow and Met. 1983; S457.
3) Jöbsis FF. Science 1977; 198: 1264-1267.
4) Takatani S, Cheung P.W., Ernst E.A. Ann. Biomed Eng 1980; 8:1-15.
5) Ferrari M, Giannini I, Carpi A, Fasella P Physiol Chem Phys 1983.

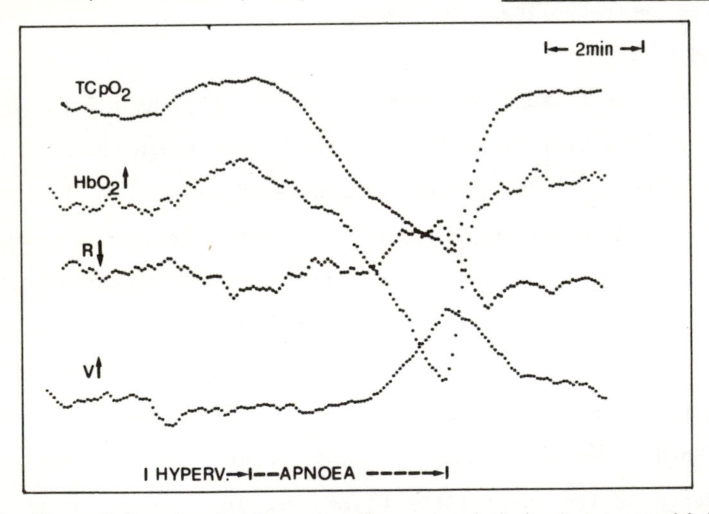

Fig. 1. Near I.R. absorption signals recorded during a rapid hyperventilation followed by apnoea. Fibers are firmly applied to frontal area. Hb oxygen saturation (HbO_2) follows the related trace of transcutaneous pO_2 measured on the forearm ($TCpO_2$). After about two minutes of apnoea an increase of blood volume content (V) and a decrease of redox level (R) of cyt a,a_3 were recorded.

PROCEEDINGS OF THE II INTERNATIONAL CONFERENCE ON APPLICATIONS OF PHYSICS TO MEDICINE AND BIOLOGY
edited by Ž. Bajzer, P. Baxa & C. Franconi

PROTON NUCLEAR ACTIVATION FOR FERROKINETIC INVESTIGATION IN PLASMA

M.C. Cantone, N. Molho, L. Pirola
Dipartimento di Fisica, Università di Milano
INFN Sezione di Milano - via Celoria 16 20133 Milano

Ch. Hansen, E. Werner
Gesellschaft fur Strahlen - Umweltforschung, Frankfurt/Main
Paul-Ehrlich-St 15-20 D-6000 Frankfurt/Main

1. Introduction

The iron bound to the transferrin of the blood plasma represents less than 0.1 per cent of the total body iron. Despite of this low value, however, the plasma iron plays a key role in iron methabolism.

The rate at which iron leaves the plasma is an index of the level of erythropoietic activity and varies significantly in different pathological conditions.

One deduces immediately how important is the determination of plasma iron clearance.

Radioactive tracers are usually utilized for such measurements, but the intrinsic risks connected with this techniques are not negligeable and in some cases (pregnant women, newborns etc.) prevent the utilization. For this reason it is important to found non damaging methodologies of investigation.

We have developed a method based on proton nuclear activation (PNA) for trace elements analysis in biological samples.

Due to the fact that PNA allows the single stable isotopes determination, it seems very suitable for iron clearance determination.

2. The Method

The PNA consists in a bombardment of a sample by a proton beam of appropriate energy to induce (p,xn) reactions on the nuclei of the target. The measure of the intensities of the gamma-rays coming from the radioactive nuclei obtained from the nuclei of the element of interest allows the analysis of the content of the element itself.

For the quantitative determination, a known amount of a reference element is added to the sample and comparison is made with a standard sample containing known amounts of the reference element and of the element of interest.

3. Preliminary results

To study the clearance of iron in plasma one injects intravenously or orally administrates a compound enriched with stable 57-Fe or 58-Fe; in this way the isotopes ratios of 57-Fe/56-Fe or 58-Fe/56-Fe can be changed without significant change of total iron concentration.

After injection or oral administration of iron, one withdraws at different times a series of blood samples.

With proton nuclear activation one can determine the content of the single isotopes 57-Fe, 58-Fe, 56-Fe in plasma samples.

The decrease of the artificially added 57-Fe or 58-Fe gives a direct measure of the plasma iron clearance which in turn is an essential parameter for the assessment of bone marrow activity in ferrokinetic investigations.

The most convenient reaction to be utilized for the determination of the iron isotopes in plasma is a (p,n) reaction.

The Table shows: the isotopes of interest, their relative natural abundance, the radioactive nuclei obtainable via a (p,n) reaction, their mean life and the most intense gamma-lines coming from the decay.

Table

56-Fe	91.66 %	56-Co	77.3 d	846.75 keV
57-Fe	2.19 %	57-Co	270. d	122.06 keV
				136.47 keV
58-Fe	0.33 %	58-Co	71.3 d	810.76 keV

To determine the detection limits of the PNA for 57-Fe and 58-Fe in plasma, samples enriched with known amounts of these isotopes have been analyzed.

The two samples to be analyzed were prepared adding to 1ml of plasma respectively:

0.90 + 5 % μg of 57-Fe

4.56 + 5 % μg of 58-Fe

In both cases the natural content of 57-Fe and of 58-Fe was negligeable. From the gamma-spectra we get the intensities of the transitions of interest:

Area (122.06 keV) = 28827 ⟶ 0.90 μg 57-Fe
Background = 11215
Detection limit ($3\sqrt{\text{background}}$) ⟶ 0.01 μg 57-Fe

Area (810.76 kev) = 26757 ⟶ 4.56 μg 58-Fe
Background = 3152
Detection limit ($3\sqrt{\text{background}}$) ⟶ 0.03 μg 58-Fe

For comparative studies with radiotracer techniques: 4.6 μCi of 55-Fe and 108.6 μg of 58-Fe have been injected intravenously in rabbit, 2.7 μCi of 59-Fe and 4.49 mg of 57-Fe have been administrated by a tube in the stomach.

The clearance data obtained with radiotracers allow to estimate that the amounts of 57-Fe and 58-Fe supplied are sufficient for PNA measurements.

References

1) M.M. Gupta, P. Roth, E. Werner, J.P. Kaltwasser, Eur. J. Nucl. Med., 4, 17, 1979

2) Ch. Hansen, K. Wittmaack, E. Werner, Iron Club Meeting, Bergen/ Norwegen, 1982

3) M.C. Cantone, N. Molho, L. Pirola, Clin. Phys. Physiol. Meas., 3, 67, 1982

PROCEEDINGS OF THE II INTERNATIONAL CONFERENCE ON APPLICATIONS OF PHYSICS TO MEDICINE AND BIOLOGY
edited by Ž. Bajzer, P. Baxa & C. Franconi

ON THE EFFECT OF SELENIUM METHIONINE UPON PROTON MAGNETIC RELAXATION IN THE LIVER TISSUE

Valéria Kovács
Department of Atomic Physics, Eötvös University,Budapest
H-1088 Puskin u. 5-7 Hungary

The present paper reports on investigations, as an important biophysical characteristic, of the influence of aminoacid selenium-methionine /se-me/ upon cell metabolism, the spectra of proton magnetic relaxation in hepoticites of Balb o mice irradiated with X-rays.

EXPERIMENTAL PART AND DISCUSSION

Animals investigated were white Balb ♂ mice of 20 g average weight, the number of animals in each group amounting to ten. Administration of selenium-methionine was carried out in amounts of 1.5 mg/kg by interpertionial injection of 0.3 ml of aqueous se-me solution. This was performed 10 minutes before irradiation with doses 200, 450 and 600 r, respectively,of X-rays. Sampling was done 3, 18, and 24 hours after irradiation; the mice were decapitated and dissected, their liver taken out and frozen in liquid nitrogen. The measurement of proton magnetic relaxation times T_1 in liver tissues at the same low temperature was performed using the Varian-E-4 radiospectrometer. The intensities of the signals in the spectra are shown in fig 1 as ratios of signal amplitudes for sampled probes and control probes, respectively.

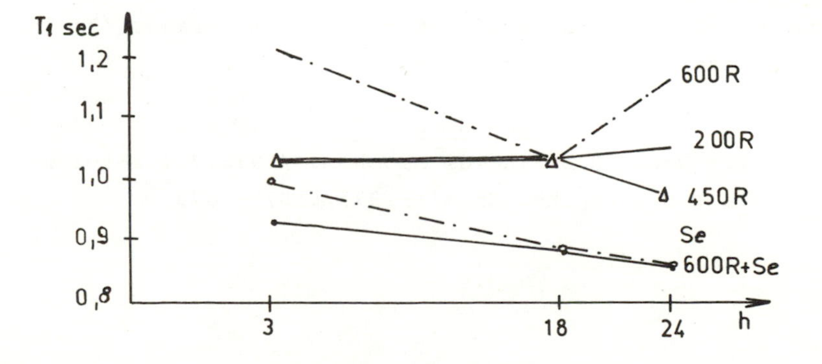

On the basis of our results we suppose that the administration of se-me to a healthy animal leads to a decrease of free water in its cells. This is in accordance with previous/1/ data on the increase in quantities of proteins containing selenium which must have an influence on T_1 values. This is corroborated by results obtained with animals treated with se-me irradiated with 600 r doses. In this case it may be supposed that if se-me had only radioprotective functions, then T_1 values would reassume normal values. On the contrary, if conformational changes do occur, then the animals' reaction to irradiation should be more sensitive, and,specifically,for the largest irradiation doses,T_1 should remain much lower than normal.

This led us to conclude that the administration of se-me causes certain conformational changes in the molecules of proteins containing selenium. This, in turn, results in changes in the micro-viscosity of the proteins and in a change of the hidrophobic action of proteins and lipids, leading to structural changes of membranes.

This effect may become of importance in pathologic states of the organism. It is known that irradiation leads to changes in the micro-viscosity of the proteins of membranes, and further, to changes in the conditions of structural changes in the membrane /6/.

The radiation protective role of se-me may be explained by supposing that this amino acid may become a regulator of mivroviscosity of protein membrane components through the appearance of molecules of altered structure containing selenium. This seems to be an important momentum in the conservation of the structural integrity of membranes /2/ in pathological states of the organism.

REFERENCES

1./ Kovács V.: Revue Roumaine de Biochim. /1976.7.4.281-286/
2./ Burlakova E.B.: Uspekhi Khimii/1975/.1871-1879.

PROCEEDINGS OF THE II INTERNATIONAL CONFERENCE ON APPLICATIONS OF PHYSICS TO MEDICINE AND BIOLOGY
edited by Ž. Bajzer, P. Baxa & C. Franconi

ON THE USE OF KALMAN FILTERS IN EMG SIGNAL PROCESSING

Tommaso D'Alessio — INFOCOM Dept. and Centro Ingegneria Biomedica
Universita di Roma — La Spienza

1. Premise

A well assessed model [1] for the surface emg signal e(t) states that:

(1) $e(t) = w^a(t) \cdot n(t)$ o, in sampled form: (1') $e_k = e(kT_c) = w_k^a \cdot n_k$

where: w(t) is the signal connected to the muscular active state, n(t) is a (possibly) white ergodic gaussian process, and T_c is the sampling interval.

The task of a emg processor (myoprocessor) is to estimate the signal w(t) following some optimality criterion. The methods of maximum likelihood [1] or of minimum root mean square error [2] lead to design a processor with a pre-whitening filter, a square law detector (in general a m-th order detector), an integrator (generally of moving average type) and a relinearizer. Due to the non additivity of noise n(t), a Kalman filter cannot be directly used. However, if we perform a suitable (logarithmic) transformation, we can reduce to the case of signal plus additive noise. In these conditions, if enough information on the signal model is available, a Kalman filter can be designed. In the following we report some results of a preliminary study on the feasibility of this approach.

2. Method

The signal of the eq.(1') has been simulated on a computer and submitted to a logarithmic transformation. Then a scalar Kalman filter has been applied. In fig.1 the general scheme of the simulator is reported. For the signal w(t) a first order autoregressive model, followed by an exponentiation, has been hypothesized, while n(t) has E n = 0 and $\sigma_N^2 = 1$. Therefore:

(2) $w_k^a = \exp[ax_k]$ and (3) $x_k = \alpha x_{k-1} + \eta_k$

where η_k is a sample of ergodic gaussian white process, and α is the correlation coefficient. After the logarithmic transformation, we have:

(4) $y_k = a\,x_k + v_k$ where: (4') $v(k) = \lg|n_k| - E\{\lg|n_k|\}$

is now a sample of a white additive noise.

Eq.(4) and (4'), together with eq.(3), are suitable for the Kalman filtering, for which we suppose to know the parameters a, α and the variance of the white processes n_k and η_k. The recursive estimation formulas are:

(5) $\hat{x}_k = \alpha\hat{x}_{k-1} + b_k[\,y_k - \alpha a\hat{x}_{k-1}]$ (6) $b_k = \dfrac{a[\alpha^2 p_{k-1} + \sigma_\eta^2\,]}{\sigma_v^2 + \alpha^2\sigma_\eta^2 + a^2\alpha^2 p_{k-1}}$

where the m.s. error p_k is obtained by means of the recursive formula:

(7) $p_k = [\alpha^2 p_{k-1} + \sigma_\eta^2](\,1 - ab_k]$ with $p_o = \sigma_x^2 = \sigma_\eta^2/(1-\alpha^2)$

The whiteness of the residuals has been checked by applying the portmanteau test, following which the quantity:

(8) $$Q = N\sum_{k=1}^{M}(\varphi_{rr}(k)/\varphi_{rr}(0))^2$$

(where M = 30 and N is the number of samples processed) is evaluated. If the residuals are white, this quantity has approximately a chi-square distribution with M degrees of freedom. Therefore, the whiteness of the residuals is checked by comparing Q with the 90% limits in the chi-square distribution.

3. Results and conclusions

From the simulation runs implemented it emerged that:

a) the filter behaves well with respect to a quadratic processor but only for some set of parameters of the simulation;

b) however, it is not completely effective in general, that is the residuals r_k are not white unless the "signal-to-noise" ratio, SNR, that is the ratio:

(9) $$SNR = 10\log_{10}(a^2\sigma_x^2/\sigma_v^2)$$

is not high enough (fig.2 a);

c) r.m.s. error and residuals power reduce with the increase of this SNR (fig. 2);

d) the filter requires the knowledge of many parameters in principle not accessible;

e) conclusions b) and c) may represent a preliminary criterion of goodness of filtering when a few information is available on the process to be estimated. In fact, even if some parameters are not known, the model inserted in the Kalman processor can be updated by looking for those values of parameters which tend to make the residuals white and with minimal power.

References

1) N. Hogan, R. W. Mann, "Myoelectric Signal Processing: Optimal Estimation Applied to Electromyography. Part I: Derivation of the Optimal Myoprocessor", IEEE Trans. Biomed. Eng., Vol.BME-27, 1980, pp.382-395.

2) G. C. Filligoi, P. Mandarini, "Some Theoretical Results On A Digital Emg Signal Processor", to be published on Trans.BME.

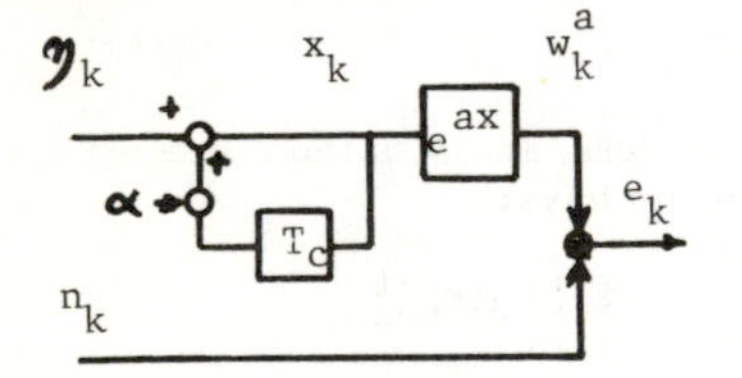

Fig.1 a) Signal generating model

Fig. 1b) Kalman filter

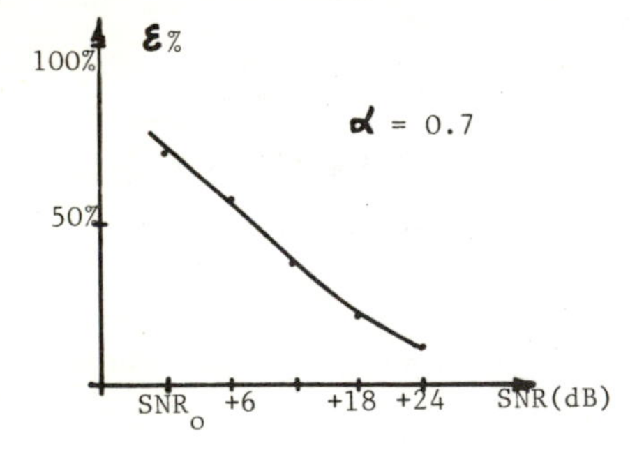

Fig.2. R.m.s. error

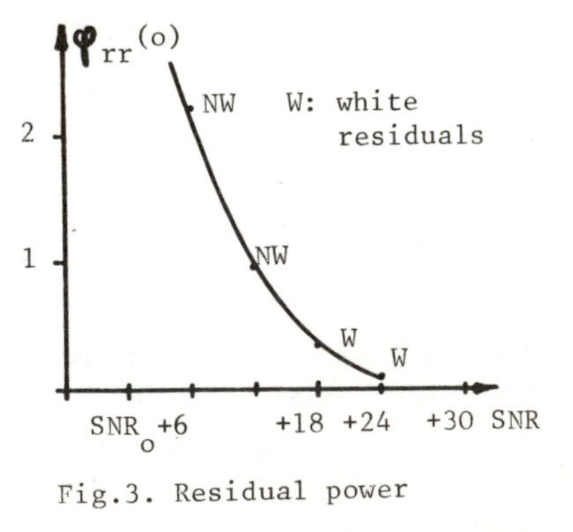

Fig.3. Residual power

PROCEEDINGS OF THE II INTERNATIONAL CONFERENCE ON APPLICATIONS OF PHYSICS TO MEDICINE AND BIOLOGY
edited by Ž. Bajzer, P. Baxa & C. Franconi

LARYNX DISEASE EVALUATION BY VOCAL SIGNAL ANALYSIS

G.Banci [+], M.Bonori [++], O.Adriani [+]

++ Dipartimento di Fisica I Università di Roma
+ Istituto di Fisica Medica II Università di Roma
Via Orazio Raimondo, 00173 Roma ITALY

Introduction

In this paper we expose a method used in an experimental study aiming at the detection of possible larynx diseases, by digital acoustic analysis of voiced speech signals. This analysis has allowed the extraction of some quantitative physiologically meaningful features, which can be used in a diagnostic two classes classifier.

Method and techniques

For the speech signal processing we have used a system based on a microcomputer, equipped with specific hardware and software.

The elaboration of the signal (a steady vowel "a") was performed either in the frequency domain or, prevalently, in the time domain.

The frequency domain analysis, carried out by an FFT algorithm, has enabled us to extract a quantitative feature, F_1F_oQ, that is the quotient between the spectrum amplitudes corresponding to the first formant and to the fundamental frequency.

In the time domain analysis we have used the "inverse filtering" tecnique to obtain the so-called "residue signal", which is highly related to the vocal cords vibratory pattern and therefore to larynx pathologies. From the analysis of the "residue" and from its further elaboration we have taken out seven quantitative physiologically meaningful features, four of which (PPQ, APQ, NRN, IPI) are extracted directly from the "residue", and three (P_o, PA, AWM) from its autocorrelation. The parameters taken out in this study are partly drawn from previous studies (1,2,3,4) and partly defined ex novo (5). Parameter P_o gives the value of the fundamental period of the vocal signal (pitch); it is measured by the distance from the origin of the first peak of the "residue" autocorrelation (see fig.1). Feature PA, given by the value of the first peak after the origin of the "residue" autocorrelation (see fig.1), is an index of the periodicity of the analysed signal and therefore of the normal/pathologic status. Feature AWM (see fig.1) is also meaningful of the speech periodicity and

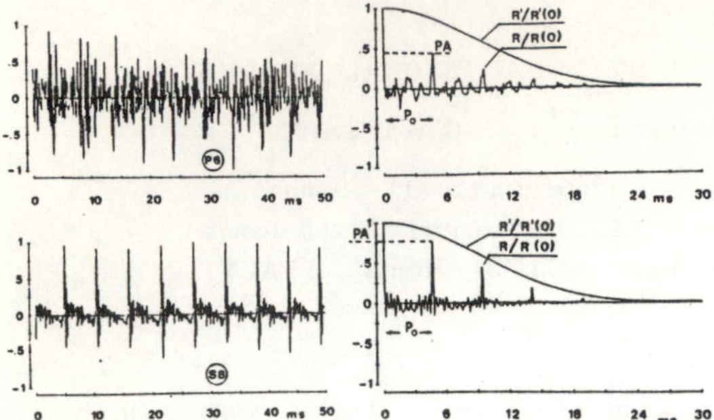

Fig.1 - Pathologic (P6) and normal (S8) residue signals and corresponding autocorrelation function (R).

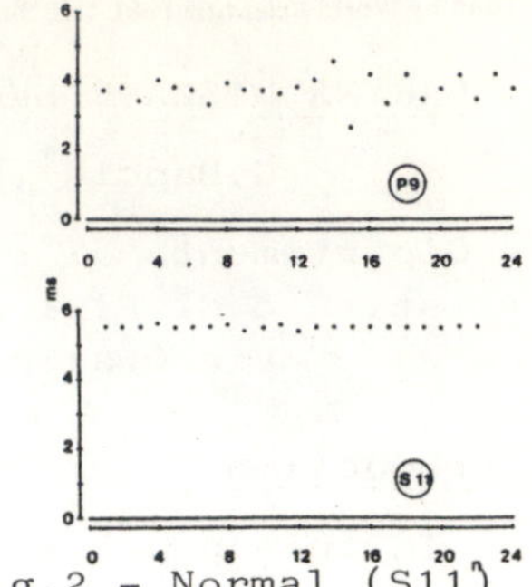

Fig.2 - Normal (S11) and pathologic pitch period sequences.

gives the measure of how far the peaks of the "residue" autocorrelation (R) are from the autocorrelation of the Hamming window (R'), used to weigh the incoming sequence. Features PPQ and APQ quantify the rapid perturbations respectively of the pitch period and of the pitch peak amplitudes, which are larger in pathologic than in normal steady voiced sound. Feature IPI, which is characteristic of pitch perturbations too, considers the number of algebraic sign inversions of the difference between each pitch period and its previous one, normalized to the number of pitch periods in the frame of the analysed residue signal (see fig. 2). Feature NRN quantify the noise-like energy between subsequent pitch peaks of the "residue". This aspect is evident in fig.1, where the residue signals of a pathologic (P6) and a normal (S8) sustained vowel "a" are plotted.

Results

The features obtained have been submitted to a series of tests in order to verify their capacity of discrimination and their fitness to use in a diagnostic classifier. 52 subjects have been experimented (18 healthy and 34 pathologic). After carrying out a direct diagnostic application of the features to the sampling above mentioned, a whole error of about 11% has been obtained (error in judging a healthy subject as a pathologic one and vice versa).

References

1) Koike Y.,Markel J. Ann.Otol.84,117,1975. 2) Kitajima K. Gould W. Ann.Oto.85,377,1975. 3) Davis S.B. SCRL Inc.1976. 4) Koike Y.,Calcaterra T. Acta Otöl.84,105,1977. 5) Bonori M., Banci G., Adriani O., Monini S. Riv.Orl.Aud.Fon.1,59, 1983.

PROCEEDINGS OF THE II INTERNATIONAL CONFERENCE ON APPLICATIONS OF PHYSICS TO MEDICINE AND BIOLOGY
edited by Ž. Bajzer, P. Baxa & C. Franconi

PAPILLARY FUNCTION DEFINED BY OPERATIVE FLOWMETRY IN BILIARY SURGERY

M. Cherubini, P. Baxa, A. Mattiussi, F. Tonelli, F. De Prosperis
University of Trieste, ITALY

The Brucke apparatus was used from 1973 till 1980. This apparatus is useful in operative perfusion of the biliary tract at a pressure of 30 cm/H_2O and measures the flow / minute rate. If the value of flow/minute is less than 10 ml/ min, the flow is abnormal which indicates a lesion of the papilla or of the common bile duct.

Mean, Standard Deviation (S.D.), Range and Significance of the flow obtained in 100 patients with Brücke apparatus:

FLOW					
	N° of PATIENTS	MEAN ml/min	S.D.	RANGE MIN.MAX. (ml/min)	SIGNIFICANCE
ABNORMAL	22	3.27	±1.98	1 - 10	$p < 0.01$
NORMAL	78	20.65	±8.89	11 - 49	

Perfusional fluid may hold 20 % iodine without modifying flow measure. Therefore it is possible to obtain a perfusion of the biliary tract at a constant pressure (30 cm H_2O), flow measure and X-ray control.

Accuracy and per cent of error obtained with cholangiography, flowmetry and thelecolangioflowmetry:

METHODS	ACCURACY %	% OF ERROR
CHOLANGIOGRAPHY	89.53	10.47
FLOWMETRY	91.43	8.57
T. C. F.	98.10	1.90

Correlation between X-ray and flowmetric data in 100 anicteric patients before biliary operation:

N° OF ANICTERIC PATIENTS EXAMINED WITH T.C.F.		100
ABNORMAL FLOW		22
FLOWMETRY/CHOLANGIOSCOPY	AGREEMENT	88
	DISAGREEMENT	12
CAUSES OF FLOWMETRY/CHOLANGIOSCOPY DISAGREEMENT	1) NON-OBSTRUCTIVE LITHIASIS	4
	2) DIFFICULTY OF NEEDLE INSERCTION	3
	3) FALSE POSITIVE	3
	4) MICROLITHIASIS OF CHOLEDOCUS	2

To reduce the false positives of the flowmetric exams an electronic apparatus was constructed (1980) in collaboration with the Physical Institute of University of Trieste(Italy).

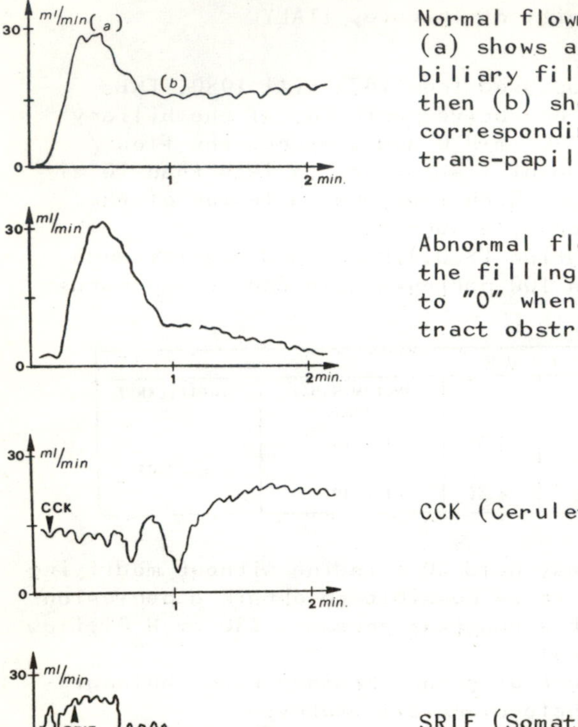

Normal flowmetric curve: (a) shows a first phase of biliary filling, then (b) shows a plateau corresponding to trans-papillary flow.

Abnormal flowmetric curve: the filling phase declives to "0" when there is biliary tract obstruction.

CCK (Ceruletide) test.

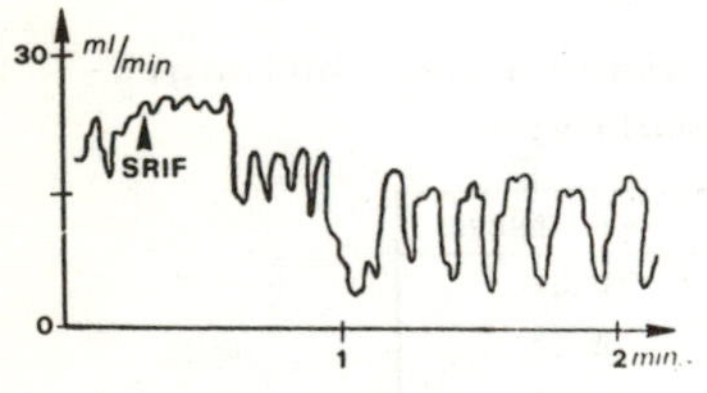

SRIF (Somatostatin) test.

1. It is possible to perfuse the biliary tract at a constant pressure (30 cm/H_2O),and to determine the flowmetric curve,and to compare it with X-ray images.The flowmetric curve is very useful in the determination of papillary or biliary diseases and in the "post-cholecistectomy syndrome" prevention.

2. It is possible to evaluate pharmacological modifications on flowmetric curve and to study the papillary function. CCK test determines an increase of flowmetric curve, and SRIF tends to reduce the flow and causes a prevalence of phasic waves.

PROCEEDINGS OF THE II INTERNATIONAL CONFERENCE ON APPLICATIONS OF PHYSICS TO MEDICINE AND BIOLOGY
edited by Ž. Bajzer, P. Baxa & C. Franconi

NEW MODEL OF RADIOCOLLOID KINETICS IN THE LIVER

G. Izzo, M. Guerrisi, S. Di Luzio*, and A. Magrini
Istituto di Fisica Medica, II Università, Rome, Italy
*Istituto di Fisica Medica, Università di Chieti, Italy

1. INTRODUCTION

Modifications were made to our latest model of radiocolloid kinetics in the liver[1,2] in order to improve the performance in non-invasive studies of the liver with the γ-camera.

Double tracer experiments[3] suggested that intrahepatic vascular system can be approximately represented with finite delay compartments. Intrahepatic diffusion processes were neglected in this instance.

2. MATERIALS AND METHOD

Tracers Two radiocolloids, both labelled with 99m-Tc but of different particle size, were used: AlbuColl (2-20 nm) and AlbuRes (0.2-3 μ).

Detector A large-field-of-view computerized γ-camera equipped with low-energy parallel-hole collimator was used.

Patients Non-liver-disease patients, submitted to liver scintigraphy for metastatic control, were included in the normal patient groups (NP) upon negative scintigraphic and ecographic findings. Histologically proven cirrhotic patients were divided in 2 groups, cirrhosis (C) and advanced cirrhosis (AC), according to the severity of the disease.

Protocol 3-4 mCi in 0.5 ml were rapidly injected into the right antecubital vein. Collection of images and selection of the regions of interest has been described in a previous paper[2].

Data analysis Data were analyzed on the basis of the model shown in Fig. 1. The following parameters were evaluated by means of non-linear least-square procedure: extraction efficiency of the Kupffer cells (E), transit time (T), arterial (AF) and venous (VF) flows through "functional" hepatic parenchima, transit time (TS) and flow (SAF) through intrahepatic artero-venous shunts, and clearance (C=E(AF+VF)). Flows and clearance were espressed with an homogeneous arbitrary unit (au).

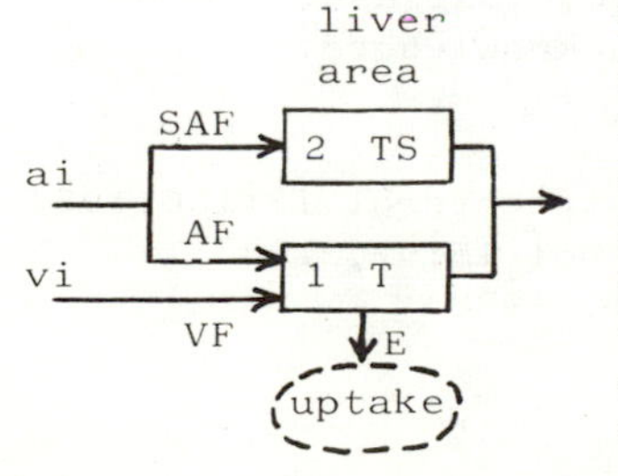

Figure 1. Block diagram of the model. ai= arterial input activity monitored over left ventricle, vi= venous input activity monitored over intestine, 1= intrahepatic "functional" compartment, 2=intrahepatic shunts (mainly representing artero-venous anastomoses). Both compartments are of the finite delay type.

3. RESULTS AND DISCUSSION

Preliminary results (Table Ia)are in agreement with physiopathological data in the literature. Comparison of results obtained with different tracers (Tables Ia,b)shows good reproducibility of tracer-independent parameters (vascular transit times and flows). Possible clinical applications include diagnosis, prognosis and follow-up (Table II) of liver diseases.

Table Ia. Results° with Albucoll (particle size 2-20 nm)

	no. cases	E(%)	AF(au)	VF(au)	T(s)	SAF(au)	TS(s)	C(au)
NP	16	50±15	24±30	351±136	24± 9	12±15	24±9	175±51
C	10	46±13	25±23	237±104	33±12	25±17	20±6	114±47
AC	10	39±17	34±23	64± 39	27± 6	38±20	16±6	35±19

Table Ib. Results° with AlbuRes (particle size 0.2-3 μ)

	no. cases	E(%)	AF(au)	VF(au)	T(s)	SAF(au)	TS(s)	C(au)
NP	5	71±15	27±31	361±155	32± 7	26±27	20±5	259±53
C	4	53±15	26±19	167±124	26±12	27±23	17±4	101±75
AC	7	41± 8	42±18	45± 27	33± 6	36±24	15±6	36±13

° mean ± 1 SD

Table II. Example of follow-up of a cirrhotic patient

Date	Tracer	E	AF	VF	T	SAF	TS	C
22.6.81	AlbuColl	28	32	76	23	28	14	30
23.6.83	AlbuRes	36	47	48	24	40	10	34
14.7.83*	AlbuRes	31	49	33	31	42	8	25

*After withdrawn of 12 Kg ascitic liquid causing decompression of the abdomen (decreased venous pressure)

4. ACKNOWLEDGEMENTS

The authors are grateful to Prof. R. Picardi, Dr. A. Favella and Dr. L. Valeri for helpful advice and for allowing access to their patients. Invaluable technical assistance of Giuliana Garzitto is also gratefully acknowledged.

5. REFERENCES

1. Magrini A and Favella A. Ital J Gastroenterol 13:136,1981
2. Magrini A, et al. Ital J Gastroenterol 15:97, 1983
3. Izzo G, et al. Eur J Nucl Med 8:101, 1983

PROCEEDINGS OF THE II INTERNATIONAL CONFERENCE ON APPLICATIONS OF PHYSICS TO MEDICINE AND BIOLOGY
edited by Ž. Bajzer, P. Baxa & C. Franconi

TWO-COMPARTMENT MODEL FOR QUANTITATION OF Kr-81m LUNG VENTILATION STUDIES

M. Zadro[+], Ž. Bajzer[+*], M. Ivanović[*], M. Kveder[+], and P. Baxa[&]

[+]Rudjer Bošković Institute, POB 1016, 41001 Zagreb, Yugoslavia
[*]Clinic for Nuclear Medicine and Oncology, "Dr Mladen Stojanović" Clinical Hospital, Vinogradska c. 29, Zagreb, Yugoslavia
[&]Istituto Geodesia Geofisica, Università di Trieste, Via del Università 7, Trieste, Italy

Predictions of single compartment model with thorough mixing for Kr-81m behaviour in the lungs were recently questioned by experiments on dogs[1)]. These experiments have shown that Kr-81m concentration is more linearly related to the ventilation/FRC ratio than single compartment model predicts. On the other hand, 177 dynamic Kr-81m lung studies performed on healthy subjects and patients indicate that wash-out time activity curves are monoexponential[2)]. Such behaviour is usually connected with well mixed single compartment model. To resolve this contradictory situation we propose a two-compartment model which describes very well Kr-81m experimental data on dogs and predicts monoexponential behaviour of the wash-out curve.

Our model consists of parallel compartments C_A and C_B which are not mutually connected with the gas flow (Fig. 1). The compartment C_A describes the part of the lung volume in which thorough mixing of inspired and resident gas is achieved. It is related to the FRC volume and the part of tidal volume. The compartment C_B is related to the

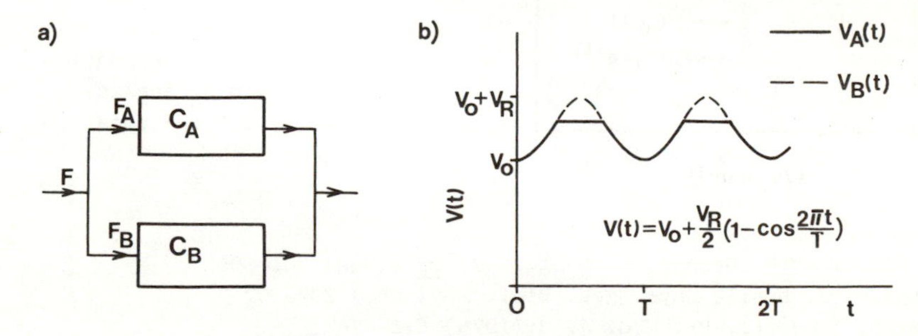

Fig. 1. a) Shematic representation of two-compartment model.
b) Volumes (V_A, V_B) of compartments as a function of time t.

remainder of the tidal volume which contains only the inspired gas and could be connected with anatomical dead space. The volumes $\bar{V}_A$, $\bar{V}_B$ and flows F_A, F_B in the model are taken to be averaged values over breathing period. $V_o + V_R/2 = \bar{V}_A + \bar{V}_B$

The compartment C_A is described by a simple model[3] which includes thorough mixing and constant input Kr-81m concentration C_o. The average activity B in C_B can be expressed as follows

$$B = \frac{\lambda}{T}\int_o^T V_B(t)\ C(t)\ dt = \lambda\bar{V}_B\ \tilde{C}, \quad e^{-\lambda T} < \tilde{C}/C_o < 1 \tag{1}$$

where C(t) is the Kr-81m concentration in C_B. The total lung activity A in steady state normalized to V_o (FRC) is given by

$$A/V_o = \lambda C_o\{(x-x_o)/[\lambda+(x-x_o)V_o/\bar{V}_A]+(1+Tx/2-\bar{V}_A/V_o)\tilde{C}/C_o\} \tag{2}$$

where $x=F/V_o$, $T=V_R/V_o x$ and $x_o=F_B/V_o$. The first term in eq. (2) contains the essential correction to the single compartment model[3]. This term causes more linear behaviour of A/V_o as a function of specific ventilation x (see Fig. 2). The wash-out curve in the present model is monoexponential: $A_{out}(t)=A\exp[-(\lambda+F_A/\bar{V}_A)t]$.

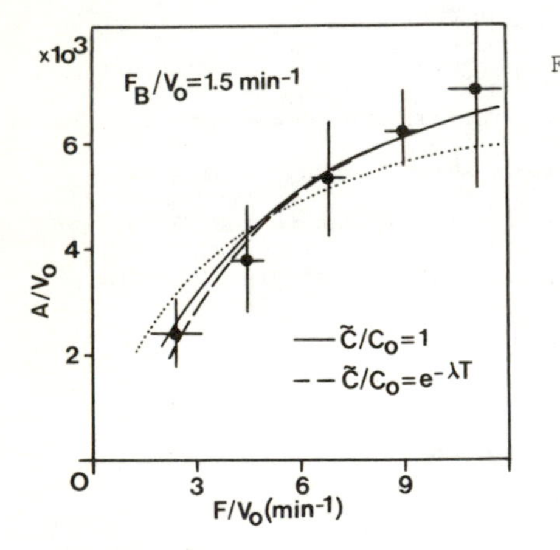

Fig. 2. Total lung activity of dogs in steady state normalized to V_o(FRC). Data are taken from ref. 1). Dotted line is a prediction of a single compartment model (with thorough mixing). Solid and dashed lines are predictions of our model. $x_o=F_B/V_o$ and C_o are adjusted to fit the data.

References

1) H.I. Modell, M.M. Graham, J. Nucl. Med. 23 (1982) 301-305
2) Ž. Bajzer, J. Nosil, Phys. Med. Biol. 25 (1980) 293-307
3) F. Fazio, T. Jones, Br. Med. J. 3 (1975) 673-676

PROCEEDINGS OF THE II INTERNATIONAL CONFERENCE ON APPLICATIONS OF PHYSICS TO MEDICINE AND BIOLOGY
edited by Ž. Bajzer, P. Baxa & C. Franconi

TECHNETIUM-99M PSEUDOGAS FOR ACCURATE ASSESSMENT OF LUNG PATHOLOGY

W.M. BURCH & I.J. TETLEY
Dept Medicine and Clinical Science
John Curtin School of Medical Research
Royal Canberra Hospital
ACTON A.C.T. AUSTRALIA

We have discovered a simple process for dispersing the rare earth radionuclide, Technetium-99m, into sufficiently small dimensions in air for it to behave as a gas. This is of great importance in the investigation of lung disease because there is no completely suitable radioactive gas available universally for examination of the airways - especially the terminal alveolated airways where gas exchange with the blood occurs.

Our "Pseudogas" overcomes most of the limitations of radioactive aerosols which are common alternative ventilating agents [1), is effectively radioactive water vapour and therefore is an ideal physiological agent for monitoring all the airways of the lung.

Pseudogas is produced by the combustion of a flammable aerosol of Technetium-99m in air. This radionuclide is soluble in alcohol, and acetone inter alia which allows a simple procedure to be devised in any Nuclear Medicine unit for creating a flammable radioactive solution. Technetium-99m is routinely available as sodium pertechnetate in physiological saline. Evaporation to dryness of this solution followed by addition of a small volume of ethyl alcohol will produce a flammable solution containing up to 80% of the original radioactivity leached from the salt residue. Ignition of a mist of this solution will generate radioactive pseudogas.

Particle size estimation with a diffusion battery [2) showed that Pseudogas was a monodisperse aerosol of 0.06u radius particles, which makes it an ideal agent for convective diffusion onto the walls of the terminal bronchi and alveolar sacs in the lung.

Internal dose calculations [3) give figures for dose equivalent of 400uSv to the lungs and 320uSv whole body for a typical inhalation of 75MBq activity. These are low and very acceptable figures for a diagnostic test.

To examine uniformity of dispersion of the radionuclide in the combustion products (CO_2, CO[0.1%] and water vapour) autoradiographs were made by diffusing the gas in saturated water vapour onto a photographic emulsion (Kodak 6556). By way of comparison a "standard" aerosol was similarly treated. The representative microscopic fields (Fig.1) confirm the striking difference in dimensions of the two agents.

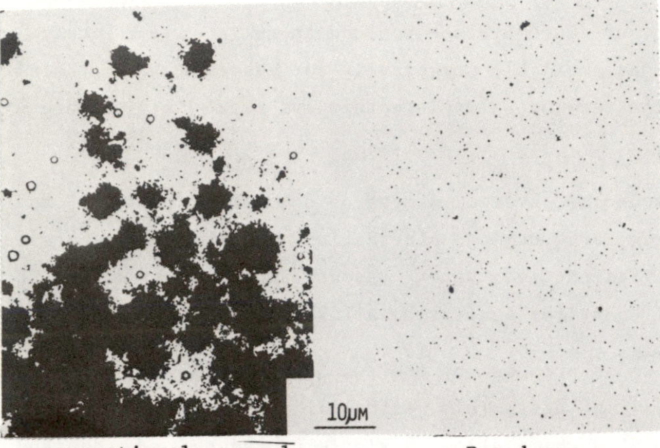

conventional aerosol **Pseudogas**

Figure 1 Autoradiographs of Technetium-99m after falling/diffusing down a 30cm column of saturated water vapour at 37°C. magnification x 500.

A typical diagnostic combination of 75MBq dose of Technetium-99m in 0.5ml ethyl alcohol will generate $2x10^{12}$ radioactive atoms dispersed in $2x10^{22}$ molecules of water vapour or a dilution of 1 in 10^{10}. However, any practical concentration is achieved easily, and it is no problem to generate Pseudogas concentrated enough for good single breath studies. Indeed a whole range of valuable information on regional lung function will now be possible.

References

1. Taplin, G.V. and Chopra, S.K., **Radiol. Clin. North Am.**, 16, 491-513 (1978).

2. Sinclair, D. and Hoopes, G.S., **Am. Ind. Hyg. Assoc. J.**, 39-42 (1975).

3. Kereiakas, J.G. and Rosentein, M. [Eds], **Handbook of Radiation Doses in Nuclear Medicine and Diagnostic X-ray**, CRC Press, Florida. U.S.A. (1980).

PROCEEDINGS OF THE II INTERNATIONAL CONFERENCE ON APPLICATIONS OF PHYSICS TO MEDICINE AND BIOLOGY
edited by Ž. Bajzer, P. Baxa & C. Franconi

STATUS REPORT ABOUT CYCLOTRON PRODUCTION OF SHORT-LIVED RADIONUCLIDES FOR MEDICAL USES AT MILAN CYCLOTRON LABORATORY

Mauro Bonardi, Claudio Birattari and Mario Cominetti

Istituto di Scienze Fisiche, Laboratorio Ciclotrone, Universita' di Milano, via Celoria 16, 20133 Milano

1. Introduction

In the last decade several studies have been carried out at the Cyclotron Laboratory of Milan University, about production of "neutron-deficient" short-lived radionuclides both for medical and for metallo-biochemical studies[1-13]. With the financial support of the Istituto Nazionale di Fisica Nucleare (INFN), of the Consiglio Nazionale delle Ricerche (CNR): Progetto Finalizzato "Biomedical Technology", as well as of the Università di Milano: Dipartimento di Fisica and in collaboration with the Radiochemistry Division of the JRC Euratom, Ispra, Varese, the following facilities have been installed at the Milan Cyclotron Laboratory:

1.1. Radioactivity Measurement Laboratory: high-resolution X- and γ-ray spectrometry (Si(Li), Ge(Li), HPGe detectors; LABEN and CANBERRA multichannel analyzers; CANBERRA/DIGITAL PDP 11-34 computer system interface; general electronics).

1.2. Chemistry Laboratory (type C): for cold preparative and analytical chemistry (thin-layer-chromatography (CAMAG); liquid-chromatography (LKB); microprocessor-gas-chromatography (VARIAN); flame atomic absorption spectroscopy (PERKIN-ELMER).

1.3. Radiochemistry Laboratory (type B): for low activity (MBq) manipulation (PVC glove-boxes) and radioanalytical chemistry (two dimensions thin-layer-radiochromatography (BERTHOLD); liquid-(NaI) radiochromatography (LKB/ORTEC)).

1.4. Radiochemistry Laboratory (type A): for medium activity (GBq) manipulation (lead-shielded telemanipulator; dose-calibrator (CAPINTEC); exposure and air contamination monitors (HERFURTH)).

2. Automatic systems for radiochemical processing of radionuclides

In order to ensure the maximum degree of safety in manipulation of medium-high activities, the following facilities have been developed:

2.1. Internal irradiation probe for remote-controlled irradiation and recovery of high-melting-point metallic targets.

2.2. Remote-controlled system for 81-Rb/81m-Kr generators production.

2.3. Microprocessor-controlled system for radiochemical processing of radionuclides (67-Ga, 111-In) by liquid/liquid extraction.

3. Conclusions

Important medical radionuclides such as 123-I, 201-Tl and 81-Rb/81m-Kr generators have been produced, with original methods, on a semiroutine basis and employed in medical research institutions of Northern Italy (Milano, Pisa).

Other radionuclides such as 67-Ga and the up-date 195m,195-Hg/195m-Au and 118-Te/118-Sb generators of ultra-short-lived radionuclides have been studied, in order to optimize the irradiation parameters, the targetry, and to set up suitable radiochemical separations and quality control routes (see Table I).

Table I - Status of short-lived radionuclides for medical uses produced at the Milan Cyclotron Laboratory

Radio-nuclide	$t_{1/2}$	E_γ (MeV)	Target	Reaction	Range (MeV)	t^{opt} (h)	A^Y (MBq/h)	Radionuclidic impurities	Status (1983)
123-I	13.3h	159	124-Te 91.86%	(p,2n) (p,2n)	28-17 25-20	0 0	463 250	124-I,0.9% 124-I,0.7%	Thyroid Scintigraphy, 4-I-antipyrine and ω-I-hexadecenoic acid.
201-Tl	74h	167	nat-Tl	(p,3n)201-Pb	27-19	0.32	4.9	200, 202-Tl,0.3%	Myocardial Scintigraphy.
		135	203-Tl 81%	(p,3n)201-Pb	27-19	0.32	14.3	200, 202-Tl,0.3%	
			202-Hg 98.6%	(p,2n)	19-10	50	17.6	199,200, 202-Tl,4.5%	No clinical tests. Radio chemical processing and quality control.
				(p,2n)	19-13	60	14.2	199,200, 202-Tl,3.4%	
				(p,2n)	19-16	90	7.1	199,200, 202-Tl,2.3%	
67-Ga	78h	93 184 300	nat-Zn	(p,xn) (p,xn) (p,xn)	34-16 28-18 24-22	85 83 81	7.3 5.1 1.1	66-Ga,1.0% 66-Ga,1.0% 66-Ga,1.0%	No clinical tests. Radio chemical processing and quality control.
81-Rb/ 81m-Kr	4.58h/	190	nat-Kr	(p,xn)	40-30	0	165	tot,0.0001%	Lung ventilation and Brain perfusion studies.
195m, 195-Hg/ 195m-Au	40h, 9.5h/ 30.6s	262	197-Au	(p,3n)	33-19	0	34	197m-Au, 0.3%	No clinical tests. Radio chemical processing in progress.

4. References

1) E. Sabbioni, L. Goetz, C. Birattari and M. Bonardi, Sci. Total Env., 17, 257 (1981).
2) M. Bonardi, Radiochem. Radioanal. Letters, 42, 35 (1980).
3) L. Goetz, E. Sabbioni, E. Marafante, C. Birattari and M. Bonardi, Radiochem. Radioanal. Letters, 45, 51 (1980).
4) D. Basile, C. Birattari, M. Bonardi, L. Goetz, E. Sabbioni and A. Salomone, Int. J. Appl. Radiat. Isotopes, 32, 403 (1981).
5) E. Acerbi, C. Birattari, M. Bonardi, C. De Martinis and A. Salomone, Int. J. Appl. Radiat. Isotopes, 32, 465 (1981).
6) C. Birattari, M. Bonardi and M. C. Gilardi, Radiochem. Radioanal. Letters, 49, 25 (1981).
7) L. Goetz, E. Sabbioni, E. Marafante, J. Edel-Rade, C. Birattari and M. Bonardi, J. Radioanal. Chem., 67, 193 (1981).
8) C. Birattari and M. Bonardi, Funzioni di eccitazione per le reazioni nucleari (p,xn) e (p,pxn) su targhette d'oro e studio del generatore del radioisotopo a vita ultrabreve 195m-Au, Report INFN/TC-80/17 (1980), Frascati, Roma.
9) C. Birattari and M. Bonardi, Ottimizzazione delle condizioni di irraggiamento per la produzione del radioisotopo 67-Ga con il ciclotrone di Milano, Report INFN/TC-80/9 (1980), Frascati, Roma.
10) M. Bonardi, C. Birattari and A. Salomone, 201-Tl production for medical use by (p,xn) nuclear reactions on Tl and Hg natural and enriched targets, Proceedings of the Inter. Symp. on Nuclear Data for Science and Technology, Anversa, Belgio, 1982, Reidel Publisher Company, Dordrecht, Holland, 1982.
11) C. Birattari and M. Bonardi, Excitation function for 195m,195-Hg/195m-Au generator production, Proceedings of the 3rd World Congress of Nuclear Medicine and Biology, Paris, 1982, Pergamon Press, Paris, 1982.
12) M. Bonardi and C. Birattari, J. Radioanal. Chem., 76, 311 (1983).
13) M. Bonardi, Produzione di radioisotopi ad elevata attività specifica con il ciclotrone AVF dell'Università di Milano, Report INFN/TC-82/15 (1982), Frascati, Roma.

PROCEEDINGS OF THE II INTERNATIONAL CONFERENCE ON APPLICATIONS OF PHYSICS TO MEDICINE AND BIOLOGY
edited by Ž. Bajzer, P. Baxa & C. Franconi

EXCITATION FUNCTIONS, YIELD AND CONTAMINATION CALCULATIONS FOR THE PRODUCTION OF THE NEW 107-Cd/107m-Ag AND 109-Cd/109m-Ag GENERATORS FOR MYOCARDIAL FIRST-PASS FUNCTIONAL STUDIES

Mauro Bonardi and Claudio Birattari
Dipartimento di Fisica, Laboratorio Ciclotrone, Università degli Studi di Milano, via Celoria 16, 20133 Milano, ITALY.

1. Single Photon Ultra-Short-Lived Radionuclides

Several Single photon gamma-emitting ultra-short-lived radionuclides (USRN) have been proposed in recent years as radiodiagnostics of various human phatologies [1], as well as tracers for functional studies (in dynamical conditions) of the myocardium and the cardiovascular system [2].

These radionuclides, showing half-lives from a few seconds to a few minutes, are commonly "generator" produced, via the decay of a parent radionuclide, with half-life from a few hours to a few months. In Table I are reported the most useful cyclotron produced USRN, with their main decay data.

Table I - Cyclotron Produced Ultra-Short-Lived Radionuclides

PARENT	t1/2	Abundance	DAUGHTER	t1/2	Main γ-Emission, keV	
^{77}Br	57 h	2.07 %	^{77m}Se	17.4 s	162	52.5 %
^{81}Rb	4.6 h	97.2 %	^{81m}Kr	13.3 s	190	67 %
^{89}Zr	78.4 h	100.0 %	^{89m}Y	16.1 s	909	99.1 %
^{90}Mo	5.8 h	93.5 %	^{90m}Nb	18.8 s	122	64 %
^{107}Cd	6.5 h	99.9 %	^{107m}Ag	44.3 s	93	4.6 %
^{109}Cd	453. d	100.0 %	^{109m}Ag	39.8 s	88	3.6 %
^{167}Tm	9.3 h	98.4 %	^{167m}Er	2.3 s	208	41.7 %
^{178}W	21.5 d	100.0 %	^{178}Ta	9.3 m	93	33.7 %
^{190m}Ir	3.2 h	95.0 %	^{190m}Os	9.9 m	616	98.6 %
^{191}Pt	2.9 d	1.0 %	^{191m}Ir	4.9 s	129	26 %
^{191}Os	15.4 d	100.0 %	^{191m}Ir	4.9 s	129	26 %
^{193m}Hg	11 h	100.0 %	^{193m}Au	3.9 s	weak	
^{195m}Hg	41 h	50.0 %	^{195m}Au	30.6 s	262	67 %
^{195}Hg	9.5 h	2.6 %	^{195m}Au	30.6 s	262	67 %
^{197m}Hg	23.8 h	6.5 %	^{197m}Au	7.7 s	278	73 %
^{201}Bi	1.8 h	98.0 %	^{201m}Pb	61 s	628	54.5 %
^{203}Bi	11.8 h	23.0 %	^{203m}Pb	6.1 s	825	70 %

With the financial support of the Italian "Consiglio Nazionale delle Rirerche", CNR, Progetto Finalizzato: BIOmedical Technology, several studies have been carried out at the Milan University Cyclotron Laboratory about cyclotron production of such generators [3-6].

Among them, the 81-Rb/81m-Kr generator has been produced for a few years in Milan and employed on a semi-routine basis in medical institutions (Milano, Pisa) for clinical and research purposes [3].

The 195m,195-Hg/195m-Au generator, has been recently employed as an agent for rapid sequential first-pass myocardial studies and determination of the important functional index: "left-ventricular ejection fraction", LVEF [2]. Some preliminary studies about its production have been carried out at the Milan Cyclotron Laboratory [4-6], in order to obtain this radiopharmaceutical in a chemical form suitable for in-human clinical tests.

2. It is the purpose of this work to present the production data of two new generators of USRN, the 109-Cd/109m-Ag and the 107-Cd/107m-Ag, which have been recently proposed for myocardial studies, in spite of the low emission gamma-energies and abundances, at 88 keV (3.6 %) and 93 keV (4.6 %) respectively. Both these generators can be produced by proton bombardment on silver targets in a low energy cyclotron, via Ag(p,xn) nuclear reactions, whose excitation functions have been previously measured at the Milan Laboratory in the proton energy range from 4 to 43 MeV [7]. In Table II are reported the main production data, as well as the optimal activation energies and "thin-target yields".

Table II - Optimal Production Data for 107-Cd and 109-Cd

NUCLEAR REACTION	Q (MeV)	CB (MeV)	Ep^{max} (MeV)	$Yield^{max}$ (μCi/μAhMeV)	(MBq/CMeV)
109-Ag(p,n)109-Cd	- 0.94	8.1	9.5	.5	5.1
109-Ag(p,3n)107-Cd	- 18.66	( )	30	3.3(E3)	3.4(E4)
107-Ag(p,n)107-Cd	- 2.20	8.2	11	1.4(E3)	1.4(E4)
107-Ag(p,3n)105-Cd	- 21.00	( )	33	2.1(E4)	2.2(E5)
107-Ag(p,4n)104-Cd	- 29.46	( )	?	?	?

3. References

1) Y. Yano, Radionuclide generators. In G. Subramanian et al.(Ed.), Radiopharmaceuticals. The Soc. Nucl. Med., Inc. N. Y., 1975, p. 236-245
2) International Symposium on Single Photon Ultra-Short-Lived Radionuclides, The Soc. Nucl. Med., Washington, D. C., 1983
3) C. Birattari and M. Bonardi, Funzioni di Eccitazione per le Reazioni Nucleari (p,xn) e (p,pxn) su targhette d'oro e studio del generatore del radioisotopo a vita ultrabreve 195m-Au, Report INFN/TC-80/17, Roma, 1980
4) E. Acerbi, C. Birattari, M. Bonardi et al., Int. J. Appl. Rad. Isotopes, 32, 465 (1981)
5) C. Birattari and M. Bonardi, Excitation functions for 195m,195-Hg/195m-Au generator production, Proc. 3rd World Congr. Nucl. Med. Biol., Paris, 1982
6) M. Bonardi and C. Birattari, Excitation functions for a 195m,195-Hg/195m-Au generator production, 1st Int. Conf. Appl. phys. Med. Biol., Trieste, Italy, 1982, World Scientific Publish. Co., Singapore, 1983
7) L. Goetz, E. Sabbioni, E. Marafante, C. Birattari and M. Bonardi, Radiochem. Radioanal. Letters, 45, 51 (1980)

PROCEEDINGS OF THE II INTERNATIONAL CONFERENCE ON APPLICATIONS OF PHYSICS TO MEDICINE AND BIOLOGY
edited by Ž. Bajzer, P. Baxa & C. Franconi

BIOMEDICAL APPLICATIONS OF MWPCs AT THE UNIVERSITY OF PISA

R.Bellazzini[+], A.Del Guerra[+], M.M.Massai[+], V.Perez-Mendez[*], and G.Spandre[+]
(+) Dipartimento di Fisica and INFN, Sezione di Pisa (Italy)
(*) Lawrence Berkeley Laboratory, Berkeley, California (USA)

We present the results of several applications of MultiWire Proportional Chambers to the biological and medical field which are currently developed at Pisa University, namely: 1) Digital Autoradiography, 2) Bone Densitometry, 3) Positron Emission Tomography.

1. Digital Autoradiography

We have built an experimental facility equipped with MWPCs, a PDP 11/23 mini-computer and a video processor for the digital imaging of ^{14}C and ^{3}H two dimensional distributions in biological and medical applications.

The Detector. The typical MWPC has an active area of 25x25 cm^2 with an anode-cathode gap of 4 mm. A cathode coupled delay line read-out system[1)] is used for both x- and y-coordinate. For the imaging of the ^{14}C β^- rays the MWPC was constructed with a very thin mylar window (9μm) and a minimized non active gas volume (1 mm thick). A spatial resolution of 4.5 mm (FWHM) has been obtained[2)] with a detection efficiency of ~ 20% and a sensitivity of ~ 1 Bq/cm^2. For the imaging of tritium distributions we have built two "windowless" MWPCs[3)]. The sample is positioned inside the chamber at ~ 200 μm from the cathode plane. The first chamber has an anode plane with a 2 mm pitch and operates at atmospheric pressure and in gas flow conditions. A spatial resolution of 1.5 mm (FWHM) has been measured along the wire direction, with an efficiency of ~ 10%, an uniformity of ~ 4% over the whole area of the detector and a sensitivity of ~ 10^{-1} Bq/cm^2. The anode plane of the second "windowless" chamber has a wire spacing of 1 mm, at 45° with the cathode planes. The MWPC is positioned inside a steel box pressurized up to 4 atm. A spatial resolution of 800μm (FWHM) has been measured along both directions for ^{3}H sources.

The Biomedical Applications. The first series of experiment was related to the study of variations in the ability of cell clones to incorporate a radioactive precursor of DNA biosynthesis[4)]. Mammalian cell mutant with defective repair of UV-induced DNA damage can be identified by the lack of unscheduled incorporation of ^{3}H-thymidine into DNA after UV irradiation. With the MWPC all colonies with absent or low incorporation of ^{3}H-thymidine can be precisely located within a population of normally labelled colonies. Several benchmark experiments have been succesfully performed to test the ability of the MWPC to discriminate between mutant and normal colonies: imaging of UV-damaged and non UV-damaged colonies, imaging of cells with a reduced DNA repair synthesis capability (UVS-20 line) and of cells with a severely impaired excision repair (Xeroderma Pigmentosum). In 20 minutes of data taking an activity as low as a few Bq/clone can be easily identified.

A second series of experiments was related to the use of the MWPC for the study of the regional carbohydrate consumption in myocardial tissue[5)] with a deposit tracer of glucose metabolism (^{3}H-deoxiglucose). Typically, $8x10^7$ Bq of ^{3}H-DG is injected intravenous to a dog. After two hours the animal is killed with an overdose of anesthetic and the heart is excised. Ultra thin heart slices of about 40μm are then obtained by means of a microtome. The slices are then placed inside the MWPC for the β^- radioactivity measurement.

2. Bone Densitometry

We have built a MWPC specially suited for bone densitometry studies. The main parameters of the MWPC are 128x128 mm^2 active area, 1 mm anode wire spacing, 3 mm anode-cathode gap, 2.7 mm wide cathode strips with fast delay-line readout (a spatial resolution of 1 mm is expected for both coordinate). The chamber is filled with Xenon-CO_2 and pressurized up to 4 atm to increase the efficiency and to reduce the mean-free-path of the photoelectron emitted by the impinging X or γ rays (an efficiency of $\sim$ 10% is expected for 45 keV radiation at 2 atm). Due to the small anode-cathode gap, the parallax error is minimized. A pile-up inspector is used to reject double events or events with a double hit (absorption within the chamber of the Xenon fluorescent line), and a pulse shape analyzer is used to discriminate events with too long ionization track (expected energy resolution of $\sim$ 20%). Two TDCs with 500 ns conversion time are utilized in conjunction with a dedicated two-dimensional (128x128 pixels) histogramming CAMAC memory. A collection data rate of 1 MHz is allowed. The MWPC is now under test.

3. Positron Emission Tomography

We originally[6-7] proposed a multiplane Positron Camera, made of six modules arranged to form the lateral surface of a hexagonal prism. Each module consisted of one MWPC equipped with two(2 cm thick)converter planes made of lead glass capillaries (6.2 g/cm^3, 80% PbO by weight). This design was based on our experimental work with glass tubing of 0.91 mm inner diameter, 0.09 mm wall thickness.This scheme was very appealing both for simplicity of construction and for the reduced number of readout channels. However, it had some limitations, both in count rate and in accidental contamination. The new design[9] for the High Spatial Resolution Positron Emission Tomograph (HISPET) has been improved in the following way:each module of HISPET will now have two MWPCs with two 1 cm thick converter planes for each chamber. We will use smaller tubes (0.48 mm I.D., 0.06 mm wall thickness) and four separate converter planes. We have increased the efficiency of each module by 1.5 and reduced the coincidence resolving time of a pair of modules by a factor of 2. This results in an increase of 4.5 in volume sensitivity of the tomograph. The parallax error is also conveniently reduced by a factor of 2, thus improving the intrinsic spatial resolution of the system. HISPET will have a volume sensitivity of 100000 c/s per 0.1 μCi/ml, a signal to noise (true to accidental coincidences) ratio of 3:1, and a spatial resolution of less than 4.5 mm (FWHM). The tomograph is now under construction.

1) R.Bellazzini et al.,Nucl.Instr. and Meth. 190, 627 (1981).

2) R.Bellazzini et al.,Nucl.Instr. and Meth. 204, 517 (1983).

3) R.Bellazzini et al.,"Biomedical Applications of MWPCs for Digital Imaging of Soft β^- Emitters",Nucl.Instr.and Meth.(1983) to be published.

4) A.Abbondandolo et al.,Radiat.Environm.Biophys. 21,109(1982).

5) R.Bellazzini et al.,IEEE Trans.Nucl.Sci. NS-30,686(1983).

6) A.Del Guerra et al.,IEEE Trans.Med.Imaging MI-1,4(1982).

7) A.Del Guerra et al.,IEEE Trans.Nucl.Sci.NS-30,646(1983).

8) A.Del Guerra et al.,Proc. of the Intern. Conf. on Applications of Physics to Medicine and Biology,Trieste,30 March-3 April 1982,World Scientific Book, p.355(1983).

9) R.Bellazzini et al.,"Some Aspects of the Construction of HISPET:a HIgh Spatial Resolution Positron Emission Tomograph",IEEE Trans.Nucl.Sci. NS-31, Vol.1 (1984) to be published.

PROCEEDINGS OF THE II INTERNATIONAL CONFERENCE ON APPLICATIONS OF PHYSICS TO MEDICINE AND BIOLOGY
edited by Ž. Bajzer, P. Baxa & C. Franconi

DECAY DATA OF RADIOTHERAPEUTICS $^{144}Ce/^{144}Pr$ USED AS BETA-RAY APPLICATORS

J. B. Olomo
Department of Physics
University of Ife, Ile-Ife
NIGERIA

Introduction

Beta-emitting isotopes such as $^{144}Ce/^{144}Pr$ have become effective radiotherapeutics for the treatment of superficial cancers especially benign conditions including selected cases of naevi in infants and various diseases of the eye by means of β-ray spectrometry. ^{144}Ce, a long-lived parent, decays by emitting low-energy β-particles to the exited states of its short-lived daughter, ^{144}Pr, which emits the required high energy β-particles to the ground state of ^{144}Nd. These high energy β-transitions are usually emitted together with weak γ-radiations which must have very low absolute values of emission probability, I_γ, if the absorbed dose to the healthy tissues around the tumour region is to remain as low as reasonably achievable. Absolute emission probability is one of the crucial nuclear data widely used in the calculations of photon penetration and energy deposition in human organs and tissues. Existing I_γ data for the most important γ-transitions of the β-emitting radiotherapeutics $^{144}Ce/^{144}Pr$ lack the accuracy sufficient to fulfil this requirement.

Experimental Method and Results

Absolute γ-ray emission probabilities in the decay of radiotherapeutics $^{144}Ce/^{144}Pr$ have been determined by a method which was similar to that employed by Debertin et al.[1] and it involved two separate techniques, (a) measurement of the absolute disintegration rate, N_o, of a radioisotope source using a high precision 4πβ-γ coincidence system and, (b) measurement of γ-ray emission rate, A_γ, by means of a well calibrated Ge(Li) spectrometer.

The absolute emission probability, I_γ is then given by A_γ/N_o.

The system and method used for measuing, and evaluation of, A_γ and N_o for each of the six sources prepared from a solution of $^{144}Ce/^{144}Pr$ obtained from TRC Amersham in the form of $CeCl_3$ in HCl acid has been fully described by Olomo[2]. The total activity measurements were performed over a period greater than one half-life of ^{144}Ce. Least-squares analyses of the decay curves yielded a mean ^{144}Ce half-life of 284.893 ± 0.008 d representing an order of magnitude improvement in precision over earlier reported values from which Tuli[3] evaluated the published half-life of 284.9 ± 0.2 d, a decay constant in close agreement with the present measurement.

The results of I_γ obtained at ±1% relative uncertainty at 68% confidence level in the present work is listed in the table below, and are compared to the data of Debertin _et al._[1] being the only earlier work that has the precision similar to that of the present work.

Absolute Gamma Ray Emission Probabilities (%) in $^{144}Ce/^{144}Pr$.

Nuclide	Photon Energy (KeV)	This work	Debertin _et al_.[1]
^{144}Ce	133.5	10.69 ± 0.12	11.09 ± 0.16
^{144}Pr	696.5	1.484 ± 0.012	1.342 ± 0.013
	1489.2	0.277 ± 0.003	0.279 ± 0.003
	2185.7	0.768 ± 0.009	0.70 ± 0.04

For the 133.5 KeV of ^{144}Ce, the present I_γ value is 4% lower than that of Debertin et al.[1]. The present I_γ values for the energies 696.5 and 2185.7 KeV of ^{144}Pr are 10% higher than that of Ref. 1. However, for 1489.2 KeV of ^{144}Pr excellent agreement exists between the present I_γ and the only earlier value of Ref. 1.

Acknowledgement

The author wishes to thank the Director of Reactor Centre, Imperial College at Silwood Park, Ascot for use of the facilities and Dr. T. D. MacMahon for the series of valuable discussions and assistance.

References

1. K. Debertin _et al._ Ann. Nucl. Energy, 2 (1975) 37.
2. J. B. Olomo. Ph. D Thesis, University of London, (1979).
3. J. K. Tuli. A = 144, Nucl. Data Sheets, 20 (1979) 97.

PROCEEDINGS OF THE II INTERNATIONAL CONFERENCE ON APPLICATIONS OF PHYSICS TO MEDICINE AND BIOLOGY
edited by Ž. Bajzer, P. Baxa & C. Franconi

MEASUREMENTS OF PERIPHERICAL BONE MINERAL CONTENT USING COHERENT AND COMPTON SCATTERED PHOTONS: FIRST "IN VITRO" RESULTS

G.E. GIGANTE and S. SCIUTI

Centro per l'Ingegneria Biomedica dell'Università di Roma
e Istituto Discipline Biologiche dell'Universitá dell'Aquila

The Bone Mineral Content (B.M.C.) can be evaluated on the forearm by the simple technique proposed by Cameron 1). This technique became of large clinical use in the last decade, especially in the assessment of osteoporosis and in the control of effectiveness of therapies. Unfortunately, the B.M.C. measured on the forearm is not always representative of skeleton mineralization. In fact, in the forearm, the bone mineral fraction (B.M.F.) exchange rate is sometimes slowed and delayed in respect to the other sites more subject to mechanical stress, as spine, calcaneum etc. Consequently, B.M.C. measurements have been attempted with dual energy absorptiometric technique in the spine 2) and with Compton scattered technique in other sites 3),4). More recently, a technique has been proposed in which both fractions of scattered spectrum (Compton and Rayleigh) have been used. This technique (referred as R/C in the following) has in principle, the advantage of confining the B.M.F. measurement in a well defined volume of bone (i.e. only trabecular bone). However, R/C technique needs a long measuring time, hundreds of seconds, due to the low intensity of elastic scattered peak. Moreover the γ source low energy (60-80 Kev), requested in order to have reasonable high elastic cross section, limits this technique to measurements on peripherical bone sites only. The R/C technique turns out to be very efficient and accurate if used at large scattering angles (ϑ), even if the elastic peak intensity decreases strongly with ϑ (i.e. as $\vartheta^{-2.3}$). Moreover, also the system sensitivity improves with ϑ, for angles bigger than 45°, as can be demonstrated by means of theoretical considerations 5). In any case, it is not possible to operate at very low ϑ angles, due to intrinsic limitations of the technique.

The above-mentioned theoretical results have been experimentally verified by us. A measuring head at a scattering angle ϑ of 135° has been expressly built characterized by a circular crown of well collimated point sources of Americium 241 surrounding a hp Germanium X-ray detector. The performances of all the equipment have been experimentally investigated. A 1% precision in the B.M.F. measurement can be reached with a measuring time of ~ 400 sec. and a dose of 300-400 mR. The parameters which affect the system response have been extensively studied, showing that a good level of accuracy can be obtained for bone sites placed at a depth of 1-2 centimeters from the skin and with a thickness of trabecular bone of 3-4 g*cm-2.

1) J.R. Cameron and J. Sorenson, Science 142,230 (1963)
2) M. Madsen, Investigative Radiology 12,2,185 (1977)
3) C.E. Webber and T.J. Kennet, Radiology 106,209,212 (1973)
4) P. Puumalainen et al. Radiology 120,723 (1976)
5) G.E. Gigante and S. Sciuti, C.I.B. Internal Report (1983) to be sent for publication.

Biological Materials

Technology, Instrumentation and Prosthetic Devices

Non-Ionizing Radiation

PROCEEDINGS OF THE II INTERNATIONAL CONFERENCE ON APPLICATIONS OF PHYSICS TO MEDICINE AND BIOLOGY
edited by Ž. Bajzer, P. Baxa & C. Franconi

LOW FIELD NMR FOR TOPICAL MEDICAL DIAGNOSIS

Georges J. BENE, Bernard BORCARD and Patrick MAGNIN
Section de Physique de l'Université
CH - 1211 - Geneva - 4
SWITZERLAND

1. Introduction

We show the capability of T_1 and T_2 relaxation times measurements on protons of physiological water in a low field range to obtain detection of pathological state of a given living tissue, identification of the precise nature of that pathology and quantitative evaluation of its importance.

We studied some pathologies of pregnant women by accompanying pollutions of the amniotic fluid (A.F.).

2. Pathologies and material investigated

We explored the foetal distress of the not yet born child who leads to a staining of the AF by the meconium, mainly constituted by mucopolysaccharides and the placenta detachment which produces a staining of the AF by fresh or degraded blood. Precisely, the material to be identified includes: healthy A.F., water solutions of meconium, fresh and degraded blood.

3. Physical principles and application

All the investigated fluids are mainly dilute solutions of diamagnetic mineral ions and macromolecules. In these solutions, molecular dynamics of water molecules may be characterized by correlation times τ_c. Such correlation times, and the energy exchanges between magnetic sublevels and the surrounding, associated to each of them, give a lot of parameters able to describe precisely such water solutions.

These parameters may be easily extracted from a curve giving T_1 of water protons as a function of the Ho field (or Larmor frequency γ_o, proportional to Ho). In the range where $\tau_c \approx \gamma_o^{-1}$, we observe a drastic change of T_1, and the T_1 variation at this place is directly connected to the energy exchanged in this process. Moreover, the actual value of relaxation times decreases with the increasing of macromolecular concentration. Measurements of T_1 or (and) T_2 in a good range of Ho values are then able to distinguish clearly a given macromolecular water solution, by determination of τ_c 's values and associated T_1 variation. Note that for pathological concentrations, we observe in T_2 measurements only a negligible difference between AF and water solutions of blood or meconium.

4. Experimental procedure and results

Two methods are now used to measure T_1 in a large range of Ho values. The field cycling [1], and the free precession of protons in the earth's field after prepolarization [2]. We were able to measure τ_c's values in the four explored solutions, and we note in meconium solutions a large τ_c value ($> 10^{-3}$s) not present in other samples and, consequently, able to lead to an unambigous distinction of that pollution. Such a dispersion may be seen for example, by comparizon of T_2 values measured in the earth's field and T_1 values measured in the polarizing field ($\sim$ 100oe) or in a field of $\approx$ 5 moe, in the free precession method. Experiments are in progress to extend that discrimination in the spin-imaging Ho field range ($\sim$ 1 Koe).

After this discrimination, the actual value of T_2 measured in the low field range gives the content of the polluting material. More precisely, the smaller value of T_2 observed in the earth's magnetic field for the healthy AF is 1.8 s and T_2 values for pathological concentrations of polluting products are in the range 1.5 to 0.2 s. Note that all measurements were made at the body temperature in view of "in vivo" application.

5. References

1) Noack, F., (1971) NMR Principles and Progress 3, 84-144

2) Florkowski, Z., et al. (1980) Nukleonika XIV, 563-569

PROCEEDINGS OF THE II INTERNATIONAL CONFERENCE ON APPLICATIONS OF PHYSICS TO MEDICINE AND BIOLOGY
edited by Ž. Bajzer, P. Baxa & C. Franconi

$^{19}F^-$ AS NMR PROBE FOR SUPEROXIDE ION AND RELATED ENZYMES

Adelio Rigo*+, Marina Scarpa*+, Roberto Stevanato+, Paolo Viglino+
* Institute of General Pathology, University of Padua, Padua, Italy
+ Institute of Physical Chemistry, University of Venice, Venice, Italy

The longitudinal nuclear magnetic relaxation rate (T_1^{-1}) of $^{19}F^-$ is very sensitive to Cu and Mn containing superoxide dismutases (SOD)which are present in the aerobic organisms and protect them from the damages of superoxide ion (O_2^-). The $^{19}F^-$ high relaxation rate is a general and specific property of superoxide dismutases[1] which further show different relaxivities in the oxidized and reduced forms. These characteristics enlarge the field of the possible applications beyond the study of the structure and dynamic of the active site[2]. Some these applications are here considered.

1. Determination of Cu and MnSOD in biological systems

The linear relationship between the relaxivity and the concentration of these enzymes, in the range 10^{-8}-10^{-3}M SOD and the high relaxivity values, which are orders of magnitude higher than those of other metal containing proteins, permit the determination of SOD in tissue homogenates, etc. To discriminate the relative contribution of Cu and MnSOD to $^{19}F^-$ relaxation, the measurements were carried out in the presence and in the absence of CN^- ions. In fact CN^-, at concentrations 10^{-3}M, inhibits the CuSOD completely and has no effect on the MnSOD relaxivity. By this method we measured the Cu and MnSOD in human fibroblasts grown in vitro to study the gene expression. The amount of Cu and MnSOD, which are about 10 and 100 pmole per mg of protein respectively, was found to depend on the culture passage and was about 10 and 100 pmole respectively.

2. Detection of superoxide ion fluxes

The determination of O_2^- fluxes is based on the equal rate of reaction of superoxide with the oxidized and the reduced form of CuSOD.This means that equal concentrations of the two redox states of the enzyme are present[3] at the steady-state.We demonstrated that, starting both from the oxidized or the reduced CuSOD, the steady-state is reached according to a first-order process, which can be easily followed by $^{19}F^-$ NMR, being the kinetic rate constant proportional to the rate of O_2^- generation[4]. The sensitivity of this method, which permits the detection of O_2^- generation as low as 10^{-10} M s^{-1}, and the possibility of operating with high absorbance media make it particularly suitable for biological systems where other methods usually fail. By this method we measured the rate of O_2^- production in red blood cells (RBC) lysates. The rate, never measured before, resulted about 2×10^{-8} M s^{-1} in RBC of healthy people.

3. Determination of redox state of superoxide dismutases

Since the molar relaxivities of the oxidized form of Cu and MnSOD are much higher than those of the reduced form and because the cell membranes are freely permeable to $^{19}F^-$, the redox state of these enzymes can be measured inside the intact cells. It must be remarked that $^{19}F^-$ NMR probes the oxidation state of SOD with a sensitivity of orders of magnitude higher than those shown by other spectroscopies. On these basis the cells must be incubated with $^{19}F^-$ and the T_1 of this ion, can be measured in packed cells. By this method we measured the ratio between the oxidized and reduced form of CuSOD in RBC. Since the relaxivity of the native human CuSOD is not exactly known we took as a reference the relaxation values measured in the presence of an intracellular flow of O_2^- stimulated by the addition of a proper compound to RBC. In particular in the case of healthy people the ratio was found to be one which indicates that in the intact RBC the CuSOD is in turnover condition as we found in RBC lysates[5].

References

1) Rigo, A., Viglino, P., Argese, E., Terenzi, M. and Rotilio, G. (1979) J. Biol. Chem. 254, 1759
2) Viglino, P., Rigo, A., Stevanato, R., Ranieri, G., Rotilio, G. and Calabrese, L. (1979) J. Magn. Res., 34, 265
3) Viglino, P., Rigo, A., Argese, E., Calabrese, L., Cocco, D. and Rotilio, G. (1981) Biochem. Biophys. Res. Commun. 100, 125
4) Rigo, A., Ugo, P., Viglino, P. and Rotilio, G. (1981) FEBS Letters 132, 78
5) Scarpa, M., Rigo, A., Orsega, E.F. and Viglino, P., in press.

PROCEEDINGS OF THE II INTERNATIONAL CONFERENCE ON APPLICATIONS OF PHYSICS TO MEDICINE AND BIOLOGY
edited by Ż. Bajzer, P. Baxa & C. Franconi

DETERMINATION OF FACTORS AFFECTING THE SPIN-LATTICE RELAXATION TIME T_1 OF PATHOLOGICAL BLOOD BY FOURIER TRANSFORM NMR SPECTROMETER

A.YILMAZ and K.BALCI
Departments of Physics and Haematology
University of Dicle, Diyarbakır, TURKEY

In recent years, Singer and his collaborators determined that $1/T_1$ rates of pathogenic blood are proportional to the hemoglobin concentration(Hb).[1] Furthermore, on the contrary of the results in the previous studies, McLACHAN has measured that proton relaxation rates in plasma from patients tend to be different than those for healthy people.[2] In this research, We have investigated the dependence of $1/T_1$ values of pathological blood on the Hb concentration and $1/T_1$ rates of plasma and resuspended cells by using FT-NMR Spectrometer at 60MHz. T_1 measurements were carried out by Inversion Recovery Method. 180-τ-90 pulse sequences were used and τ is altered within (0.3-1.8)s. Data is given in fig(1).

According to the simple NMR theory of aqueous protein solutions, $1/T_1$ values of solutions are proportional to the solute concentration and $1/T_1$ of solvent.[3] In the case of blood, least squares fitting of $(1/T_{1b} - 1/T_{1p})$ versus Hb comes to an agreement with this findings (b:blood , p:plasma). Regression equation of the straight lines in fig(2) can be expressed as

$$1/T_{1b} - 1/T_{1p} = \alpha_i + k_i \mathrm{Hb} \quad (i: 1,2) .$$

Where 1 and 2 numbers denote the line segments of (8-16)gmHb/100mL interval and (5-8)gm/100mL interval respectively (α_1:-0.003, k_1:0.0152, α_2:-0.126 and k_2:0.0303). α_1 can be taken as zero and the regression equation gives an equality such as $1/T_{1b} \simeq 1/T_{1p} + k_1\mathrm{Hb}$. On the other hand, NMR study of blood supports the existence of a fast exchange mechanism between plasma and erythrocyte phases.[4] So if the protons of water in blood are classified as the water protons in plasma and water protons bounded to the Hb molecules irrotationally, then the result found above may be reasonable.[3] Thus, the relaxation mechanism of pathogenic blood with higher Hb values can be analyzed in terms of the fast exchange mechanism between intracellular and extracellular phases. The correlation between Hb and blood T_1 values has been calculated as 0.96 and 0.83 for interval(1) and interval(2) respectively. Therefore, the change of slopes about 8 gm/100mL may be related to decreasing of interaction between Hb and water molecules. In addition to these, it has been shown that $1/T_1$ of 0.5mL of pathogenic cells readded in 1mL isotonic solution is constant on average

REFERENCES

1. SINGER J.R. et al, J.Clin.Eng., 3: 237-243,1978
2. McLACHAN L.A. Phys. Med. Biol., 25: 309-315, 1980
3. DASKIEWICZ O.K., Nature, 200: 1006-1007, 1963
4. CONLON T. and OUTHRED R., Biochim.Biophys.Acta,228:354-361,1972

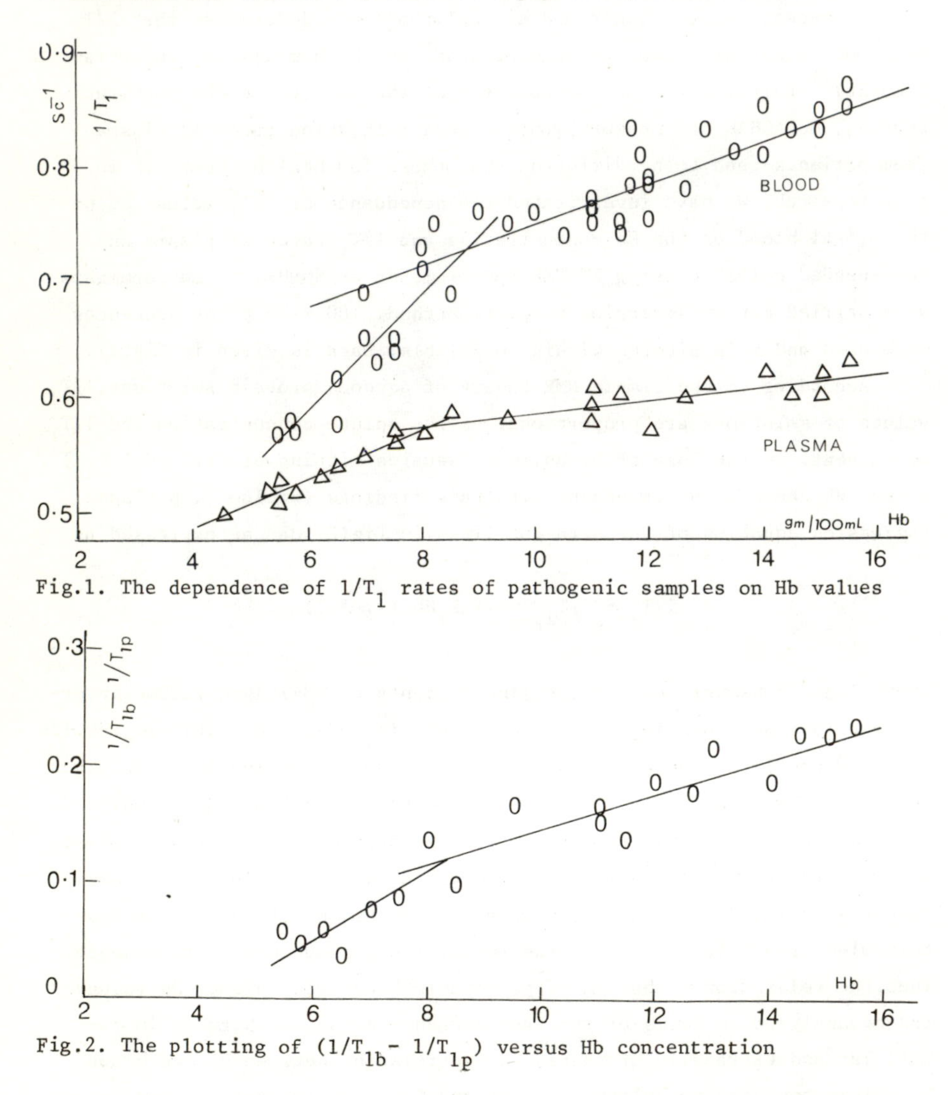

Fig.1. The dependence of $1/T_1$ rates of pathogenic samples on Hb values

Fig.2. The plotting of $(1/T_{1b} - 1/T_{1p})$ versus Hb concentration

PROCEEDINGS OF THE II INTERNATIONAL CONFERENCE ON APPLICATIONS OF PHYSICS TO MEDICINE AND BIOLOGY
edited by Ż. Bajzer, P. Baxa & C. Franconi

MEASUREMENTS AND CALCULATIONS OF NMR SPIN-LATTICE RELAXATION TIME IN HUMAN BLOOD

A.YILMAZ and B.TURAN
Deparments of Physics, University of Dicle
Diyarbakır, TURKEY

Although the explanation of the relaxation mechanism of whole blood has been examined by many researchers, a satisfactory result has not been obtained yet.[1] In this work, we have suggested a relaxation mechanism for normal blood and plasma by considering the effect of paramagnetism in plasma. Normal blood samples diluted by its own plasma have been studied. The measurements were performed by computerized 60MHz FT-NMR Spectrometer at 30^oC. Inversion Recovery Method was used and τ was altered within (0.4-1.8)s. The least squares fitting of $1/T_1$ rates versus hemoglobin(Hb) concentration gives a regression equation as

$$1/T_{1b} = 0.573 + 0.006 \text{ Hb} \simeq 1/T_{1p} + k \text{ Hb} \qquad (1)$$

Where $1/T_{1b}$ and $1/T_{1p}$ are the relaxation times of blood and plasma respectively. On the other hand, the relaxation mechanism of paramagnetic solutions can be written as[2]

$$1/T_1 = \frac{16\pi^2 (g g_N \beta \beta_N)^2 \, S(S+1)\, N \eta}{15\, \hbar^2\, kT} . \qquad (2)$$

Where N and η denote the paramagnetic ion concentration and the viscosity of solution respectively. Since the paramagnetic iron is distributed in the extracellular plasma, we can calculate the concentration of the iron in blood from equality $N_b \simeq N_p(1\text{-HCT})$. Also, for the viscosity of blood, we can use an expansion formula by neglecting the second and higher order terms.[3] Thus $\eta_b \simeq \eta_p(1 + \text{HCT})$ is obtained. If these are substituted in eq(2) and necessary cancellations are made, a similar result to eq(1) is derived from eq(2).

In order to apply eq(2) to the plasma, it was assumed that there is a fast exchange mechanism between free water and water protons bounded to the transferrin in plasma. The concentration of paramagnetic iron in the plasma was determined as 1.47μgm/mL. By using of Ostwald Viscometer, relative viscosity was calculated as 1.75cp at 30^oC. S and g-value are

5/2 and 4.3 respectively.[4] By using of these values, $1/T_{1p}$ is calculated as 0.55 $sec.^{-1}$ Experimental value was 0.58 $sec.^{-1}$ On the other hand, the relaxation rates of diluted blood can be calculated from eq(2) by using of η_b values in fig(1).As it is seen from fig(2), there is a good agreement between the experimental and theoretical values. Finally, we have concluded that the paramagnetic iron in plasma plays a dominant role for the relaxation mechanism of blood and plasma.

REFERENCES

1. BROOKS R.A. et al, IEEE Trans. Biomed. Eng., BME-22: 12-18,1975
2. CARRINGTON A. and McLACHAN A.D., Introduction to magnetic resonance, HARPER and ROW Publisher, 1967
3. GIORDANO R. et al, Physics Letters, 70A: 64-66,1979
4. DODD N.J.F., Br.J.Cancer, 32: 108-119, 1975

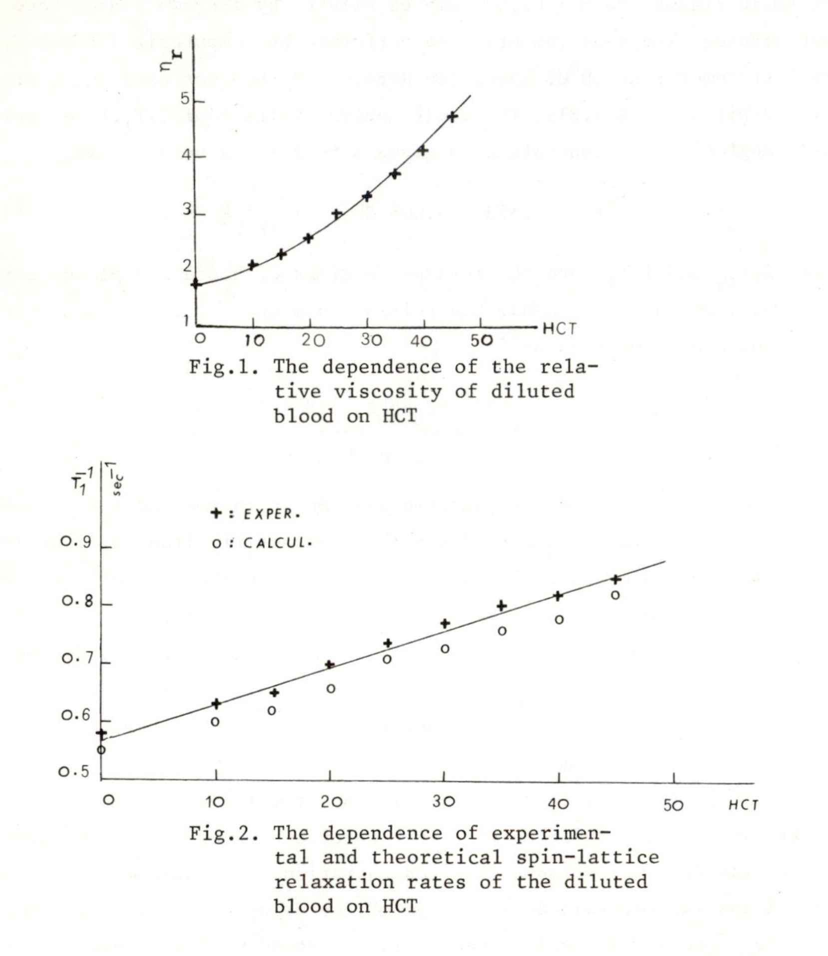

Fig.1. The dependence of the relative viscosity of diluted blood on HCT

Fig.2. The dependence of experimental and theoretical spin-lattice relaxation rates of the diluted blood on HCT

PROCEEDINGS OF THE II INTERNATIONAL CONFERENCE ON APPLICATIONS OF PHYSICS TO MEDICINE AND BIOLOGY
edited by Ž. Bajzer, P. Baxa & C. Franconi

THE SYSTEMATIC EFFECT IN HUMAN SERA OBSERVED BY ^{1}H NMR RELAXATION TIMES

L.Guidoni, P.L.Indovina, V.Viti - Lab. di Fisica, Ist. Superiore di Sanità, Roma (Italy).

1. Introduction

The NMR properties of tissue water in healthy organs of tumor bearing animals and humans are affected by the presence of the desease. This effect, referred as systemic effect, has been investigated by measuring the NMR relaxation times $T_1$and T_2 of water protons in distant tissues, including serum (For a review on this subject, see ref. 1). These parameters,in fact, are considerably affected in sera of tumor bearing animals and, at a lesser extent, in sera of humans. Despite the interest on this matter, no definite conclusion can be drawn on the origin of the effect nor on a possible diagnostic utilization of the phenomen.

The present study has been therefore undartaken with the aim of: i) further clarifying the origin of the increase of NMR relaxation time T_1 in cancer; ii) providing a better methodology to discriminate between sera from halthy and tumor affected individuals.

2. Results

All measurements were carried out at temperatures below the freezing point of the water in sera. In fact, in frozen samples only the unfrozen water associated with the biological entities is observable with the high resolution NMR spectroscopy.

The experiments were run on frozen samples under identical thermal treatment, mainly at -6°C, that is the highest temperature attainable without defrosting the sample. At this temperature the area of bound water peak is at maximum, and the relaxation times are longer than at lower temperatures, thus amplifying possible variations.

The water peak area decreases by decreasing temperature, exibiting a binary phase behavior, on the kind already observed in frozen protein solutions.

The relaxation times T_1 for water protons in frozen samples, measured by the inversion-recovery technique, reveals at least two components, but more markedly the longer one, are lenghtened by the presence of a cancer in the patient. This finding rules out the explanation, proposed for T_1 increases in cancerous tissues and also suggested for sera, that the observed increase in relaxation times is produced by an increase in bulk water concentration.

The analysis of a large number of sera of patients affected by neoplasms and different diseases is now under investigation, possibly providing new diagnostic and/or prognostic tools.

(1) P.T. Bell,D.Medina, andF. Hazlewood , cap 3 in "NMR in Medicine" R. Damadian Editor.

PROCEEDINGS OF THE II INTERNATIONAL CONFERENCE ON APPLICATIONS OF PHYSICS TO MEDICINE AND BIOLOGY
edited by Ž. Bajzer, P. Baxa & C. Franconi

STRUCTURAL MODIFICATIONS INDUCED IN ERYTHROCYTE MEMBRANES BY THE LIPOPHILIC VITAMINS A, D_3, E AND K_1: a ^{31}P NMR AND SPECTROFLUORIMETRIC STUDY

V.Viti, L.Guidoni - Lab. di Fisica, Ist. Superiore di Sanità, Roma (Italy).
R.Cicero - Ist. di Biologia Generale, Fac. di Medicina, Università di Bari, Bari (Italy).
D.Callari, A.Billitteri - Ist. di Patologia Medica, II Cattedra, Fac. di Medicina, Università di Catania, Catania (Italy).
G.Sichel - Ist. di Biologia Generale, Fac. di Medicina, Università di Catania, Catania (Italy).

1. Introduction

The mechanism of action of the lipophilic vitamins has not yet been fully understood. A correlation between the physical perturbations induced on the plasmatic membrane by some lipophilic vitamins and their fusogenic activity has been suggested (1).

In the present study we correlate morphological modifications produced by lipophilic vitamins on erythrocyte membranes, chosen as a model for natural membranes, with changes in the static and dynamic structure of the lipid bilayer. Changes in the bilayer organization were studied by means of ^{31}P NMR and fluorescence anisotropy of the DPH probe. The occurrenec of cell fusions was checked by light microscopy.

2. Results

Vitamins A, E, K_1, which produce cell fusion, also induce the formation of configurational phases other than the bilayer. They also increase the membrane fluidity.

On the contrary, Vitamin D_3 induces aggregation of the erythrocytes not followed by fusion. This behavior is accompanied by bilayer phase stabilization and microviscosity increase in erythrocyte ghosts.

When Vitamin A is added to the erythrocytes and their membranes in the presence of Vitamin E or K_1, two different behaviors can be observed:

- at low Vitamin A to Vitamin E (or K_1) ratios, the effects of the latter predominate;
- at high Vitamin A to Vitamin E (or K_1) ratios, the former Vitamin affects the bilayer, at a lesser extent than alone.

When Vitamin D_3 is added in the presence of Vitamin A, a very strong protective effect is observable.

In conclusion, the fusogenic activity of the lipophilic Vitamins seems to be very strongly correlated to the modifications of the bilayer organization and to changes of fluidity of the membrane.

(1) V.Viti,R.Cicero,D.Callari,L.Guidoni,A.Billitteri,G.Sichel, FEBS Letters 158, 36-40 (1983).

PROCEEDINGS OF THE II INTERNATIONAL CONFERENCE ON APPLICATIONS OF PHYSICS TO MEDICINE AND BIOLOGY
edited by Ž. Bajzer, P. Baxa & C. Franconi

RADIOWAVE CONDUCTIVITY DISPERSIONS IN NORMAL AND HOMOZYGOUS β-THALASSEMIC ERYTHROCYTE SUSPENSIONS

C. Ballario[*], A. Bonincontro[*], C. Cametti[*], A. Rosi[1], L. Sportelli[2]

[*]Dipartimento di Fisica, Università di Roma "La Sapienza", Italy

[1]Laboratorio di Fisica, Istituto Superiore di Sanità
Viale Regina Elena, 299 - Roma, Italy

[2]Dipartimento di Fisica, Gruppo di Biofisica Molecolare,
Università della Calabria, Arcavacata di Rende - Italy

The conductometric behaviour of human erythrocyte suspensions both in normal and pathological state (homozygous β-thalassemia) was studied in a frequency range from 5KHz to 100MHz at various temperatures in the range 5÷45°C.

The cells were dispersed in saline physiological solution (5mM Na-phosphate, pH=7.4, 0.15M NaCl), and two different hematocrits (30 and 15%) were examinated.

A well defined dispersion was observed characterized by a relaxation time of about $3\div4\ 10^{-8}$ sec depending on temperature. These dispersions are attributed to the Maxwell-Wagner mechanism, considering the cell as a conducting particle covered with a less conducting membrane.

The analysis of the experimental results, carried out using an expression given by Hanai et al. for the conductivity of a suspension of ellipsoidal particles covered with a shell, allows to estimate the capacitance and conductance per unit surface of the cell membrane.

Normal cells display a surface capacitance C_M of about 1.1 $\mu F/cm^2$, roughly independent of temperature, in good agreement with the mean value reported in literature for most biological cells.

On the other hand, a value of about 0.5 $\mu F/cm^2$ is obtained for β-thalassemic erythrocyte cell membrane.

These values approach those observed in a lipid bilayer, thus suggesting a reduction in the protein content of the hydrocarbon region of the membrane.

Also, a different protein interaction with the membrane lipids, by means of hydrophobic associations of the non polar residues with the lipid bi layer, may be considered to justify the low values of C_M in the pathological membrane.

In the normal samples, the membrane conductance G_M assumes values ranging from 0.5 to 3mho/cm^2 depending on temperature. On the contrary, a value lower than about one order of magnitude, with less pronounced dependance on temperature, must be taken into account for the β-thalassemic cell membranes. It may be supposed that this lower membrane conduc tance may be due to a different lipid distribution or, perhaps, to the presence of inclusion bodies, which attach themselves to the membrane and occlude the microchannels in the lipid bilayer, thus affecting the ionic transport equilibrium between cell and medium.

In conclusion, this work supports evidences that conductometric measurements in the frequency range, where β-dispersions occur, may give informations on the structural state of the red blood cell membrane in pathological state.

PROCEEDINGS OF THE II INTERNATIONAL CONFERENCE ON APPLICATIONS OF PHYSICS TO MEDICINE AND BIOLOGY
edited by Ž. Bajzer, P. Baxa & C. Franconi

ROTATIONAL TUMBLING OF PD-DTBN IN CHOLESTEROL-LECITHIN MEMBRANES

Feride Severcan
Phyics Department, Hacettepe University
Beytepe 83, ANKARA
TURKEY

and William Z. Plachy
Chemistry Department, S.F.S.U.
San Francisco, CA, 94132
U.S.A.

1. Introduction

Rotational mobility of the perdeutero di-t-butyl nitroxide (PD-DTBN) spin probe in dimyristoyl lecithin (DMPC) bilayer membrane containing various amounts of cholesterol has been investigated by ESR technique. Above the phase transition temperature of pure DMPC, when the mol percent of cholesterol exceeds 12, two hydrophobic subphases present in cholesterol-phospholipid mixtures[1,2,3]. The spectral parameters obtained by computer deconvolution are directly related to the sub-phase properties of interest. The linewidths in each sub-phase are dominated by the rotational tumbling motion of the probe. The linewidth also gives the effective rotational viscosity of each subphase using Stokes-Debye relationship and the known tumbling radius of this probe[4].

2. Materials and Methods

Samples were prepared using standart methods[5]. A gas permeable teflon sample holder[6] was used to exclude the dissolved oxygen. ESR spectra were taken at X and K bands with a Varian E-12 spectrometer. Spectra were digitized using an HP 3437A digital voltmeter interfaced to an HP 9825A computer. The temperature was around 37°C for all experiments. ESR signal does not give good resolution and computer fit because of overlapping with the aqueous site(water line);Therefore samples were prepared in a nearly nonaqueous environment by using Aquacide(Calbiochem-San-Diego,CA). Overlapping lines were deconvoluted using standart regression methods which fit hybrid Gaussian-Lorentzian lineshapes, including ^{13}C satellites to the spectra[7].

3. Results and Discussion

Lorentzian fraction is dominated in ESR line shape since we used PD-DTBN as spin probe. In addition, we assumed isotropic motion of the probe. For this reason rotational correlation times, τ_r, can be obtained using the equation in refs[8,9,10]. From the line width analysis we obtained three different rotational correlation times of the probe for two lipid sites(one of which is called "cholesterol poor", the other "cholesterol rich") and an aqueous site. τ_r for aqueous site is in the order of $5.5\ 10^{-12}$sec.(±%10). As mol percent of cholesterol increases from 15 to 25, τ_r for cholesterol poor site varies from $9.5\ 10^{-11}$sec to $2.5\ 10^{-10}$sec(±%5). However, τ_r for cholesterol rich site varies from $1.5\ 10^{-10}$ to $6.5\ 10^{-10}$sec(±%20). Increasing the cholesterol content of the bilayer tends to decrease the probe mobility, especially in the "cholesterol rich" site.

Effective rotational viscosity of each sub-phase was calculated by using Stokes-Debye relationship and $r=2A^o$ for tumbling radius of this probe[4]. Corresponding values are 0.0007 $N.s/m^2$ for aqueous site, between 0.0121 $N.s/m^2$ and 0.0255 $N.s/m^2$ for cholesterol poor site, and between 0.019 $N.s/m^2$ and 0.083 $N.s/m^2$ for cholesterol rich site depending on the cholesterol concentration.

4. References

1. D.Recktenwald and H.M.McConnel, Biochemistry, 20,4505 (1981).
2. W.Plachy, J.Frank, C.Morse and G.Krowech, Bulletin of Magnetic Res., 2,399 (1981).
3. F.Severcan, J.Frank, C.Morse and W.Plachy, X^{th} International Conference on Magnetic Resonance in Biological Systems, Stanford, CA, U.S.A. (1982).
4. W.Plachy, D.Kivelson, J.Chem.Phys. 47,3312 (1967).
5. E.J.Shimshick, H.M.McConnel, Biochem.Biophys. Res.Commun., 53, 446 (1973).
6. W.Plachy,D.A.Windrem, J.Mag.Res., 27,237 (1977).
7. D.A.Windrem, W.Plachy, Biochim.Biophys. Acta 600,655 (1980).
8. R.J.Stone, T.Buckman, P.L.Nordio and H.M.McConnell, Proc.Natl.Acad. Sci. U.S.A. 54,1010 (1965).
9. B.Connon, C.F.Polnaszek, K.W.Butler, L.E.G.Eriksson and I.J.P.Smith, Arch. Biochem.Biophys. 167,505 (1975).
10. L.R.Brown, C.Bosch, K.Wuthrich, Biochim.Biophys. Acta 642,296 (1981).

PROCEEDINGS OF THE II INTERNATIONAL CONFERENCE ON APPLICATIONS OF PHYSICS TO MEDICINE AND BIOLOGY
edited by Ž. Bajzer, P. Baxa & C. Franconi

BOUND WATER IN FROZEN AQUEOUS SOLUTIONS OF BIOLOGICAL MATERIALS: A DIELECTRIC STUDY BY THE DTC TECHNIQUE

P.Pissis, L.Apekis, G.Boudouris, D.Diamanti and C.Christodoulides
National Technical University, Physics Laboratory A,
Zografou Campus, Athens 624
GREECE

We report in this paper on the dielectric behaviour of frozen aqueous solutions of several biological materials (oligosaccharides, aminoacids, proteins) studied by the depolarization thermocurrent (DTC) technique in the temperature range of 85-250 K and over a wide range of concentrations. Our aim is to obtain information on the state of water in the solutions, i.e. the fraction of the non-influenced water (free or bulk water), the fraction of the water influenced by the solute molecules (hydration or bound water), the hydration sites and the hydration mechanism. Apart from their theoretical interest, such studies are important also from a practical point of view, since many biological phenomena depend on the amount and state of water present.

The use of dielectric methods in the study of the state of water in hydrated materials is based on the fact that the dielectric relaxation time is different for bound than for free H_2O molecules, due to the difference in their surroundings. The DTC technique consists of studying the thermally activated release of stored dielectric polarization[1]. For our purposes, the main advantage of the DTC technique as compared to the conventional ac techniques, is that DTC can experimentally resolve overlapping relaxation processes.

In the DTC technique, the measurements have to be carried out on frozen solutions, at subzero temperatures. For interpretation these results have to be compared with those of DTC measurements in pure ice. An example is shown in figure 1. The low-temperature DTC peak in pure ice at about 120 K is attributed to polarisation of H_2O molecules in ice, while the peaks at about 120 and 135 K in the maltose solution are attributed to polarization of free and bound H_2O molecules respectively. The most significant quantities obtained from our measurements are the fractions of free and bound water and their activation energies for reorientation.

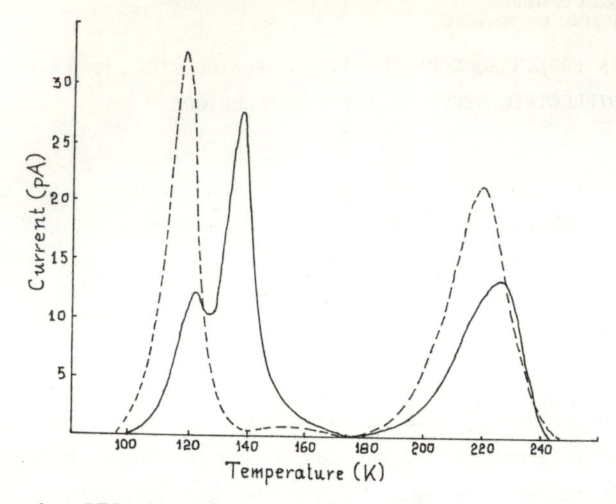

Figure 1. DTC plots for a 0.003 M maltose solution (—) and polucrystalline pure ice (---).

We studied nine oligosaccharides: the monosaccharides glucose, mannose, galactose, ribose and arabinose, the disaccharides cellobiose, lactose and maltose and the trisaccharide raffinose. Our results showed that in the glucose, mannose, galactose and raffinose solutions there is a continuous transition from bound to free H_2O molecules in dilute solutions and that there are only bound H_2O molecules in concentrated solutions. In the ribose, arabinose, cellobiose, lactose and maltose solutions, there are two discrete kinds of H_2O molecules, namely free and bound molecules. The fraction of bound water increases in the above sequence from ribose to maltose.

We studied eight aminoacids: glucine, alanine, valine, norleucine, β-alanine, γ-aminobutyric acid, serine and proline. In the α-aminoacid solutions the fraction of bound water was found to increase with increasing chain length. This result suggests the mechanism of hydrophobic hydration. Also the fraction of bound water was found to be significantly higher in the β- than in the α-alanine.

Finally, the solutions of the proteins studied (bovine hemoglobin and bovine serum albumin) differed from each other in their hydration behaviour. This is in contrast to experimental results reported for measurements above 0° C.

References

1) C.Bucci, R.Fieschi and G.Guidi, Phys. Rev. 148, 816 (1966).

PROCEEDINGS OF THE II INTERNATIONAL CONFERENCE ON APPLICATIONS OF PHYSICS TO MEDICINE AND BIOLOGY
edited by Ž. Bajzer, P. Baxa & C. Franconi

STUDY OF CRYSTALLURIA BY X-RAY DIFFRACTION

M. Baldassarri*, S. Barocci^, R. Caciuffo^, P. Mariani°, S. Melone^, G. Muzzonigro*, M. Polito*, F. Rustichelli°

* Istituto di Patologia dell'Apparato Urinario (Direttore Prof. M.Polito) Facoltà di Medicina e Chirurgia, Università di Ancona

^ Sezione Fisica, Dipartimento Scienze dei Materiali e della Terra, (Responsabile Prof. S.Melone), Facoltà di Ingegneria, Università di Ancona

° Istituto di Fisica Medica (Direttore Prof. F.Rustichelli), Facoltà di Medicina e Chirurgia, Università di Ancona.

Crystals formation is an obbligatory step between chemical and physical alterations in urine and the growth of renal stones[1]. In fact, crystals can represent the initial form of the central nucleus around which the growing of the renal stones takes place. The structure and chemical composition of human urinary calculi has also studied by means of X-ray diffraction on pre-powdered samples. Anyhow, the most up-to-date studies have a limitation in the solely determination of the position and relevant intensities of the diffraction peaks: in fact such data are generally sufficient to determine structural properties of detected samples[2,3]. By means of different techniques, as chemical methods of inorganic analysis, optical microscopy with polarizing light and scanning electron microscopy, some authors carred out several studies on crystalluria, in order to find some possible correlation between crystal occurring and calculosis[4,5]. When a pathological situation arises, it seem possible to make some correlations with the number and size of the "grains", which crystalluria is made of[6]. In the presents notes, we report some preliminary results obtained trough X-ray diffraction on crystalluria samples gathered by urine filtration. A study of the shapes of diffraction peaks as well as an analysis of the signal-noise ratio can actually supply information about crystal defect characteristics: in fact it is possible for such defects to be connected with aggregation phenomena of crystallites and hence with the growing ability of renal stones[7]. The first results obtained, enabled us to achieve a sure determination of chemical composition of crystalluria, even where classical methods, chemical or optical, could not succeed in leading reliable data. The figure shows some of the diffraction profiles obtained. This respectively refer to samples of Uric Acid Hydrate, Calcium Ortho Phosphate Hydrate and Calcium Oxalate Hydrate, provided by patients stoneformers.

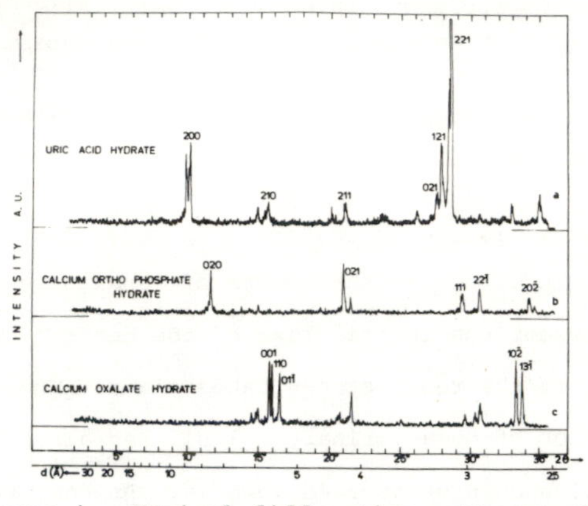

Figure 1 : Typical diffraction patterns of some crystalluria samples.

This study is a part of a systematic work, whose future task is a comparison of results on a statistical population of patients, both formers and non-formers of renal stones.

1. H.Fleisch, in Advances in Nephrourology, vol.9, Plenum Press, (1981) p.275.
2. D.J.Sutor and S.Scheidt, Brit.J. of Urol., 40, (1968) 22.
3. G.Brien, G.Schubert and C.Bick, Eur.Urol., 8, (1982) 241.
4. A.Krajewski, R.Mongiorgi, P.Sabatino and A.Castellano, Miner.Petrog. Acta, 21, (1976) 101.
5. P.C.Hallson and G.Alan Rose, Brit.J.of Urol., 48, (1976) 515.
6. A.Martelli, V.Pulini and P.Buli, in International School of Urology and Nephrology 3° Corso, Plenum Press (1980).
7. M.Polito, III Settimana Bianca Urologica, Cortina (1983).

PROCEEDINGS OF THE II INTERNATIONAL CONFERENCE ON APPLICATIONS OF PHYSICS TO MEDICINE AND BIOLOGY
edited by Ž. Bajzer, P. Baxa & C. Franconi

EPICADMIUM NEUTRON ACTIVATION ANALYSIS OF ELEMENTS PRESENT IN TRACE QUANTITIES IN BIOLOGICAL MATERIALS BY $k_{e,o}$-METHOD

T. Elnimr

Physics Department, Faculty of Science, Tanta Univ., Tanta, EGYPT

In modern medical and biological investigations a frequent necessity is the simultaneous determination of a number of trace elements in low weight biological materials.This problem is solved by the application of various analytical methods, including radioactivation ones.

The present paper deals with a method of increasing the selectivity and sensitivity of the determination of trace elements in biological materials, by means of sample irradiation with epithermal neutron using a new standardization technique called $k_{e,o}$- method[1-3]. The deviation of the epithermal neutron flux distribution from 1/E law, true coincidence effects of cascade γ - rays, the cadmium transmission factor for epithermal neutrons and the efficiency of the Ge(Li) detector were taken into consideration.

The present work examines the possibility and the accuracy of the developed $k_{e,o}$ - method, in the analysis of a standard Bowen's Kale powder, Gold(^{197}Au) was used as a comparator.

Experimental:

Samples. Six samples of roughly 50mg. wet weight were prepared The sample were wrapped in Mylar foil.

Irradiation. The samples together with monitors were placed in a cooled cadmium cylinder h/d =2 and 1mm. wall thickness. The cylinder was transferred to a standard plastic container, which was inserted in a cryostat in the reactor. The sample were irradiated for 5 min. and others for 30 hr.

Measurements. The isotopes activities induced in the samples were evaluated by means γ- spectrometric analysis with Ge(Li) detector connected to 4096 MCA.

Analytical results of Bown's Kale obtained in this work, as well as, the values reported in the literature are presented in table(1) . The elements Ba, Br, Cl, Cs, Fe, Ga, K, Na, Rb, Se, and W can be determined with good precision, for As, Co, Cr, Cu, La, Mo, Zn and Zr our result are probably low which might be due to poor counting statistics and to the high compton background.

Table (1) Element concentrations in Bowen's Kale

Element	ρ_{ppm}(%) This work	ρ_{ppm} lit.data	Element	ρ_{ppm}(%) This work	ρ_{ppm} lit.data
As	0,18(6)	0,15(4	K	25845,7(1,2)	24615(5
Ba	4,41(3,1)	4,55(5	La	o,o98(2,5)	0,098(9
Br	28,37(5,2)	26,7(7	Mo	3,35(3,1)	2,28(5
Cl	3210(3,0)	3415(5	Na	2556,1(0,8)	2506(5
Co	$7,8.10^{-2}$(3,3)	$8,3.10^{-2}$(6	Ni	0,99(1,8)	1,08(6
Cr	0,43(1,8)	0,31(5 &6	Rb	47,5(2,1)	52,2(5
Cs	0,0708(3,0)	0,0738(5	Se	0,13(1,1)	0,131(4
Cu	6,1 (1,2)	4,99(5	W	0,056(6,0)	0,059(4
Fe	126,8(2,3)	123(8	Zn	39,5(5,3)	33,2(5
Ga	0,07(6)	0,069(5	Zr	0,78(4,1)	0,2 (5

It can be conclude that the present procedure allows a very fast non-destructive determination in biological materials. The major advantage is obviously the absence of pre-irradiation treatments of the samples and of any chemical separation after irradiation.

References

1) T.Elnimr, F.De Corte, L.Moens, A.Simonits, J.Hoste
J.Radioanal.Chem. 67(1981)421

2) T.Elnimr, A.Alian, I.I.Bondouk
Proc.Math.Phys.Soc.Egypt in press

3) T.Elnimr, I.I.Bondouk
J.Phys.D.Appl.Phys. 16(1983)1407

4) E.Steinnes" Trace Elem.in Rel.to Cardi. Disease, Tech.Rep. IAEA, Vienna, (1973) p.149

5) H.J.M.Bowen,
J.Radioanal.Chem. 19(1974)215

6) B.Maziere, et.al. J.Radioanal.Chem. 24(1975)279

7) J.Kucera , J.Radiochem.Radioanal.Letters 38(1979)229

8) L.O.Plantin, Nucl.Activ.Tech.in the life science ,IAEA,Vienna(1972) p.73

9) R.A.Nadkarni, et.al. J Radioanal.Chem. 3(1969)175

PROCEEDINGS OF THE II INTERNATIONAL CONFERENCE ON APPLICATIONS OF PHYSICS TO MEDICINE AND BIOLOGY
edited by Ž. Bajzer, P. Baxa & C. Franconi

THE OXFORD SCANNING PROTON MICROPROBE AND ITS APPLICATIONS IN MEDICAL DIAGNOSTICS

J. Takacs, F. Watt, G.W. Grime
University of Oxford
Department of Nuclear Physics
Keble Road, Oxford, U.K.

1. Introduction

Although in many pathological conditions specific trace elements play an important role, the mechanism of their involvement is often obscure. Most analytical techniques currently used in histopathology provide quantitative information on the presence of specific trace elements but they can rarely provide information simultaneously on the local distribution of the various trace elements in the sample. A new instrument called the Scanning Proton Microprobe (SPM) is capable now of providing this information simultaneously(1). This instrument, which uses the analytical technique of proton induced X-ray emission (PIXE) is now becoming recognised as a highly sensitive device for elemental analysis in medical and biological fields(2).

2. Scanning Proton Microprobe

When a beam of energetic protons (2 MeV or above) impinges on matter there is a high probability that inner shell electrons are ejected from the sample atoms. Then electrons from outer shells cascade inwards to fill the vacancies and in doing so X-rays are produced with energies characteristic of the parent atom. By using solid state X-ray detectors (eg. Si(Li), Ge) these X-rays can be identified thereby allowing the elemental character of the sample to be obtained.

By focusing down the proton beam to small spot sizes and rastering the beam across the sample, elemental maps can be constructed by observing the induced characteristic X-rays as a function of beam position. The SPM system developed in Oxford Nuclear Physics Laboratory has a spatial resolution of approximately 1μm. With the facility of mapping of all elements above Z=10 simultaneously, it also offers direct visual correlation between elementary distributions.

The SPM with the capability of detecting elements in samples at about 10^{-17} g/μm^2 concentration level makes a valuable contribution to trace element analysis of biological and medical samples.

3. Applications

The fields of application of the SPM are numerous and interest has been shown in the Oxford SPM by workers in fields as diverse as archaeology, geochemistry, zoology, metallurgy, pathology, pharmacology, botany and medicine. Two medical problems which are currently under investigation are described below.

a. Alzheimer's disease (senile dementia)

Recent research into this disease has indicated its possible link with an increase of aluminium present in the brain. Preliminary SPM studies have indicated that the presence of parts per million of aluminium can be detected in the brain tissues. Efforts are being made to correlate the Al distribution with the characteristic neuro pathological features of the disease.

b. Primary biliary cirrhosis (PBC)[2,3]

Primary biliary cirrhosis is a terminal disease characterised by progressive destruction of small intrahepatic bile ducts, cholestasis and high level of copper in the liver.

The elemental maps constructed by the Oxford SMP at several different magnifications show that the copper is deposited on the periphery of diseased portal tracts in small deposits of approximately 5μm in size. These deposits also contain sulphur with a 1:1 Cu/S atomic ratio. Work is continuing on other heavy element related diseases of the liver.

REFERENCES

1) F.Watt, G.W.Grime, G.D.Blower, J.Takacs and D.J.T.Vaux
Nucl.Instr.and Meth. 197, 1, (1982) 65.

2) F.Watt, G.W.Grime, J.Takacs and D.J.T.Vaux
Proceedings of the "Third Int. Conf. on Particle Induced X-ray Emission" Heidelberg 1983.

3) D.J.T.Vaux
Oxford Medical School Gazette Vol.XXXIV No.2, (1983) 30.

PROCEEDINGS OF THE II INTERNATIONAL CONFERENCE ON APPLICATIONS OF PHYSICS TO MEDICINE AND BIOLOGY
edited by Ž. Bajzer, P. Baxa & C. Franconi

TRACE ELEMENT DISTRIBUTION IN LIVER : A SCANNING PROTON MICROPROBE ANALYSIS*

B. Gonsior, W. Bischof, M. Höfert, B. Raith, A. Stratmann
Ruhr-Universität Bochum, Institut für Experimentalphysik 3
Postfach 102149, D-4630 Bochum 1, Fed. Rep. Germany

and

E. Rokita
Jagellonian University, Institute of Physics, Reymonta 4, Cracow, Poland

In two different research projects we investigated time dependent fluctuations of trace element distributions in liver samples with a spatial resolution of a few micrometers. As analytical method we used the proton induced X ray emission (PIXE)[1] in combination with a scanning proton microprobe [2]. The basic principle of this technique is the following: The sample is irradiated with protons that are accelerated to an energy of a few MeV. The protons create vacancies in the inner shells of the sample atoms. These vacancies are filled by electrons from outer shells, followed by emission of X rays which energies are characteristic of the element the radiation originates from. Using an energy-dispersive Si(Li)-detector the radiation of all sample elements with $Z \geq 14$ (Silicon) can be detected simultaneously. The detection limits, calculated on the base of a statistical definition [3] are between 1 and 10 ppm. They depend on the radiation background - which is due to bremsstrahlung and Compton scattering of γ rays - that superimposes the peaks of the characteristic X rays in the energy spectrum and are optimum for a proton energy of 3 MeV [4].

To apply PIXE to microanalysis the diameter of the proton beam from the accelerator is reduced to 30 μm by an object diaphragm. Four magnetic quadrupole lenses are used to further reduce the beam diameter down to 2 μm at the sample position. A deflection coil between the lens system and the specimen chamber enables beam deflection for raster or line scan operation. Beam movement and data acquisition are controlled by a PDP 11/ CAMAC data processing system.

The main emphasis of our work is to study the role of trace elements in metabolic processes at the level of cooperative cell colonies. In our case the liver lobuli constitute these functional units. Our investigations revealed significant trace element profiles in the case of samples of rabbit liver. We could show the time dependent changes of the distribution of bromine after general anaesthesia supplemented by halothan. We also find very distinctive distributions for other trace elements and for the macro elements. In contrast to experimental results obtained so far elsewhere it turns out that in the liver the distribution of the trace elements is indeed significant.

On the other hand investigations on samples of human liver do not result in similar significant distributions. The reason for this, as we believe, is connected to the more complicated sample taking under clinical conditions of post-mortem excision. The main problem is the unavoidable time delay until deep freezing of the sample is possible. The element composition of the microstructure seems to decay very rapidly, so that analytical results are not of comparable reliability.

References

1) S.A.E. Johansson, T.B. Johansson, Nucl. Instr. and Meth. 137(1976)473

2) W. Bischof, M. Höfert, H.R. Wilde, B. Raith, B. Gonsior and K. Enderer, Nucl. Instr. and Meth. 197(1982)201

3) F.S. Goulding and J.M. Jaklevic, Ann. Rev. Nucl. Sci. 23(1973)45

4) B. Raith, M. Roth, K. Göllner, B. Gonsior, H. Ostermann and C.D. Uhlhorn, Nucl. Instr. and Meth. 142(1977)39

*Supported in part by Bundesminister für Forschung und Technologie, D-5300 Bonn, Fed. Rep. Germany, under the auspices of Umweltbundesamt.

PROCEEDINGS OF THE II INTERNATIONAL CONFERENCE ON APPLICATIONS OF PHYSICS TO MEDICINE AND BIOLOGY
edited by Ž. Bajzer, P. Baxa & C. Franconi

AN APPRAISAL OF K-ESCAPE EFFECT AND CROSS SECTION VALUES IN HPGe DETECTORS

Agostino TARTARI, Gina NAPOLI, Claudio BARALDI and Ernesto CASNATI
Istituto di Fisica dell'Università I-44100 Ferrara (Italy)

A method is proposed for evaluating the photon-atom collision parameters useful in interpreting information obtained with intrinsic germanium detectors which are increasingly used in the biomedical field (radiodiagnosis, elemental analysis). This method is based on the well known Axel model[1] for analytical evaluation of K-escape probability $\varepsilon_{\alpha,\beta}$ which is expressed as:

$$\varepsilon_{\alpha,\beta} = \frac{A^i}{a^i_{\alpha,\beta}+A^i} = \omega_k \, P_{\alpha,\beta} \, \frac{\tau_k(E)}{\mu(E)} \left[1 - \frac{\mu(E_{\alpha,\beta})}{\mu(E)} \ln\left(1 + \frac{\mu(E)}{\mu(E_{\alpha,\beta})}\right)\right] \qquad 1)$$

where ω_k is the K-shell fluorescence yield; τ_k the K-photoelectric attenuation coefficient; μ the total attenuation coefficient for the energy E of the incident photon and $E_{\alpha,\beta}$ of the photon escape; $P_{\alpha,\beta}$ is the relative $K_{\alpha,\beta}$ x-ray emission rate[2]; A^i the registred peak area corresponding to the E energy and $a^i_{\alpha,\beta}$ the area corresponding to the escape peak α or β respectively.

Analyzing the escape fraction $a^i_{\alpha,\beta}/A^i$ by means of an ORTEC 16325 HPGe-LEPS detector the following objectives are proposed: 1) to verify the absolute value and the trend of escape fractions vs incident photon energy; 2) to select the $\tau_k(E)$ and $\mu(E)$ sets for Ge.

The escape fraction calculated in the present work is supplied by a computer program which analytically evaluates each individual peak registred. The validity and accuracy of the results are checked by calculating the total escape fraction $K_T = (a^i_\alpha + a^i_\beta)/A^i$ for each incident photon energy E from the ratio between escape peak integration "in toto" and peak A^i integration.The results are reported in Table 1.

Table 1 . Present experimental results **(%)**

E (keV)	K_α	K_β	$K_\alpha + K_\beta$	K_T
17.44	9.04 ± .05	1.54 ± .02	10.58 ± .05	10.60 ± .06
19.65	7.36 ± .09	1.24 ± .03	8.60 ± .09	8.47 ± .08
23.11	5.40 ± .04	.861 ± .006	6.26 ± .04	6.30 ± .09
26.16	4.20 ± .09	.723 ± .007	4.92 ± .09	4.86 ± .09

Once $K_{\alpha,\beta}$ and K_T are experimentally obtained (and thus $\varepsilon_{\alpha,\beta} = K_{\alpha,\beta}/(1+K_T)$) it is possible to draw ω_k from equation 1)using self-consistent sets for τ_k and μ values. Noting that, to date, the least uncertain of all collision parameters in equation 1) is ω_k[3),4)] (= 0.553) the value inferred with the above-mentioned method could supply a powerful test. Focusing attention on the two most highly acredited and updated sets of data (the one based on Veigele's experimental data [5)] and those derived from the theoretical results of the single electron model [6),7)]) calculations carried out on the present experimental data has shown that while the results with the sets of ref. 5) essentially do not pose any problems ($\omega_k = 0.540 \pm 0.011$) those obtained with sets deduced from ref. 6) clearly understimate the value ($\omega_k = 0.506 \pm 0.011$). An accurate statistical analysis seems to indicate that these conclusions are not accidental and, therefore, one can reasonable presume that the sets in ref. 5) are more representative.

In conclusion, for practical use, equation 1) can prove to be complicated due to the fact that one must know τ_k and μ for each E. It was found that the expressions:

$$K_\alpha(E) = \exp(-6.82 + 5.64\,w - 1.75\,w^2 + 0.112\,w^3) \qquad 2)$$

$$K_\beta(E) = \exp(-9.15 + 5.47\,w + 1.55\,w^2 + 0.081\,w^3) \qquad 3)$$

with $w = \ln E$, yields results within 3% agreement of the experimental and analytical data and for incident photon energy up to 60 keV.

References

1) Axel P 1953 BNL report 271; see also Seltzer S M 1981 Nucl. Inst. Meth. 188, 133
2) Khan Md R , Karimi M 1980 X-Ray Spectrometry 9, 32
3) Casnati E, Tartari A, Baraldi C, Napoli G (to be published)
4) Langenberg A, van Eck J 1979 J. Phys. B: At. Mol. Phys. 12, 1331
5) Veigele W J 1973 At. Data 5, 51
6) Hubbell J H 1982 Int. J. Appl. Rad. Is. 33, 1269 (and references therein)
7) Scofield J H 1973 Report UCRL 51326

PROCEEDINGS OF THE II INTERNATIONAL CONFERENCE ON APPLICATIONS OF PHYSICS TO MEDICINE AND BIOLOGY
edited by Ž. Bajzer, P. Baxa & C. Franconi

DEVELOPMENT OF A 50/60Hz UNIVERSAL HARMONIC ELIMINATOR

Nicolosi, D.E.C.; Farias, R.A.C.
CTPE, Instituto "Dante Pazzanese"
de Cardiologia - São Paulo, Brazil

Introduction

In the medical instrumentation field, one of the most serious problems of working is to eliminate the AC power source interference from biological signals. Our purpose is to clean with a filter, AC interference from ECG signals received from traditional biological amplifiers. The use of passive RLC network makes a filter that is dependent on the inductor's quality. Thus we have the active filter's option but its sensitivity and instability are high when we need a very selective filter. Considering these problems, we decided to develop a filter wich utilizes a technique known as "n-path filter", where the complex poles are generated by switching of a sequence of capacitors in parallel arrangement. This filter has the advantage of obtaining high Q in low frequencies, to be little sensitive, to be sinchronized with the AC power source signal and to be simpler than the digital process of similar performance[1]

The N-Path Filter[2,3]

The general N-path filter theory was developed by Franks & Sandberg[2]. In our application, the element Z_{ij}[2] is a capacitor, as we see in figure 1, in a practical circuit with successively grounded capacitors and its transfer function is:

$$E_2(jw)/E_1(jw) = [\sin(\tau/N)/(\tau/N)]^2 [G(jw-jw_o)+G(jw+jw_o)] \quad (1)$$

where

$$G(jw)=(NR_2/jwC)[(NR_1+(jwC)^{-1})(NR_2-(jwC)^{-1})-1/(wC)^2]^{-1} \quad (2)$$

as we see in Fig.1 $R_2 >> |(jwC)^{-1}|$

$$E_2(jw)/E_1(jw)=(\sin(\tau/N)/(\tau/N))^2[NR_1C(jw-jw_o)+1)^{-1}+(NR_1C(jw+jw_o)+1)^{-1}]. \quad (3)$$

This relation is equivalent to a 2nd order band-pass filter. Thus we see that the central frequency is independent of the discret components, and it depends only on w_o and its sensitivity is pratically null.

The diagram shown in figure 1 is an 8-path filter, which is syncronized with the AC power source through a Phase-Locked-Loop, PLL. As the 8-path switching frequency needs to be $8.w_c$, the PLL has a "divide by eight" circuit on its feedback to multiply the w_o AC frequency source. Then, the signal e_2 (t) tuned by the 8-path is inverted and added with the original input signal $e_i(t)$. In this way the signal $e_o(t)$ is obtained and its spectrum does not have the AC w_o frequency. Thus this circuit is a notch with high selectivity and stability.

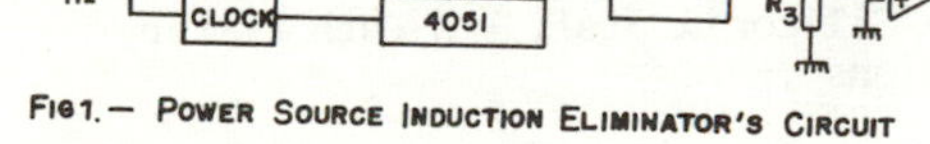

FIG1.— POWER SOURCE INDUCTION ELIMINATOR'S CIRCUIT

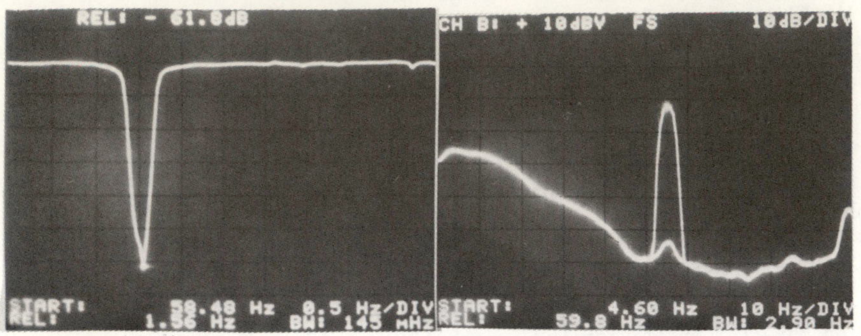

Figure 2.a Figure 2.b

Figure 2.a shows the 60Hz eliminator's frequency response. Figure 2.b shows the frequency spectrum of an ECG before and after passing by eliminator. As we see, the circuit does not alter the signal spectrum only rejecting the 60Hz frequency. Figure 3 illustrate the ECG impregnated with the 60Hz induction in a real time sampling, and the ECG after passing by de harmonic eliminator, where one notices the process effectiveness.

Conclusions:- The method's efficacy is superior to that obtained with traditional analog circuits, is equivalent to digital methods[1], and the electronic circuit is simpler to construct. A problem of this method is that it does not reject significantly the harmonics of the AC power source. The n-path was studied without using the Band-Pass-Filter. It allows to obtain notches in the AC power source harmonics (120 and 180Hz), however, due to the fact that n-path output is a sampled signal we do not manage to obtain a degree of rejection bigger than 25 dB for 60, 120 and 180Hz. As the circuit is very simple,it can be reproduced in any biologic amplifier for cardiology or for biological signals. If this band-width of 240Hz is a limiting factor, it is possible to double the capacitors'number and to obtain a 480Hz bandwidth. With the appearing of integrated circuits filters with switching techniques like the MF-10, this technique will be substituted in the future by simpler circuits.

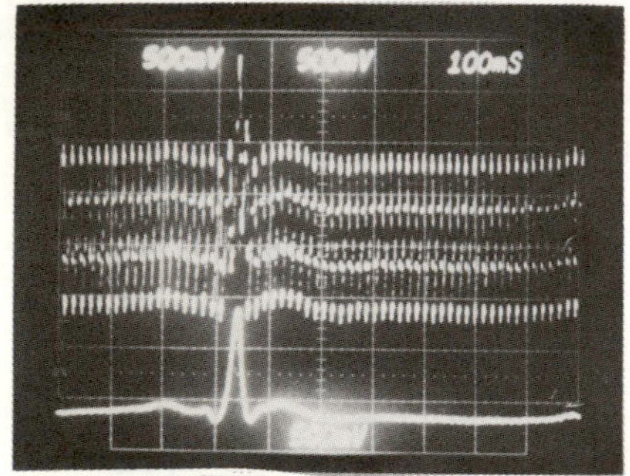

Figure 3

Bibliography

1- Paul B. Brown and Lavi Malhotran, "A 60Hz Harmonic Eliminator", IEEE Trans. Bio. Med.Eng. BMG - 25, 392-397, 1978.

2- L.E.Franks and I.W.Sandberg, "An alternative Approach to the realization of Network Transfer Functions: The N-Path Filter", The Bell System Technical Journal, September 1960 - pg. 1321-1350.

3- Siliconix Staff - N-Path Filter in "Analog Switches and Their Applications".

PROCEEDINGS OF THE II INTERNATIONAL CONFERENCE ON APPLICATIONS OF PHYSICS TO MEDICINE AND BIOLOGY
edited by Ž. Bajzer, P. Baxa & C. Franconi

PRELIMINARY RESULTS OF A HOME BUILT WHOLE-BODY NMR IMAGING APPARATUS

F. De Luca, B.C. De Simone, B. Maraviglia, C. Casieri and F. Vellucci*

Dipartimento di Fisica, Università di Roma, P.le A. Moro, 2
00185 Rome - Italy

A whole-body NMR imaging apparatus was built as a technological and biomedical research prototype in our physics dept. The scheme of our apparatus was based on the experience of a previous small scale (5cm) system, also built in our laboratory, which started operating in spring 1981, the whole-body apparatus can operate with the projection-reconstruction[1] and spin-warp[2,3] methods, with all the common pulse sequences used to obtain ρ, T_1, T_2 contrast. Other imaging methods like planar-echo[4] are feasable, but due to their low efficiency are still under investigation.

The whole software is home built and also most of the hardware was projected and built at home to get as much insight and control of the machine as possible. The system can also provide 3D images[5].

The project began in summer 1979 after a proposal from "Progetto Finalizzato Tecnologie Biomediche". From 1982 a collaboration with ANSALDO of Genoa has started. This collaboration has brought to the project of a third prototype for clinical evaluation, now under construction in a hospital.

In our apparatus the main functions are controlled by two microprocessors one of which controls the dynamics of the pulsed gradients and the other determines the shape and the sequence of the R.F. pulses. The R.F. pulses are linearly amplified up to a required power. The NMR signal is quadrature detected digitized along two different channels averaged and stored.

Our resistive magnet is at moment implemented to produce head scans or limb scans. The image shown in Fig. 1 represents a transverse section of human head obtained at 4.2 MHz. The thickness of the slice is about 2 cm and the width about 25 cm.

The image clearly shows the cerebellum, the skull, the nose with nostrils and many other anatomical details.

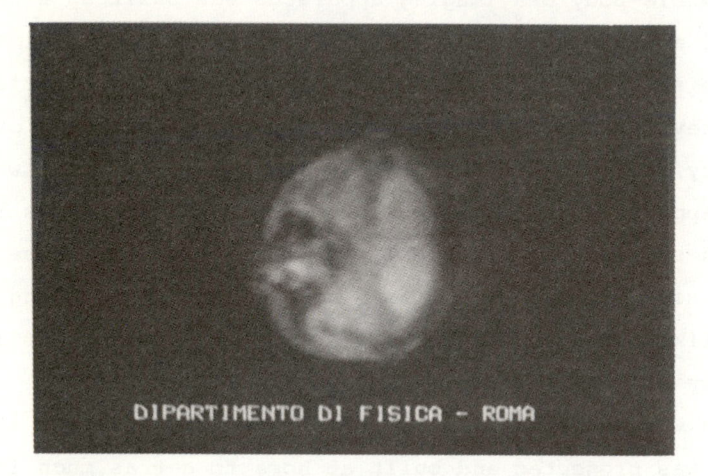

Fig. 1 - 2D image of a transverse section of human head obtained at 4.2 MHz. The slice thickness is 2 cm.

References

1) P.C. Lauterbur, Nature 242, 190 (1973).
2) A. Kumar, J. Welty and R.R. Ernst, J. Magn.Res. 18, 69 (1975).
3) I.M.S. Hutchinson, R.J. Sutherland and J.R. Mallard, J. Phys. E, 11, 217 (1978).
4) P. Mansfield and P.G. Morris, Supp. 2, Adv. Magn. Res. Academic Press, New York 1982.
5) F. De Luca, B.C. De Simone and B. Maraviglia, Magn. Res. Imaging, 1, 205 (1982).

*On leave of absence from ANSALDO biomedicale.

PROCEEDINGS OF THE II INTERNATIONAL CONFERENCE ON APPLICATIONS OF PHYSICS TO MEDICINE AND BIOLOGY
edited by Ž. Bajzer, P. Baxa & C. Franconi

"P.A.D.A.S." A SYSTEM TO MONITOR AND EVALUATE FACTORS RELATING TO BACK PAIN

E.O. Otun, I. Heinrich, J. Crooks, H. O'Hare, V. Moss, J.A.D. Anderson
A.R.C. Occupational Health Research Unit,
Department of Community Medicine,
Guy's Hospital Medical School,
London. SE1

BACK PAIN is a medical disorder significantly affecting 80% of people sometime during their lives[1)], causing serious socio-economic problems of which industrial absenteeism is just one. Attempts have therefore been made to study the occupational factors relating to back pain in industry. One of the problems encountered in studies of this sort is the creation of objective and reproducible measures of the various tasks within a job at work. We have developed a Physiological Ambulatory Data Acquisition System (P.A.D.A.S.) to monitor jobs and in particular their component tasks.

METHOD

"P.A.D.A.S." System description

The system is small, robust, unobtrusive and light causing minimal interference with work and is capable of continuously recording 8 separate non-invasive inputs of objective physiological data related to back stress and work load such as electromyogram (E.M.G.) of the lower back, antero-posterior (A/P)/lateral flexion of the spine[2)], and intra-abdominal pressure (I.A.P.)[3,4)], without requiring direct observation by a third party.

The main components of the system are, two miniature 4 channel analog cassette tape recorders, three miniature inclinometers to measure A/P and lateral flexion of the spine at levels T12 and S1, a miniature pressure sensitive radiotelemetry pill and receiver to measure I.A.P., 4 Ag/AgCl electrodes to record the electromyogram of the back at levels L4 and L5 and a timer to synchronise both tape recordings.

Validation Study

As part of the validation experiment a series of 6 full stoop lifts, lift height being 80 cm, were performed by volunteers with weights ranging from 10 kg to the volunteers individual maximum. Each series was preceded by a rest period and each sequence of rest and lift series was replicated 9 times.

All data recorded by P.A.D.A.S. was digitized and anaysed by a motorola 6800 based microcomputer.

RESULTS

As an example of the P.A.D.A.S. system output the results of one volunteer performing the validation experiment are summarized in Figs. 1-4 showing the means and standard errors over 9 replicates for E.M.G. spike counts/second, I.A.P., A/P and lateral flexion of the spine against the weights lifted by the volunteer.

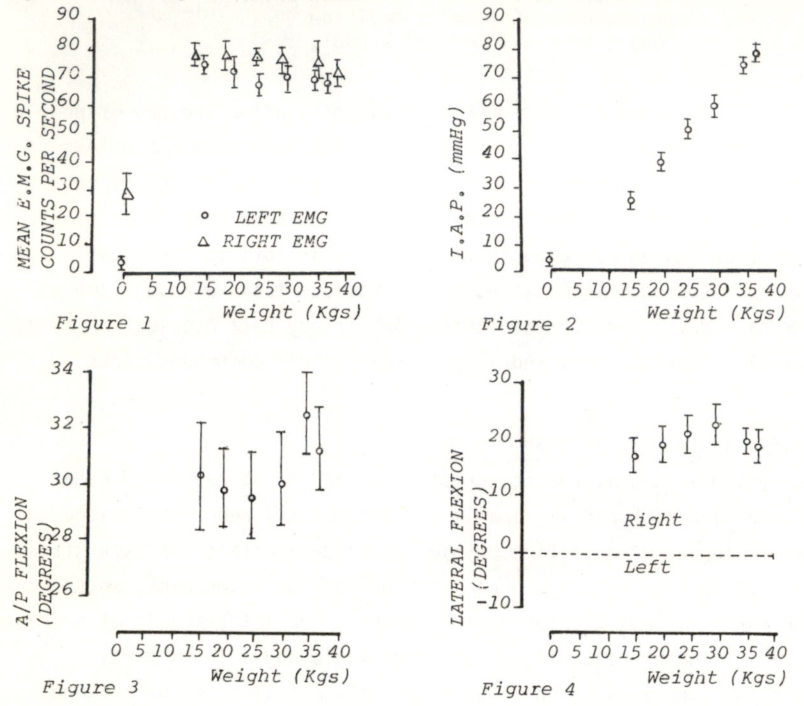

Figure 1

Figure 2

Figure 3

Figure 4

CONCLUSION

This project sets out to measure in a simple manner just a few physiological response variables which have been shown by previous research[2,3,4] to be related to stress on the spine. In proposed industrial surveys we hope to discriminate between various response patterns within a job and between jobs considering age, sex differences and body build with the ultimate aim of seeking correlations between occupational factors and back pain.

REFERENCES

1) KELSEY, J.L. and WHITE. Epidemiology and impact of low back pain (1980).
2) ANDERSON, J.A.D. Occupational aspects of low back pain. Clinics in Rheumatic Diseases, Vol. 6, No. 1, April 1980.
3) DAVIS, P.R., STUBBS, D.A. & RIDD, J.E. (1977). Radio pills: their use in monitoring back strains. Journal of Medical Engineering Technology, 1, 209-215.
4) ANDERSSON, G.B., ORTENGREN, R. & NACHEMSON, A. (1977). Intradiskal pressure, intra-abdominal pressure and myoelectric muscle activity related to posture and loading. Clinical Orthopedics, 129, 156-164.

PROCEEDINGS OF THE II INTERNATIONAL CONFERENCE ON APPLICATIONS OF PHYSICS TO MEDICINE AND BIOLOGY
edited by Ž. Bajzer, P. Baxa & C. Franconi

SYMMETRIC PURE INDUCTION 27 MHz APPLICATOR

C.A. Tiberio[+], L. Raganella[o], C. Franconi[o]

+ Physics Dept., I University of Rome
o Internal Medicine Dept., II University of Rome

Induction applicators of the pancake type exhibit "hot spots" near the feeding point and a cool region around the loop center. As shown theoretically by Morita and Bach Andersen (1) a linear radiofrequency current parallel to the surface of the medium to be heated gives a better power distribution with a maximum located in front of the current line. An inductive applicator employing an array of parallel equiverse currents has been designed and experimented by Bach Andersen et al (2).

This note describes a 27 MHz applicator made of 3 parallel lines arranged in a balanced electrical resonating structure. Fig. 1 shows an outline of the prototype and Fig. 2 the electrical line equivalent of the resonating circuit. As the line length is only 0.025 of one wavelength, the resonance current is constant within 1% along the lines, and voltage has a null at the centre.

As seen in Fig. 1 the excitation is made by a separate adjustable loop for easy matching between the power generator and the load. Suitable shields have been added at the ends of the line to minimize the stray capacitive field E_c without affecting the induction contribution E_i.

The power deposition rate has been measured with the temperature/time gradient method. The temperature rise distribution was mapped over a 2x2x2 cm grid using a Bailey digital thermocouple thermometer switched over 10 PT6 sensors. Temperature steps measurements were found to be reproducible within 10%.

Fig. 3 shows the relative positions of the lines and the phantom in a cross section of the applicator drawn in the YZ plane.

Experimental data describing the power deposition pattern in a phantom are reported in the diagrams of Fig. 4. Due to electrical balancing of the device power deposition curves at various sections show a fairly symmetrical distribution with respect to ZX and ZY plane, absence of sidelobes and single elliptical beam centered around the Z-axis. The width in the Y (transverse) direction can be increased by increasing the number of parallel strip lines.

(1) N. Morita and J. Bach Andersen. Bioelectromagnetics 3:253-274 (1982)
(2) J. Bach Andersen, Aa. Baun, K. Harmark, L. Heinzl, P. Raskmark and J. Overgaard. A hyperthermia system using a new type of inductive applicator. To be published in IEEE Transactions on Biomedical Eng.

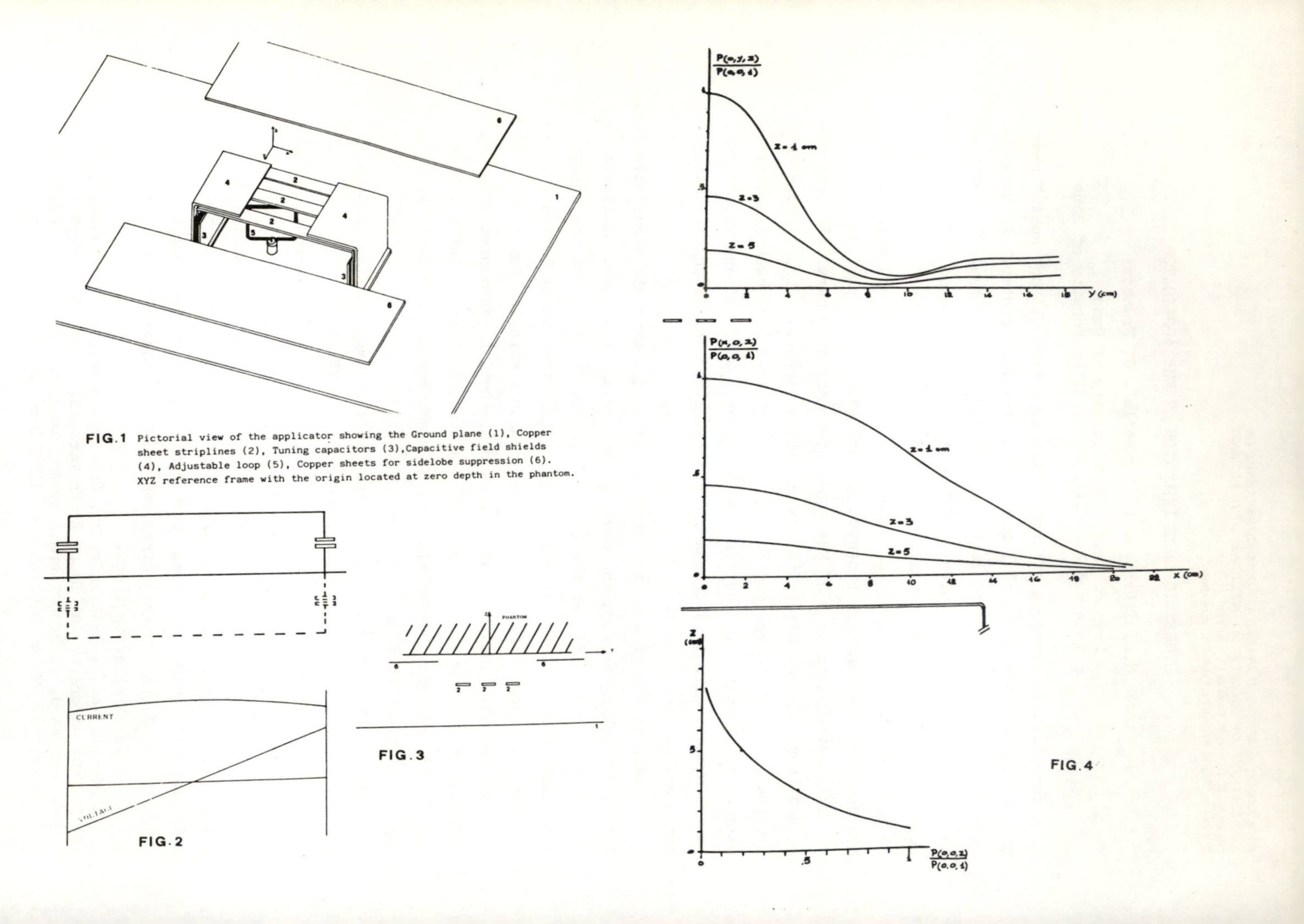

FIG.1 Pictorial view of the applicator showing the Ground plane (1), Copper sheet striplines (2), Tuning capacitors (3), Capacitive field shields (4), Adjustable loop (5), Copper sheets for sidelobe suppression (6). XYZ reference frame with the origin located at zero depth in the phantom.

FIG.2

FIG.3

FIG.4

PROCEEDINGS OF THE II INTERNATIONAL CONFERENCE ON APPLICATIONS OF PHYSICS TO MEDICINE AND BIOLOGY
edited by Ž. Bajzer, P. Baxa & C. Franconi

A MULTIFREQUENCY WATER-FILLED WAVEGUIDE APPLICATOR: THERMAL DOSIMETRY IN VIVO WITH AND WITHOUT COOLING

G.A. Lovisolo[1], M. Adami[1], G. Arcangeli[2], A. Borrani[3], G. Calamai[3], A. Cividalli[1] and F. Mauro[1]

(1) Laboratorio di Dosimetria e Biofisica, ENEA, 00060 Roma

(2) Istituto Medico e di Ricerca Scientifica, 00184 Roma

(3) S.M.A., Via del Ferrone, 50124 Firenze

Efficient microwaves applicators should offer the following performance characteristics:

a) deep heating and variable surface cooling;
b) matching and mismatching when in contact with the skin or not, respectively;
c) best fitting to irregular skin surface;
d) large band-width for applications at different frequencies.

According to these requirements, the design and construction of a ridged horn, filled with liquid dielectric and operating in the bandwidth 300 to 1000 MHz, have been carried out. In the first attempt, deionized water has been selected as dielectric. Such a choice can be considered as a fair compromise between thermic and electric characteristics. Other liquids or liquid and gas emulsions would represent a better solution, but have been temporarily discarded because of the technical problems of mixing and the potential toxic hazard.

The applicator return losses as a function of the frequency are displayed in Fig. 1 for different contacts. The 400 MHz frequency has been selected for the initial tests on a meat phantom (beef thigh) and then on two sheep (weighing 35 and 25 kg. respectively). In both cases, 5 thermocouples were used to monitor the temperature at 5 test points: 3 mounted (in axis with the applicator) on the same probe (model IT17), at the tip, and at 1.5 cm and at 3 cm from the tip. The two other thermocouples were inserted for surface monitoring, on separate probes. All probes were connected to a clinical monitoring thermometer (Bailey model, TM10A). In the various tests, the temperature of the circulating deionized water was maintained around 15°C, 20°C or 35°C by means of a Haake F3 thermocryostat.

Fig. 2 and 3 show temperature profiles for different surface cooling and applied power. Fig. 4 has been obtained by changing surface cooling and power so that a homogeneous and deep heating could be achieved.

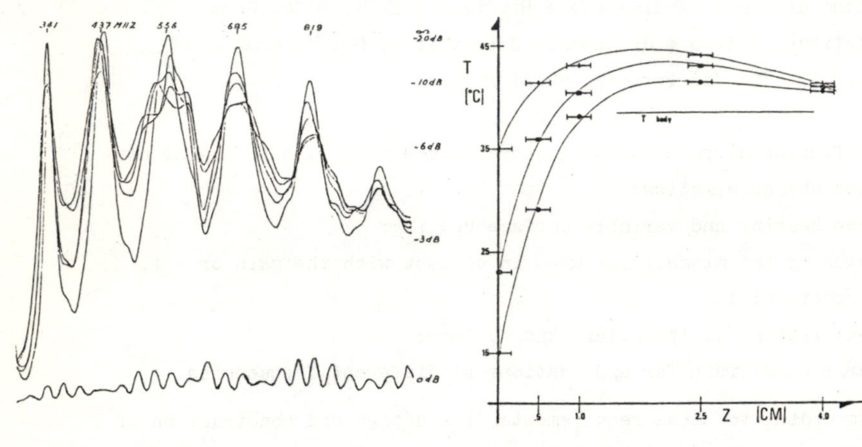

Fig. 1

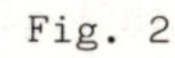

Fig. 2

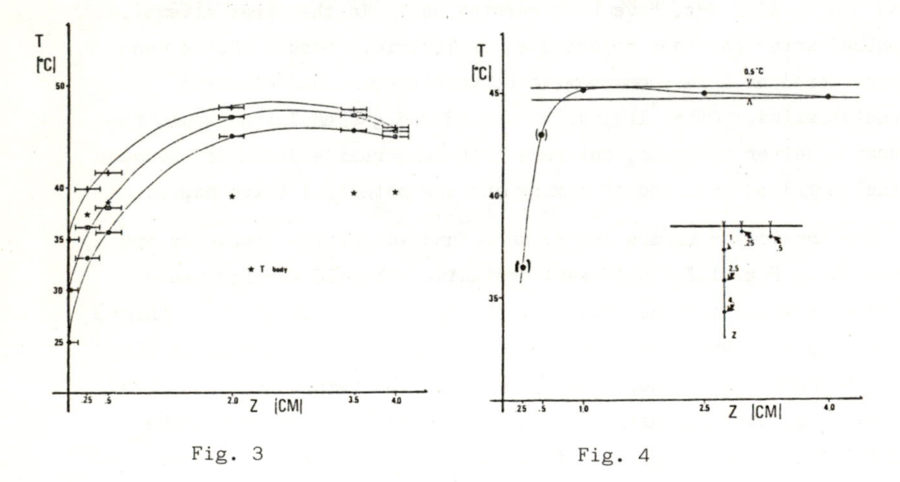

Fig. 3

Fig. 4

PROCEEDINGS OF THE II INTERNATIONAL CONFERENCE ON APPLICATIONS OF PHYSICS TO MEDICINE AND BIOLOGY
edited by Ž. Bajzer, P. Baxa & C. Franconi

JOINT FOURIER TRANSFORM CORRELATOR FOR ECHOSTRUCTURE RECOGNITION

P. Sirotti * G. Rizzatto ** E. Mumolo *

* Istituto di Elettrotecnica e di Elettronica
Università di Trieste - 34127 TRIESTE - ITALY

** Reparto di Radiologia
Ospedale Civile di Gorizia - 34170 GORIZIA - ITALY

Ultrasonic parenchymal textures can be considered as an hystologic macrostructure of an organ, modified by the acoustic pulse-tissue interaction and various performance factors of the instrument [1] . We have already demonstrated that a multiple holographic filtering proves satisfactory for recognition and classification of liver echostructures [2] . In this paper we present another coherent optical correlation method based on the Joint Fourier Transform (JFT) [3] .

Textures to be correlated are located at the input plane (whatever fore plane) of a convergent lens. Their JFT is formed at the back focal plane, quadratically recorded (f.i. on a photographic film with suitable properties) and retransformed, by placing it on the input plane again. If a set of reference images h_i (i = 1 .. n) has to be correlated with an unknown one, g, the output pattern can be expressed as:

$$O(r) = \sum_{1}^{n}{}_{i,j\ j\geq i} h_i(r) * h_j(r \pm 2s_{ij}) + g(r) * g(r) + \sum_{1}^{n}{}_{i} h_i(r) * g(r \pm 2a_i)$$

where r is the pair of spatial coordinates (x,y), a_i the distance between g and h_i, s_{ij} is the distance between h_i and h_j (j = 1 .. n) and the symbol * indicates correlation.

The only significant term is the third, that is a sum of the correlation functions between reference and unknown textures. The brightness of correlation peaks and correlation symmetry individuate the reference texture more similar to the unknown one (fig. 1). In fact it can be proved that if the input functions are very like one another, the correlation is symmetrical around the correlation peak (fig. 2).

The preliminary results demonstrate the effectiveness of JFT correlator for the recognition of diffuse liver diseases; further refinements in transform recording will probably allow as good results as with the above mentioned method, that allowed a diagnostic accuracy of about 91% [2] .

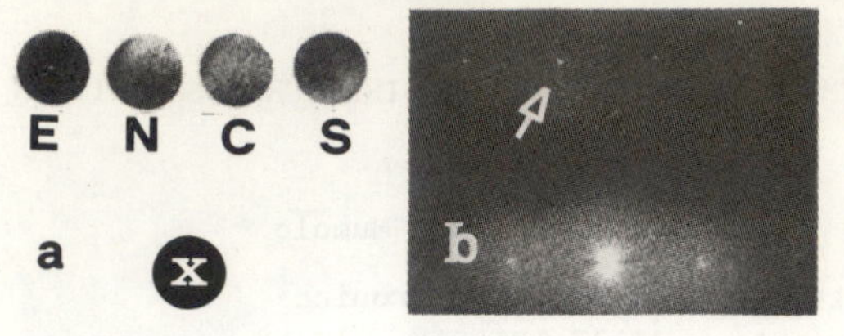

E - hepatitis
N - normal
C - cirrhosis
S - fatty liver

Fig. 1: Multiple correlation among four reference images and an unknown image (x).
a) Input plane. The location of the images was choosen to simplify the separation of useful terms in b).
b) Output pattern. The brightest correlation peak recognizes x as normal.

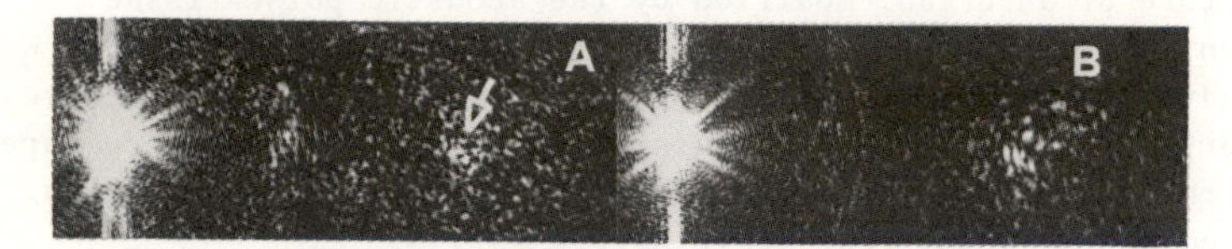

Fig. 2: Correlation (→) between two ultrasonic scans of
A) normal livers B) normal and cirrhotic livers

JFT correlator is advantageous because it is very insensitive to position imprecisions and does not require vibration-isolation tables; nevertheless it is not a real time method, requiring an intermediate recording of the transform on a photographic film. To overcome this limitation we have just realized an optical-digital correlator: the JFT is acquired by a TV camera and reconstructed by a computer. In this way the flexibility of the system is greatly improved, allowing on line filtering on the transform itself. The effectiveness of such optical-digital correlator in classifying echostructures has now to be tested.

REFERENCES

[1] J.C. Birnholz: Ultrasound Evaluation of Diffuse Liver Disease - CDU 1979, 1 : 23÷33.

[2] P. Sirotti, G. Rizzatto: Holographic Filtering for Echostructure Analysis - EUSIPCO 83, Erlangen 12-16 September 1983.

[3] D. Casasent: Coherent Optical Pattern Recognition - Proc. IEEE, Vol. 67, May 1979, pp. 813-825.

PROCEEDINGS OF THE II INTERNATIONAL CONFERENCE ON APPLICATIONS OF PHYSICS TO MEDICINE AND BIOLOGY
edited by Ž. Bajzer, P. Baxa & C. Franconi

ULTRASONIC PRESSURE PROFILES MEASURED BY AN OPTICAL METHOD

Franz Holzer, Paul Wach, Stefan Schuy
Institute of Biomedical Engineering, TU Graz
A-8010 Graz, Inffeldgasse 18
AUSTRIA

1. Introduction

Ultrasonic beams emitted by medical devices can be imaged by the Schlieren method which is based on the diffraction of light by ultrasound. This method is very helpful in analysing the geometrical structure of an ultrasonic beam.

The aim of this paper is to show that in addition to this qualitative aspect of the method, the spatial and temporal distribution of ultrasonic pressure can be measured by detecting the light intensity distribution in the Schlieren pattern.

2. Spatial distribution of ultrasonic pressure

If the ultrasonic beam is illuminated at right angles by coherent light and all but the zero diffraction order contribute to the Schlieren pattern, the normalized light intensity in the pattern is defined by

$$I = 1 - J_0^2(v) \quad , \tag{1}$$

where J_0 is the zero order Bessel function and v is the integrated phase shift or Raman-Nath parameter[1], a magnitude which is proportional to the pressure amplitude. For ultrasonic beams with rotational symmetry, one scan of the light intensity profile is sufficient to reconstruct the pressure profile[2]; an example is given in Fig. 1.

3. Temporal distribution of ultrasonic pressure

If only the zero and first diffraction orders contribute to the Schlieren pattern, the normalized light intensity is given by

$$\begin{aligned} I &= |J_0(v) + J_1(v)\exp(i\Omega t)|^2 \\ &= J_0^2(v) + J_1^2(v) + 2J_0(v)J_1(v)\cos\Omega t \quad , \end{aligned} \tag{2}$$

where Ω is the angular frequency of ultrasound.

For sufficiently low pressure amplitudes, corresponding to a level of v ≤ 0.3, the light intensity signal as detected by a fast photodetector represents a good approximation to the actual ultrasonic signal.

High amplitude pulses as emitted by diagnostic equipment can also be detected in the same way by placing a thin membrane of known reflectivity in the beam section of interest and performing the measurement in the reflected ultrasonic beam at a suitable pressure level. A comparison with a hydrophone output is given in Fig. 2.

4. Conclusion

The method described in this paper yields ultrasonic beam profile and pulse shape data at an accuracy which is sufficient in most cases of practical interest, thus completing the well-known Schlieren method used in transducer characterization.

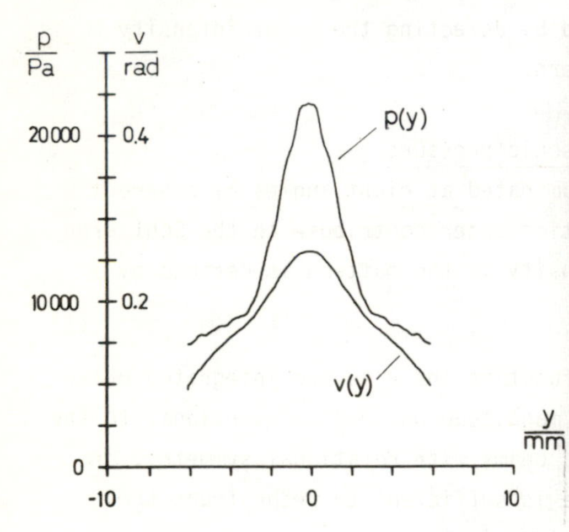

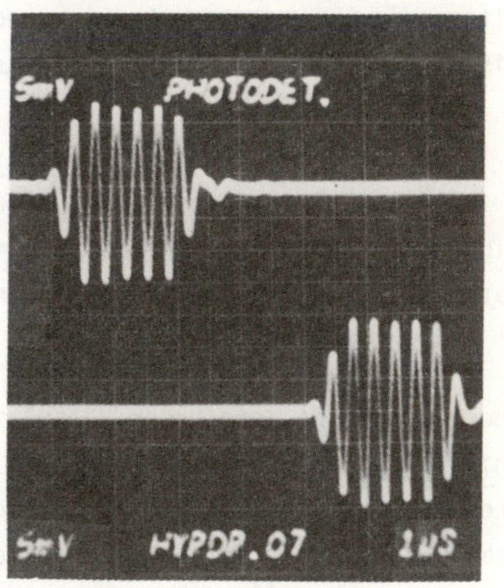

Fig. 1 (left): pressure profile p(y) obtained from the corresponding v(y) profile in the farfield of a 2.25 MHz beam.

Fig. 2 (right): the signal at the bottom corresponds to the output of a PVDF hydrophone, the other was optically detected in the Schlieren plane; the pressure amplitude is 0.3 bar.

5. References

1) C. V. Raman, N. S. N. Nath: Proc. Ind. Acad. Sc., A2 (1935), 406.

2) F. Holzer, P. Wach, S. Schuy: Proc. FASE/DAGA'82, Göttingen (1982), 783.

PROCEEDINGS OF THE II INTERNATIONAL CONFERENCE ON APPLICATIONS OF PHYSICS TO MEDICINE AND BIOLOGY
edited by Ž. Bajzer, P. Baxa & C. Franconi

LONG-TERM AVERAGE SPECTRAL CHARACTERISTICS OF VOICES AFTER THYROIDECTOMY

G.Kouroupetroglou*, I.Orfanos** and M.Katiri-Stefanou***

* Athens Univ.,Electr. Lab.,TYPA,Ilisia, Athens,GREECE
** National Techn. Univ.,Physics,Lab.A',Athens,GREECE
*** Hosp."Agia Olga",1st Surg. Unit,N.Ionia,Athens,GREECE

INTRODUCTION: Long-term average speech spectra reflect several important characteristics of the speech mechanism of a man so they have been applied successfully for speaker recognition [1], as well as objective acoustical methods in phoniatric diagnostics of speech organ disorders [2-3]. The present paper is a preliminary report dealing with the possible use of these spectra for measuring the changes of the voice after thyroidectomy. The study emphasize the performance of the phonatory mechanism rather than the development of diagnostic procedures.

METHOD: Subjects were 12 patients selected among 211 who had been surged for goiter with the following critiria: the patient had to be an adult of at least average intelligence, speaker of Greek, reasonable healthy except for his goiter pathology. 33 patients met the above criteria. All these at the previous day of the thyroidectomy and at the 7th post-operatively day recorded the same phrase: "To nero rei sto riaki". Recordings were made in a quite room using a high-quality magnetic tape recorder. Precautions were taken for all subjects to remain at the same recording conditions. From these patients we selected finally only 12 with the following critirion: 4 of them had not any recurrent paralysis before or after the thyroidectomy (group 1), other 4 had recurrent paralysis only before (group 2) and the rest 4 only after the operation (group 3). (The case of group 2 is very rare and we chose equal number of patients for each group). Although the number of speakers is small, this study has the advantage of being restricted to groups with well documented homogeneous lesions of the vocal folds.

The recordings analysed on a digital narrow-band spectrum analyser (Bruel & Kjaer model 2031) which essentially consists of a digitizing system and an FFT algorith.This instrument was connected to a computer system (HP 9825) to obtain automatic data analysis and storage.

A laryngoscopy examination of the vocal folds' mobility for possible recurrent laryngeal nerve paralysis was made for each subject one day pre-operatively and the 7th post-operative day.

RESULTS AND DISCUSSION: An average smoothed spectral envelope of up to 5 KHz was derived for each patient before and after the thyroidectomy. Composite average spectra for the three groups were obtained by taking the mean of the individual smoothed spectra and the results before and after the operation were plotted on the same figure for each group. The maximum level associated with peaks appropriate to the first formant energy concentration in the average spectra was used as a reference (0 dB) for normalization.

The results indicate that patients who get a reduced energy spectrum above about 1.5 KHz during the recovery had recurrend paralysis only after the operation and patients who get an increased spectrum energy at the same region had recurrent paralysis only before the thyroidectomy. There were not uniform and significant changes of the spectral shape for the patients who had not any recurrent paralysis before or after the operation. There was also a strong correlation of the long-term average spectral analysis compared with the auditory evaluation of the qualities of the patients' voices by a group of listeners. The observed differences indicate the possible use of long-term average spectral analysis for the subjective measurement of the voice quality.

CONCLUSIONS: Long-term average spectra before and after thyroidectomy indicate a strong relation with recurrent paralysis before and after the therapy.Also these spectra indicate a correlation with voice quality. It appears that readings of speech spectrograms may be used by a computer spectrogram expert for automatic subjective voice quality measurements

REFERENCES:

[1].M.Kuhn:"Access control by means of automatic speaker verification", J.Physics E,vol.13,p.85,1980.

[2].B.Weinberg,Y.Horii and B.Smith:"Long-time spectral and intensity characteristics of esophageal speech", J.Acous. Soc. Amer., Vol.67(5),p.1781,1980.

[3].C.Formby and R.Monsen: "Long-term average speech spectra for normal and hearing-impaired adolescents", J. Acous. Soc. Am., Vol.71(1),p.196,1982.

PROCEEDINGS OF THE II INTERNATIONAL CONFERENCE ON APPLICATIONS OF PHYSICS TO MEDICINE AND BIOLOGY
edited by Ž. Bajzer, P. Baxa & C. Franconi

VARIATION IN TIME OF PHONOCARDIOGRAPHIC FREQUENCY POWER SPECTRA: PRELIMINARY RESULTS AND EVALUATION

E. M. Staderini
Medical Physics Institute, II University of Rome, 00173 ROME

1. Introduction

Spectral phonocardiography is now gaining more and more positions in cardiovascular research work thanks to computer technology which is less expensive and more user friendly than in the past. In the study of heart tones proper frequency analysis of signal is very useful as heart tones are known to contain infromation on mechanic and viscoelastic state of the cardiac structures emitting them[1]. Such information are of the utmost importance in physiological research as well as in clinical practice. As variations in mass and/or constant of elasticity in a mechanical system are expected to produce variations in natural resonation frequency, so spectral phonocardiography is a useful tool in evaluating, in vivo and not invasively, the anatomo-pathological and physiological state of the cardiac structures involved in sound generation[2].

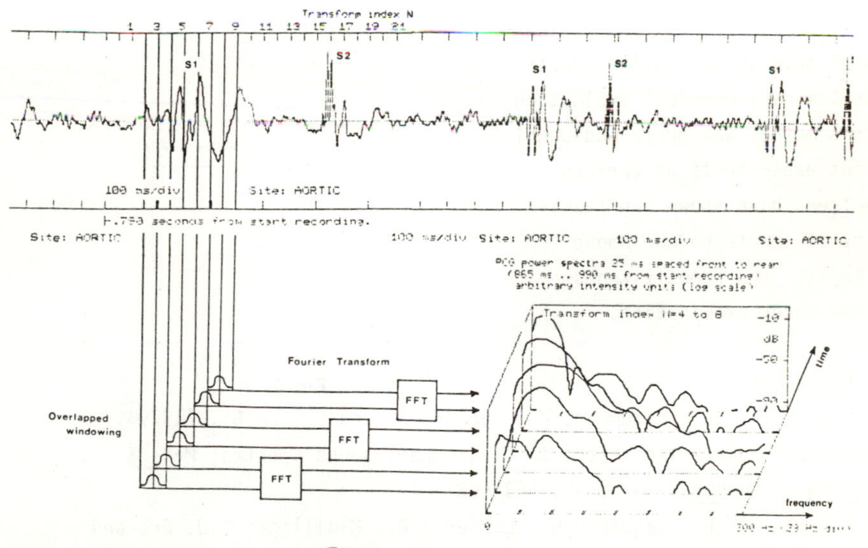

Fig. 1

2. Method of analysis

In this research successive frequency spectra are taken from successive overlapping time intervals of the sound as in Fig. 1. PCG signal has very low power content above 800 Hz so a simple 6th order low pass active filter with 1000 Hz corner frequency was considered just adequate to overcome aliasing with a sampling frequency of 2557.5 Hz. Data were windowed with a 128 points 50 msec wide Hamming function. Picket-fence effect was made less relevant adding zeroes to fill 512 points. 512 points radix-2 decimation in time FFT is used to compute two spectra at a time as input signal is real. Finally the log value of the squared modulus of the transform is taken. Frequency resolution is about 5 Hz but window bandwidth increases such value to about 20 Hz and this was always taken into account in spectra evaluation. Data selection for analysis was made as depicted in Fig. 1. Hamming window was applied on 50% overlapping signal segments and a spectrum was obtained every 25 msec of signal. Overlapping was used as main purpose of PCG frequency analysis is to detect short-duration tone-like signals which might go undetected without overlapping as Hamming window has very low values at the boundaries.

In Fig. 2 second heart sound (S2) and associated spectra are drawn. A clear 160 Hz peak in the spectrum is visible due to frequency content of the signal. This peak is clearly visible only during S2 and varies in amplitude on successive spectra. No preliminary clinical evaluation of this method was performed until now but experiments on various volunteers have shown the feasibility of the technique whose reliability and repetitiveness are now under examination.

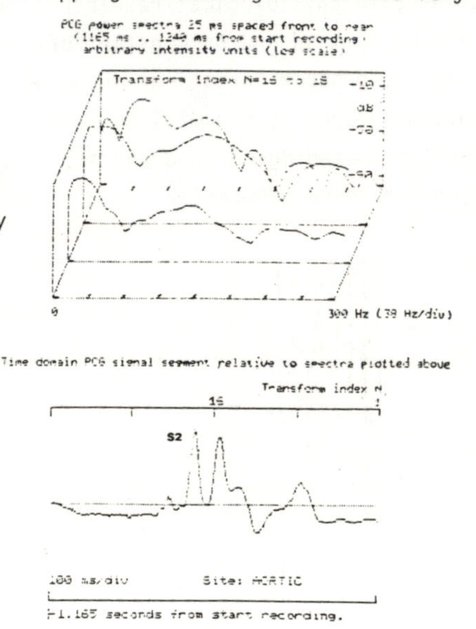

Fig. 2

3. References

(1) Hearn T.C., Mazumdar, J., Hubbard R. and Eyster G., (1979). Temporal and heart-size effect in first-heart-sound spectra, Med. & Biol. Eng. & Comput., 1979, 17, 563-568.

(2) Stein P.D., Sabbah H.N., Lakier J.B., Magilligan D.J. Jr. and Goldstein S., (1981). Frequency of the first heart sound in the assessment of stiffening of mitral bioprosthetic valves. Circulation 63, No. 1, 1981.

PROCEEDINGS OF THE II INTERNATIONAL CONFERENCE ON APPLICATIONS OF PHYSICS TO MEDICINE AND BIOLOGY
edited by Ž. Bajzer, P. Baxa & C. Franconi

IMPLANT MATERIALS: STABILIZATION OF SHAPE MEMORY NiTi ALLOYS

G. Airoldi[+], G. Bellini[+], C. Di Francesco[o]
[+] CISE - C.P. 12081 - Milano
[o] CESNEF - Politecnico di Milano

The interest[1-5)] to use NiTi as implant material arises from its unique property to develop a stress state induced by the recovery of a formerly imprinted shape, which is appealing to find out quite new solutions in orthopaedic or maxillofacial surgery. NiTi availability lacks till now, stimulating their production and characterization.

A NiTi alloy is suited to be considered as implant material if its transition temperatures M_f, M_s, A_s, A_f are properly allocated respect to body temperature: A_s, A_f must be higher than body temperature (T_b) and M_f, M_s lower than T_b. A proper choice of chemical composition is the first but not unique prerequisite to reach this goal. Thermal cycling between martensite $\rightleftarrows$ parent phase can shift transition temperatures[6,7)].

The rôle of thermal cycling has here been considered on different NiTi alloys: the ones here examined, prepared as already described[8,9)], correspond to different chemical composition and/or different thermomechanical treatment: initial M_s transition temperatures span +76°C ÷ -36°C. The shape memory behaviour has previously been checked[10)] both applying a uniaxial tension stress or a bending stress: 90% and 81% recovery deformation have been obtained respectively after uniaxial strains of 2,5% and 4%.

Transition temperatures have been determined from electrical resistance v/s temperature curves, as already suggested[7,11)]. All the alloys examined and their transition temperatures as a function of the number of thermal cycles are given in Table 1.

The major shift in transition temperatures is found for alloys with initial $M_s \geq T_R$ where T_R is the transition temperature to a rhomboedral phase, recently[11)] pointed out to play a distinct rôle from martensite phase. No shift at all is found for alloys with $M_s \ll T_R$. Alloys as C12 and C14, suitable for biomedical applications, initially displaying $M_s \approx T_R$, though showing a shift of M_s temperature with cycling, appear rather well stabilized after twenty cycles.

We did not till now succeed to obtain alloys with M_s slightly lower than room temperature which might be, most probably, the true ideal to be used in biomedical application.

Table 1

Specimen Code Starting Composition	% Reduction	cycle	M_s (°C)	T_R (°C)	A_f (°C)
C 10	69	1	+75.5		+112.0
50 at % Ni		2	+73.0		+105.5
		8	+56.5		+ 96.5
		20	+52.5	+67	+ 93.5
C 12	85	1	+51.5		+ 68.0
50 at % Ni		8	+29.5		+ 62.0
		20	+25.0	+45	+ 51.0
C 14	80	1	+33.0	+40	+ 64
51 at % Ni		20	+29.0	+40	+ 53
C 13	70 + 90	1	-36.0		+ 30
50 at % Ni		3	-36.0	- 7	+ 30
		20	-36.0	- 7	+ 30

% Reduction = $h_o - h/h_o$ (h_o initial height, h final height of the plate after several hot rolling steps)

1) M.A.Schmerling et al., in "Shape Memory Alloys" pag. 563, 1975, Ed. Perckins - Plenum Press.
2) L.S.Castleman et al., J. Biom. Mater. Res. 10, 695 (1976).
3) G.Bensmann et al., Techn.Mitt.Krupp Forsch.Ber. 37(1), 21(1979).
4) D.Schettler et al., J. of Maxillof. Surg. 7, 51 (1979).
5) G.Airoldi, L.Parrini, Arch. Ortop. Reum. 94(I-II), 17 (1981).
6) G.D.Sandrock, R.F.Hehemann, Metallography 4, 451 (1971).
7) H.C.Ling, R.Kaplow, Met. Trans. 11A, 77 (1980).
8) ADR-K7/E32 - R.A. 1/9/1981 - 30/4/1982 (1982).
9) G. Airoldi et al., CISE 1848 (1982).
10) ADR-K7/E32 - R.A. 1/5/1982 - 31/12/1982 (1983).
11) M.C.Ling, R.Kaplow, Metal. Trans. 12A, 2101 (1981).

Work partially supported by CISE ADR K7/E32.

PROCEEDINGS OF THE II INTERNATIONAL CONFERENCE ON APPLICATIONS OF PHYSICS TO MEDICINE AND BIOLOGY
edited by Ž. Bajzer, P. Baxa & C. Franconi

THERMOLUMINESCENCE OF ZnO:CU,La PHOSPHORS UNDER THE EXCITATIONS OF UV AND γ-RAYS

S. Bhushan, Deepti Diwan and S.P. Kathuria*
Department of Physics, Ravishankar University,
Raipur, India.
*Physics Health Division, BARC, Bombay, India

1. Introduction

Recent work on the electroluminescence (EL) and photoluminescence (PL) of undoped and rare earth (RE) doped ZnO[1,2)] have shown that these present quite interesting systems for display technology. However, so far not a single paper has been published on the thermoluminescence (TL) of rare earth doped ZnO. The present work reports the TL studies of ZnO:Cu,La under the excitations of UV and γ-rays.

2. Experimental

ZnO doped with Cu and La were prepared by firing a mixture of pure ZnO, copper sulphate and lanthanum nitrate at 980°C for 1 hr. Low pressure Hg lamp and Co^{60} were used as excitation sources. The recording system consisted of a photomultiplier (EMI-9514), a high voltage unit, a dc amplifier, a temperature programmer (Hewlett Packerd F and M model 240) and a strip chart recorder (Rekadenki 10 mV).

3. Results and Discussions

Undoped ZnO shows two peaks at 346° and 511.25°K and those at 300° and 576°K for ZnO:Cu. ZnO:Cu,La shows peaks at 340°, 564°, 574°, 592°, 604° and 610°K after 1 min UV excitation. Also decrease in intensity of low temperature peak is observed, which shows that it may be due to lattice-defects forming traps. Due to γ-irradiation of 1.2×10^{3} rad only one peak at 590°K is seen which becomes pronounced at higher irradiation times. At higher irradiation times a broad peak at 420°K along with some sharp peaks are also observed (Fig.1). Increasing concentration of La also produces these sharp peaks along with suppression of intensity of different peaks (Fig.2). Annealing at 400°C has resulted in the disappearance of smaller peaks at higher temperatures (Fig.1). Since ZnO:La shows poor TL compared to ZnO:Cu,La thus recombination phenomena may be due to transfer of energy from Cu to La. From Fig. 3 it is seen that the intensity of higher temperature peak continuously increases up to 2.6×10^{6} rad and linear up to 2.4×10^{5} rad. Since the intensity of the higher temperature peak is quite good thus it is expected that it may find application in dosimetry. Due to increase in exposure rate more of centres are produced and hence recombination increases.[3]

Activation energy for higher temperature peak determined by heating rate method, peak shape and iso-thermal decay methods is found to be ~1.7 eV. The order of kinetics is found to be 2 and escape frequency factor of the order of 10^{13} sec^{-1}.

References

1. Bhushan S., Pandey A.N. and Kaza B.R., J.Lum. 20 (1979) 29
2. Takata S., Minami T., Manto M. and Kawamura T. Phys. Stat. Sol. 65 (a) (1981) K83.
3. Kathuria S.P., Bapat, V.N. and Jain V.K. Fourth Intl. Conf. on Lum. dosimetry, Poland, Aug. 19, 1974, 1177.

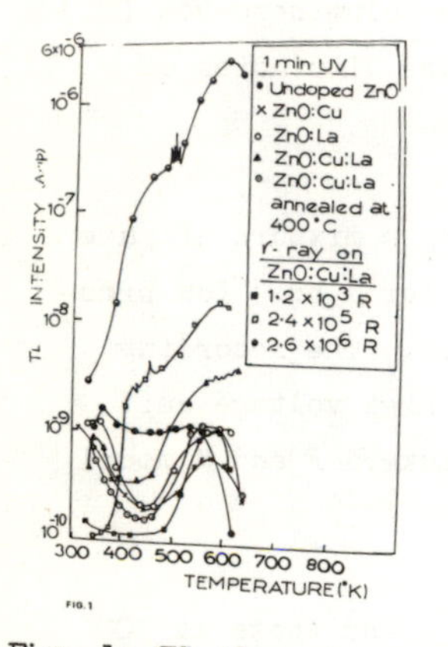

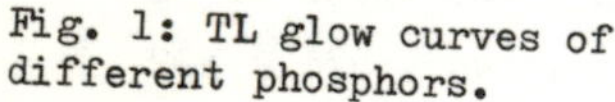
Fig. 1: TL glow curves of different phosphors.

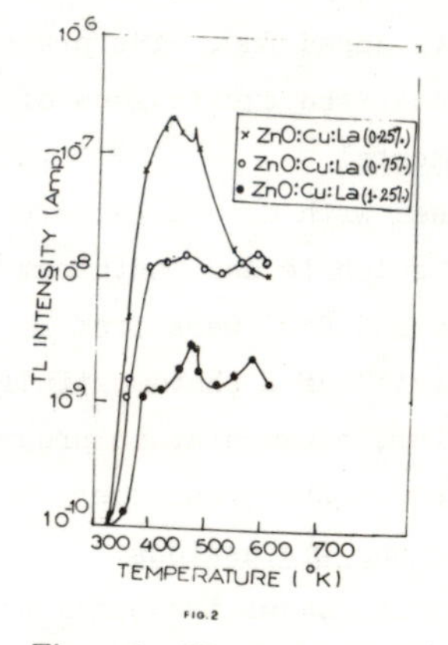

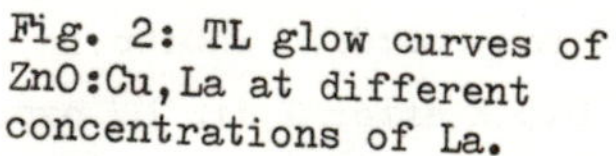
Fig. 2: TL glow curves of ZnO:Cu,La at different concentrations of La.

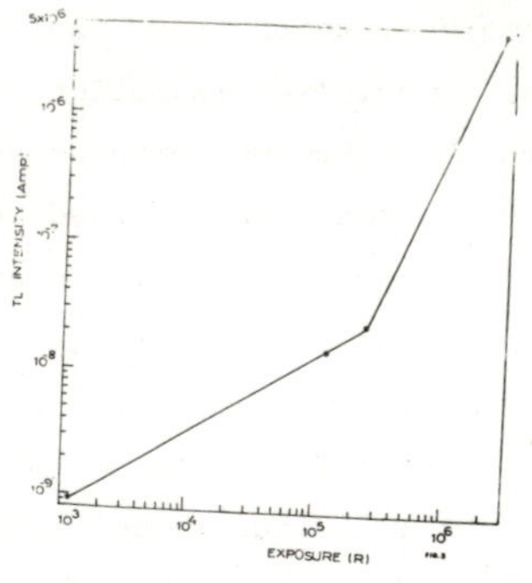

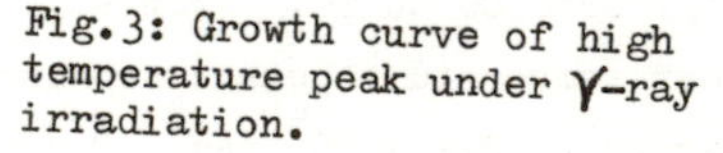
Fig.3: Growth curve of high temperature peak under γ-ray irradiation.

Ionizing Radiation

Risk Assessment and Control

Hyperthermia

Education and Training

PROCEEDINGS OF THE II INTERNATIONAL CONFERENCE ON APPLICATIONS OF PHYSICS TO MEDICINE AND BIOLOGY
edited by Ž. Bajzer, P. Baxa & C. Franconi

DOSE COMPUTATION OF SCATTERED DIAGNOSTIC X-RAY: A MATHEMATICAL MODEL

Giuseppe Gnani, Luigi Alberti, Maria Morelli
Health Physics Department, Ravenna Hospital - Italy

1. Introduction

The scattered radiation around an object provides an important contribution to the stray radiation field. The Monte Carlo method is rarely required in medical radiology, because of the complexity of the approach. The purpose of this work was to develop a simple mathematical model for the computation of scattered dose from diagnostic X-ray beams.

2. Calculation

The beam phantom geometry is illustrated in Fig.1 . The spectral energy fluence of the primary beam is given[1] by:

$$\Psi(E) = f_1(E,V) + f_2(E,V) + f_3(E,V) \tag{1}$$

where E is the photon energy and V is the tube voltage.

The mathematical model utilized for the determination of $\Psi(E')$ (spectral energy fluence of scattered beam), is:

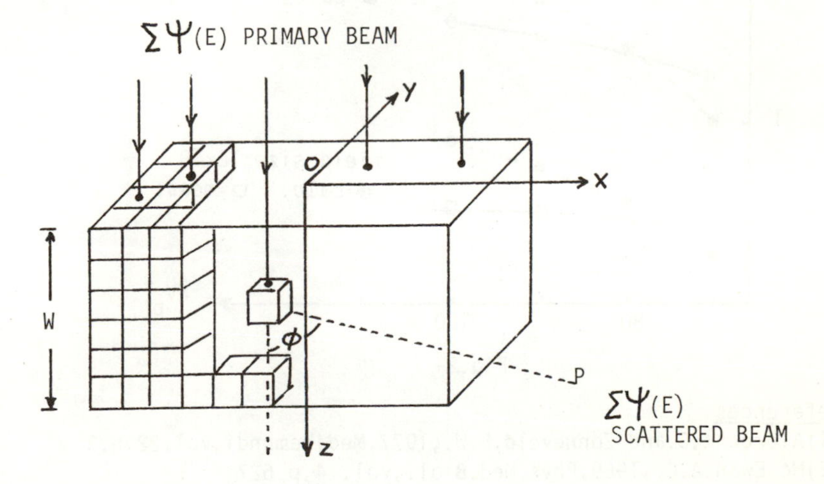

- Fig.1 -

$$\Psi(E') = \Psi(E) \cdot a_E(E, \emptyset_E, \mu(E), \mu(E'), L). \quad (2)$$

The value of $\sum \Psi(E')$ (i.e., the scattered dose) is given by:

$$\sum \Psi(E') = \sum \Psi(E) \cdot d\sigma(E, \emptyset_E)/d\Omega \cdot n_e \cdot (W - 1/\mu(E)) \exp(-\mu(E')L) \quad (3)$$

where $d\sigma/d\Omega$ is the Klein-Nishina differential cross-section, n_e is the electron density, W is the phantom thickness, $\mu(E)$ and $\mu(E')$ are the linear attenuation coefficients of primary and scattered photons, L is the scattered photon path lenght and $\emptyset_E$ is the photon scattering angle, with the following simplifications[2,3] :

-the phantom(cuboid phantom)is made of unit-density material;

-the incident photons are monodirectional and uniformly distributed over a plane perpendicular to the central axis of the radiation field

-multiple scattering is not taken into account.

3. Results

Comparison between computation and experimental measurements is show in Fig.2 ($\emptyset$=90°,1 m from the central ray of the primary beam).

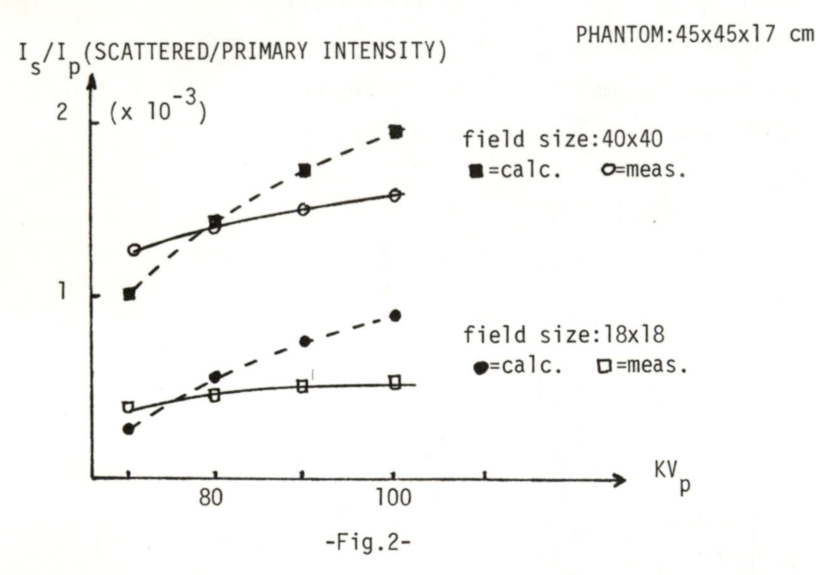

-Fig.2-

References

(1)Albrecht,C.and Zonneveld,F.W.,1977,Medicamundi,vol.22,n.3

(2)Mc Ewan,A.C.,1969,Phys.Med.Biol.,vol.14,p.627

(3)Battista,J.J.and Bronskill,M.J.,Phys.Med.Biol.,1978,vol.23,p.1

PROCEEDINGS OF THE II INTERNATIONAL CONFERENCE ON APPLICATIONS OF PHYSICS TO MEDICINE AND BIOLOGY
edited by Ž. Bajzer, P. Baxa & C. Franconi

FINITE VOLUME CORRECTION IN IONIMETRIC EXPOSITION MEASUREMENTS FOR BRACHITHERAPY DOSIMETRY

G. Arcovito, A. Piermattei, G. D'Abramo & F. Andreasi

Istituto di Fisica
Università Cattolica S.C. - Roma
ITALY

The dosimetry of the radiation sources used in brachitherapy must be realized by high sensibility point detectors, like solid state detectors. The ionimetric method, however, must be used in order to measure the absolute values of the exposition in the calibration of the isodose curves. Ionization chambers generally employed cannot be considered point detectors because their volume ranges between 0.1 and 0.6 cm^3.

We present a method to correct the exposition measurements obtained using an ionization chamber 0.22 cm^3, in the dosimetry of the high activity radiation sources of Ir 192 (740 GBq) and Cs 137 (148 GBq) employed in the endocavitary afterloading Buchler system. This system can operate with the oscillating Ir 192 source controlled by program-disks. The cylindrical sources have a volume of 3.1 mm^3 (Ir 192) and 199.3 mm^3 (Cs 137).

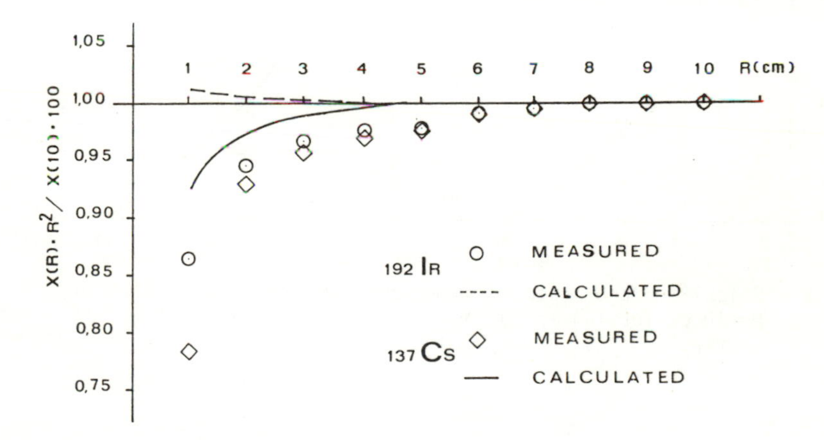

Fig. 1 - Ir 192 and Cs 137. Calculated and measured air exposure rate, multiplied R^2 and normalized to R = 10 cm against R.

Exposure rate was calculated by using the "quantization method". The data obtained clearly show that a difference does exist between the experimental and computed air exposure rate as far as both Ir 192 and Cs 137 are concerned, as reported in Fig. 1. This difference depends upon the finite volume of the ionization chamber.

On this ground we were able to derive a correction factor F(R) for the exposure measurements as a function of the distance R from the center, along the lateral axis of the source:

$$F(R) = 0.90 \exp(-2.8 \cdot R) + 0.10 \exp(-0.37 \cdot R) \quad .$$

The water experimental measurements multiplied by the F(R) factor agree, within 2%, the calculated data.

In Fig. 2 this agreement is shown for (a) one of the possible moving configurations of the Ir 192 source (program disk no. 9) and (b) for the Cs 137 source.

A third order polynomial was used to describe the attenuation in water[2].

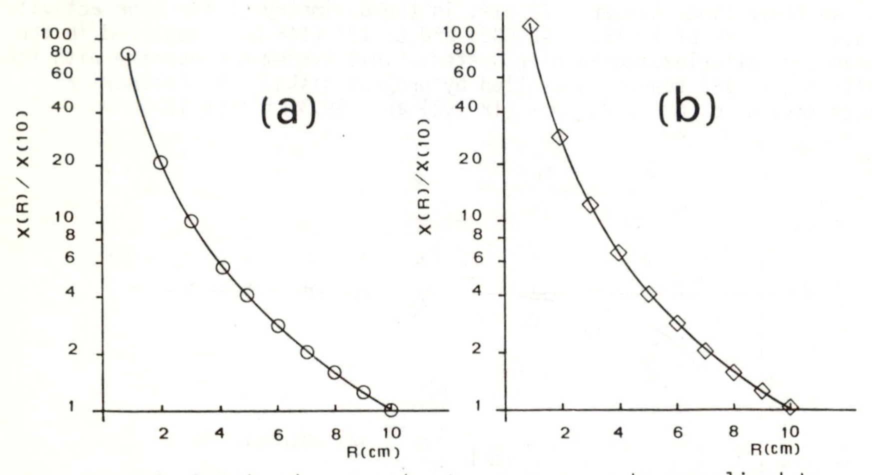

Fig. 2 - Calculated and measured water exposure rate normalized to R = 10 cm for (a) Ir 192 oscillating source and (b) Cs 137 source.

References

1. K.J. Cassell, Brit. J. Radiol. 56, 113, 119, (1983).
2. L.L. Meisberger, R.J. Keller, R.J. Shalek. Radiology, 90, 953, 957, (1968).

PROCEEDINGS OF THE II INTERNATIONAL CONFERENCE ON APPLICATIONS OF PHYSICS TO MEDICINE AND BIOLOGY
edited by Ž. Bajzer, P. Baxa & C. Franconi

STATISTICAL ANALYSIS ON ISOEFFECT IN CURIETHERAPY

S.Belletti, P.Feroldi, R.Moretti, L.Verzeletti
Medical Physics Dept.,General Hospital,Brescia-Italy

Isoeffect curves are fundamental in curietherapy,as in radiotherapy in general,when people intend to optimize the therapeutic ratio with various treatment regimes.In fact, radiobiological and clinical data have shown the necessity to formulate isoeffect models (1)(2)(3)(4),with a high specificity on hystology,volume and other characteristics.

The aim of this paper is to propose a quite simple isoeffect formula widely valid in curietherapy and to consolidate it through a statistical analysis in a suitable sample.

From a bestfitting of radiobiological and clinical data (5), we derived this isoeffect formula:

$$\text{Total Dose} = D \times (0.56 + 0.75 \exp(-1.5 \times d))$$

where d = dose-rate ranging from 0.20 to 2.50 Gy/hr and D=reference total dose at 0.357 Gy/hr clinically stated in order to suit different tissues,sites,volumes and treatment regimes.

Between Jan 1973 an June 1982 55 patients, affected by oral tongue cancer (32 T1 cases and 23 T2 cases), were treated with exclusive Ir 192 hairpins interstitial at the General Hospital of Brescia, Italy.
The dose-rate ranged from 0.3 to 2.6 Gy/hr: in 42 cases the total dose was computed according to our formula (72.5 Gy mean total dose referred at 0.357 Gy/hr) whilst in 13 no biased cases no isoeffect correction was applied (102.2 Gy mean total dose referred at 0,357 Gy/hr), thus resulting in two subgroups normo and hyper dosed.

Standard variance analysis was used,considering recurrences and radionecroses incidence.

Results are shown in table I. As recurrences, no significant difference exists between doses and the difference is at a limit value (P=0.05) between stages.

As radionecroses, there is an expected highly significant difference ($P < 0.025$) between stages, but an even higher difference ($P < 0.005$) between doses.

In conclusion, to ignore an isoeffect relationship causes an increase in normal tissue damage with no benefit in recurrences reduction and no improvement of therapeutic ratio.

Tab.1 - Recurrences and radio necroses in oral tongue cancer treated with Ir 192 according to Belletti's isoeffect relationship

Stage	Dose normo	Dose hyper	Dose normo	Dose hyper
T1	3/26 (12%)	1/6(17%)	0/26 (-)	1/6 (17%)
T2	6/16 (37%)	2/7(29%)	2/16 (12%)	4/7 (57%)
Total	9/42 (21%)	3/13(23%)	2/42 (5%)	5/13(38%)

Recurrences
Between stages P = 0.05

Radionecroses
Between stages $P < 0.025$
Between doses $P < 0.005$

1) Barendsen G.W. -Dose fractionation, dose-rate and isoeffect relationships for normal tissue responses - Int.J. Rad.Oncol.Biol.Phys. 8,1981-1997, 1982

2) Ellis F. - Dose, tissue and fractonation: a clinical hypotesis. Clin.Radiol. 20,1-7, 1969

3) Hall E.J. - Radiation dose-rate: a factor of importance in radiobiology and radiotherapy-Brit.J.Radjol.45,81-97, 1972.

4) Liversage W.E. - A general formula for equating protracted and acute regimes of radiation-Brit.J.Radiol.42,432-440,1969.

5) Goitein M. - Review of parameters characterizing response of normal connettive tissue to radiation-Clin.Radiol. 27,389-404, 1976.

PROCEEDINGS OF THE II INTERNATIONAL CONFERENCE ON APPLICATIONS OF PHYSICS TO MEDICINE AND BIOLOGY
edited by Ž. Bajzer, P. Baxa & C. Franconi

EFFECT OF FRACTIONATED DOSES ON HEMOPOIETIC TISSUE OF MICE

Sergio Belletti, Rosina E. Gallini, Ugo Giugni
Università degli Studi, Facoltà di Medicina
Brescia, Italy

The aim of our research is to study the effect of gamma rays on the hemopoietic tissue of mice, after the administration of daily fractionated doses, varying from the typical values of segmentary radiotherapy to the values of total body irradiation(TBI), namely from 3.0 to 0.1 Gy/die.

Now in this paper we illustrate the results obtained with daily doses up to 0.75 Gy, by studying the stem cells compartment, through the spleen colonies technique, the cellularity of femurs and spleens and also by studying the peripheral blood, if the treatment exceeds 10 days.

The experiments were carried out on 1200 hybrid mice, 3-4 months old; we used for each experimental point males (from 3 to 12 animals) as bone marrow and blood donors and 10 females as recipients, previously irradiated with 9 Gy.

The hematological data, obtained by sampling the blood of 3 mice were: leucocytes, erythrocytes and platelets, analized by electronic counter. The donors were sacrificed immediatly after the end of each irradiation, while the receivers were killed on the 9th day after the bone marrow transplantation.

For the irradiation we utilized a ^{60}Co source at dose rate of 0.25 Gy/min and the dosimetry was performed with a thimble ionization chamber PTW of 1.1 cm^3.

The daily doses studied were 3.0-2.0-1.5-1.0-0.75 Gy with with times of total treatment ranging from 4 to 16 days.

The survival of bone marrow stem cells, expressed as surviving fraction of CFU(Colony Forming Unit) is given in Fig 1;the experimental errors don't exceed ± 15%.The straight line represents the results obtained through acute dose.There is no difference between acute and fractionated dose up to a total dose of 4 Gy;after this point it is possible to evidence the repair.With higher doses the surviving fraction remains nearly constant,because the radiation damage is balanced by the proliferation.

The test of the peripheral blood shows at first a deep reduction of the elements and then a return to normal values. The normal values(mean value ± 2 S.D.) of the elements of the blood were determined in 40 mice of the same sexe and age.

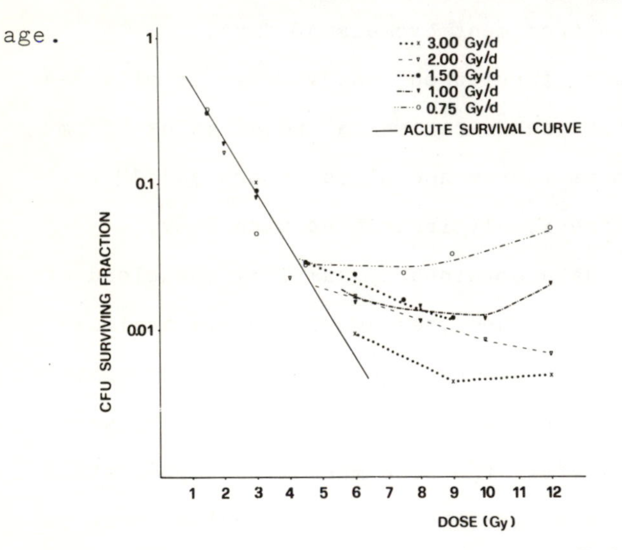

Fig.1-CFU Survivng Fraction for the 5 doses studied

Research supported by CNR,Special Project"Control of Neoplastic Growth",grant N 82.00229.96 and 81.01297.96

PROCEEDINGS OF THE II INTERNATIONAL CONFERENCE ON APPLICATIONS OF PHYSICS TO MEDICINE AND BIOLOGY
edited by Ž. Bajzer, P. Baxa & C. Franconi

PENCIL BEAM DOSE CALCULATION

Leopoldo Conte , Roberto Brambilla
Raffaele Novario , Mosé Visconti
ISTITUTO DI SCIENZE RADIOLOGICHE DELL' UNIVERSITA' DEGLI STUDI DI MILANO - SERVIZIO DI FISICA SANITARIA DELL'OSPEDALE REGIONALE DI VARESE.
21100 Varese, viale Borri n.57
ITALY

A computer program has been devised wich processes CT images to obtain electronic densities in areas where the minimum dimensions are one pixel.

The images obtained allow computing doses in radiotherapy with photons, taking into account variations in electronic density, even in small regions of the irradiated volume.

A calculation algorithm of the dose in a point is suggested, based on the laws describing interactions of pencil beam with the matter.

This approach is comparable with the method suggested by Wong, Henkelman, Fenster and Johns 1).

Preliminary applications concern 60-Co source in homogeneous media of different densities as water,lung,bone.

The dose absorbed in a point comes from electrons and photons scattered by volume elements situated along the path of the pencil beam (fig.1.).

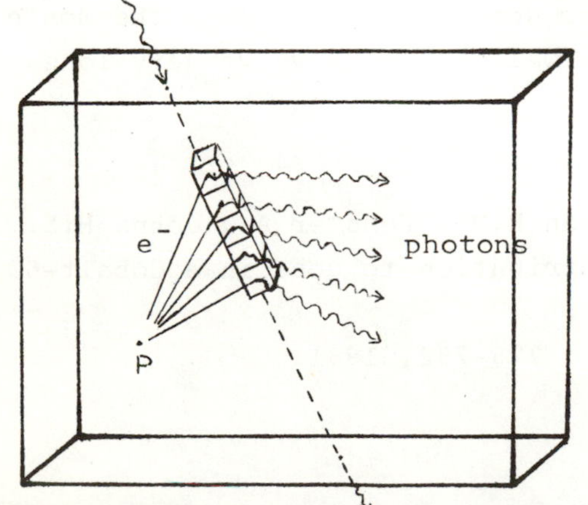

Fig. 1: model for the calculation of the dose in a point.

In this way transmission curves, dose profiles and iso doses curves were obtained for pencil beam in density 1 (water), 1,45 (bone), 0,30 (lung) media.

One can see a good agreement between the transmission curve due to electron contribution and the transmission curve of the zero area TAR (fig.2).

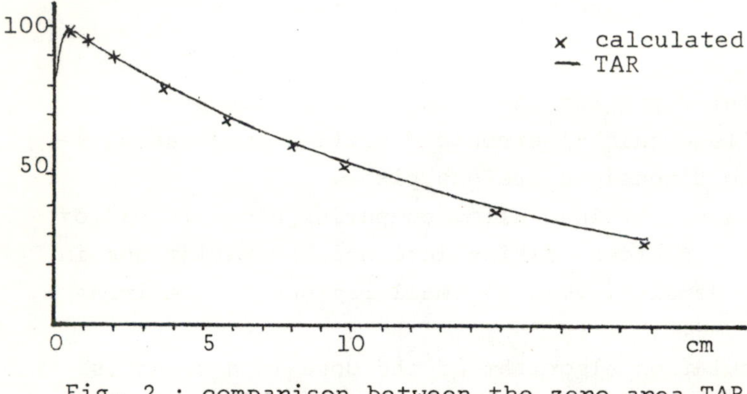

Fig. 2 : comparison between the zero area TAR transmission curve and the calculated transmission curve (electron contribution)

It was also possible, with integration procedure, to obtain the contributions to the transmission curve in water given by scattered electrons and photons in the case of fi= nite dimensions field.

The method proposed requires further verifications.

However it could offer some advantages over the Monte Carlo methods wich, as is well known, involve very long times.

References

1) Wong J.W., Henkelman R.M., Fenster A., Johns H.E. Second scatter contribution to dose in a Cobalt-60 beam.
Med. Phys. 8 (6) : 775-782, 1981.

PROCEEDINGS OF THE II INTERNATIONAL CONFERENCE ON APPLICATIONS OF PHYSICS TO MEDICINE AND BIOLOGY
edited by Ž. Bajzer, P. Baxa & C. Franconi

FETAL DOSES IN DIAGNOSTIC X RAY EXAMINATIONS

Elisabeth Mateus Yoshimura and Emico Okuno
Instituto de Física da USP - Departamento de Física Nuclear
C.P. 20516 - São Paulo - SP - BRASIL

Introduction

Occasionally women are submited to radiological examinations without knowing about their pregnancy. It is known that X rays are harmful to growing tissues, like human embryo and fetus. According to the absorbed dose, even abortion may be recommended.[1-4]

Unfortunately, these examinations are not always performed by following the best procedures, leading to poor quality X ray images and excess of dose to patients.

In the present work doses to patients undergoing abdominal X ray exposures were experimentally determined.

Experimental Method

Dose determinations were carried out with thermoluminescent dosemeters (TLDs)of lithium fluoride (TLD-100) and Brasilian natural calcium fluoride.[5]

Patients were simulated by an Alderson phantom with TLDs inserted insuitable holes. Abdomen exposures were made in different hospitals, and evaluation of skin and uterus doses were performed.

For comparison with theory, in some cases, the X ray beam exposure and effective energy were determined, also with TLDs.

Results

Experimentally obtained absorbed doses ranged from 30 to 2500 mrads to the uterus and from 120 to 9500 mrads to the skin (Table I). Compared to theoretical values from tables based on Monte Carlo calculations of tissue air ratios[6],experimental doses showed good agreement (±20%), providing a simple method of dose checking.

Dose variation with X ray voltage (kVp) and charge (mAs) was also sudied. Skin and uterus doses showed a linear response with mAs. As expected, skin dose increased with the second power of kVp, but the uterus dose followed a third power law.

TABLE I: Mean experimental values and range (brackets) of uterus and skin doses per abdomen radiograph of thin, standard (stdr) or fat patient. Techniques (kVp, mAs and Focus to Skin Distance (FSD)) were suggested by hospitals staf.

PATIENT	kVp	mAs	FSD(cm)	SKIN DOSE (mrads)	UTERUS DOSE (mrads)
thin	60 (50-66)	92 (64-115)	(109-180)	436 (123-706)	76 (22-135)
stdr	78 (60-85)	130 (80-250)	(109-180)	1323 (268-3900)	285 (54-685)
fat	95 (70-117)	158 (100-320)	(109-180)	2850 (475-9240)	760 (100-2600)

Discussion and Conclusions

The high values of fetal doses that can be reached in one radiograph, together with the great number of X ray exposures during a routine examination of a patient causes concern. Some recommendations to radiologists are worthwhile, although not new:

- X ray examinations of women in the reproductive age must be avoided, unless necessary. Non emergency X rays should be performed according to the "ten days rule".
- Radiation fields must be restricted to the area of interest. This includes the use of protective shielding if required.
- All available improvements should be used to minimize the patient dose, without loss of medical information.
- X ray units and imaging forming systems must have adequate quality control and maintenance.

References

1) MOLE,R.H. - "Radiation effects on pre-natal development and their radiological significance" Brit.J.Radiol. 52,89-101,1979.
2) BAKER,M.L. et al - "Fetal exposure in diagnostic radiology" Health Phys. 37,237-239,1979.
3) JACOBSEN,A. & CONLEY,J.G. - "Estimation of fetal dose to patient undergoing diagnostic X ray procedures" Radiol.120,683-685,1976.
4) HAMMER-JACOBSEN,E. - "Therapeutical abortion on account of X ray" Dan.Med.Bull. 6,113,1959.
5) WATANABE,S. & OKUNO,E. - "Thermoluminescent properties of Brazilian natural calcium fluoride" Proc. 3rd Int. Conf. on Luminescence Dosimetry,Riso,Denmark,11-14 Oct 1971,p 380-391.
6) ROSENSTEIN,M. - "Organ doses in diagnostic radiology" HEW Publication (FDA) 76-8030, 1976.

PROCEEDINGS OF THE II INTERNATIONAL CONFERENCE ON APPLICATIONS OF PHYSICS TO MEDICINE AND BIOLOGY
edited by Ž. Bajzer, P. Baxa & C. Franconi

PATIENT EXPOSURE AND QUALITY ASSURANCE IN X-RAY DEPARTMENTS IN PAPUA NEW GUINEA

Indravadan C. Patel
University of Technology
Lae, Papua New Guinea

INTRODUCTION A Quality Assurance program in radiology is essential for achieving highest quality radiographs with the lowest possible retakes and minimum radiation exposure to patients and personnel. In Papua New Guinea, there are over 150 medical x-ray units. However, there is no physicist employed in the health care services, and there is no QA program either. To assess the situation, a few x-ray departments were surveyed and some of the results are presented here.

METHOD & RESULTS The main features of each x-ray department together with the radiological safety factors are summarised in Table 1. The

TABLE 1. EQUIPMENT CHARACTERISTICS AND SAFETY FACTORS

X-ray Department	Type of Equipment	Structural Shielding	Output mR/mAs	HVL mmAl	Light and x-ray beam Alignment	Exposure at operator's position-mR
A	Fixed	Yes	3.9	2.1	Fair	0.5
B	Fixed	Yes	4.4	1.8	Bad	0.1
C	Fixed	Yes	2.9	2.3	Good	0.3
D	Mobile	No	2.6	2.7	Bad	7.5
E	Mobile	No	2.1	-	Fair	-
F	Mobile	No	2.0	2.8	Fair	-
G	Mobile	No	4.7	2.4	N/A	10
H	Fixed	Yes	2.2	2.7	Fair	NIL
I	Fixed	Yes	2.3	2.1	V.Bad	0.3
J	Mobile	No	2.9	2.4	Bad	9
K	Fixed	Yes	3.7	2.3	Good	NIL
L	Fixed	Yes	3.4	2.8	Fair	0.1
M	Fixed	Yes	3.2	2.4	Bad	0.1
N	Fixed	Yes	3.5	3.0	Fair	NIL

tube output was measured at one metre from the focus using a NBS calibrated 15c.c. ionization chamber. The last column showing the exposure at operator's position was measured using a 900c.c. calibrated ionization chamber, and refers to the most frequent radiological examination performed in that department. All the other factors were measured by the common methods [1]. Table 2. shows the variations in exposure factors and patient skin exposures for some typical radiological examinations.

TABLE 2. VARIATIONS IN SKIN EXPOSURES

Examination	kVp	mAs	Skin Exposure (mR)
Skull AP	65-80	20-80	64-512
Cervical Spine	60-70	6-80	25-368
Pelvis	65-75	24-80	136-508
Abdomen Pregnancy	75-100	30-160	240-1424
Arm	50-65	3-16	8-112
Chest	65-70	5-15	10-144

DISCUSSION The results of this survey should be considered with caution, since it is incomplete. Some important parameters like kVp and film processing technique were not investigated due to lack of suitable equipment. In spite of these drawbacks, some general comments are in order. This study has shown that the situation in the x-ray departments in the country is not alarming. Most x-ray units are in reasonable working state. However, the radiological safety in most departments could be improved. For a given radiological examination, there is a large range of exposure factors used, resulting in large variations in patient exposures to x-rays. The radiographs in all cases are stated to be satisfactory. To understand the causes of these variations, and to reduce patient exposure, it is hoped that in the very near future a detailed survey of all x-ray departments will be carried out. This survey will include measurement of kVp as well as evaluation of film processing techniques.

REFERENCES

1. "Basic Quality Control in Diagnostic Radiology", Ed. Siedband et al. A.A.P.M. Report NO. 4, 1978.

PROCEEDINGS OF THE II INTERNATIONAL CONFERENCE ON APPLICATIONS OF PHYSICS TO MEDICINE AND BIOLOGY
edited by Ž. Bajzer, P. Baxa & C. Franconi

RADIATION EXPOSURE IN MEDICAL DIAGNOSTIC PROCEDURES IN FRIULI VENEZIA GIULIA REGION

V. Barbina, G. Contento, M.R. Malisan, C. Omet, R. Padovani
G. Gozzi (*)
Servizio di Fisica Sanitaria, Ospedale Civile, Udine, Italy
(*) Istituto di Radiologia dell'Università, Ospedale Maggiore, Trieste, Italy

This research aims at assessing the radiation exposure due to diagnostic radiology in Friuli Venezia Giulia region (FVG), North-East Italy. In this area the whole population is about 1.5 million: there are 22 National Health Service hospitals and a few private medical centers. The research is carried out with the collaboration of the Gruppo Regionale SIRMN-Friuli Venezia Giulia.

The final goal is the evaluation of the Collective Somatic Effective Dose Equivalent (CSEDE) and the Genetically Significant Dose (GSD). To assess the CSEDE, which is a weigthed sum of organ doses, both the numbers of each type of examination as well as the relative mean organ doses are to be considered. The GSD is defined as the dose which, if received by all members of the population with an age capable of reproduction, would produce the same genetical effects of the doses effectively received by the gonads of the single members.

In order to compute the GSD a number of different factors must be considered. The first one is the number of individuals in each age-sex group who underwent a particular X-ray examination, the second one is the mean gonadal dose received by an individual, in each age-sex group, for a particular examination. Finally the age-sex structure of the population and the pattern of child expectancy must be evaluated.

This paper presents the preliminary results of a survey of the radiological activity in FVG for a period of two weeks, resulting in the collection of the data of over 40000 examinations, out of which 19000 have been processed.

Table 1 shows the overall frequencies of the different types of examinations, compared to the recent results of similar surveys in UK and USA. It is noted that CT-scanning procedures are becoming a relevant contributor to GSD and CSEDE in FVG.

As for the age-sex distribution of all examinations, it seems that the age group below 15 in FVG is less radiologically exposed, regardless of the examination type, than the equivalent age group in UK and USA (FVG 7.8%, UK 13.7%, USA 11.9%).

Moreover, sample organ dose measurements have been performed, in two radiological departments, allowing the refinement of the dose measurements protocols for most of common X-ray procedures.

The dosimetry system consists of LiF ribbons (TLD 100) in polythene sachets. The sensitivity figure of LiF over the range of X-ray qualities found in diagnostic radiology has been determined as a function of kVp and total filtration of the beam.

The doses were measured according to the type and location of the organ in the body. For compact organs close to the surface as breasts, thyroid and testicles the doses were directly measured on patients with TLDs attached to the skin. For other organs of relevant interest, doses were derived from the measurements of entrance skin doses using Monte Carlo simulation.A comparison of the measured organ doses with those derived from the Monte Carlo simulation has shown that, for compact organs close to the edge of the beam, the doses measured with TLDs are much greater, see Table 2.

Table 1. Frequency of the different types of examinations in FVG

Examination type	Percentage distribution		
	FVG 1983	UK 1977	USA 1970
Chest, heart and lungs	39.9	32.8	49.5
Limbs	18.4	28.3	17.0
Cervical, dorsal, lumbar spine	10.0	8.7	5.1
Head	5.5	7.8	7.4
Pelvis	3.2	4.0	2.0
Abdomen	3.0	4.1	3.3
Barium meal	2.9	2.9	4.4
CT scan	2.1	0.3	0.0
Cholecystography, cholangiography	1.7	1.6	3.1
Intravenous pyelography	1.7	1.9	3.9
Barium enema	1.3	1.2	3.3
Others	10.3	6.4	1.0

Table 2. Organ doses in chest PA projection.
A - 65 kV, 2.5 mm Al tot. filtr., SFD = 180 cm, Field = 40x32 cm^2
Mean entrance skin dose in 50 examinations: 610 uGy
B - 120 kV, 3.0 mm Al tot. filtr., SFD = 180 cm, Field = 40x32 cm^2
Mean entrance skin dose in 50 examinations: 680 uGy

Organ	TLD uGy		Monte Carlo uGy		TLD/Monte Carlo	
	A	B	A	B	A	B
THYROID	46.5	71	5.5	28	8.45	2.54
BREASTS	49	77	36	85	1.36	0.91
OVARIES			0.2	1.5		
LUNGS			154	330		
RED MARROW			34	110		

CEC contract n. BIO 516/I(S)

PROCEEDINGS OF THE II INTERNATIONAL CONFERENCE ON APPLICATIONS OF PHYSICS TO MEDICINE AND BIOLOGY
edited by Ž. Bajzer, P. Baxa & C. Franconi

TIME RESPONSE IN RADIOMETRIC MEASUREMENTS OF BIOLOGICAL T

G. Marsiglia, G.P. Preti, S.M.A. Firenze ITALY

P.P. Lombardini, University Torino ITALY

A stratified plane model of 5 parallel layers of idealized biological tissues (in Fig. 1- from left to right: skin, fat, muscle, tumor, muscle) of which thickness, density (ρ), thermal conductivity (k), specific heat (c), metabolic heat (W), blood cooling rate (C) are known, is assumed to be exposed to a 10 minute pulse of an externally controlled diathermic heater of rate Q. The following expression, derivable from energy conservation considerations:

$$\rho c \frac{\partial T}{\partial t} - K \frac{\partial^2 T}{\partial x^2} = Q + W - C$$

has been numerically solved. In Fig. 2, the calculated temperature at a depth of 5 mm under the skin is displayed vs. time, counted from the end of the 10 minutes theating pulse.

Curve A refers to the case in which the tumoral layer is present, and curve B to the case in which the tumoral layer is not present.

Clearly, the presence of the tumoral layer enhances the temperature (as already experimented by [1]). Furthermore, the time delay in the maximum of the thermal response varies in the two curves, indicating a change in the spacial distribution of heating for the two cases.

A delay in thermal response related to a difference in position of heated regions has been observed in an experiment performed at S.M.A. (See Fig. 3). In figure, a slab of necrotic biological tissue, laid on an aluminum heatable plate, is observed by two radiometers, one operating in the microwave band (2 cm), the other in the IR band (about 1 micron). The temperature of the tissue is also monitored at different depths by two thermocouples.

The readings of the two radiometers during a period of switching

tching off of the heater, are plotted vs. time, in Fig. 4. It is parent that the microwave radiometer reaches a maximum reading approximately 2 minutes after the switching-off of the heater, while the maximum of the IR radiometer occurs more than 2 minutes later. This delay is attributable to the fact that, the penetrating power of the two radiometers being different, they probe two different regions of the heated tissue, each having its own thermal history.

1) K.L. Carr, A.M. El-Mahdi, J. Shaeffer (1981), IEEE MTT-29, No 3

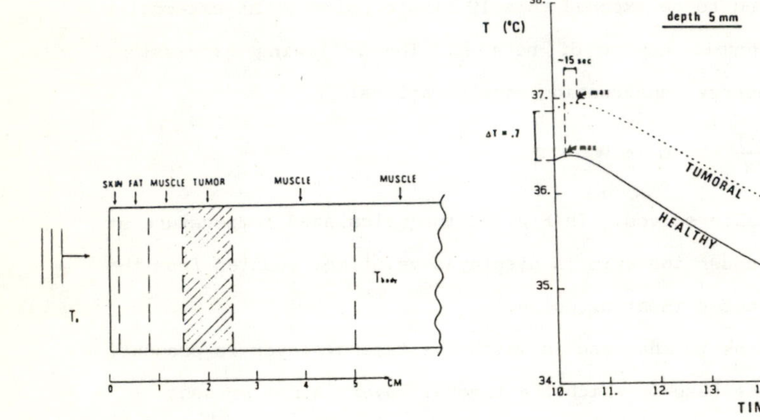

Fig. 1 Schematic of biological tissue

Fig. 2 Time response since heating switched-off

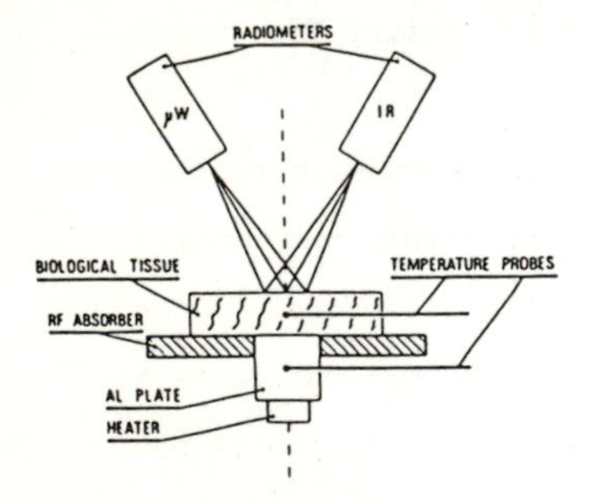

Fig. 3 Measurement schematic diagram

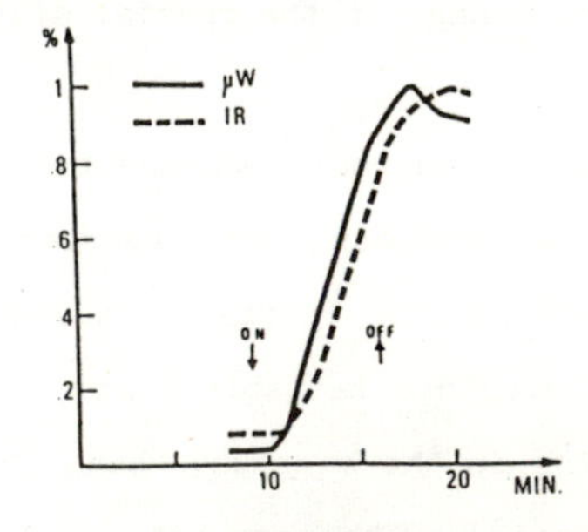

Fig. 4 MW versus IR radiometric response comparison

PROCEEDINGS OF THE II INTERNATIONAL CONFERENCE ON APPLICATIONS OF PHYSICS TO MEDICINE AND BIOLOGY
edited by Ž. Bajzer, P. Baxa & C. Franconi

MICROWAVE-INDUCED HYPERTHERMIA AND IONIZING RADIATION-TECHNIQUE AND CLINICAL RESULTS

Bertil Persson, Per Nilsson, Sven Hertzman
Clas-Ebbe Lindholm*, Elisabeth Kjellén and Torsten Landberg*
Lund University Hospital, S-221 85 Lund
*Malmö Allmänna Sjukhus, S-214 01 Malmö
SWEDEN

1. Introduction

Hyperthermia in combination with ionizing radiation offers many potential advantages in tumour treatment. A fractionation schedule using microwave induced hyperthermia and low-dose radiation therapy has significantly increased the treatment results of superficial recurrances of tumours or metastases in areas which were previously given full dose radiotherapy[1].

2. Method and material

For the induction of hyperthermia in clinical practice a 2450 MHz microwave system has been used since August 1980[2]. During 1982 a system using the frequency 915 MHz has also been constructed. A control system based on a microcomputer adjusts the temperature in the tissue by alternately temperature registration and microwave irradiation. Conventional thermistor probes (Ø=0.6 mm) which are placed in the tissue by using intravenous cannulas are used for the temperature reading.

For the treatment of superficial tumours different waveguide applicators have been designed. A temperature increase with acceptable homogeneity can be reached in tumours with 4-6 cm diameter and to a depth of 2-3 cm below the skin surface. Radiation therapy has been given both by photon and electron irradiation.

The combined treatment has so far been offered to about 30 patients. At the time of treatment these patients had superficial, local recurrances of tumours or metastases, in spite of earlier used established treatment of recurrances. The radiation absorbed dose (10x3.00=30.0 Gy) as well as the hyperthermia treatment (41-44 °C during 45 minutes for each treatment) have been constant. In those cases where the patients have had more than one superficial tumour, at least one tumour, usually the smallest, has been treated only with radiation therapy to the same absorbed dose and following the same fractionation schedule as in the combined treatment.

3. Results

The treatment of 29 patients with a total of 56 tumour regions has been evaluated. Complete response (CR) and partial response (PR) have been estimated within the treated regions. Of the 43 tumours which were given the combined treatment 29 have responded (14 CR and 15 PR). The median follow up time is 8 months (range 1+-31+ months). The corresponding result for 13 tumours treated with radiotherapy only is 4 responders (2 CR and 2 PR). In this material 13 of the patients have had 2-5 different tumour regions each. This has offered the possibility to compare the different treatment modalities on the same patient. The combined treatment were given to 24 tumours out of which 19 responded (10 CR and 9 PR), based on a median follow up time of 8 months (range 1-29 months). Radiotherapy alone resulted in 4 responders out of 13 treated tumours (2 CR and 2 PR).

4. Conclusion

The results obtained with the present equipment for the induction of hyperthermia in superficial tumours have increased the interest for a further development of the system. This is now made more flexible with possibilities to use radiofrequencies down to 100 MHz. To obtain better temperature distributions and increase the homogeneity of the heating pattern new applicators will be designed.

5. References

1) Acta radiol. Oncology 21 (1982), 241.
2) Acta radiol. Oncology 21 (1982), 235.

PROCEEDINGS OF THE II INTERNATIONAL CONFERENCE ON APPLICATIONS OF PHYSICS TO MEDICINE AND BIOLOGY
edited by Ž. Bajzer, P. Baxa & C. Franconi

MICROWAVE HYPERTHERMIA IN SIMULATION MODELS : CORRELATION AMONG SOURCE, ELECTROMAGNETIC ABSORPTION, AND TEMPERATURE .

S.Caorsi°, G.Scielzo°°, G.Ogno°, B.Brunelli°°, F.Bistolfi°°°

° Biophysical and Electronic Engineering Division-Institute of Electrical Engineering - University of Genoa.
Via all'Opera Pia 11 A , 16145 Genoa , Italy

°° Health Physics Dept. and °°° Radiotherapy Dept.
Galliera Hospital , Via Volta 19 , 16128 , Italy .

In this paper, the problem of controlling the various elements that contribute to characterize a microwave hyperthermic treatment |1| is de= veloped on the basis of experimental models and tests.

In particular, we consider : the characteristics of the e.m. source; the applicator; the e.m. absorption distribution and the temperature di= stribution in the heated medium; the e.m. power density that is present in the environment during the hyperthermic treatment.

For this purpose, we have prepared the following set of electromag= netic and thermal measurements:

a) impedence matching between the source and the applicator, and power transmission to the applicator, both under loaded conditions (the me= dium that is to be heated) and under free-space conditions;
b) electromagnetic field distribution, in the heated medium, using a small microwave probe;
c) temperature distribution, by means of a series of thermistors control= led by a microprocessor;
d) electromagnetic fields in the environment .

The schematic diagram of the whole apparatus is shown in Fig.1 .

Following this diagram, under the assumption that the simulation me= dium has linear characteristics, we carry out, as a first step, the measu= rements described in the previous points a) and b) . In the case b) the measurement is performed at low values of available electromagnetic po= wer and by means of a microwave probe |2| realized according to the Chen work |3| .

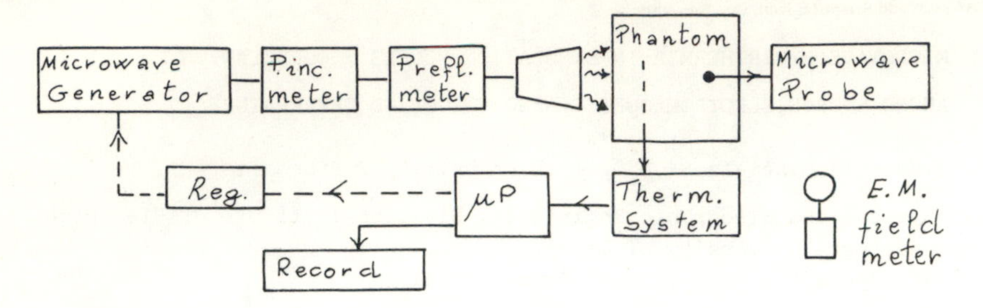

Fig.1 Schematic diagram of the measurement apparatus.

The temperature profiles are recorded using 16 thermistors controlled by an AIM 65 Rokwell microprocessor |4,7| .

Moreover,in order to use in the next future non-invasive thermometric tecnhique based on microwave radiometry |5| , a microwave radiometer is now being developed |6| .

Finally,the first experimental results have been obtained in the case of heating by microwave irradiation at the frequency 433 MHz .

References

|1| F.Bistolfi, (Ed.): "L'ipertermia in terapia oncologica " Piccin Padova,1981 .

|2| S.Caorsi,G.Ogno: "A probe for electromagnetic field measurements in a microwave-heated phantom" Institute of Electrical Engineering University of Genoa ; Rep. May 1983 .

|3| Kun-Mu Chen,S.Runkspollmuang and D.P. Nyquist:"Measurement of induced fields in a phantom model of man". Radio Science, Vol.17 55S; 1982, pp. 49S-59S .

|4| G.Scielzo,B.Brunelli,M.Carso:"Ipertermia mediante radiofrequenza e microonde:allestimento di un sistema a microprocessore per la misura di temperatura" Convegno A.I.R.P. Genova 1982,Atti pp.347-354.

|5| S.Caorsi:"Il riscaldamento elettromagnetico e il controllo radiometrico di temperatura in ipertermia oncologica",Convegno Nazionale su "L'ipertermia a microonde in terapia oncologica",invited paper , Trento 30-31 March 1983 .

|6| S.Caorsi:"Temperature distributions by microwave radiometric measurements" Third Int. Congr. of Thermology ; Bath England 29th March 2nd April 1982 .

|7| G.Scielzo:"Esperienze di termodosimetria nell'impiego di radiofrequenze e microonde in terapia oncologica",Convegno Nazionale on "L'ipertermia a microonde in terapia oncologica" invited paper, Trento 30-31 March 1983

PROCEEDINGS OF THE II INTERNATIONAL CONFERENCE ON APPLICATIONS OF PHYSICS TO MEDICINE AND BIOLOGY
edited by Ž. Bajzer, P. Baxa & C. Franconi

MICROWAVE HYPERTHERMIA OF BIOLOGICAL SYSTEMS WITH NON-LINEAR THERMAL PROPERTIES: PREDICTION OF TEMPERATURE DISTRIBUTIONS.

S.Caorsi, L.Laderchi, P. Lazzereschi, R.Merlo

Biophysical and Electronic Engineering Division
Insitute of Electrical Engineering - University of Genoa (ITALY).

In this paper temperature distributions are studied in biological systems exposed to the microwave hyperthermic treatment. The biological system is simulated by easy geometrical models that can be assembled according to a modular technique. Each elementary block (cubes or plane layers) is characterized by constant values of the complex dielectric permittivity, specific heat, density, and thermic conductivity; these values can be deduced from the bibliographic references.

Furthermore, besides the cooling of the external surfaces, we consider the thermoregulatory properties of the body both of the chemical type (i.e., metabolic heat generation) and of the physical type (i.e., variations in the blood flow, and the radiative and convective combined exchange between the body and the boundary medium).

In order to approximate as much as possible the actual behaviour of the biological system, the thermoregulatory properties are assumed to be non-linear functions of the local temperature.

Under these conditions, the temperature distributions inside the biological system $v(\bar{r},t)$ must satisfy the following equation of the heat transfer:

$$c(\bar{r})g(\bar{r}) \frac{\delta}{\delta t} v(\bar{r},t) = \nabla \cdot [K(\bar{r}) \nabla v] + Qtr(v) + Qem(r)$$

where:

c,g,K are the specific heat, the density, and the thermic conductivity, respectively;

Qtr represents the combined effect of metabolic heat generation and heat dissipation, and it can be written as follows:

$$Qtr(v) = A(v).v(\bar{r}) + B(v)$$

Qem represents the effect of the absorption of the electromagnetic field inside the body.

In this paper, (as well as in previous ones |1,2,3|) the tansient temperature distribution in the biological system is obtained by applying the well-know method of finite differences. In this case, it may be suitable, even if not necessary, to perform a segment-linearization of the Qtr functions, thus pointing out various threshold temperature that diversify the thermoregulatory behaviour of each region into linear ranges. An example of application is represented by the case of a layered biological model as shown in Fig.1.

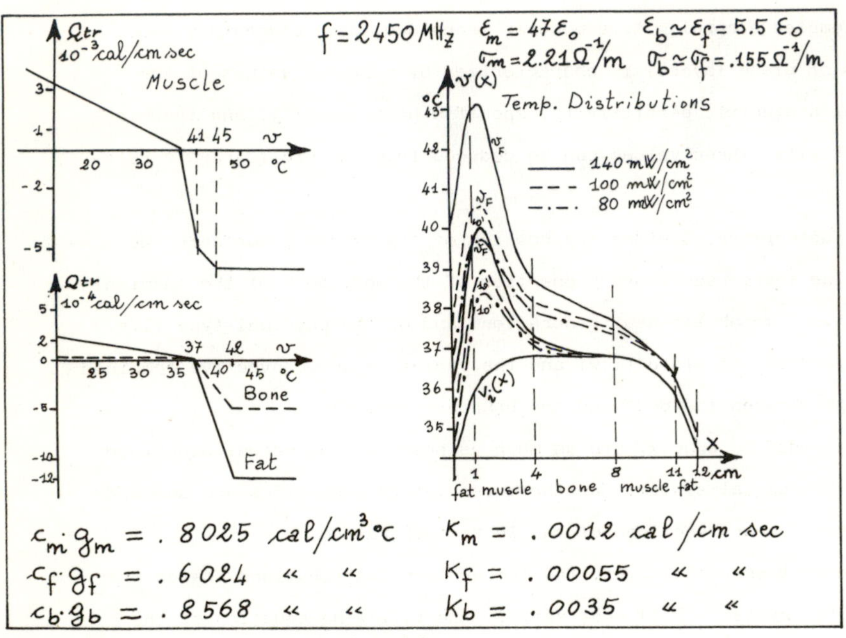

Fig. 1 Five layered biological model.

References

|1| S.Caorsi, A.Gialdini: "Riscaldamento Elettromagnetico di modelli a strati piani di tessuti biologici: influenza dei fenomeni termoregolatori" IV° Riun. Naz. "Elettromagnetismo applicato", Firenze 1982, proc. pp. 105-107.

|2| S.Caorsi: "temperature distributions in biological plane layers: microwave heating and thermoregulatory effects". "Melecon '83", Conf. IEEE, Athens May 1983. Proceendings pp.B4.10.

|3| S.Caorsi: "Electromagnetic heating and non-linear thermoregulatory response of biological systems" U.R.S.I. 1983 Conf. on "Electromagnetic Theory" Santiago de Compostela, Spain, August 1983. Proc. pp.673-676.

PROCEEDINGS OF THE II INTERNATIONAL CONFERENCE ON APPLICATIONS OF PHYSICS TO MEDICINE AND BIOLOGY
edited by Ž. Bajzer, P. Baxa & C. Franconi

THEORETICAL AND EXPERIMENTAL RESEARCH ON MICROWAVE HYPERTHERMIA

F. Bardati*, G. Gerosa**, P. Tognolatti***

*Università di Roma "Tor Vergata"

**Università di Roma "La Sapienza"

***Telespazio S.p.A.

With reference to microwave localized hyperthermia for cancer treatment, the problem of the electromagnetic heating of living tissues has been considered both from theoretical and experimental viewpoints.

In order to theoretically investigate the space and time temperature distribution inside biological bodies irradiated electromagnetically, a mathematical model of the thermal behavior of the living tissue must be assumed.

There is experimental evidence [1), 2)] that a linear model, even if the heat exchange with blood flow is taken into account, is not completely adequate to describe such a thermal behavior, because local thermoregulating effects are disregarded. In this work non-linear effects have been taken into account, by assuming the blood flow convective effects and the metabolic heat generation to be represented in each point by a non-linear function $F(\theta)$ of the local temperature θ ; moreover a temperature dependent effective thermal tissue conductivity $K(\theta)$ has been assumed. The effect of a cooling fluid forced past the surface of the living tissue has been also taken into account.

The non-linear operator equation

$$Q(\theta) = \rho C \frac{\partial\theta}{\partial t} - \nabla\cdot(K\nabla\theta) - F - Q_{EM} = 0,$$

where $Q_{EM}(\underline{r}, t)$ is the electromagnetic mean power dissipation rate, and ρ and C are density and specific heat respectively, has been linearized and solved by using a variant of the Newton iterative method.

The functions $F(\theta)$ and $K(\theta)$ have been represented as Taylor series expansions in a neighborhood of the arterial blood temperature θ_B.

In particular the case of planar stratified structures simulating living bodies, irradiated by plane electromagnetic waves, has been considered [3]. The temperature distribution has been determined for various values of the coefficients of the series expansions of F(θ) and K(θ), which have been suitably truncated. From the numerical results it appears that the non-linearity associated with F mainly reduces (with respect to the linear case [4), 5)] the mean heating temperature, while non-linearity associated with K flats the temperature distribution, without a significative reduction of the mean temperature. From a proper combination of both non-linearities a matching of the theoretical temperature distributions with experimental evidences is expected.

With regard to the experimental part of the research, L-band direct-contact applicators for microwave hyperthermia for cancer treatment have been developed and tested [6]. In order to evaluate the electromagnetic power deposition within biological bodies, the various experimental set-up for both electrical probes measurements and cholesteric plate thermography have been developed.

References

1) H.F. Cook, "A physical investigation of heat production in human tissues when exposed to microwaves", Brit. J. Appl. Phys., vol. 3, pp.1-6, 1952.
2) J.L. Guerquin-Kern, L. Palas, M. Samsel, M. Gautherie, "Hyperthermie micro-ondes: influence du flux sanguin et des phénomènes thermorégulateurs", Bull.Cancer (Paris), vol 68, pp. 273-280,1981.
3) F. Bardati, G. Gerosa, "An improved model describing the thermal behaviour of electromagnetically irradiated living tissues", Proceedings 1983 U.R.S.I. Symposium on Electromagnetic Theory, Santiago de Compostela, 23-26 agosto 1983, pp. 677-680.
4) F. Bardati, G. Gerosa , P. Lampariello, "Temperature distribution in simulated living tissues irradiated elctromagnetically", Alta Frequenza, vol. XLIX, pp. 61-67, 1980.
5) F. Bardati, "Time-dependent microwave heating and surface cooling of simulated living tissues", IEEE Trans. Microwave Theory Techn., vol. MTT-29, pp. 825-828, 1981.
6) F. Bardati, P. Tognolatti, "Experimental techniques for direct-contact applicators for microwave hyperthermia", Alta Frequenza, vol. LII, maggio-giugno 1983, pp. 173-175.

PROCEEDINGS OF THE II INTERNATIONAL CONFERENCE ON APPLICATIONS OF PHYSICS TO MEDICINE AND BIOLOGY
edited by Ž. Bajzer, P. Baxa & C. Franconi

DETERMINATION OF TEMPERATURE DISTRIBUTION IN LIVING TISSUES BY MICROWAVE RADIOMETRIC MEASUREMENTS. A STUDY OF FEASIBILITY.

F. Bardati and D. Solimini

Dipartimento di Ingegneria Elettronica
II Università degli Studi di Roma "Tor Vergata"
Via Orazio Raimondo, 00173 Roma, Italy

Measurement of microwave emission from living tissues yields information on the thermal state of the subcutaneous biological medium[1]. In addition, it has been suggested that the spatial distribution of the temperature con be determined quantitatively from a convenient set of measurements, provided suitable algorithms are available to extract the seeked data from the measurements[2),3)].

The feasibility of reconstructing the bidimensional distribution of temperature in the biological structure from radiometric measurements at microwave frequencies is discussed here. A three-layer skin-fat-muscle model of the living tissues has been considered, and the subcutaneous temperature has been assumed to undergo bidimensional variations in a plane perpendicular to the tissues layers. By using a plane wave spectral representation of the thermal emission field, an expression which relates the emitted radiation intensity to the temperature distribution whithin the tissues has been worked out, so that the temperature distribution in the biological structure can be determined by solving an integral equation, provided a suitable set of measurements be available.

The case of radiometric measurements at five different locations (Fig. 1) and at seven microwave frequencies has been considered. The relevant aspect of the detection of localized thermal anomalies has been examined. The retrieval technique has been the Kalman filtering, which, besides its god retrieving characteristics, exhibits the advantage of recursivity, which is useful whenever the temporal evolution of the thermal distribution in the tissues is of interst, as is the case, for instance, in hyperthermical treatments. Figs. 2a) and b) show two examples of numerical retrieval of the excess temperature when a localized thermal anomaly with $\Delta T=2.2$ °K is assumed to be present at two depths in the muscular tissue.

References

1) P.C. Myers, N.L. Sadowsky, A.H. Barrett, "Microwave thermography: principles, methods and clinical applications", J. Microwave Power, vol. 14, pp. 105-115, 1979.
2) F. Bardati, D. Solimini, "Radiometric sensing of biological layered media", Radio Science, vol. 18, 1983.
3) F. Bardati, U. Conventi, D. Solimini, 'Determination of temperaure profiles in biological media by microwave radiometry", U.R.S.I. Symposium on Electromagnetic Waves, Santiago de Compostela, pp. 681-683, agosto 1983.

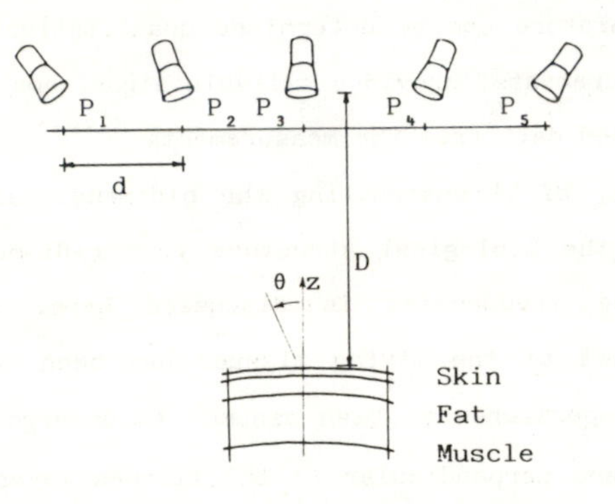

Fig.1 - Model of radiometric sensing of biological tissues by using five equidistant antennas.

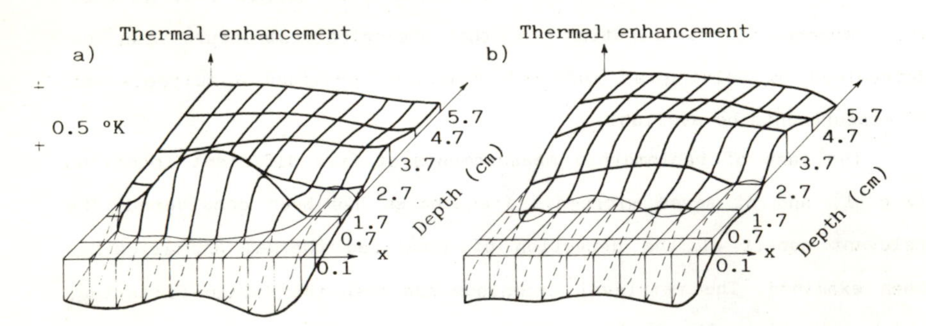

Fig.2 - Retrieved enhancement of temperature when a thermal anomaly $\Delta T_0 = 2.2$ °K is centered on depth 1.7 cm (a) and at 3.7 cm (b)

PROCEEDINGS OF THE II INTERNATIONAL CONFERENCE ON APPLICATIONS OF PHYSICS TO MEDICINE AND BIOLOGY
edited by Ž. Bajzer, P. Baxa & C. Franconi

SURVIVAL OF HUMAN RED BLOOD CELLS IN MICROWAVE HEATING

Alessandro Checcucci e Riccardo Vanni

CNR-IROE
Firenze, via Panciatichi 64
ITALY

Radiofrequency and microwave irradiation has been proposed as a faster and aseptic method for warming refrigerated blood before massive transfusions[1) 2)]. Blood heating is also involved in electromagnetic hyperthermic treatment in cancer therapy. It is consequently useful to know the thermostability of Red Blood Cells (RBC's) as well its dependence on heating profile.

Among different features characterizing RBC spoiling (fine morphologic alterations, function impairment, membrane breaking) the hemolysis extent has been chosen by us as a first step in RBC survival evaluation. However, further studies are in progress as regards the other end points. Samples of whole blood of human healthy donors have been heated in an especially designed apparatus[3)], by means of which a linear growth of temperature can be obtained at different velocities. The temperature of blood samples, with anticoagulant (EDTA) added, was raised from 4°C (cold banking temperature) to 37-40-45-50-55°C with different warm-up velocities (2-5-10-15-20°C/min). In a first series of experiments the plasmatic hemoglobin content was measured immediately after heating, while in a second series, the samples were kept 15 min at constant temperature after its prefixed value was reached. Plasmatic hemoglobin content was determined in heated and control samples by the Drabkin's method[4)].

The result of our experiments can be so summarized : a. no significant increase of hemoglobin content in the plasma is measured in samples heated up to 55 °C, whatever warming velocity is used; b. equally, no significant hemolysis occurs when samples were heated up to 37-40-45°C and kept at these temperatures for 15 min; c. a relevant hemolysis is observed only at 50 and 55°C.

Our results, indicating an approximative hemolysis threshold around

$50^{o}C$,are somewhat at variance with respect to the findings of Baranski et al.[5] and Peterson et al.[6] who observe a significant hemolysis at 41-42°C.

However,the higher thermostability we find can be explained on the basis of different methodologies used,mainly as regards : a. cell conditions (buffered suspensions or whole blood); b. warming procedures (microwave far field or cavity); c. initial temperature (banking or room temperature); d. profile and velocity of temperature growth (uncontrolled or linear).We could hypothesize,for instance,that plasma protects erythrocytes,or that in far field exposure significant temperature gradients are present vhereas in cavity exposure substantially uniform heating of the whole sample occurs[3].Furthermore,in our exposure conditions only the electric field is acting on material.Other conditions seem to be less effective.In fact,we did not observe significant differences among uncontrolled and linear temperature growth,although this second method is highly preferable for experiment reproducibility.Likewise,we did not measure strong differences by varying the warming velocity or the initial temperature.

It is,finally,to be noted that both in the case of massive transfusions and in hyperthermic therapy,whole blood only is of interest.

REFERENCES

1) DU PLESSIS,J.M.E.,BULL,A.B.,BESSELING,J.L.N.,*Anesth.Analg.* 46,96-100, 1967

2) RESTALL,C.J.,LEONARD,P.F.,TASWELL,H.F.,*Anesth.Analg.* 50,302-306,1971

3) BACCI,M.,BINI,M.,CHECCUCCI,A.,IGNESTI,A.,MILLANTA,L.,RUBINO,N.,VANNI, R.,*Proc.IMPI Micorw.Power Symp.Monaco*,1979,pp.42-44

4) VAN KAMPEN,E.J.,ZIJLSTRA,W.G.,*Clin.Chim.Acta* 6,538-541,1961

5) PETERSON,D.J.,PARTLOW,L.M.,GANDHI OM P.,*IEEE Trans.Biomed.Engin.* BME-26,428-436,1979

6) BARANSKI,S.,SZMIGIELSKI,S.,MONETA,J.,in : *Biologic Effects and Health Hazards of Microwave Irradiation, Proc.Int.Symp.Warsaw,Poland*, P. Cserski et al.Eds.,Polish Medical Publ.,pp.173-177

PROCEEDINGS OF THE II INTERNATIONAL CONFERENCE ON APPLICATIONS OF PHYSICS TO MEDICINE AND BIOLOGY
edited by Ž. Bajzer, P. Baxa & C. Franconi

EVALUATION OF THERAPEUTICAL EFFICIENCY OF WAVEGUIDE APPLICATORS

R. Antolini - Istituto di Fisica, Università di Trento - Povo (TN)- Italy.
G. Cerri, R. De Leo - Dipartimento di Elettronica ed Automatica, Università di Ancona - Ancona - Italy.

1. Introduction

Open-ended waveguides are beeing employed in the realization of microwave applicators for hyperthermia.

The dissipated power pattern in a biological body radiated by waveguide applicators is strongly affected by the characteristics of the dielectric body itself; therefore an accurate analysis of the interaction between the applicator and the biological body is essential in order to assess the efficiency of an applicator for a particular clinical situation.

In this paper the e.m. characteristics of apertures fed by standard rectangular waveguides are considered; the aperture admittances along with the radiation pattern in the biological medium are evaluated as a function of the tissue parameters.

2. Network formulation of the e.m. coupling

In Fig. 1a the formulation of the problem is shown: a waveguide radiates into a biological medium through a bolus of distillated water (ε_w=81); in general $\underset{\sim}{\varepsilon}_b = \underset{\sim}{\varepsilon}_b(r)$.

In this near-field situation, as the sources are strongly coupled to the human tissues, the field distribution and the radiation characteristics of the aperture array are obtained by the solution of coupled integro-differential equations that is performed numerically by means of the method of moments.

A simple circuit model of the hyperthermia process is derived (see Fig. 1b):

- the applicator is described by a n-port current generator (n beeing the number of subsecution in which the whole applicator aperture is subdivided) and by a n n "source" admittance matrix. The elements of the n-port current generator $[I]$ and of the n n "source" admittance matrix $[Y_{wg}]$ may be evaluated as a function only of the waveguide dimensions and of the excitations (magnitude and phase) at every waveguide;
- the body to be threated is also described by a n n "load" admittance matrix $[Y_b]$ whose elements depend only upon the local clinical situation [1][2]

The voltage matrix:

$$[V]=[Y_{wg} + Y_b]^{-1}[I] \qquad (1)$$

is the tangential electric field at every subsection of the applicator.

In this way a simple algebrical procedure permits the evaluation of the electric field distribution at the applicator aperture and hence the dissipated power distribution in the body.

As a first step, this approach may be used in order to select the applicator that is the most appropriate for a specific clinical situation. Moreover, the procedure will be used in the future for the synthesis of applicators to realize a prescribed power distribution into a biological body.

It is noticeable that the formulation in (1) may be easily extended to arrays of several applicators [3].

The efficiency of the applicator for a particular clinical situation is deduced by the dissipated power distribution in the biological body that must fit the temperature distribution imposed by therapeutical protocol.

3. Numerical and experimental results

In Fig. 2a a typical dissipated power distribution for a rectangular waveguide applicator is shown (f=434 MHz; ε_b=62.11-j 93.19) with the corresponding electric field at the aperture (Fig. 2b).

The power distribution is experimentally verified by infra-red thermographics on phantoms with dielectric characteristics equal to that of the biological body. As short as possible power pulses (typical P~100W and Δt~1') are used in order to reduce the effects of thermal diffusion inside the phantoms when an increase of temperature of some degrees must be achieved in order to get a sufficient resolution in the thermal profile.

References

|1| R. De Leo et al. - Radio Science 1981 pp. 1217-1222.
|2| R. De Leo et al. - Alta Frequenza 1983 pp. 211-213.
|3| R. De Leo et al. - Proc. of VII Microcoll - Budapest pp. 483-486.

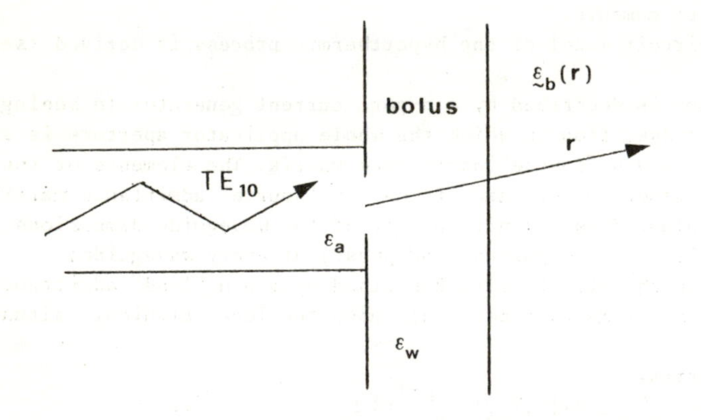

Fig. 1a) Electromagnetic model of applicator coupling to a biological body through a distilled water bolus

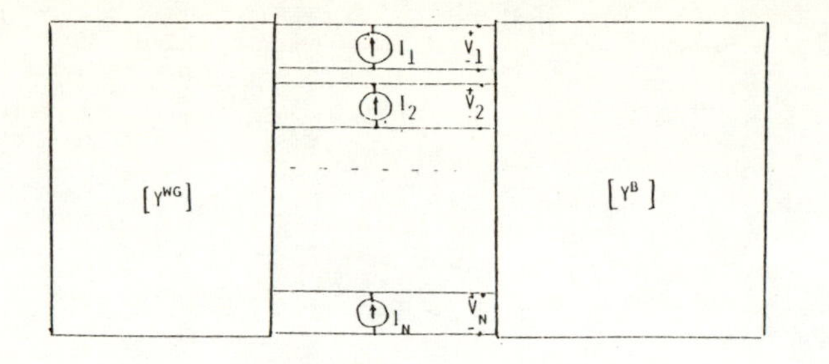

Fig.1b) Network formulation of the e.m. coupling

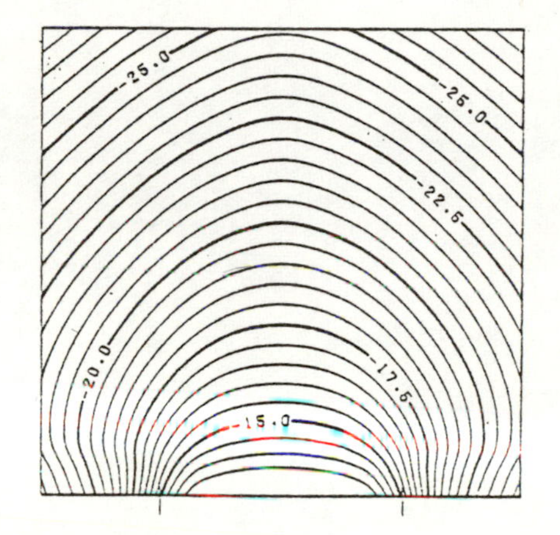

Fig. 2a) Normalized dissipated power distribution (in Log_{10}

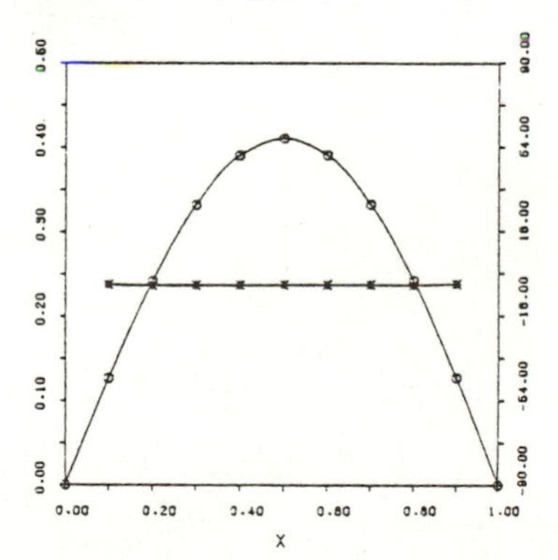

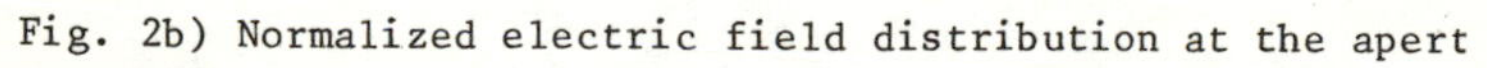

Fig. 2b) Normalized electric field distribution at the apert

PROCEEDINGS OF THE II INTERNATIONAL CONFERENCE ON APPLICATIONS OF PHYSICS TO MEDICINE AND BIOLOGY
edited by Ž. Bajzer, P. Baxa & C. Franconi

PHYSICS IN MEDICINE IN PARADISE

Indravadan C. Patel
University of Technology
Lae, Papua New Guinea

INTRODUCTION An university in a developing country cannot be just a seat of learning. Through its highly qualified staff, it must assist directly in the national development, by way of providing experts and specialized services. A dedicated university will run not only the stereotyped courses found internationally, but develope special courses tailor made to the needs and ambitions of the nation.

Papua New Guinea, a young and developing country commonly advertised as the "Paradise" because of its rich flora and fauna, and great number of species of the Birds of Paradise, is fortunate in having the P.N.G. University of Technology. It is dynamic and pledged to the development of the country. This paper describes the role played by its Department of Applied Physics in the medical services of the nation.

GENERAL There are over 150 diagnostic x-ray units scattered all over the country. There is also a central radiotharapy unit consisting of a cobalt-60 unit, a superficial x-ray machine, Cs - 137 needles and Sr-90 plaques. However, there is no physicist employed in the health services. The Department of Applied Physics of the University of Technology, with its strong bias towards medical physics, is making significant contribution in the smooth running of the radiology and radiotherapy services in the country, and in providing specialist advice and services in this field. Figure 1 shows the involvement of the Department of Applied Physics in the field of medical physics.

TRAINING There are no therapy radiographers in the country. After failure to train some overseas, the University finally mounted its own program to produce these much needed radiographers. This course is unique in many ways and meets the needs of the country. It is a four year course leading to the degree of Bachelor of Technology. The qualified radiographers will be able to perform many of the duties of a physicist, treatment planning being one of them. The Department

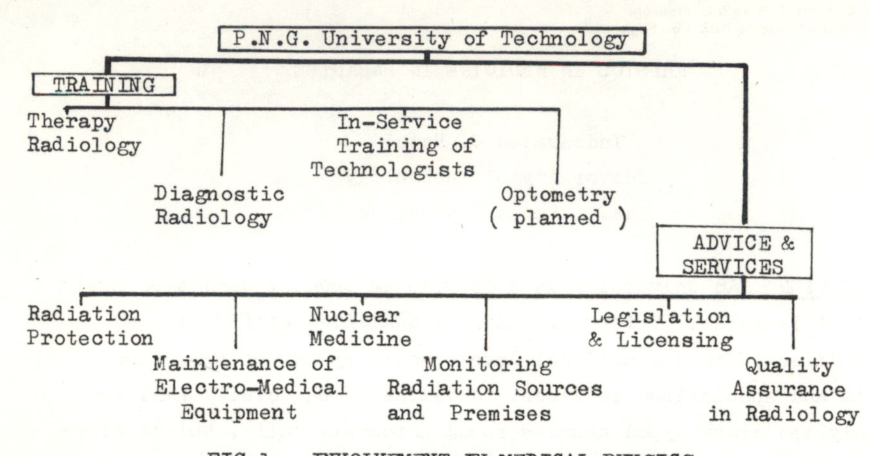

FIG.1 INVOLVEMENT IN MEDICAL PHYSICS

is assisting in the running of a certificate course in diagnostic radiology. However, plans are in progress to raise the course to a degree level. The University is also considering mounting a course in optometry to meet the urgent need of the country in this field.

ADVICE & SERVICES As in education and training, the University is making significant contributions in providing expert advice and services in the field of medical physics. The author has been appointed by the government as its honarary medical physicist. With advice and assistance from the Department, legislation is about to be introduced to control the import and use of sources of ionizing radiations, and to control the overall radiological hygine in the country. Plans have been drawn up to establish a radiation protection service based at the University. The University is also helping in the routine monitoring of radiation sources and premises, and in quality control in diagnostic radiology which would soon be extended to therapy radiology also.

CONCLUSION One of the greatest obstacle in implementing full medical services is lack of trained manpower. The University of Technology, through its Department of Applied Physics, is trying hard to train the required manpower. In the meantime, it is providing the essential services within the constraints of time and manpower available to it.

CLOSING REMARKS

L. Dalla Palma
Istituto di Radiologia
Università di Trieste
Trieste

It is somewhat embarrassing for me as a physician, to comment on this conference on medical physics as I stand in awe of what I have heard regarding the possibilities that the future offers us.

I conclude a meeting between scientists who have expanded their field of research from the theoretical or the pure experimental physics to that of physics applied to medicine.

Man with his suffering and distress, i.e. the patient, has become the main theme of your research. Many physicists still working in theoretical and experimental high energy physics are rather sceptical about this evolution.

I have observed this attitude first hand at the University of Trieste, over the past 15 years, where I have worked trying, without success, to institute an academic medical physics group.

This conference, held in memory of a Triestine physicist who understood the message coming from Man, confirms the high level of modern medical physics, which already includes several Nobel Prize Laureates. As a radiologist, let me recall the great progress promoted by medical physicists in diagnostic imaging: conventional radiology, computed tomography, ultrasonography and, more recently, digital radiography and nuclear magnetic resonance. These represent extraordinary discoveries to assist man afflicted by disease.

The physicists approaching medicine have laboured with a purpose that transcends the goals of traditional research. When Professor Cormack was asked, while in Trieste, what was the motivation leading to his discovery of computed tomography, he answered: "To help the cancer patients". His intuition occurred during a visit to a Department of Radiotherapy in South Africa working together with physicists who were measuring isodose curves on anatomical axial sections in cadavers.

Professor Mistretta had similar goals when he realized the technique of digital subtraction angiography as a substitute for invasive techniques.

Therefore, medical physicists not only carry out research in the laboratory but also on man for the health of mankind. This is significant, in this moment of severe crisis in medicine, when the physician, overwhelmed by sophisticated medical instruments, is losing the sense of personal contact with the patient. The physicist is coming closer to the physician, and therefore to the patient, favouring the human aspect in addition to the technical one. This is a consideration which must be added to the cultural one which is also certainly of great importance.

Some of the results that have been presented during this conference are extraordinary. I hope they will stimulate the Italian academic physicists and strengthen their motivation to contribute more to medical physics.

The organizing of this international conference is a sign of a sensitivity that is changing. I wish to thank Professor Abdus Salam, Professor Franconi and all the organizers for their initiative; their message has a deep meaning, and I close by asking them: "for the advancement of application of physics to medicine and biology, please organize the Third Conference in the near future."

AUTHOR INDEX

Joint Meetings

2nd INTERNATIONAL CONFERENCE ON APPLICATIONS OF PHYSICS TO MEDICINE AND BIOLOGY

«Giorgio Alberi Memorial»

and

2nd Annual A. I. F. B. Meeting

WORK IN PROGRESS IN MEDICAL PHYSICS IN ITALY 1983

organized by

The International Centre for Theoretical Physics

and

Associazione Italiana di Fisica Biomedica

November 7 - 11, 1983 - I. C. T. P. - Miramare - TRIESTE, Italy

FINAL PROGRAMME

Monday, 7 November

MORNING SESSION — Chairmen: R.A. RICCI, I. LERCH

8.30 - 9.00	Registration in the Lobby
9.00 - 10.00	Opening
10.00 - 10.40	G. FANT - Human speech and communication aids
10.40 - 10.50	Discussion
10.50 - 11.10	Break
	Symposium TECHNOLOGIES FOR THE CARDIOVASCULAR SYSTEM
11.10 - 11.50	D. OLSEN - Implantation of total artificial heart on humans. Past, present and future developments
11.50 - 12.30	E. HENNIG - Technologies of the total artificial heart
12.30 - 12.40	Discussion
12.40 - 14.00	Lunch

FIRST POSTER SESSION — Including "WORK IN PROGRESS IN MEDICAL PHYSICS IN ITALY 1983"

14.00 - 15.30	No. 1 Physiological models and measurements
	No. 2 Medical informatics
	No. 3 Radioisotopes

AFTERNOON SESSION — Chairman: M. DELLA CORTE

15.30 - 16.10	E. BELARDINELLI - Vascular system modelling fundamentals
16.10 - 16.50	N. WESTERHOF - Functional evaluation of the cardiovascular system
16.50 - 17.00	Discussion
17.00 - 17.40	H. REUL and (M.M. BLACK)* - The design, development and assessment of heart valve substitutes
17.40 - 18.20	T. D'ALESSIO - Ultrasound techniques for the vascular system
18.20 - 18.30	Discussion

* Did not attend conference

Tuesday, 8 November

MORNING SESSION	Chairmen: B. MARAVIGLIA, L. DALLA PALMA
	Symposium NMR IN BIOMEDICINE
9.00 - 9.40	D. GADIAN - NMR studies of metabolism in vivo
9.40 - 10.20	R.R. ERNST - Pulse technology. A basic requisite for biomedical applications of NMR
10.20 - 11.00	S. SYKORA - An introduction to the different measuring techniques in NMR imaging
11.00 - 11.10	Discussion
11.10 - 11.30	Break
11.30 - 12.10	P. LAUTERBUR - New techniques and new applications of NMR zeugmatography
12.10 - 12.50	W.R. BRODY - Clinical considerations in the design and application of NMR imaging systems
12.50	Discussion
SECOND POSTER SESSION	Including "WORK IN PROGRESS IN MEDICAL PHYSICS IN ITALY 1983"
14.00 - 15.30	No. 4 Biological materials
	No. 5 Technology, instrumentation and prosthetic devices
	No. 6 Non-ionizing radiation
	No. 7 Education and training
AFTERNOON SESSION	Chairman: P. LAUTERBUR
15.30 - 16.10	S. KOENIG - Magnetic fields dependence of T_1 in tissues: implications for contrast enhancement in NMR imaging
16.10 - 16.50	D. SHAW - Technical problems connected with imaging NMR and TMR equipment. Performance and quality assessment of NMR imaging systems
16.50 - 17.00	Discussion
17.00 - 17.40	M.M. TER POGOSSIAN - The latest development in ECT and PET
17.40 - 18.00	Discussion
19.00	Reception
20.15	Choral concert

Wednesday, 9 November

	Chairmen: S. LIN, S. MASCARENHAS
	Symposium NUCLEAR METHODS & TECHNIQUES
9.00 - 9.40	K.V. ETTINGER - Cyclotrons for nuclear medicine
9.40 - 10.20	M. DELLA CORTE - Functional imaging in nuclear medicine
10.20 - 11.00	R. CESAREO - Nuclear analytical techniques in medicine
11.00 - 11.10	Discussion
11.10 - 11.30	Break
11.30 - 12.10	S. MASCARENHAS and J. CAMERON - New techniques in radiation dosimetry
12.10 - 12.50	F. SAULI - Advanced radiation detectors

AFTERNOON FREE

Thursday, 10 November

MORNING SESSION	Chairman: J. CAMERON
	Symposium HYPERTHERMIA AND CANCER
9.00 - 9.40	B. MONDOVI' - Biological and physiological basis of hyperthermia
9.40 - 10.20	J. HAND - Microwaves and ultrasound in clinical hyperthermia. Some physical aspects of heating and thermometry
10.20 - 11.00	J. BACH ANDERSEN - Electromagnetics of hyperthermia
11.00 - 11.10	Discussion
11.10 - 11.30	Break
11.30 - 12.10	T.T. CETAS - Planning and dosimetry in thermal therapy
12.10 - 12.30	Discussion
12.30 - 14.00	Lunch
THIRD POSTER SESSION	Including "WORK IN PROGRESS IN MEDICAL PHYSICS IN ITALY 1983"
14.00 - 15.30	No. 8 Ionizing radiation
	No. 9 Risk assessment and control
	No. 10 Hyperthermia
AFTERNOON SESSION	Chairman: C. FRANCONI
15.30 - 16.10	J.J.W. LAGENDIJK - Experimental aspects of hyperthermia
16.10 - 16.20	Discussion
16.20 - 17.00	F. DI FILIPPO and (R. CAVALIERE)* - Clinical hyperthermia and chemotherapy

* Did not attend conference

17.00 - 17.40 G. ARCANGELI and (F. MAURO)* - Biological basis for clinical applications of combined hyperthermia and radiation

17.40 - 18.00 Discussion

18.30 Reception at the Trieste City Hall

Friday, 11 November

Chairmen: F. RUSTICHELLI, D. SOLIMINI

NEW DEVELOPMENTS IN MEDICAL PHYSICS

9.00 - 9.40 C. CORSI - Computerized radiometry

9.40 - 10.20 A. BARKER - Some biological effects of low-frequency magnetic and electric fields

10.20 - 11.00 I. LERCH - Progress in telemedicine

11.00 - 11.10 Discussion

11.10 - 11.30 Break

11.30 - 12.10 G.L. ROMANI - Advances in clinical evaluation of the biomagnetic method

12.10 - 12.20 Discussion

CLOSING REMARKS: L. DALLA PALMA, C. FRANCONI

END OF CONFERENCE

* Did not attend conference